W9-AVN-487

Applied Simulation Modeling

Duxbury titles of related interest

Albright, Winston & Zappe, *Data Analysis & Decision Making*

Albright, *VBA for Modelers: Developing Decision Support Systems with Microsoft Excel*

Berk & Carey, *Data Analysis with Microsoft Excel*

Berger & Maurer, *Experimental Design with Applications to Management, Engineering & the Sciences*

Clemen & Reilly, *Making Hard Decisions with DecisionTools*

Devore, *Probability & Statistics for Engineering and the Sciences*

Devore & Farnum, *Applied Statistics for Engineering and the Sciences*

Farnum, *Modern Statistical Quality Control and Improvement*

Hart & Hart, *Statistical Process Control for Health Care*

Hayter, *Probability & Statistics for Engineers and Scientists*

Kao, *Introduction to Stochastic Processes*

Keller & Warrack, *Statistics for Management & Economics*

Kenett & Zacks, *Modern Industrial Statistics: Design of Quality and Reliability*

Kirkwood, *Strategic Decision Making: Multiobjective Decision Analysis with Spreadsheets*

Lapin & Whisler, *Quantitative Decision Making with Spreadsheet Applications*

Lattin, Carrol & Green, *Analyzing Multivariate Data*

Lawson & Erjavec, *Engineering and Industrial Statistics*

Middleton, *Data Analysis with Microsoft Excel*

Minh, *Applied Probability Models*

Savage, *Decision Making with Insight*

Shapiro, *Modeling the Supply Chain*

Vardeman & Jobe, *Basic Engineering Data Collection and Analysis*

Weida, Richardson & Vazsonyi, *Operations Analysis Using Excel*

Weiers, *Introduction to Business Statistics*

Winston, *Introduction to Mathematical Programming*

Winston, *Introduction to Probability Models*

Winston, *Operations Research: Applications & Algorithms*

Winston & Albright, *Practical Management Science*

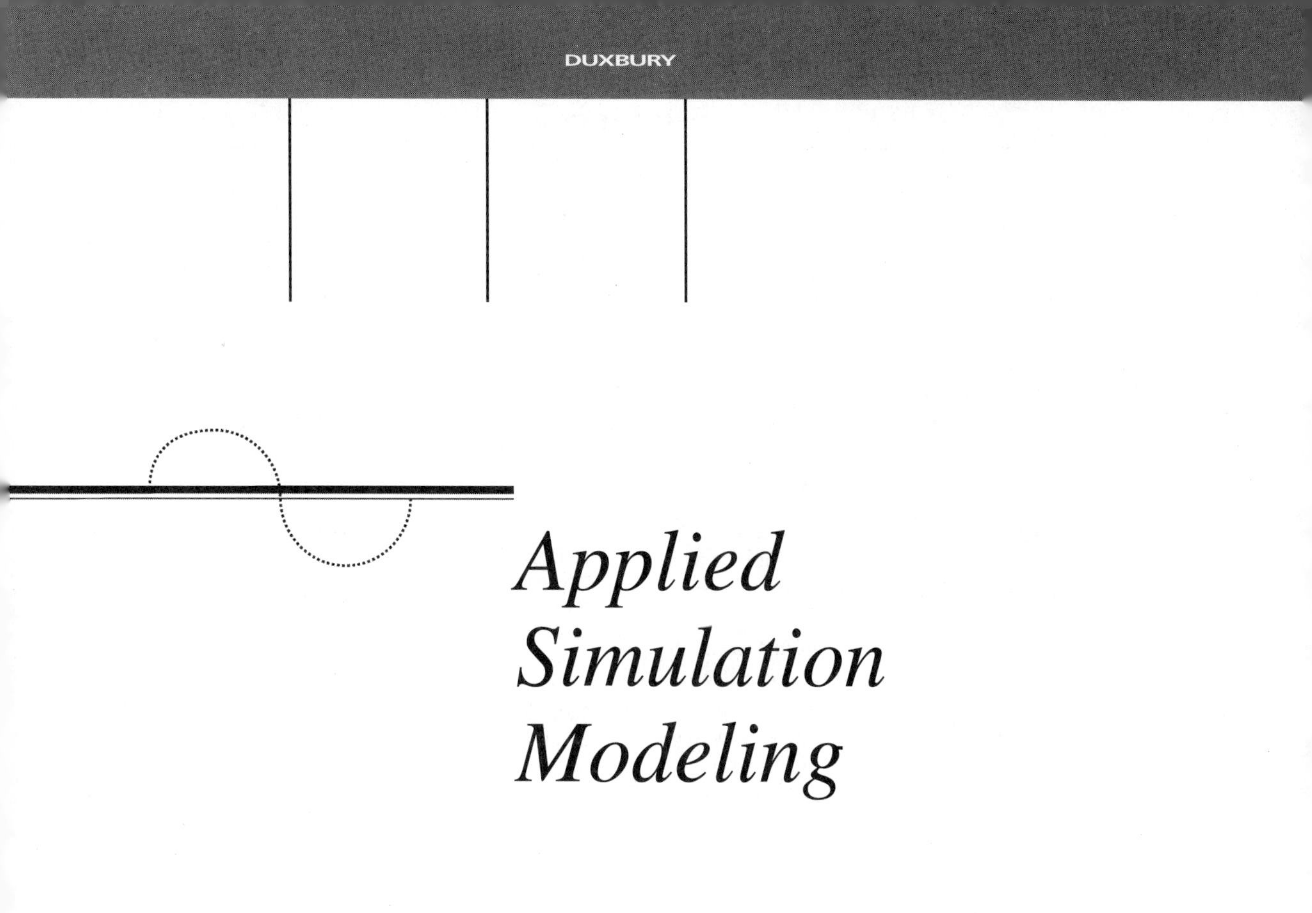

Applied Simulation Modeling

Andrew F. Seila
Terry College of Business, University of Georgia

Vlatko Ceric
University of Zagreb

Pandu Tadikamalla
University of Pittsburgh

THOMSON
BROOKS/COLE

Australia • Canada • Mexico • Singapore • Spain
United Kingdom • United States

THOMSON
BROOKS/COLE

Sponsoring Editor: Curt Hinrichs
Assistant Editor: Ann Day
Editorial Assistant: Katherine Brayton
Technology Project Manager: Burke Taft
Marketing Manager: Joseph Rogove
Marketing Assistant: Jessica Perry
Advertising Project Manager: Tami Strang
Print/Media Buyer: Jessica Reed
Permissions Editor: Joohee Lee
Production, Copyediting, Illustration, & Composition: Summerlight Creative
Text Designer: Andrew Ogus ■ Book Design
Cover Designer: Lisa Langhoff
Cover Image: PhotoDisc
Text and Cover Printer: Maple-Vail Binghamton

Printed in the United States of America

1 2 3 4 5 6 7 06 05 04 03 02

For more information about our products, contact us at:
Thomson Learning Academic Resource Center
1-800-423-0563

For permission to use material from this text, contact us by:
Phone: 1-800-730-2214 **Fax:** 1-800-730-2215
Web: http://www.thomsonrights.com

ISBN 0-534-38159-6

Brooks/Cole—Thomson Learning
10 Davis Drive
Belmont, CA 94002
USA

Asia
Thomson Learning
5 Shenton Way #01-01
UIC Building
Singapore 068808

Australia/New Zealand
Thomson Learning
102 Dodds Street
Southbank, Victoria 3006
Australia

Canada
Nelson
1120 Birchmount Road
Toronto, Ontario M1K 5G4
Canada

Europe/Middle East/Africa
Thomson Learning
High Holborn House
50/51 Bedford Row
London WC1R 4LR
United Kingdom

Latin America
Thomson Learning
Seneca, 53
Colonia Polanco
11560 Mexico D.F.
Mexico

Spain/Portugal
Paraninfo
Calle/Magallanes, 25
28015 Madrid, Spain

Dedication

To the memory of my father, Claude F. Seila.
Andrew F. Seila

To my family.
Vlatko Ceric

To the memory of my father, Sri Tadikamalla Satyanarayana.
Pandu Tadikamalla

Brief Contents

1 Introduction to Simulation 1

2 Static Simulations 21

3 Financial Models and @RISK 77

4 Dynamic Simulations 117

5 System Modeling Concepts for Discrete-Event Simulation 154

6 System Modeling Paradigms 209

7 Visual Interactive Simulation and Arena 245

8 Graphical Simulation Modeling 279

9 Problem Solving Using Simulation 326

10 Simulation in Practice 390

Appendix: Review of Probability and Statistics 418

Contents

Preface **xix**

List of Figures **xxiv**

List of Tables **xxxiii**

List of Models **xxxv**

About the Authors **xxxvii**

1 Introduction to Simulation 1

1.1 Decisions and Decision Models **2**

1.2 Getting Answers from Models **5**

1.3 A General Definition of Simulation **7**

1.3.1 Gaming Simulations 9

1.3.2 *Static Simulations* ***9***
1.3.3 *Dynamic Systems Simulations* ***10***

1.4 Components of a Simulation Study **11**

1.5 Reasons for Using Simulation as an Analysis Tool **14**

1.6 Summary **15**

References ***17***
Problems ***18***

2 **Static Simulations** **21**

2.1 Introduction **22**

2.2 A Model for Profit on a Special Sale Promotion **23**

2.2.1 *The Spreadsheet Model* ***23***
2.2.2 *Setting Up the Simulation* ***25***
2.2.3 *Running the Replications* ***27***
2.2.4 *Analyze the Output Data* ***29***
2.2.5 *Comments about Spreadsheet Simulations* ***30***

2.3 Simulation for Sensitivity Analysis **31**

2.3.1 *A Financial Model for an Office Building* ***31***
2.3.2 *Sensitivity Analysis for Office Building Model* ***32***
2.3.3 *What-If Calculations* ***33***
2.3.4 *Simulation for Sensitivity Analysis* ***35***

2.4 Simulation: Sampling on the Computer **38**

2.4.1 *A Model to Estimate* π ***38***
2.4.2 *An Experiment to Estimate* π ***39***
2.4.3 *Analysis of the Data* ***40***

2.5 Some Techniques for Generating Random Variates **41**

2.5.1 *Bernoulli Random Variates* ***42***
2.5.2 *Uniform Random Variates* ***43***
2.5.3 *Triangular Random Variates* ***43***
2.5.4 *Normal Random Variates* ***44***
2.5.5 *Exponential Random Variates* ***44***

2.5.6 Discrete Integer-Valued Random Variates **45**
2.5.7 Other Discrete Random Variates **45**
2.5.8 The Inverse Transform Method **47**
2.5.9 Special Considerations **48**

2.6 Evaluating Decisions: A One-Period Inventory Model **48**

2.6.1 The Inventory Model Revisited **48**
2.6.2 Optimal Order Quantity Using Simulation **49**
2.6.3 Other Cost Functions **50**
2.6.4 Sensitivity Analysis **52**

2.7 Data Analysis for Static Simulations **55**

2.7.1 Performing Independent Replications **55**
2.7.2 Estimating the Mean from Independent Replications **57**

2.8 Simulation with Spreadsheets: A More Complex Model **58**

2.8.1 A More Realistic Real Estate Model **58**
2.8.2 Risk Analysis for the Real Estate Model **60**
2.8.3 Fitting Distributions for Uncertain Variables **60**
2.8.4 Simulation Results **63**
2.8.5 Alternative Decisions to Improve Performance **64**

2.9 An Insurance Model **66**

2.9.1 A Model for Loss Ratio for an Insurance Agency **67**
2.9.2 The Simulation **69**
2.9.3 Simulation Results **70**

2.10 Summary **71**

References **73**
Problems **73**

3 Financial Models and @RISK 77

3.1 Introduction **78**

3.2 A Model for the Price of a Stock **79**

3.3 Options, Futures, and Simulation **88**

3.3.1 Options and Futures **88**

3.3.2 Estimating the Price of a Call Option Using Simulation ***89***
3.3.3 Hedging Using Put Options ***91***

3.4 Dynamic Financial Models of Stock Prices **93**

3.4.1 A Worksheet to Sample Stock Prices over Time ***93***
3.4.2 Estimating the Price of an Asian Option ***96***

3.5 Correlated Asset Values **99**

3.5.1 Estimating the Correlation Matrix ***99***
3.5.2 Sampling Correlated Variates ***101***

3.6 Fitting a Distribution to Data **104**

3.6.1 Using BestFit to Fit a Specific Distribution to the Data ***106***
3.6.2 Finding the Distribution with the Best Fit ***110***

3.7 Summary **113**

References ***114***
Problems ***115***

4 Dynamic Simulations 117

4.1 Introduction **118**

4.2 Waiting Times in a Single-Server Queueing System **119**

4.2.1 Lindley's Formula ***119***
4.2.2 Taxonomy of Queueing Systems ***120***
4.2.3 A Spreadsheet Simulation of M/M/1 Queue Waiting Times ***121***

4.3 Characteristics of Data from Dynamic Simulations **122**

4.3.1 The Initial Transient Period ***123***
4.3.2 Autocorrelated Observations ***126***

4.4 Batch Means to Estimate the Mean from Stationary Data **126**

4.4.1 Batch Means Computation ***127***
4.4.2 Some Guidelines to Applying the Batch Means Method ***128***

4.5 Simulating Discrete-Time Markov Chains **129**

4.5.1 A Markov Chain Inventory Model ***129***
4.5.2 A Markov Chain Queueing Model ***131***
4.5.3 A Markov Chain Reliability Model ***133***

4.5.4 *Performance Measures for Markov Chains* ***135***

4.6 The Regenerative Method for Estimating the Mean **136**

4.6.1 *Regenerative Processes* ***137***
4.6.2 *The Regenerative Method for Estimating the Mean* ***138***

4.7 An Advanced Queueing Model **142**

4.8 A Marketing Model **143**

4.9 Summary **149**

References ***150***
Problems ***151***

5 **System Modeling Concepts for Discrete-Event Simulation** **154**

5.1 Introduction **155**

5.1.1 *Static and Dynamic Model Descriptions* ***156***

5.2 Events and Event Sequencing **156**

5.2.1 *Events and State Changes* ***157***
5.2.2 *Event Scheduling* ***158***
5.2.3 *Discrete-Event Simulation Model Development* ***159***
5.2.4 *A Generic Simulation Language* ***160***

5.3 Example: A Single-Server Queue **161**

5.3.1 *State Variables for the Single-Server Queue* ***162***
5.3.2 *Event Routines for the Single-Server Queue* ***162***
5.3.3 *A Detailed Description of Event Execution and State Changes* ***163***
5.3.4 *Review: Event Scheduling* ***167***
5.3.5 *Estimating the Mean Number in the Queue* ***169***

5.4 Event Graphs **175**

5.4.1 *Simplifying an Event Graph* ***176***
5.4.2 *Conditional Events* ***178***
5.4.3 *Event Routines and State Changes* ***178***

5.5 Event Parameters **178**

5.5.1 *Passing Event Parameters to Event Routines* **179**
5.5.2 *Using Event Parameters in Event Routines* **180**
5.5.3 *Event Parameters in Event Graphs* **181**
5.5.4 *Event Parameters and Output Data Analysis* **181**
5.5.5 *Event Canceling* **187**

5.6 Static Model Description **189**

5.6.1 *Entities and Attributes* **189**
5.6.2 *Sets, Lists, and Queues* **190**

5.7 A Message-Processing Model that Uses Entities and Lists **191**

5.7.1 *Top-Down Design for Model Building* **192**
5.7.2 *Top-Level Model Description* **193**
5.7.3 *Model Details* **194**
5.7.4 *Model Output* **195**

5.8 A Single-Server Queue Using Entities and Events **198**

5.9 Summary **204**

References **205**
Problems **206**

6 System Modeling Paradigms 209

6.1 Introduction **210**

6.2 Review of Discrete-Event Dynamic Systems Concepts **211**

6.2.1 *Temporary and Permanent Entities* **211**
6.2.2 *Attributes and Entity States* **212**
6.2.3 *System Dynamics* **213**
6.2.4 *Conditional and Scheduled Events* **214**
6.2.5 *Activities and Processes* **215**

6.3 The Activity View **216**

6.3.1 *Original Version* **217**
6.3.2 *The Three-Phase Approach* **220**
6.3.3 *An Example of Simulation Execution Using the Three-Phase Approach* **222**

6.4 The Process View **225**

6.4.1 *The Process View of the Single-Server Queue* **226**
6.4.2 *Process Interaction* **228**
6.4.3 *Processes, Events, and Activities* **230**

6.5 A Manufacturing Example **231**

6.5.1 *The Event View Model* **233**
6.5.2 *The Activity View* **236**
6.5.3 *The Process View* **236**

6.6 Summary **241**

References **242**
Problems **243**

7 Visual Interactive Simulation and Arena 245

7.1 Visual Interactive Simulation **246**

7.1.1 *Simulation Hardware and Graphical User Interfaces* **247**
7.1.2 *Software for VIS* **248**

7.2 A First Look at Arena **249**

7.2.1 *Build the Model* **250**
7.2.2 *Preparing a Simulation Run* **257**
7.2.3 *Interactive Model Runs* **259**
7.2.4 *Reports* **260**
7.2.5 *Modeling Costs in Arena* **262**
7.2.6 *Modeling Elements* **262**
7.2.7 *Modeling Blocks* **263**

7.3 A Bank Lobby Model in Arena **264**

7.3.1 *Bank Lobby System* **264**
7.3.2 *Model Building* **265**
7.3.3 *Simulation Run and Results* **271**

7.4 Visual Interactive Simulation: A Recap **274**

7.5 Summary **276**

References **277**
Problems **277**

8 Graphical Simulation Modeling 279

8.1 Graphical Modeling **280**

8.2 Some Graphical Modeling Techniques **282**

8.2.1 *Event Graphs* ***282***
8.2.2 *Activity Cycle Diagrams* ***283***
8.2.3 *Petri Nets* ***285***

8.3 Graphical Modeling with Arena **288**

8.3.1 *The Batch and Separate Modules* ***290***
8.3.2 *A Copy Center Model* ***291***
8.3.3 *The Arena Model* ***291***
8.3.4 *Running the Simulation* ***299***

8.4 Hierarchical Modeling **303**

8.4.1 *Basic Concepts* ***303***
8.4.2 *Hierarchical Modeling with Arena* ***304***
8.4.3 *A Fuel Depot Model* ***304***

8.5 Animation **309**

8.5.1 *Animation with Arena* ***310***
8.5.2 *The Fuel Depot Model and Facility-Based Animation* ***310***
8.5.3 *Station Sequences* ***317***
8.5.4 *Named Views and Navigation* ***320***

8.6 Summary **321**

References ***323***
Problems ***323***

9 Problem Solving Using Simulation 326

9.1 Introduction **326**

9.2 Service Systems **328**

9.2.1 *Characteristics of Service Systems* ***328***
9.2.2 *Resources* ***331***
9.2.3 *Multiple Servers and Multiple Queues* ***332***

*9.2.4 Networks of Queues **337***
*9.2.5 Performance Measures for Service Systems **339***
*9.2.6 Cost Parameters **341***

9.3 Manufacturing System Models **345**

*9.3.1 Characteristics of Manufacturing Systems **346***
*9.3.2 A Manufacturing System Model **347***
*9.3.3 Conveyor Modeling **348***
*9.3.4 Building the Model in Steps: Version 1 **350***
*9.3.5 Modeling Blocking **353***
*9.3.6 Adding the Rework Loop: Version 2 **355***
*9.3.7 Adding the Loop Conveyor **357***
*9.3.8 Special Characteristics: Transporters and Batch Processing **360***
*9.3.9 A Manufacturing Decision Support System **363***

9.4 Transportation Systems **365**

*9.4.1 Phase 1: The Truck Unloading Process **369***
*9.4.2 Phase 2: Refining the Model **374***
*9.4.3 Phase 3: Data Collection and Experimental Design **378***

9.5 Summary **382**

*References **384***
*Problems **384***

10 Simulation in Practice 390

10.1 A Simulation Project **391**

*10.1.1 Project Origination and Systems Analysis **392***
*10.1.2 Data Collection **392***
*10.1.3 The Preliminary Model **394***
*10.1.4 Project Conclusion **395***

10.2 The Science of Simulation Project Management: Problem Analysis and Solution Process **396**

*10.2.1 Problem Formulation and Statement of Objectives **397***
*10.2.2 Project Plan and Schedule **397***
*10.2.3 System Analysis **398***
*10.2.4 Model Conceptualization and Formulation **398***

10.2.5 *Data Collection and Analysis* ***399***
10.2.6 *Model Building* ***400***
10.2.7 *Model Validation* ***400***
10.2.8 *Experimentation and Analysis* ***401***
10.2.9 *Reports and Presentations* ***401***
10.2.10 *Implementation* ***402***

10.3 The Art of Project Management **402**

10.3.1 *Problem Identification and Formulation* ***403***
10.3.2 *Understand the Client's Objectives* ***403***
10.3.3 *Collect Data to Verify and Document Problems* ***404***
10.3.4 *Consider Alternative Models* ***405***
10.3.5 *Project Planning and Management* ***405***
10.3.6 *Team Management* ***407***
10.3.7 *Communicating with Clients* ***408***

10.4 Data Collection **409**

10.4.1 *Coordinating Model Development and Data Collection* ***409***
10.4.2 *The Amount and Availability of Data* ***410***
10.4.3 *Data Quality* ***410***

10.5 Model Management **411**

10.5.1 *Model Complexity* ***411***
10.5.2 *Hierarchical Model Building* ***412***
10.5.3 *Start with a Simple Model* ***413***

10.6 Summary **415**

References ***417***

Appendix: Review of Probability and Statistics 418

A.1 Probability Concepts: Some Definitions **418**

A.1.1 *Probability Rules* ***419***

A.2 Random Variables, Probability Distributions, and Expectations **421**

A.2.1 *Discrete Random Variables* ***421***
A.2.2 *Some Discrete Distributions* ***423***

A.3 Continuous Random Variables **426**

A.3.1 *Some Continuous Distributions* ***427***

A.4 Inferential Analysis **434**

A.4.1 *Sampling Distribution of a Statistic* ***435***
A.4.2 *Examples: Sampling Distribution of* $\overline{X}$ ***437***
A.4.3 *Estimation* ***439***
A.4.4 *Confidence Intervals* ***440***
A.4.5 *Sample Size Determination* ***442***
A.4.6 *Hypothesis Testing* ***443***
A.4.7 *Examples: Hypothesis Testing* ***445***
A.4.8 *Testing on the Difference of Two (Population) Means* ***447***

Index 449

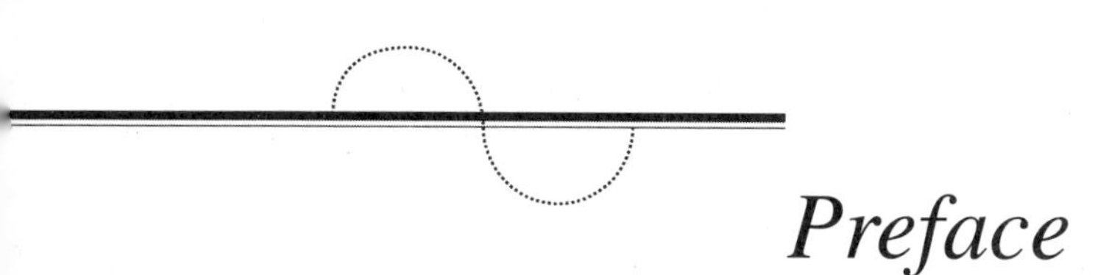

Preface

The idea for this book began when two of the authors (Seila and Tadikamalla) were discussing problems in teaching simulation to students in business and other fields such as forestry, social sciences, and agricultural engineering. At the time, the few available simulation texts were written primarily for students in industrial engineering (IE) and operations research (OR). We had found that business students wanted to use simulation to analyze stochastic financial problems as well as discrete-event models. We also found that they primarily wanted to learn how to build simulation models rather than delve deeply into statistical methodology for simulation. This book is an attempt to serve these students and others with similar interests. Thus, it is directed at a broad range of students, including IE and OR students who wish to adopt the practical approach we take.

Applied Simulation Modeling takes the perspective of the consultant whose job is to analyze stochastic decision problems by building a simulation model and using it to understand the behavior of the system and explore the effects of alternative decisions. Our approach is balanced. Two modeling methodologies are presented:

(1) models, both static and dynamic, that can be implemented as mathematical expressions in a spreadsheet or similar interactive software tool; and (2) discrete-event models that utilize the event scheduling formalism. Since spreadsheets are ubiquitous, we use this software to demonstrate how to build and execute the first type of simulation model. Arena, from Rockwell Software, is the tool we use to demonstrate how to build discrete-event simulation models.

Chapters 2 through 4 focus on general simulation ideas and use a strong spreadsheet orientation to model and simulate various financial and managerial problems. In Chapter 2, the reader is introduced to simulation in a gradual and deliberate series of tutorial examples that demonstrate how to create the models, execute the simulation, analyze the output data, and interpret the results. Some of these examples involve models for which the "correct" answers are known. In these cases, we show that simulation can produce an answer that is close to the "correct" answer. The purpose of these examples is to develop confidence in simulation as a tool for problem analysis.

Chapters 5 and 6 present the conceptual foundations of discrete-event simulation using diagrams and an artificial simulation language. The objective of these two chapters is to provide a good understanding of the concepts and underlying structure of discrete-event simulation. We believe that if the analyst understands how discrete-event simulation works, he or she will be a more effective modeler and will find that problems and errors in simulation models are avoided and corrected much more easily.

Chapters 7 through 9 use Arena to demonstrate how to build and execute discrete-event simulations. These chapters introduce Arena by building rather simple models and then proceeding to increasingly complex models that use more of Arena's extensive collection of modeling tools. The material in these chapters is presented in a tutorial manner. Once completed, the reader should be well prepared to create moderately complex models using Arena.

In all chapters, our approach is to start with simple models first and then proceed to more complex models. We have found that many students fail to master basic modeling concepts well enough to build simple models reliably. We believe that an analyst cannot possibly develop a correct model of a complex system if he or she cannot develop a correct model of a simple system. Many of the problems at the end of the chapters ask the reader to build simple models that have various new features. If these models are mastered, the reader has the basic tools and building blocks for creating correct models of realistic systems.

The past decade has seen an extensive change in the way simulation has been taught, both in business schools and in IE and OR departments. Business schools have changed from teaching dedicated simulation courses to teaching decision modeling courses that include simulation as a module. Both business and IE and OR programs have begun to expect students to build more realistic models because powerful interactive simulation software is available. We believe that in order for a simulation modeler to use this software competently, he or she must have a full understanding of the underlying simulation methodology, both the modeling and statistical methodologies. Throughout this book, we try to carefully explain the underlying concepts and constructs.

This book takes a more managerial perspective than other simulation texts with which we are familiar. In most models, we take the reader through the entire process of designing the model, implementing it in the appropriate software, executing the simulation, collecting and analyzing the output data, and using the results of the analysis to evaluate alternative decisions. It is this last phase that managers are ultimately concerned with: decision making.

The only background we assume on the part of the reader is an introductory course in probability and statistics. Of course, more knowledge of probability and statistics is better. The reader should understand the basic concepts of probability (including random variables and probability distributions) and sampling, estimation, and hypothesis testing in statistics.

The modeler has many alternative paths to take in building complex system models. Some of these paths will lead more quickly to a correct model than others. In the more advanced modeling examples, especially in Chapter 9, we discuss modeling strategy—that is, we first develop a plan for our model-building process and then use it to actually create the model in stages that ultimately lead to a working and reliable model. This strategy incorporates a hierarchical approach to model building that we think is necessary to manage the complexity of large models. In our experience, students often have difficulty planning and starting the model-building process. We hope this material will provide some useful guidance to them.

Although some of the models in this text are relatively complex, they still fall short of the level of complexity of real-world models. Readers who attempt to use the techniques and methods in this book to solve real-world problems will find a gap between the relatively clean modeling situations in this book and the messy, ill-defined modeling problems in the real world. Chapter 10 is an attempt to bridge this gap by alerting readers to some of the problems they will encounter and providing advice based on the authors' consulting experience.

The appendix has a brief review of probability and statistics concepts. This is not intended to be a course on basic probability and statistics but a reference in which the reader can quickly look up concepts and techniques that might have been forgotten.

This text is appropriate for courses that deal with general simulation models in business and other contexts as well as courses that focus on discrete-event simulation. The chapters are modular and in many cases can be skipped. A course on simulation using spreadsheets can use Chapters 1 through 4. Chapters 5 through 10 constitute an appropriate collection for a course on discrete-event simulation. A course that teaches decision modeling and analysis could use Chapters 1 and 2 and 7 through 10, omitting Chapters 4 through 6. We recommend that any course in simulation include the material in Chapters 1 and 10.

A CD-ROM containing the academic versions of two commercial software packages—@RISK from Palisade Corporation and Arena from Rockwell Software—is included with this book. Instructions for installing these software packages are provided at the end of the book. The reader also needs to have a spreadsheet program available to follow the tutorial examples in Chapters 2 and 4. Microsoft Excel, Lotus 123, Quattro Pro, and StarOffice Calc will all suffice for these two chapters. Chapter 3 requires Microsoft Excel since @RISK is only available as an add-in for Excel. The CD-ROM also contains files with the spreadsheets and Arena models in the text.

ACKNOWLEDGMENTS

Many people have contributed to this text both directly and indirectly. Richard Fenton at West Publishing encouraged us to write this book and worked with us to get the project under way. Nancy Hill-Wilton worked with us during our early drafts. At Duxbury, we would like to thank Curt Hinrichs, whose support and encouragement have been vital to the completion of this project. Nathan Day and Hal Humphrey at Duxbury have also provided valuable help. We want to thank S.M. Summerlight at Summerlight Creative for his patience and cooperation during the copyediting and typesetting processes.

Many colleagues have influenced our view of simulation. We want to give a special thanks to George Fishman, who introduced Andy Seila to discrete-event simulation and has been a mentor, friend, and colleague over the years. John Ramberg not only taught Pandu Tadikamalla simulation but also inspired him to do research in the simulation area—in particular, on random variate generation. David Goldsman and Christos Alexopoulos have engaged in many stimulating discussions and provided numerous helpful comments. Deb Sadowski reviewed the material on Arena and gave us valuable insights. Other colleagues we would like to recognize are David Kelton, Lee Schruben, John Miller, Russell Cheng, Bruce Schmeiser, James Wilson, Alan Pritsker, Bob Sargent, Mike Pidd, Kathryn Brohman, and Patrick McKeown. Students who have reviewed various parts of the text and provided feedback about the problems are Kevin Burns, Kisu Kim, and Kathryn Kotzan.

We also thank the following reviewers who helped us improve our work: Veena Adlakha, University of Baltimore; Mohammad Amini, University of Memphis; Hossein Arsham, University of Baltimore; Paul Bobrowski, University of Oregon; David Fritzsche, University of Washington; David Goldsman, Georgia Institute of Technology; David Grimmett, Austin Perry State University; Ronald Jensen, Emory University; Tarja Joro, University of Alberta School of Business; Cem Karacal, Rochester Institute of Technology; Gary Kochenberger, University of Colorado at Denver; Ronald McPherson, James Madison University; Jason Merrick, Virginia Commonwealth University; Barry Nelson, Ohio State University; David Pentico, Duquesne University; J. W. Schmidt, Virginia Polytechnic Institute and State University; Thomas J. Schriber, University of Michigan; Carlton Scott, University of

California-Irvine; Wei Shih, Bowling Green State University; Ashok Soni, Indiana University; Patrick Thompson, University of Florida; and Walt Wheatley, University of West Florida.

Finally, we would like to thank our families for their love, support, and patience during the writing of this book. Their contributions are just as real and important as the others named above, and we could not have completed this project without them.

Andrew F. Seila
Vlatko Ceric
Pandu Tadikamalla

List of Figures

2.1	Profit calculations for special furniture promotion	**24**
2.2	Distributions of (a) demand and (b) initial price	**25**
2.3	Furniture promotion model simulation	**26**
2.4	Twenty observations of demand, price, and profit	**28**
2.5	Spreadsheet for office building cash-flow analysis	**31**
2.6	Summary of calculations for the office building model	**33**
2.7	Density functions of uncertain parameters in office building spreadsheet	**35**
2.8	Histogram of after-tax cash flow	**36**
2.9	Simulation data for ATCF	**37**
2.10	Cumulative distribution function for office building model	**38**
2.11	Quarter-circle inscribed in a square	**39**
2.12	Plot showing convergence of estimates to π	**41**
2.13	Triangular density function with most likely value on extreme right	**43**
2.14	Asymmetric triangular density function	**44**
2.15	Use of a lookup table to sample from a discrete distribution	**46**
2.16	Binomial distribution cdf	**47**

2.17 Results of multiple runs for inventory model **50**
2.18 Results of multiple runs for inventory model with nonlinear costs **51**
2.19 Results of simulation runs with normally distributed demand **52**
2.20 Results of simulation runs with Erlang demand **53**
2.21 Results of a long simulation run with Erlang demand **54**
2.22 Histogram for actual inventory cost **54**
2.23 Analysis of inventory level decision using one replication **54**
2.24 Repeat of 2.23 four times **56**
2.25 *Random number generation* **56**
2.26 Multiyear office building cash-flow model **59**
2.27 Empirical distribution of building cost **61**
2.28 Empirical distribution with triangular distribution **61**
2.29 Sample distribution of present value of ATCF **63**
2.30 Sample cdf of present value of ATCF **64**
2.31 Plot of scenarios in Table 2.7 **66**
2.32 Frequency plots for four scenarios of claims ratios **71**

3.1 Stock price model in Excel **80**
3.2 @RISK toolbar **80**
3.3 Simulation Settings dialog **81**
3.4 Sampling tab on Simulation Settings dialog **82**
3.5 @RISK output windows **83**
3.6 Summary Statistics window **84**
3.7 Detailed Statistics window **85**
3.8 Reports to Worksheet dialog **86**
3.9 Copying simulation results to Excel **86**
3.10 Confidence interval calculation for stock price **87**
3.11 Histogram of future prices for stock **87**
3.12 Cumulative distribution of future prices for stock **88**
3.13 Option pricing worksheet **90**
3.14 Worksheet to evaluate hedging using put options **91**
3.15 @RISK output for evaluating hedging with put options **92**
3.16 Model of stock price over time **94**
3.17 Plot of price series for stock **95**
3.18 @RISK window for dynamic simulation of stock price series **96**
3.19 Graphs of stock price after 13 and 26 weeks **97**
3.20 Summary graph for stock price series **98**
3.21 Worksheet for pricing an Asian option **98**
3.22 Logs of weekly price changes for six stocks in portfolio **100**
3.23 Sample correlations between six log stock price changes **101**

3.24 Full correlation matrix for six stock price changes **101**
3.25 Parameter portion of the portfolio model **102**
3.26 Computations to sample prices **102**
3.27 Portfolio simulation assuming uncorrelated stock prices **103**
3.28 Output from simulation of a portfolio of six correlated stock prices **104**
3.29 Comparison of percentage returns with and without correlation **105**
3.30 Differences of logs of daily closing prices for Daimler-Chrysler **106**
3.31 Distribution-fitting dialog **107**
3.32 Distribution-fitting output window **107**
3.33 Data for effective average interest rates by state **110**
3.34 Output window for fitting a distribution to interest rate data **111**
3.35 Goodness-of-fit statistics for interest rate data fitted to the logistic distribution **112**
3.36 P–P plot for logistic distribution and interest rate data: P–P **112**

4.1 Spreadsheet for simulating a M/M/1 queue **122**
4.2 Results of 20 replications of queueing system **124**
4.3 Twenty replications of stationary single-server queue **125**
4.4 Sample from Markov chain inventory model **131**
4.5 Waiting times from a regenerative M/M/1 queue **138**
4.6 Plot of mean waiting time estimate versus threshold **144**
4.7 Initial transient portion for scenario 1 **147**
4.8 Stationary portion of marketing model output: Market share for scenario 1 **147**
4.9 Initial transient portion for scenario 2 **148**
4.10 Stationary portion of marketing model output: Market share for scenario 2 **148**

5.1 Plot of number of customers in system versus time **157**
5.2 Event routines for single-server queue **163**
5.3 Intialization routine **163**
5.4 State variables and event list for queue before initialization **163**
5.5 State variables and event list for queue after initialization **164**
5.6 State variables and event list for queue after first *arrival* event **165**
5.7 State variables and event list for queue after first *begin service* event **165**
5.8 State variables and event list for queue after second *arrival* event **166**
5.9 State variables and event list for queue after first *end service* event **166**
5.10 State variables and event list for queue after second *begin service* event **167**
5.11 Summary of 20 events in M/M/1 queue model **168**

5.12 Plot of $Q(t)$ with $A(0, T)$ **170**

5.13 Event routines for single-server queue with data collection **173**

5.14 Results of simulation run for single-server queue **174**

5.15 A simple event graph with two events **175**

5.16 Event graph for a simple queueing system **175**

5.17 Event graph with time delays included **176**

5.18 Event graph with *begin service* event removed **176**

5.19 A sample event graph with unnecessary event B **177**

5.20 The event graph in 5.18 with event B removed **177**

5.21 Single-server queue event graph with event conditions added **178**

5.22 Completed event graph for single-server queue **179**

5.23 Spreadsheet for office building cash-flow analysis **180**

5.24 Trace of actions in heterogeneous server queue **182**

5.25 Event graph with formal and actual parameters **183**

5.26 Event graph of two-server queue with nonidentical servers **183**

5.27 Event routines for single-server queue with data collection for waiting times **184**

5.28 Trace of 20 events in single-server queue with event parameters **185**

5.29 Results of run of single-server queue to estimate mean waiting time **187**

5.30 Representation of event canceling in event graphs: (a) event cancelling; (b) with condition and time **188**

5.31 Event graph model of a multiple identical servers shop with closing time **188**

5.32 *Initialize* **194**

5.33 *Receive messages* **194**

5.34 *Process messages* **195**

5.35 Sample simulation run for message-routing model **196**

5.36 Results of a long simulation run to estimate mean cycles for message-routing model **199**

5.37 Structure of customer and server entities **200**

5.38 Event routines for single-server queue with entities **200**

5.39 *Routine initialize* **201**

5.40 *Routine initialize* **201**

5.41 Sample trace of M/M/1 queue simulation with entities **202**

5.42 *Event graphs* **207**

6.1 Relationship among conditional events, scheduled events, and activities **215**

6.2 Activity routines for single-server queueing model **217**

6.3 Timing routine for activity view **219**

6.4 Initialization routine for activity view **220**

6.5 Event routines for three-phase approach **221**

6.6 Event routines for three-phase approach **221**

6.7 Three-phase approach showing event list and state variable values for a single-server queue execution **223**

6.8 Process routine for single-server queue **226**

6.9 Process routine for single-server queueing model **227**

6.10 Process interaction view of single-server queueing model **228**

6.11 Process interaction version of single-server queueing model without labels **230**

6.12 Combined process and event view of single-server queueing model **231**

6.13 Manufacturing system with two coupled subsystems **232**

6.14 *Event routines for 10 events* **234**

6.15 *Event routines for 10 events* **236**

6.16 *Process view model* **239**

7.1 Arena model containing several flowchart elements (modules) **249**

7.2 Arena's modeling environment **251**

7.3 Simple queueing system flowchart **252**

7.4 Placing the Create module in the model window **253**

7.5 Model flowchart after placing the Process and Dispose modules in the model window **253**

7.6 Create module property dialog with the data for the simple queueing system with a single server **254**

7.7 Resource dialog with the data for the simple queueing system with a single server **255**

7.8 Process module property dialog with the data for the simple queueing system with a single server **256**

7.9 Dispose module property dialog with the data for the simple queueing system with a single server **256**

7.10 Resource spreadsheet with the data for the simple queueing system with a single server **257**

7.11 Simulation module property dialog with the data for a simulation run **258**

7.12 Completed Arena model of the simple queueing system with a single server **258**

7.13 Animation of the simple queueing system with a single server **259**

7.14 Category Overview Report for the simple queueing system with a single server **260**

7.15 Bank lobby flowchart **265**

7.16 Defining resources in the bank lobby model **266**

7.17 Defining set (of resources) in the bank lobby model **266**

7.18 Initiating the arrivals schedule in the bank lobby model **266**

7.19 Numerical definition of the arrivals schedule in the bank lobby model (edited via spreadsheet) **267**

7.20 Graphical view of the arrivals schedule in the bank lobby model **267**

7.21 Defining the probability of simultaneous arrivals in the bank lobby model **268**

7.22 Defining the probability that the customer will request specific number of banking transactions **269**

7.23 Defining the balking decision in the bank lobby model **269**

7.24 Defining the service in the bank lobby model **270**

7.25 Defining the balking statistics in the Record module of the bank lobby model **271**

7.26 Defining the clock animation in the bank lobby model **271**

7.27 Clock animation picture inserted in the bank lobby model (screen snapshot during simulation run) **272**

7.28 Simulation reports by replication for the bank lobby model (Replication 1, Process, WIP) **273**

8.1 Event graph for multiple identical server queue **282**

8.2 Elements of activity cycle diagrams **283**

8.3 Activity cycle diagram of a single-server queue **284**

8.4 Elements of elementary Petri nets **285**

8.5 Firing of Petri net transitions: (a) transition T_4 enabled and (b) after firing of transition T_4 **286**

8.6 Timed transitions in extended Petri nets and their firing mechanism: (a) timed transitions, (b) transition T_4 firing at $t = t_0$, and (c) state changes at $t = t_0 + t_s$ caused by finalizing the firing of transition T_4 **287**

8.7 Petri net model of a single-server queue **288**

8.8 Arena model of single-server queueing system **289**

8.9 Arena model for bank lobby **289**

8.10 Copy center model **293**

8.11 Decide dialog for slow copier **296**

8.12 Dialog for Separate module in copy center model **296**

8.13 Process dialog for slow copier **298**

8.14 Dialog for Batch module in copy center model **298**

8.15 Animation of copy center in Arena **299**

8.16 Run setup dialog **300**

8.17 Tree view of copy center simulation output **300**

8.18 Entity Time report for copy center model **301**

8.19 Waiting Time detail report for all replications **302**

8.20 Top-level hierarchical model for fuel depot **305**
8.21 Create module for trucks in fuel depot model **305**
8.22 Decide module for top level in fuel depot model **306**
8.23 Submodel for Process module Get Fuel 1 **306**
8.24 Fuel depot model replenishment Decide module dialog **307**
8.25 Resupply submodel **307**
8.26 Fuel 1 resupply Process module dialog **308**
8.27 Pump Fuel 1 submodel in fuel depot model **308**
8.28 Animate Transfer toolbar **311**
8.29 Background for fuel depot animation **311**
8.30 Stations for fuel depot animation **312**
8.31 Dialog for animation station at Entrance location **312**
8.32 Finished animation network for fuel depot model **313**
8.33 Fuel depot submodels **313**
8.34 The Submodel Properties dialog for Enter Depot submodel **314**
8.35 Modules for Enter Depot submodel **314**
8.36 Dialog for Entrance station module **315**
8.37 Dialog for Route module in Enter Depot submodel **315**
8.38 Submodel for Check In station **316**
8.39 Select a Pump Decide module dialog **316**
8.40 Assign module dialog for Pump 2 in fuel depot model **317**
8.41 Sequence spreadsheet view **318**
8.42 Assignment of sequence to trucks **318**
8.43 Pump 2 Assign module dialog **318**
8.44 Route module dialog in Check In submodel **319**
8.45 Pump Fuel submodel **319**
8.46 Navigate panel in Fuel Depot model **320**
8.47 Navigate panel with Named View added **320**

9.1 Model of a simple queueing system **332**
9.2 A tandem queue **333**
9.3 Complete tandem queueing system **333**
9.4 Resource spreadsheet **334**
9.5 Completed Resource spreadsheet **334**
9.6 Building the model **335**
9.7 Process dialog for Get a Server Process module **335**
9.8 Resources dialog for Server Set **335**
9.9 Decide module dialog **336**
9.10 Process module dialog **336**
9.11 Resources module dialog **337**

9.12 Sample network of queues **337**
9.13 Network model of an election poll **338**
9.14 Process module dialog **342**
9.15 Entity spreadsheet with cost parameters **343**
9.16 Resource spreadsheet with cost parameters **343**
9.17 Demo model to show how costs are allocated **344**
9.18 Manufacturing system with production and assembly units **347**
9.19 A model conveyor **348**
9.20 Station module dialog **349**
9.21 Spreadsheet for defining segments **349**
9.22 Station definition spreadsheet **350**
9.23 Conveyor module spreadsheet **350**
9.24 Version 1 of the manufacturing model **351**
9.25 Dialog for Access module **352**
9.26 Convey module dialog **352**
9.27 Exit dialog for Conveyor 1 in version 1 **353**
9.28 Version 1 of model with blocking for full conveyor **354**
9.29 Dialogs for new process to machine part **354**
9.30 Version 2 of the manufacturing model with rework loop **356**
9.31 Animation graphic for version 3 of manufacturing system **357**
9.32 Segment spreadsheet for version 3 **357**
9.33 Conveyor spreadsheet for version 3 **358**
9.34 Model for version 3 of manufacturing system **359**
9.35 Decide module dialog for version 3 **359**
9.36 Dialog for rework part process **360**
9.37 Example of a flexible manufacturing system using AGVs **361**
9.38 Diagram of terminal operations **367**
9.39 Resources for package depot model **369**
9.40 Phase 1 package depot model: (a) truck arrival and unloading process; (b) van arrival, loading, and package delivery process **371**
9.41 Dialog for scheduled truck arrivals Create module **372**
9.42 Dialog for unscheduled truck arrivals Create module **372**
9.43 Process dialog for unloading trucks **372**
9.44 Dialog for Process module unload truck (truck arrival and unloading) **374**
9.45 Truck unloading subprocess model **375**
9.46 Unloading process for extra workers **377**
9.47 Process module fragment for unload truck submodel **377**
9.48 Search module for finding other worker requests **378**
9.49 Series plot of daily average number of packages in the terminal **380**

9.50 Autocorrelation function plot for average number of packages in the terminal (original unbatched data) **381**

9.51 Batch means for batches of 20 observations **381**

9.52 Autocorrelation function for batch means (batch size 20) **381**

10.1 Emergency department patient flow and urgent care center model **414**

A.1 *Union of events A and B* **419**

A.2 Cdf of X = sum of points on two dice **422**

A.3 *Parent population and sampling distribution of* $\hat{\theta}$ **435**

A.4 Distribution of $\overline{X}$ **437**

A.5 Distribution of $Z = \dfrac{\overline{X} - \mu_X}{\sigma_{\overline{X}}}$ **422**

A.6 Distribution of $t = \dfrac{\overline{X} - \mu_X}{s_{\overline{X}}}$ **422**

List of Tables

2.1 Minimum, most likely, and maximum values of uncertain parameters in the office building model **33**

2.2 Results of some what-if calculations on the office building model **34**

2.3 Estimates of π **40**

2.4 Summary of methods for generating random variates from common distributions **42**

2.5 Formulas used to compute random variates in the five-year office building model **63**

2.6 Sample cdf of present value of ATCF **65**

2.7 Outputs for additional simulation runs of office building model **65**

2.8 Scenarios for loss ratio simulation **70**

2.9 *Parameters and values* **73**

3.1 Holdings in stock portfolio **100**

3.2 *Years, aggressive growth, and bonds* **115**

4.1 Daily demand distribution for computers **130**

4.2 Sample path for Markov chain **141**

4.3 Confidence intervals (95%) for mean waiting time as a function of service time threshold **143**
4.4 Parameter values for two scenarios **146**
4.5 Batch data for inventory model **151**

5.1 Randomly sampled observations for interarrival and service times for the single-server queue model **164**
5.2 System state for the model of a supermarket checkout system **191**
5.3 Parameters in the message-processing model **193**

6.1 Randomly sampled observations for interarrival and service times for the single-server queue model **222**

7.1 Probability distribution of number of transactions per customer **264**
7.2 *Replication data* **274**

8.1 Parameters in copy center model **294**

9.1 Summary of nodes in election poll network model **338**
9.2 Entity costs for cost demo model **344**
9.3 Resource costs for cost demo model **344**
9.4 Arena modules to control transporters **363**
9.5 Initial costs in the manufacturing model **364**
9.6 Four scenarios for manufacturing model **365**
9.7 Results of four scenarios **365**
9.8 *Summary statistics* **382**
9.9 *Nodes and service time distributions* **386**
9.10 *Doors and probabilities* **386**
9.11 *Steps and distributions* **386**
9.12 *Steps and distributions* **386**
9.13 *Orders and probabilities* **387**
9.14 *Processes and distributions* **387**
9.15 *Periods and arrival rates* **388**
9.16 *Numbers and probabilities* **389**

A.1 *Probability mass function distribution* **421**
A.2 *Cumulative mass function distribution* **422**
A.3 True state of nature **422**

List of Models

The following chapter sections cover the following models:

2.2 Special furniture sale promotion
2.3 Office building construction and lease
2.4 Estimation of π
2.6 Inventory order quantity
2.8 Multiyear office building construction and lease
2.9 Loss ratio for insurance agency

3.2 Stock price
3.3 Call option price
3.3 Put option price and hedging
3.4 Dynamic stock price
3.4 Asian option price
3.5 Correlated asset values

4.2.3 M/M/1 queue
4.5.1 Markov chain inventory
4.5.2 Markov chain queueing
4.5.3 Markov chain reliability
4.6 M/M/1 queue: regenerative method
4.7 Advanced queue
4.8 Market share

5.3 Single-server queue
5.7 Message processing

6.3 Single-server queue
6.4 Single-server queue
6.5 Manufacturing system with shared resources

7.2 Simple queue
7.3 Bank lobby

8.3.3 Copy center
8.4.3 Fuel depot: hierarchical structure
8.5.2 Fuel depot: facility-based animation

9.2.3 Tandem queue
9.2.3 Tandem queue with blocking
9.2.6 Demo of cost parameters
9.3 Manufacturing system with conveyors
9.4 Package delivery system

About the Authors

Andrew F. Seila is Professor of Management Information Systems at the University of Georgia's Terry College of Business. He received the B.S. degree in physics and the Ph.D. in operations research and systems analysis, both from the University of North Carolina at Chapel Hill. He worked for North Carolina Blue Cross Blue Shield, the Highway Safety Research Center at the University of North Carolina, and Bell Laboratories in Holmdel, New Jersey, prior to coming to the University of Georgia.

Professor Seila's interest in simulation began in graduate school and has continued throughout his career. At the University of Georgia, he has taught many courses on simulation, both at the undergraduate and graduate levels, to students in business, engineering, statistics, computer science, forestry, and other areas. He was one of the earliest advocates of using spreadsheets to develop simulation models. His research interest has concerned both statistical methodology for simulation and simulation modeling methodology. He is the author of SIMTOOLS, a software toolkit for simulation using Pascal. Professor Seila is a regular contributor to the Winter Simulation Conference and served as its Associate Program Chair and Program Chair in 1993 and 1994, respectively. Currently, he is interested in the application of simulation and other modeling methods to problems in health care management.

Vlatko Ceric is a Professor in and head of the Business Computing Department at the Graduate School of Economics and Business, University of Zagreb, Croatia. His research interests are simulation modeling, decision support systems, information retrieval, Internet technology, electronic commerce, and operations management. He has published more than 80 papers and several books in this field. He also led several research and application-oriented projects. He was the editor-in-chief of the international *Journal of Computing and Information Technology,* head of the International Program Committee of the international conference "Information Technology Interfaces," and a member of program committees of several international conferences. He reviews papers for several international journals and conferences.

Pandu Tadikamalla is a Professor of Business Administration at the Joseph M. Katz Graduate School of Business, University of Pittsburgh. Dr. Tadikamalla received his M.S. and Ph.D. in Industrial and Management Engineering from the University of Iowa. He teaches courses in decision technologies, statistical techniques for management, simulation, and total quality management. Dr. Tadikamalla received several awards in recognition of his dedication to and excellence in teaching. He has also consulted for several multinational corporations. Dr. Tadikamalla's research interests lie in the areas of simulation methodology and statistical techniques for business management. He is on the editorial review board of the *American Journal of Mathematical and Management Sciences*. Dr. Tadikamalla has published more than 40 research articles in several professional journals.

1

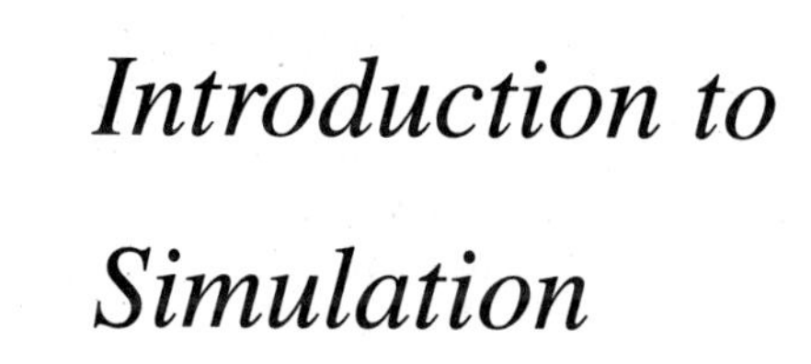

Introduction to Simulation

This chapter provides a general introduction to the area of modeling and analysis of probabilistic models using simulation. We will stress the idea that simulation methodology involves a set of techniques for representing operational aspects and relationships in a model by sampling variables (or obtaining observations) and using these observations to estimate one or more parameters of interest. It is important to understand not only the methods and techniques for developing and analyzing a model using simulation, but also where the field of simulation fits into the overall scheme of operations research (OR) and management science (MS) modeling. Modeling and analysis using simulation are two separate but related activities. Several examples will be used to solidify these ideas.

1.1 DECISIONS AND DECISION MODELS

Almost any time a decision is made, a model is used to aid the decision maker. In many, if not most, cases, this model is an implicit and ill-defined behavioral model that involves relationships and scenarios such as "I believe that if I make *this* decision, then I will get *that* outcome." On the other extreme, models can be overt and explicit—for example, a spreadsheet model that gives numerical relationships between decision variables (the quantities the decision maker can control) and the outcome of the decision, or a linear programming model in which the decision maker's objective is explicitly related to the decision variables.

Decisions can take many forms. Business decisions may involve determining warehouse capacities, production levels, staff hiring levels, or other parameters of a production system; selecting or rejecting certain investment opportunities; determining whether to increase the prime interest rate in a financial context; selecting a plan to market a new product; or determining how much money to retain for a self-insurance plan. Engineering decisions may involve such considerations as whether to accept a design for a new automobile suspension or a new bridge, or how many processors to put in a newly designed computer system.

As an example of a decision in statistics, suppose that we wish to estimate the mean cost per year for a college education in the United States using a random sample of college students in the United States. This involves computing the usual 95 percent confidence interval, obtaining lower and upper confidence limits, and announcing that we are 95 percent confident that the mean cost is some value between these two limits. In fact, we have used a model—a statistical model—to make this decision.

The field of *operations research* uses explicit mathematical models to make decisions; that is, any time a decision is made with the use of a mathematical model, the work can be classified as operations research. *Management science,* on the other hand, is a closely related field that involves the use of models to make administrative or managerial decisions. There is considerable overlap in these two fields, because most models used are mathematical in nature, and many decisions made can be considered to be managerial. In practice, the two titles are used interchangeably, as OR/MS or MS/OR.

A *system* is a set of interacting components or entities. Generally, the components operate together to achieve a common goal or objective. When the components work together, the system can perform tasks and accomplish objectives that individual components cannot. As an example, consider the system for stocking and selling goods at a grocery store. The components could be:

- the arriving customers,
- the checkout counters and attendants,
- a customer service attendant,

- the queues or waiting lines for customers,
- the baskets for carrying groceries, and
- the store with its stock.

Other systems include a hospital, with its patients, rooms, personnel, equipment, and so on; an air traffic control system, with its radar detectors, planes, runways, and controllers; a telecommunications system with its messages, communication network, and receiving stations; a manufacturing system with its machine centers, inventories, production schedule, and items produced; and a criminal justice system with its law enforcement officers, courts, jails, and probation officers. Note that the entities in a system may be physical entities, such as hospital rooms, planes, or machines, or they may be only conceptual entities, such as messages in a communication network.

Most systems that are of interest in the real world are highly complex. It is useful to divide complex systems into subsystems. A *subsystem* is part of a system that performs a specific task or achieves a specific objective for the entire system. One might divide a hospital into natural subsystems of emergency room, surgical suite, intensive care unit, maternity ward, medical wards, X-ray department, laboratory, dietary department, and laundry. Each subsystem can then be divided further into its own subsystems as necessary to describe the subsystem's operation.

A *model* is an abstract and simplified representation of a system. Generally, a model is a specification of which system components are important and of the way in which they interact. The words *abstract* and *simplified* are important in this definition. A model is not an exact re-creation of a system but a simplified description that specifies the assumed relationships between system components. For example, our model of the checkout system might be as described in the following diagram:

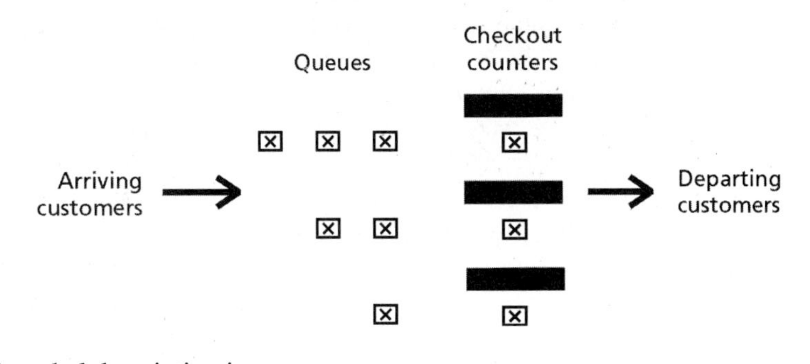

A verbal description is:

Customers arrive singly with independent and identically distributed exponential* interarrival times with mean .85 minutes. There are four checkout lines,

*See Appendix for a review of probability and statistics.

each having its own queue. Upon arrival, the customer joins the queue with the fewest customers, or, if a facility is empty, service begins immediately. Customers are served from each queue in first-come, first-served order. Service times are independent and uniformly distributed between .2 and 5.0 minutes. Upon completion of service, the customer departs the system.

We have abstracted and simplified the system in five ways:

1. We have not considered the store and its stock explicitly.
2. We have represented interarrival times as independent, identically distributed exponential random variables.
3. We have represented service times as independent, identically distributed uniform random variables.
4. We have not allowed customers to move between queues—that is, when another queue becomes shorter or empty, a customer who is waiting may not move to this queue.
5. Each queue is served in strict first-come, first-served order.

Thus, many of the details of the real system have been omitted from the model. In a later section, we will discuss models and model building and explore when and why some of these details should be omitted.

A *stochastic* or *probabilistic model* is a model in which randomness or uncertainty is inherent—that is, variables that are random or uncertain are involved in an essential way. As an example, consider the problem of determining how much of a particular soft drink to order for a basketball game. The optimal amount balances (1) the costs of having excess drink after demand has been met and (2) not having enough drink to meet demand. Unfortunately, the exact number of people attending the ball game is unknown at the time the soft drink order must be placed. The attendance depends on such factors as the public interest in the game, the weather, and numerous personal considerations applying to each potential attendee. Even if the attendance were known in advance, the sales of soft drinks would be unknown (but more predictable) because the number of drinks purchased by each attendee cannot be predicted. Thus, the demand for soft drinks must be considered a random quantity at the time the order is placed. However, by looking at similar games in the past, we can determine the relative likelihood of different values for the demand. In other words, we can fairly well specify the probability distribution of this random variable. If the demand for soft drink were known exactly, then we would order that exact amount of soft drinks. In this case, none would remain after the game and no attendee would be denied a soft drink because of lack of supply. Because the demand is not known exactly, one might argue that the mean, or expected value, of the demand should be ordered. Simple examples will show that this is not the amount that will minimize the cost (Hillier and Lieberman, 1995). Thus, the *distribution* of the demand must be considered explicitly to solve this problem. It is not sufficient to consider just the mean or any other *certainty equivalent* for the demand, and therefore this model is a stochastic model.

1.2 GETTING ANSWERS FROM MODELS

We have already alluded to the reason for developing models—to provide information for making decisions. Hence, the term *decision model* is frequently used. The model helps in making decisions by allowing us to understand how the system behaves and by predicting how the system will respond to various decisions. Then the decision that produces the most desirable response can be given stronger consideration. We say that it is given stronger consideration to recognize that *decision makers, not models, make decisions.* The considerations involved in selecting a decision include those that can be quantified and included in the model plus those that cannot be built into the model. These latter considerations usually fall into the area of political considerations.

A *parameter* is defined to be any numerical characteristic of a model or system. In other words, a parameter is a number that describes something about the model. We can classify parameters into two types: input parameters and output parameters. An *input parameter* is any parameter whose value is required as part of the model specification. In the queueing model just presented, there were four input parameters:

1. the number of counters, or servers (4);

2. the mean interarrival time between customers (.85 minutes); and

3. (and **4.**) the service time distribution parameters (.2 and 5.0).

Before a model is fully specified, all input parameters must be given explicit numerical values. In the soft drink inventory model, the mean and variance—μ and σ^2—of the demand distribution, as well as the order cost, stockout cost, and holding cost are all input parameters. They are known or can be estimated from existing data, and they must be given before the model is fully specified. Conversely, once they are all specified, the behavior of the model is determined (though that behavior may be random, or stochastic).

Some decisions are specified by the values of input parameters; others are given by policies, or rules, that govern the operation of the system. For example, in our checkout counter model, if we set the number of servers to four, then we have decided to operate the system with four checkout counters. The decision to increase the number of open checkout counters to five is represented by changing the value of this input parameter to five. As an example of a "policy" decision, the specification that customers in each queue will be served in strict first-come, first-served order represents a policy regarding the system. If this rule were changed to say that each time a customer leaves the checkout counter, the next customer to be served will be the one with the shortest service time (the smallest amount of groceries) among those currently waiting, then this would represent a different operating policy for the system. Another operating policy might be that a new checkout counter would be opened at the moment that three customers are in any queue, and remain open until the queue served by that counter is empty.

An *output parameter* is defined to be any parameter whose value is determined by the system and its input parameters. Generally, an output parameter specifies some measure of the system's performance. Examples of output parameters for the above system are:

1. the mean number of customers waiting in the system,
2. the mean time spent waiting in the queue, and
3. the utilization of checkout counters.

Each of these output parameters specifies something about how well the system performs. We want the mean number of customers in the system and the mean waiting time per customer to be minimal; however, we want the utilization of checkout counters to be as large as possible.

The relationship between input parameters, the model, and output parameters can be depicted as in the following diagram:

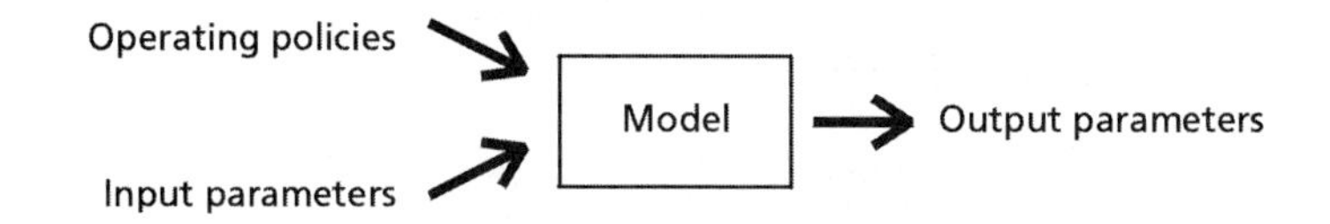

Output parameters, therefore, are the "answers" that we want the model to provide to aid us in decision making. These parameters must be obtained from the model because they are not given as part of the model specification. (Indeed, if they were part of the model specification, we would have no need for the model—we would already have the answers we want!) Thus, we must specify how the model relates the output parameters to the input parameters. There are two basic ways to do this:

1. use mathematical analysis, and
2. use simulation.

There is an important difference between these two approaches. A mathematical analysis will yield formulas or a computational procedure to produce an exact value of the model's performance measure (output parameter). A simulation, however, will yield a sample of observations that can be used to compute a confidence interval for the performance measure, and therefore to estimate the value of the performance measure from data. Thus, simulation cannot be used to compute the exact value of the performance measure. We will see this distinction more clearly in the following chapters.

The mathematical tool that is used to derive and compute output parameters for stochastic models is *probability theory.* For example, for certain simple queueing systems, probability theory can be used to derive a formula that relates the mean waiting time in the queue to the arrival rate and the service rate (Hillier and Lieberman, 1995). Unfortunately, the majority of realistic stochastic models are too complex for analysis using probability theory. This leaves simulation as the only other available method for obtaining information about the performance measure(s) in which we are interested.

1.3 A GENERAL DEFINITION OF SIMULATION

Since the early 1970s, simulation has been increasingly employed for the solution of problems of human endeavor in business, engineering, science, and social science. Most simulation applications are developed for private businesses and are not reported in the academic literature. The Society for Computer Simulation (http://www.scs.org) has many publications that concern the application of simulation to a wide range of problem areas. The Winter Simulation Conference (http://www.wintersim.org) is held annually in early December to share the latest developments in simulation applications and methodology. The complete text of published articles in previous Winter Simulation Conference Proceedings is available online at the INFORMS College on Simulation home page (http://www.informs-cs.org/wscpapers.html). In these publications you will find a wide variety of simulation applications, including:

- passenger flows at an airport terminal (Gatersleben and van der Weij, 1999),
- patient care at an outpatient obstetrical unit (Isken, Ward, and McKee, 1999),
- call center operations (Chokshi, 1999),
- public housing construction in Hong Kong (Jinsheng, Zheng, and Tam, 1998),
- product development at Timberland (Grosz, 1998),
- staffing in a semiconductor manufacturing system (Kotcher, 2001), and
- work flow in a financial services firm (Ferrin, Miller, and Giron, 2000).

Other applications of simulation in the private sector involve analyzing the operations of flexible manufacturing systems, automated guided vehicles, and automated storage-and-retrieval systems; evaluating manufacturing designs for circuit boards, designs of telecommunications systems, and designs of warehousing operations and logistics; and analyzing computer system designs. In the public sector, simulation has been applied to evaluate road design and traffic signal scheduling, mail processing systems, driver licensing systems, patient transportation, emergency room operations and radiology department services in hospitals, nuclear waste disposal, and financial planning in universities. Other applications have involved analysis of battlefield strategies, taxi fleet sizing, newspaper production, and crop management. These applications are all described in various articles in the *Proceedings of the Winter Simulation Conference,*[*] and represent only a small portion of the current applications of simulation to real-world problems and a much smaller portion of the potential applications.

The increase in the application of simulation to business and industrial problems over the past 30 years is attributable to three developments. First, fast and easy-to-use computers are available to most people employed by businesses or government agencies. Indeed, many simulations can be run on personal computers. Second, efficient software is available that allows simulation programs to be developed quickly and easily. Much of this software is available for personal computers as well as larger computers. And third, the number of students who study simulation in engineering, business, and related curricula has increased greatly in the past decade. Simulation is valuable both for its relative simplicity as compared to analytical (mathematical) techniques and for its potential to study a large number of variants of the original model without incurring the costs associated with experimenting with the real system or the difficulties in rederiving formulas for output parameters that are inherent in mathematical analysis.

Simulation is a set of numerical and programming techniques for representing stochastic models and conducting sampling experiments on those models using a digital computer. This is a broad definition and includes virtually any type of stochastic model. It also explicitly reflects the fact that a simulation produces (random) observations on a quantity of interest.

It is important to note that simulation is a set of *techniques*—analysis methodology. It is not a special type of model, as would be implied by the often-used term *simulation model.* Instead, simulation involves methodology for extracting information from a model by "observing" the behavior of the model with the use of a digital computer. The term *simulation model* actually means a model that has been adapted to be analyzed with the use of simulation.

Many excellent reference works and other textbooks on simulation are available, each presenting the authors' personal views of simulation (as does this text). Winston (1996) deals with static simulations and uses a spreadsheet as the tool to build and run the models. Pidd (1992) concentrates on discrete-event simulation and general simulation principles. Kelton, Sadowski, and Sadowski (2002) provide a general introduction to discrete-event simulation and detailed instructions on using the Arena software product. Banks et al. (2001) cover discrete-event simulation but provide more fundamental concepts and more complete coverage of the statistical

---

[*]The *Winter Simulation Conference Proceedings* are available online at <http://www.informs-cs.org/wscpapers.html> for the years 1997 through 2001; new *Proceedings* papers are added each year. Print copies for earlier years are available in many libraries.

analysis methodology. Law and Kelton (2000) have a text that has been available in various versions since 1982; it provides even more detailed material on modeling and analysis methodology. Fishman (1996) is a highly authoritative and complete reference for the statistical methodology involved in simulation. Finally, Banks's *Handbook of Simulation* (1998) has broad coverage of all important methodologies and topics that concern simulation. This is a good reference and tutorial book for readers who plan to apply simulation to actual problems.

We can categorize simulations into three broad types: gaming, static systems, and dynamic systems.

1.3.1 Gaming Simulations

Gaming simulations involve the interaction of one or more persons with the simulation program in an essential way. A familiar example is that of a video game in which a human player tries to beat a computer or another player; the game is represented by the simulation program. Other examples include war games used by the military to train officers in battle tactics, flight simulators used to train pilots to handle aircraft in bad weather, and management games that teach managers how to effectively administer various business systems. Such management games are available for marketing, finance, and other functional areas of business.

Arcade and other video games primarily serve to entertain the players; other games primarily serve to educate. (In fact, all simulations do both.) These games may or may not involve time as an essential element; however, time is usually an important part of the simulation. It is implicitly part of amusement games and aircraft simulators, which run in *real time*. In other games, decisions from the human participant are made and processed sequentially. The objective of an inventory game, for example, may be to teach a participant how to deal with lags in product delivery.

1.3.2 Static Simulations

Static simulations operate by sampling observations and transforming them according to formulas or rules that compose the model. This process is repeated independently many times to produce a set of independent, identically distributed observations that are then used to study the characteristics of the transformed random variable. For example, suppose that on a particular day, the number of gasoline sales at a convenience store is a random variable having a Poisson distribution with mean 12.3. The amount of each sale is a random variable having approximately a triangular distribution with minimum value \$4.33, most likely value \$14.10, and maximum value \$37.48. We are interested in studying the distribution of the total sale on a randomly selected day. For example, we might wish to know the probability that the total sale exceeds \$200. To study the total sale, we would first sample the number of sales using a Poisson distribution; then, given the number of sales, we would sample each sale using the triangular distribution. The total sale would be computed by summing the sales. This would produce one observation of the total sale in a day. This entire process would be repeated to produce a sample of observations of the total sale. These data would then be used to estimate the probability that the total sale exceeds $200.

Note that the output of a static simulation consists of *independent, identically distributed* observations, as opposed to observations that are ordered or produced in a particular sequence. Thus, time does not play an essential role in static simulations (although time may be involved in the transformation that is used to produce each observation). These simulations have been used a great deal by mathematicians, physicists, engineers, and mathematical statisticians to solve mathematical problems that could not be solved using mathematical analysis. They are also important to risk analysis in business.

1.3.3 Dynamic Systems Simulations

The term *dynamic* refers to the behavior of a process over time, and dynamic systems simulations observe the behavior of system models over time. Simulations of this type constitute the bulk of simulation work in management science. Examples include inventory, queueing, production, and transportation systems.

We can further divide dynamic systems simulations into two subareas: continuous and discrete. *Continuous simulations* (dynamic systems) involve models in which the quantities of interest are represented as variables in differential equations that may also be influenced by random disturbances. These variables, which are called *state variables,* change continuously as time progresses, hence the name *continuous simulation.* Examples include models of water flow in a river, steel production, and cash flow in a financial institution.

Discrete simulations, on the other hand, allow system quantities to change only at discrete points in time. Those points in time when system quantities (state variables or attributes) change are called *events,* hence the name *discrete event simulation.* Examples include manufacturing systems that produce discrete parts, inventory, transportation, computer, and communications systems. Dynamic systems simulations can be both discrete and continuous, as in the case of a parts manufacturing plant that also makes the castings for the parts.

Discrete event simulations represent the majority of simulations in applied management science for two reasons: First, most of the models of interest to management scientists are discrete event models: production and inventory, queueing, transportation, and other systems. Second, discrete simulations are somewhat more natural to program. In practice, continuous simulations can be approximated by discrete event simulations; however, the converse is not usually true—continuous simulations cannot generally be used to approximate discrete simulations.

In this textbook, we will focus our attention on static simulations in Chapters 2 and 3 and on discrete event simulations in Chapters 5 through 9. These types of simulation represent most of the simulation applications in management science. The other types—games and continuous simulation—require specialized programming techniques, are directed toward specialized audiences, or both.

1.4 COMPONENTS OF A SIMULATION STUDY

Most studies that involve modeling have the following components:

1. statement of the decision problem and objectives,
2. system analysis,
3. analysis of input distributions and parameters,
4. model building,
5. design and coding of the simulation program,
6. verification of the simulation program,
7. analysis of output data to estimate parameters,
8. validation of the model,
9. experimental design,
10. simulation production runs,
11. statistical analysis and interpretation of data,
12. recommendations for decisions and implementation of the model, and
13. final documentation of the model and simulation program.

The initial step involves determining on what decision or decisions the model will be used. It is important at this point to be comprehensive and specific and to involve all managers who will be making decisions relative to the system. The nature and the scope of the model will depend on the decisions being considered. Sometimes, one may contemplate using a single model for several different types of decisions concerning the same system. For example, a model of a production system may be used to determine the shop floor layout or the sizes of buffer inventories as well as to evaluate different scheduling policies. If all important decisions are not considered at this point, then the model, once developed, will probably be inadequate. If extraneous decisions are considered, then additional time, effort, and money will be expended unnecessarily and the model will possibly be too complex. With increasing complexity comes increasing cost and decreasing reliability, so providing for extraneous decisions could also lead to unnecessary expense, errors, and reliability problems.

In the second step, system analysis, a careful study and documentation of the system is done. At this point, the analyst wants to make sure that the system—

whether existing and operating or just being designed—is fully understood. It is important to get information from all persons that are involved with the management and operation of the system at this stage so that the information will be accurate and unbiased and those who manage the system will support the modeling effort. Sometimes, this step uncovers additional decision criteria that should be included in the statement of the decision problem, and thus we must return to step 1 to modify the objectives of the study. We will discuss this topic in greater detail in Chapter 10.

In the third step, data are gathered on the operation of the system and then analyzed to estimate model parameters. These data frequently combine explicit observations directly from the operating system or from a database associated with the system (orders, arrivals, schedules, etc.) and subjective estimates of system parameters, such as response times to schedule changes, or breakdown frequency for new equipment. In Chapter 3, we will discuss how to analyze data from an existing system in order to provide input to the model-building process.

The model-building process in step 4 involves extracting from the actual system those components and their interactions that are central to system operation, relative to the decisions and objectives determined in item 1. Model building is more art than science and requires much practice to master. It is important to develop a model that is as simple as possible while also being highly credible and having a good chance of being found valid for the anticipated decisions. Most experienced modelers agree that it is better to start with a simple model and add details later if necessary than to start with an excessively complex and realistic model. At this stage, it is not unusual to find that important information about system operation or parameters is missing, and thus that you must return to either or both of the two previous steps. In fact, this step is frequently performed simultaneously with the data-gathering and data-analysis step because the model-building process determines which system parameters and distributions must be known in order to run the simulation. Documenting the model is also important. State explicitly all assumptions and simplifications that have been made and give a detailed description of the model's structure. Model building will be discussed throughout this text.

Most simulation programs are quite complex. The actual complexity of the program depends on the complexity of the model as well as the software tools that will be used to write the simulation program. Even with the current generation of simulation languages, simulation programs usually have hundreds and frequently thousands of lines of code. Therefore, the programming process is frequently assigned to a team of programmers working under a project leader. Good programming management is required at this point to keep the project on schedule and to develop reliable software. More discussion of this point will be given in Chapter 10.

Verification is the process of making sure that the simulation program actually represents the intended model. Debugging the program is part of this process, but verification also involves carefully observing the operation of the program, comparing it with the operation predicted by the model specification and observing any discrepancies. Good documentation of the model and the simulation program greatly expedites this step. Sometimes during the verification stage, one determines that important model characteristics were not specified in sufficient detail, and therefore the model specification must be revised.

The next two steps (analysis of output data and validation of the model) are generally performed simultaneously. Simulation programs, especially discrete event simulation programs, frequently require that special statistical techniques be applied to analyze the data produced. In the first chapters, where models are developed and simulated, simple but valid methods for analyzing data will be presented. In step 7, these techniques are identified and, if appropriate, are programmed to analyze the data. Validation is the process of determining if the model is a useful representation of the real system relative to the intended purpose of the model. To ensure that the model is valid, data must be available from the real system so that it can be compared to the data produced by the simulation program. If the model is found not to be valid, then one may need to return to step 2 (system analysis) or step 4 (model building) to revise the model before modifying the program, reanalyzing the data, and revalidating the model. The potential users of the model must be included in the validation process so that they will gain confidence in the model's predictive capabilities.

Once the model has been validated, it can be used to evaluate alternative decisions or system designs. Frequently, the number of alternative decisions is so large that a simulation run cannot be made for each of them. Therefore, some decisions must be selected to be evaluated. The process of determining which decisions will and will not be evaluated is called *experimental design.* Establishing the experimental design to be used is an important but often overlooked aspect of simulation. After the experimental design has been specified, production runs can be made. Sometimes this requires that "pilot" runs be made to determine how long the production runs must be to provide acceptable accuracy in the estimates of the performance measures.

Statistical analysis of the results may be done using a commercially available statistical package, or it may be programmed as part of the simulation program, depending on the nature of the model, availability of software, and preferences of the investigators. Because the simulation can provide only estimates, and not exact values, of system performance, it is important to know their accuracy. The primary purpose of statistical analysis of the data is to provide a measure of accuracy. With the estimates of system performance and their accuracy in hand, alternative decisions can be evaluated. This may result in selecting other decisions to evaluate that were not included in the current experimental design and then modifying the model to include additional alternatives and evaluating these, or it may involve selecting one or more decisions to recommend. It is not unusual that a model will be developed so that it can be used on a regular or irregular basis to evaluate decisions. If this is the case, then the program must be installed on one or more computers, training and reference manuals must be developed, and users must be taught to use the model. This process, which might also involve adding "bells and whistles" to the user interface, is called *implementation.*

Finally, once the model has been implemented, it is important to document "permanently" both the model and the simulation program. We want to stress that documentation should be part of each step in the modeling process. If this effort is saved for last, then important details will be lost. However, it is equally important to make sure at the end that the documentation is assimilated and put into a form that is easy to access and understand for persons who might wish to understand or modify the model or the program in the future.

1.5 REASONS FOR USING SIMULATION AS AN ANALYSIS TOOL

There are many reasons for using simulation, as opposed to mathematical analysis, to get information from a model. Here are a few.

The model may be too complex to allow the output parameters to be computed using mathematical analysis, leaving simulation as the only method available. As we have discussed previously, most realistic models of actual systems are much too complex to be analyzed mathematically. This leaves the analyst with the options of either further simplifying the model and perhaps making it unrealistic or using simulation.

The system under study may yield a model that is so complex that it cannot easily be described by a mathematical model (i.e., a set of equations or inequalities); however, the operations of the model can reasonably be represented by a computer program. In this case, it is easier to *sample* the model's behavior than to analyze it mathematically because it is difficult to represent the behavior mathematically. Examples of systems in this category involve large-scale models of computer communications systems and production, inventory, and distribution systems.

The investigator's primary interest might be to either experiment with the system model to find the design that maximizes one or more performance measures or simply study the behavior of the system. Experimenting with the actual system could be impossible, if it does not exist yet, or extremely expensive if it does exist. Examples of situations in which these considerations apply include combat simulations, design of nuclear generating facilities, and large-scale production and distribution facilities. Indeed, if one were to experiment with the real system, then the time needed to accumulate enough data would be normally unreasonably long, and only one replication of the experiment would be performed.

The modeling effort is frequently useful in itself because it leads to a better understanding of the system. To run the simulation, the model must be completely specified—that is, all components of the model along with explicit rules for how they interact must be given as well as all relevant input parameters. The effort of analyzing the system for model specification usually leads to a better understanding of the system and can suggest useful changes even without the remainder of the simulation study.

As a tool, simulation carries a certain amount of credibility with management. It is easier to explain to management the efforts involved with a simulation study than to explain the process of deriving a mathematical solution for the model using the arcane language of mathematics. Many modern simulation languages include facilities for animation that present a pictorial image of the system under study. This is quite useful to show management the operation of the model and gain its acceptance. Without management's acceptance, it is unlikely that the results of the modeling efforts will be used.

For these reasons and the availability of computer hardware and software for simulation, it is now one of the most widely used analysis techniques in operations research and management science. Simulation and linear programming are generally considered to be the two most widely used tools in applied MS/OR work. These tools are widely used for two reasons:

1. They are applicable to a large number of models, and
2. computer equipment and methodology are widely available to implement models using these techniques.

1.6 SUMMARY

Most decision makers use a model of some type to help them evaluate various alternatives. Operations research and management science are fields of study concerned with the use of explicit mathematical models to analyze the effects of alternative decisions. A system is a collection of interacting components that work together to achieve a common goal. Real-world systems are usually quite complex; therefore, to study their behavior, a model—an abstract and simplified representation of a system—is created and analyzed.

If the model uses uncertainty, or probability, as an essential part, then it is a stochastic model. To fully specify a model, all input parameters as well as operating policies must be specified. These, along with the model, will determine the values of output parameters, or performance measures. There are two ways to determine the relationship between input parameters and operating policies, and output parameters: (1) mathematical analysis and (2) simulation. Mathematical analysis allows an exact or approximate value of the output parameters to be calculated. Simulation can only produce statistical estimates of the parameters.

Simulation is a set of numerical and programming techniques for representing stochastic models and conducting sampling experiments on them using a digital computer. There are three categories of simulations: (1) gaming simulations, which require the interaction of one or more persons; (2) static simulations, which produce independent observations on a random variable; and (3) dynamic systems simulations, which observe the behavior of a system over time. Dynamic systems simulations can be further subdivided into continuous simulations and discrete event simulations. Discrete event simulations, which allow the system to change state only at discrete points in time, represent the largest group of simulations in management science.

A simulation study consists of the following steps:

- stating the decision problem and its objectives,
- analyzing the system,
- analyzing data on input parameters and distributions,
- building the model,
- writing the simulation program,
- verifying the simulation program,
- analyzing output data,
- validating the model,
- making production runs,
- analyzing and interpreting data from production runs,
- making recommendations and implementing the model, and
- documenting the model and simulation.

Usually, these steps will be repeated as the project develops and later stages show the need for refinement in the results of earlier stages.

Simulation has become a popular and widely used tool in management science and operations research for several reasons. Most realistic stochastic models cannot be analyzed mathematically, so simulation is frequently the only tool available—especially when the model is so complex that it cannot easily be represented mathematically. Simulation is useful if the investigator wishes to experiment with the system but such experimenting with the actual system would be prohibitively expensive and normally would not yield a useful amount of data. Management often is more willing to accept the results of a simulation because the model can be demonstrated with the use of animation or by presenting the results of validation runs.

References

Banks, J. 1998. *Handbook of simulation.* New York: John Wiley.

Banks, J., J. S. Carson III, B. L. Nelson, and D. M. Nicol. 2001. *Discrete-event system simulation.* Upper Saddle River, New Jersey: Prentice Hall.

Chokshi, R. 1999. Decision support for call center management using simulation. In *Proceedings of the 1999 Winter Simulation Conference,* ed. P. A. Farrington, H. B. Nembhard, D. T. Sturrock, and G. W. Evans, 1634–1639.

Ferrin, D. M., M. J. Miller, and G. Giron. 2000. Electronic workflow for transaction-based work cells in a financial services firm. In *Proceedings of the 2000 Winter Simulation Conference,* ed. J. A. Joines, R. R. Barton, K. Kang, and P. A. Fishwick, 2055–2058.

Fishman, G. S. 1996. *Monte Carlo: Concepts, algorithms and applications.* New York: Springer.

Gatersleben, M., and Simon W. van der Weij. 1999. Analysis and simulation of passenger flows in an airport terminal. In *Proceedings of the 1999 Winter Simulation Conference,* ed. P. A. Farrington, H. B. Nembhard, D. T. Sturrock, and G. W. Evans, 1226–1231.

Grosz, D. 1998. Application of business process modeling at Timberland. In *Proceedings of the 1998 Winter Simulation Conference,* ed. D. J. Medeiros, E. F. Watson, J. S. Carson, and M. S. Manivannan, 1357–1361.

Hillier, F. S., and G. J. Lieberman. 1995. *Introduction to Operations Research.* 6th ed. New York: McGraw-Hill.

Isken, M. W., T. J. Ward, and T. C. McKee. 1999. Simulating outpatient obstetrical units. In *Proceedings of the 1999 Winter Simulation Conference,* ed. P. A. Farrington, H. B. Nembhard, D. T. Sturrock, and G. W. Evans, 1557–1563.

Jenshing, J. S., S. X. Zheng, and C. M. Tam. 1998. Modeling and simulation of public housing construction in Hong Kong. In *Proceedings of the 1998 Winter Simulation Conference,* ed. D. J. Medeiros, E. F. Watson, J. S. Carson, and M. S. Manivannan, 1305–1310.

Kelton, W. D., R. P. Sadowski, and D. A. Sadowski. 2002. *Simulation with Arena.* Boston: McGraw-Hill.

Kotcher, R. C. 2001. How "overstaffing" at bottleneck machines can unleash extra capacity. In *Proceedings of the 2001 Winter Simulation Conference,* ed. B. A. Peters, J. S. Smith, D. J. Medeiros, and M. W. Rohrer, 1163–1169.

Law, A. M., and W. D. Kelton. 2000. *Simulation modeling and analysis.* Boston: McGraw-Hill.

Pidd, M. 1992. *Computer simulation in management science.* Chichester, England: Wiley.

Winston, W. L. 1996. *Simulation modeling using @RISK.* Belmont, California: Duxbury.

Problems

1.1 For the following systems, identify the components and describe briefly how they interact to achieve a goal.

a. a hospital emergency room

b. a railway freight depot

c. a university

d. the sales department of a manufacturing company

e. a small farm

f. a school system

g. a telephone network in an office building

h. a small computer system

i. a harbor

1.2 The following is a model of a small copying center that might be in a business or open to the public.

The center has two copy machines. Machine 1 is a large, high-volume copier with a collator that can produce 60 pages per minute on average. The second machine is a small, low-volume copier that can produce 20 pages per minute on average and does not have a collator. Customers arrive individually. Each customer has a document with a specific number of pages to copy and a specific number of copies desired. When customers arrive, they choose copier 1 if they have more than 50 pages to copy or intend to make two or more copies of the document. Each copier has a queue of customers waiting for it, and customers are served from each queue in first-come, first-served order. On each copier, the time required to make the copies consists of setup time plus the time to actually produce the copies at the machine's rate.

Describe in what ways this model might have simplified the operation of an existing copying center. Make a list of all variables in this model, and classify each variable as to whether its value is certain or random.

1.3 The following is a sketch of a model of the deli counter in a large grocery store.

The counter has three employees who serve customers. Customers may request two types of items: prepared foods such as salads and precut meats and cheeses, and unprepared foods such as meats and cheeses that must be sliced. Waiting customers are served in the order they arrive. Service times depend on how many of the prepared foods and how many of the unprepared foods are requested. We assume that all foods requested are always available.

Describe the ways in which this model might have simplified the operation of an existing deli counter. Make a list of all variables in this model, and classify each variable as to whether its value is certain or random. Identify the input parameters of the model.

1.4 The following is a model to determine the amount of three products to produce.

Let x_1, x_2, and x_3 be the amounts of items 1, 2, and 3, respectively, to produce. The profit per item is $10 for item 1, $12 for item 2, and $23 for item 3. Labor requirements are 2.2 hours for item 1, 2.5 hours for item 2, and 3.1 hours for item 3. Total labor must not exceed 1200 hours. Storage space requirements per item are 2.0 cubic feet for each product. Total storage space available is 1100 cubic feet. Because we want to maximize profit, the model can be stated as the following linear programming model:

$$\text{maximize } 10x_1 + 12x_2 + 23x_3$$

subject to

$$\begin{aligned} 2.2x_1 + 2.5x_2 + 3.1x_3 &\leqslant 1200 \\ 2.0x_1 + 2.0x_2 + 2.0x_3 &\leqslant 1100 \\ x_1 \geqslant 0, x_2 \geqslant 0, x_3 &\geqslant 0 \end{aligned}$$

a. State all assumptions and simplifications that are included in this model relative to the real system.

b. Identify all of this model's input parameters. How would these values be determined in practice? Which parameters might actually be random instead of fixed?

c. Identify the decision variables in this model.

d. Identify the output parameter of this model. State explicitly how the output parameter is related to the decision variables.

1.5 Consider the following inventory model for paper at a newspaper printer.

Periodically, an order will be placed for a quantity Q of paper. The cost for placing an order is K, and the cost per thousand pounds ordered is c. Let h denote the inventory holding cost per day per thousand pounds, and p denote the cost per day per thousand pounds when there is a shortage. (If there is a shortage, a more expensive paper can be used in place of the usual newsprint.) Let a denote the rate of paper use per day. We want to determine the optimal amount to order, Q, and the optimal amount of shortage to allow during an order cycle, R. An analysis of this model can be found in Hillier and Lieberman (1995), pages 761–767. The total cost per day (T) is

$$T = \frac{K + cQ + \frac{hS^2}{2a} + \frac{pR^2}{2a}}{\frac{Q}{a}}$$

The optimal value of Q is

$$Q^* = \sqrt{\frac{2aK}{h}} \sqrt{\frac{p+h}{p}}$$

and the optimal value of the maximum shortage allowed per period is

$$R^* = \sqrt{\frac{2aK}{p}} \sqrt{\frac{h}{p+h}}$$

a. State all assumptions and simplifications that are included in this model relative to the real system.

b. Identify all of this model's input parameters. How would these values be determined in practice? Which parameters might be random instead of fixed?

c. Identify the decision variables in this model.

d. Identify the model's output parameters. State explicitly how the output parameters are related to the decision variables.

1.6 Consider the following investment model.

An investor has \$10,000 to invest in one of two investments. Both investments will last five years. The first investment is a bond fund that pays quarterly interest, compounded quarterly. Let r_i be the quarterly interest paid in quarter i, for $i = 1$ to 20 (because there are 20 quarters in the five-year period). Then, the value (V_1) of the investment at the end of five years will be

$$V_1 = 10{,}000\,(1 + r_1)\,(1 + r_2)\ldots(1 + r_2)$$

The second investment is a stock fund that both pays dividends and increases in value over the five years. Let the price of the stock at the end of month i be p_i, for $i = 1$ to 60 (because there are 60 months in a five-year period), and let p_0 be the initial price of the stock. Let d_i be the dividend paid at the end of quarter i, for $i = 1$ to 20. The number of shares purchased initially is $S = 10{,}000/p_0$, and the total value of the investment at the end of the five-year period, ignoring present-value considerations, is

$$V_2 = p_{60}\,S + [d_1 + d_2 + \ldots d_{20}]$$

The objective is to select the investment that has the maximum value at the end of the five-year period.

a. State all assumptions and simplifications that are included in this model relative to the real system.

b. Identify all of this model's input parameters. How would these values be determined in practice? Which parameters might be random instead of fixed?

c. Identify the decision variables in this model.

d. Identify the output parameter. State explicitly how it is related to the decision variables.

1.7 Referring to the inventory problem on page 4, which was concerned with purchasing soft drinks for a basketball game, suppose that the soft drink must be purchased by the canister. A canister contains enough to serve 100 drinks, and a partially opened canister cannot be returned. It is assumed that the demand probabilities are .20 for between 0 and 100 drinks, .40 for between 100 and 200 drinks, .30 for between 200 and 300 drinks, and .10 for between 300 and 400 drinks. For each leftover and unopened canister, there is a \$10 return fee; for each canister (or portion of a canister) that was needed to meet demand but was not available, there is a cost of \$80 in lost profit. Suppose that two canisters are ordered. Then, the mean, or expected, cost is computed as follows:

$$\text{Expected cost} = 10(.20) + 0(.40) + 80(.30) + 160(.10) = 42$$

This calculation comes from the following analysis. The cost is \$10 if the demand is less than 100 drinks and one of the two canisters must be returned. The cost is zero if the demand is between 100 and 200 drinks and no canisters must be returned. If the demand is between 200 and 300 drinks, then we will be one canister short and incur a cost of \$80. Finally, if the demand is between 300 and 400 drinks, then we will be two canisters short and incur a cost of $80 for each canister, or $160 total.

For each of the other three possible decisions—order 1, 3, or 4 canisters—compute the expected cost. Which decision minimizes the expected cost?

Suppose you decide to order the mean number of canisters demanded (or the next larger number of canisters if the mean is not an integer). How many canisters would this solution tell you to order? Is this the optimal number of canisters to order, given the costs?

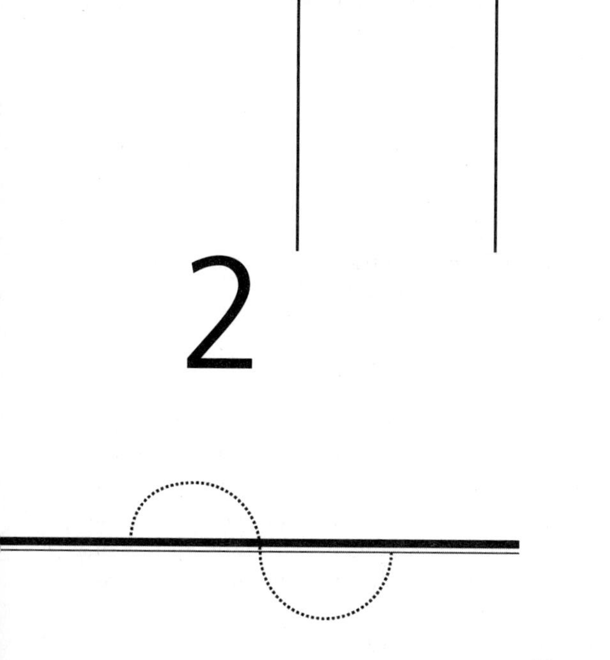

2

Static Simulations

This chapter focuses on static simulation using a spreadsheet. The topic is introduced and explored using a series of examples, including a single-period inventory model that first illustrates the basic concepts and techniques and later shows how to evaluate alternative decisions. A model to estimate the value of π demonstrates that reliable answers can be obtained if sufficient data are generated. A real estate model involving an office building illustrates the use of simulation in a more complex environment and shows how to use this technique to evaluate variation in spreadsheet models. The chapter concludes with an insurance model that is a static simulation but does not fit neatly into a spreadsheet. This model is presented not only to show that spreadsheets are not appropriate for every static model but also to make the reader aware that other useful simulation platforms are available.

The topics of random variate generation and output data analysis are also presented in an example context. Random variate generation is the technique used in simulation to sample observations from specified populations. Techniques for sampling from the most common distributions are demonstrated using a spreadsheet. Output analysis methods for static simulations use the familiar statistical techniques taught in introductory statistics courses. These are reviewed and demonstrated.

2.1 INTRODUCTION

In Chapter 1, we introduced the concept of a model as a useful representation for analyzing the possible (or likely) consequences of decisions, and the concept of a simulation as a sampling experiment concerning the outcomes of the model. In this chapter, these ideas will be developed and explored using static models. Recall that a static model is one in which we do not record observations on the system over time. The simulation consists of generating random variates and combining them according to the formulas or rules of the model to produce an observation for the output parameter, or performance measure, and then repeating this process a specified number of times—or running a specified number of *replications*—to produce a collection of independent, identically distributed observations that can then be analyzed using standard statistical and graphical methods to show the distribution of the data and estimate parameters of this distribution, such as the mean performance measure.

Models are so varied and specific to the system they represent that it is extremely difficult to present a general theory of model building. Many of the skills needed to be an effective model builder depend on the discipline from which the model comes rather than general model-building techniques. As a result, we will use a series of specific models in this chapter to teach various aspects of developing, running, and analyzing static simulations. The first model—which is a simple one concerning the relationship between profit (our performance measure) and price, cost, and sales—is used to introduce you to static simulation. Next, we look at a model involving construction and operation of an office building. This is a more realistic model that concerns complex relationships among the parameters and random variates that might begin as a spreadsheet calculation. The third model is also a simple one concerned with estimating the value of the mathematical constant π. Our objective here is to demonstrate that a simulation can provide a reliable estimate of the desired output parameter if it is allowed to produce enough data. This will hopefully increase your confidence in simulation as a tool to analyze stochastic models. For the fourth model we return to the inventory problem introduced in section 1.1. In real applications, models are used to aid in decision making. With this model, we show how simulation can be used to evaluate decision alternatives and then allow the best decision to be selected. The fifth model, which is a more complex version of the office building model in section 2.5, shows how to use simulation as an analysis technique to evaluate the risk induced by uncertain variables in a spreadsheet model. The calculations and relationships in some models are structured such that the model cannot easily be implemented in a spreadsheet and therefore must be represented using a general-purpose or simulation programming language. The sixth and last model in this chapter is such a model that involves calculating the loss ratio for a small insurance agency. Other models in the problems at the end of the chapter provide examples from other fields.

Two other aspects of simulation are generation of random variates and analysis of output data. In this chapter, we will discuss these topics from a rather utilitarian point of view. Our purpose here is to show how to generate random variates from some common distributions and how to compute confidence intervals using independent observations so that you can begin to develop and run your own simulations.

2.2 A MODEL FOR PROFIT ON A SPECIAL SALE PROMOTION

A large catalog merchandiser is planning to have a special furniture promotion a year from now. To do this, the company must place its order for the furniture now. It plans to sign a contract with the manufacturer for 3000 chairs at a cost of \$175 per unit, which the company plans to offer initially for \$250 per unit. The promotion will last for eight weeks, after which all remaining units will be offered for sale at half the initial price, or \$125 per unit. The company believes that 2000 units will be sold during the first eight weeks. We can represent the profit from this sale by defining the following variables:

P the profit from the promotion

C the per unit cost for the chairs (\$175)

R the initial price per unit for the chairs (\$250)

S the number of units ordered (3000)

V the number of units sold during the first eight weeks of the promotion (2000).

$R - C$ is the net profit per unit for units sold during the first eight weeks, and $R/2 - C$ is the net profit (or loss) per unit for units sold after the first eight weeks. Because $S - V$ is the number of units sold at the discounted price $R/2$, the profit is related to the other variables by

$$P = (R - C)\,V + \left(\frac{R}{2} - C\right)(S - V)$$

Then, if each variable has the value given above, the profit will be exactly \$100,000.

2.2.1 The Spreadsheet Model

To create a spreadsheet with these calculations, enter the following elements:

1. **Inputs.** The numerical inputs to the profit calculation—that is, the data needed to calculate profit—are S, the number of units ordered; C, the unit cost for the

FIGURE 2.1
Profit calculations for special furniture promotion

	A	B	C
1	**Model for Special Promotion Furniture Sale**		
2			
3	Stock ordered (S):	3000	
4	Unit cost for stock (C):	$175.00	
5			
6			
7	Demand within first 8 weeks (V):	2000	
8	Sales within first 8 weeks (V):	2000	
9	Initial price (R):	$250.00	
10	Sales after first 8 weeks (S-V):	1000	
11	Sale Price (R/2):	$125.00	
12			
13	Profit (P):	$100,000	
14			
15			
16			
17			
18			
19			
20			

Original Model / Distributions / Simulation Model / Replications /

chairs; *V*, the demand during the first eight weeks of the promotion; and *R*, the initial price. These are entered in cells B3, B4, B7, and B9, respectively.

2. **Computed values.** The profit could be computed from the inputs in one formula, but in a spreadsheet it is often convenient to separate computations into logical pieces.
   - Enter the sales in the first eight weeks in cell B8 using the formula

     `=B7`

   - Enter the sales after the first eight weeks in cell B10 using the formula

     `=IF(B8.B3,0,B3-B8)`

   - Enter the discounted price in cell B11 using the formula

     `=B9/2`

   - Enter the profit in cell B13 using the formula

     `=(B9-B4)*B8+(B11-B4)*B10`

   The spreadsheet will now look similar to that in Figure 2.1.

Two of the four inputs—the cost per unit and the number of units ordered—are fixed because the company will sign a contract with the manufacturer to deliver 3000 units at a price of $175 per unit. However, the other two inputs are uncertain.

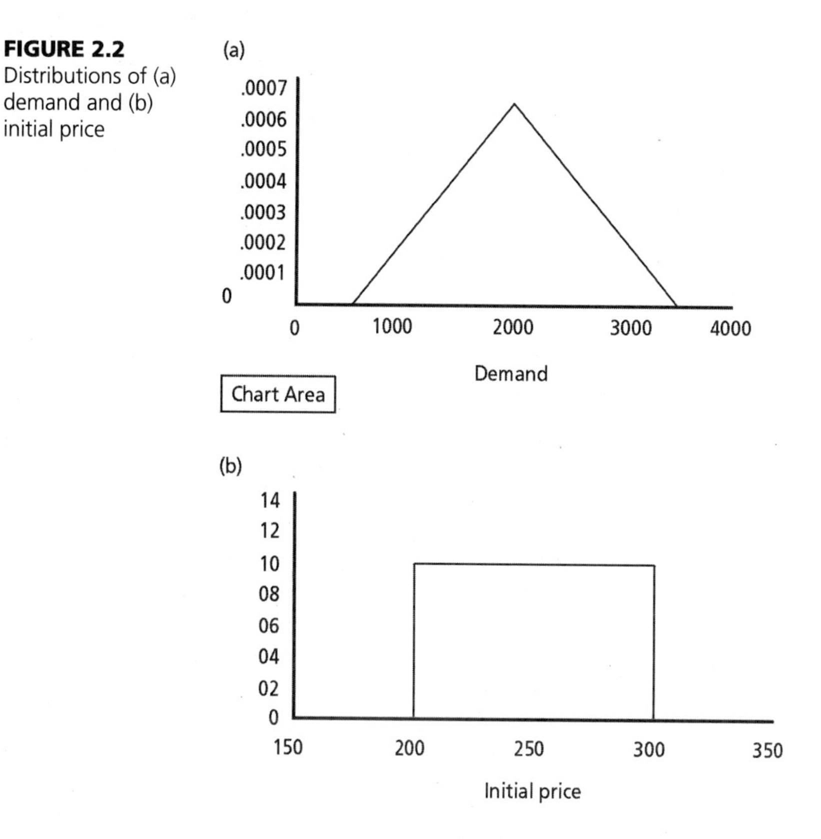

FIGURE 2.2 Distributions of (a) demand and (b) initial price

We cannot know for sure how many units will be sold during the first eight weeks. This depends on a lot of factors: the condition of the economy, the appearance of the chairs, the success of the advertising campaign, and others. In addition, the initial price of the chairs, which will be determined just before the advertising campaign begins, will depend on current economic conditions and what the company's competitors are doing.

2.2.2 Setting Up the Simulation

Research suggests that the demand for the chair during the first eight weeks has a symmetric triangular distribution between 500 and 3500 units, with a peak at 2000 units. This distribution is shown in Figure 2.2. Later in this chapter, we will discuss how to sample observations from this distribution. For now, we will assume that this can be done. Total demand for the first eight weeks can be as much as 3500 units, but we only contracted to purchase 3000 units, so the number of units demanded can be more than the number available to sell. To accommodate this, we must modify the model. Let D represent the demand for the chair in phase I of the sale. Then, $V = \min(D, S)$; that is, V is the minimum of the number of units available and the

FIGURE 2.3
Furniture promotion model simulation

	A	B	C	D	E
1	**Simulation Model for Special Promotion Furniture Sale**				
2					
3	Stock ordered:	3000			
4	Unit cost for stock:	$175.00			
5				Distribution Parameters	
6				Lower	Upper
7	Demand within first 8 weeks:	1758		500.00	3500.00
8	Sales within first 8 weeks:	1758			
9	Initial price:	$244.20		200.00	300.00
10	Sales after first 8 weeks:	1242			
11	Sale Price:	$122.10			
12					
13	Profit:	$55,941			
14					
15					
16					

Original Model / Distributions / Simulation Model / Replications

number requested. D is the variable that has a triangular distribution. Note that when a deterministic model (one that does not involve random variates) is being converted into a stochastic model to examine the variation in the performance measure, the model frequently must be modified somewhat in case certain values fall outside the bounds assumed in the deterministic model. When dealing with stochastic models, we must also consider the unlikely possibility that random quantities will have extreme values. If the model is to be replicated 1000 or 10,000 times, these extreme values are likely to occur a few times, and they must be dealt with.

The initial price, R, will be between \$200 and \$300, and we have no reason to believe one value more likely than another in this range, so we will represent the initial price as a uniformly distributed random variable between \$200 and \$300 as shown in Figure 2.2. Since these quantities are random variables and the net profit depends on them, the net profit is also a random variable.

Figure 2.3 shows the modified model to be simulated. To create the spreadsheet in Figure 2.3, first copy the spreadsheet created in Figure 2.1 to a new worksheet and change the title in cell A1. The simulation has some new inputs: the parameters of the two distributions for demand during the first eight weeks and the initial price. Now follow these steps:

1. Enter these parameters and their labels as indicated in the range D5..E9.
2. Enter the formula

   =MIN(B3,B7)

   in cell B8 to allow the possibility that demand during the first eight weeks could exceed the number of units on hand.
3. Enter the formulas for sampling the demand and initial prices from their respective distributions (see Figure 2.3).
   - Enter the formula

```
=D7+(E7-D7)/2*(RAND()+RAND())
```
(3a)

in cell B7.

- Enter the formula

```
=D9+(E9-D9)*RAND()
```
(3b)

in cell B9.

Now each time the recalculate key (**F9**) is pressed, new values are sampled in cells B7 and B9, and these values are used to calculate the profit in cell B13.

2.2.3 Running the Replications

To run the simulation, we must first specify the number of replications to be run—100 or 1000, for example. For each replication, we sample a value for each of the random variables in the model—demand and initial price, in this case—and plug these sampled values into the formulas above to compute the net profit. Then we save the resulting net profit value in an unused cell. We could do this by hand: Press **F9** to resample and recalculate, then copy the value (not the formula) in cell B13 to a cell where we want to save it. Most spreadsheets provide a command to automate this process: Excel provides the **Table** command under the **Data** menu. It can also be performed using a general-purpose programming language or a special simulation language or even by hand. Figure 2.4 shows the results of 20 replications of the experiment. When the run is completed, we will have a large number of independent observations like those in Figure 2.4.

Several authors also have written about using spreadsheets to do simulation; see especially Winston (1996) and Evans and Olson (1998). To create a spreadsheet like that in Figure 2.4, start a new worksheet and enter the titles in cells A1 and A3..D3. Now do the following:

1. Enter the sequence of numbers 1 through 20 in cells A5..A24.
2. Click on cell B4 and enter a formula by pressing "=" and then select cell B7 in the worksheet containing the simulation model (Figure 2.3). The resulting formula is

   ```
   ='Simulation Model'!B7
   ```

   where **Simulation Model** is the name on the tab of the worksheet in Figure 2.3.
3. Similarly, enter the formula

   ```
   ='Simulation Model'!B9
   ```

   in cell C4.
4. Enter the formula

   ```
   ='Simulation Model'!B13/1000
   ```

FIGURE 2.4

Twenty observations of demand, price, and profit

	A	B	C	D	E	F	G	H	I
1	Furniture Promotion Simulation								
2						Saved data from 20 replications			
3	Replication	Demand	Price	Profit		Replication	Demand	Price	Profit
4		1758.00	244.20	55.94					
5	1	1377	234.16	-12.55		1	1633	285.77	136.95
6	2	2768	235.23	153.39		2	2040	201.16	-18.06
7	3	2070	227.82	52.50		3	1785	211.25	-19.59
8	4	758	204.42	-140.87		4	2589	255.62	189.39
9	5	2337	212.01	40.77		5	1751	226.58	13.29
10	6	2002	281.53	179.06		6	2049	248.97	103.55
11	7	1472	286.97	116.74		7	872	233.58	-72.77
12	8	1983	262.64	129.38		8	1722	281.62	139.98
13	9	2110	211.07	14.28		9	1660	294.90	162.16
14	10	2020	240.65	78.97		10	2885	299.33	355.77
15	11	3021	257.31	246.93		11	2048	240.99	83.23
16	12	2463	213.94	59.32		12	3116	297.41	367.23
17	13	2075	277.87	180.08		13	2058	259.80	132.02
18	14	2442	244.81	141.07		14	2453	269.46	209.62
19	15	1299	221.12	-49.71		15	2990	293.38	353.66
20	16	2043	224.98	42.34		16	2471	214.39	61.50
21	17	2457	259.69	183.61		17	1413	270.35	71.56
22	18	1083	273.07	32.45		18	2221	224.58	61.25
23	19	1730	260.95	92.12		19	1577	231.22	4.14
24	20	1242	247.93	0.91		20	2352	221.82	68.58
25									
26						Observations	20	20	20
27						Sample mean	2084	253.11	120.17
28						Sample stdev	564.46	31.98	125.82
29						Std error	126.22	7.15	28.13
30						Half-width	218.25	12.37	48.65
31						LCL	1866	240.75	71.53
32						UCL	2303	265.48	168.82
33									
34									
35									
36									

Original Model / Distributions / Simulation Model / Replications

in cell D4.

The final formula copied the value of profit but divided it by 1000 to express the profit in thousands of dollars. Now each time the spreadsheet is recalculated, cells B4, C4, and D4 will contain the values computed in cells B7, B9, and B13 in the Simulation Model worksheet. Next, highlight the range from A4..D24 and select the **Data** menu, then the **Table** command. In the **Table** dialog, click on the lower entry labeled "Column input cell:" and then click on any unused cell, such as cell E1. Click **OK**. If you have done everything correctly, you will see the range from B5 through D24 fill with values. These are the computed (i.e., sampled) values in cells B7 (demand), B9 (initial price), and B13 (profit) from the worksheet containing the simulation. The values you get will not be identical to those in Figure 2.4 but they will be sampled from the same distributions.

If the automatic recalculation option is on, the entries in the range B4 through D24 will be recalculated whenever a cell is changed in the spreadsheet. Press the recalculate key, **F9**, to observe this. Each change in the spreadsheet effectively recalculates the entire spreadsheet so that a new set of 20 replications is performed. This is not exactly what we want. We actually want only one set of 20 replications to be performed. To capture a set of 20 replications, highlight the range from A5 through D24 and select **Copy** from the right-click menu. Then select a cell in a new area of the spreadsheet (we selected F5), select **Paste Special** from the right-click menu, click on **Values,** and click **OK** to paste the values for the 20 replications to this new area.

2.2.4 Analyze the Output Data

The final step in the simulation analysis involves using the output data to estimate or graphically show the output parameter(s) for the model. Suppose we want to know two things about this model: (1) the expected net profit and (2) a most likely range of values for the net profit. To specify the expected net profit, we will compute a 90 percent confidence interval for the mean using the data in Figure 2.4. To specify a most likely range of values for the net profit, we will use the range of values between the 5th and 95th percentiles. This range is sometimes called a *tolerance interval* or *prediction interval.* Generally, 20 observations are too few to compute reliable values for these parameters, but we will only use 20 here to demonstrate explicitly how these parameters are computed and give you an opportunity to confirm our calculations.

We will denote the observations by $X_1, X_2, \ldots, X_{20}$. To simplify things, we will express our data values in thousands of dollars. Thus, the first observation in Figure 2.4 is 136.95 in thousands of dollars. Then for our data set, the sample mean and sample standard deviation are 120.17 and 125.82 in cells I27 and I28, respectively. (These were computed using the standard Excel functions "average" and "stdev.") The 95 percent confidence interval for the mean is 120.17 ± 48.65, or from 71.53 to 168.82, in thousands of dollars. These figures are in cells I27 through I32. It follows that, although we still do not know the exact value for the mean profit from the simulation, we have learned from 20 replications that we can be 90 percent confident that the range of values from $71,530 to $168,820 contains this mean.

The 5th sample percentile for this data is .95 times the smallest observation plus .05 times the next largest observation, or .05(–72.77) + .95(–19.59) = –22.24. The 95th percentile is .05 times the next-to-the-largest observation plus .95 times the largest observation, or .05(355.77) + .95(367.23) = 366.66. Thus, the range of most likely values for net profit is between –$22,240 and $366,660. There is an approximate 90 percent probability that the net profit will be between these two limits.

Note the difference in interpretation between these two intervals. The first interval, $71,530 to $168,820, is a 90 percent *confidence interval for the mean.* If we were to repeat the entire simulation experiment—producing 20 replications and computing a confidence interval using the same methods—many times, then 90 percent of the confidence intervals would include the true, but unknown, mean profit. We actually computed only one of these intervals, but we can be 90 percent sure that the interval we computed is one of the 90 percent of the intervals that includes the true mean. The other interval, –$22,240 to $366,660, is a *prediction interval* for profit. If we were to repeat the basic experiment, which consists of sampling the demand and price and computing the profit, the probability is approximately 90 percent that the profit in a single year will fall within this interval. To state it another way, if we were to repeat the experiment many times, approximately 90 percent of the replications would show the profit within this interval. If we are going to repeat this promotion each year for many years, we are primarily interested in the mean profit, and therefore, the confidence interval for the mean. However, if we are going to do the promotion only once, we are primarily interested in the prediction interval for the mean. See if you can explain why.

In this section, we have used a very simple model to show that simulation is a means to analyze, or get information from, a model by repeatedly sampling the stochastic, or uncertain, variables in the model, plugging these values into the model calculations, and saving the model output. In the next section we will look at a model that is more complex, and therefore more realistic, than this one.

2.2.5 Comments about Spreadsheet Simulations

Three distinct areas are needed in the spreadsheet:

1. an area to contain the model (including areas with input parameters, model calculations, and output values),
2. an area to contain the observations produced by the replications, and
3. an area to contain the summary data computed from the observations.

Where these areas are located is a matter of taste; however, you should remember that a large number of contiguous cells are normally needed to contain the observations. Thus, you may put the model in the first 100 rows, the summary data in the next 20 rows, and the observations in the rows after 120. In most spreadsheets, this will leave perhaps 8000 or more cells to hold the data. Another alternative, which was used in the example in Figures 2.33 and 2.34, is to put the replication data in columns to the right of the model.

A few more words are in order about analyzing the data. Most spreadsheets have built-in functions for doing statistical calculations: mean, variance, standard deviation, minimum, and maximum. Generally, commands are also available for computing the (empirical) distribution of a set of observations. Some statistical calculations are more difficult to do using a spreadsheet (the median and percentiles, for example), but most spreadsheets also have these functions. Thus, the capabilities of the spreadsheet are frequently adequate to compute summary statistics; however, it may be preferable to import the data into a "real" statistical package for analysis.

2.3 SIMULATION FOR SENSITIVITY ANALYSIS

Simulation is a useful tool to determine the sensitivity of a model, especially a spreadsheet model, to changes in input parameters. In this section, we will learn how to use simulation to evaluate model sensitivity.

2.3.1 A Financial Model for an Office Building

Figure 2.5 presents a model of the financing of an office building. The objective of this model is to predict the after-tax cash flow (ATCF) resulting from constructing and operating this building during the first year. If the after-tax cash flow is negative, the investors will have paid out more money in principal, interest, operating expenses, and taxes than they received in rent, thus taking a loss on the investment. If the ATCF is positive, then they posted a profit.

The planned size of the building is 180,000 square feet. If the building cost per square foot is assumed to be $80, then the total cost to build will be $14.4 million. The investors plan to take out a mortgage for 85 percent of the cost of the building, and they have been guaranteed a rate of 12 percent for a term of 30 years, or 360

FIGURE 2.5

Spreadsheet for office building cash-flow analysis

	A	B	C	D
1	**CASH FLOW CALCULATION FOR PURCHASE OF AN OFFICE BUILDING**			
2				
3	**Year**	**2002**		
4	Building cost per square foot:	$80.00		
5	Original cost of building:	$14,400,000		
6	Size of building (sq. ft):	180,000		
7	Avg. rent per square foot:	$9.00		
8	Assumed Operating Exp./sq.ft.:	$1.20		
9	Assumed vacancy rate:	0.30		
10	Depreciation horizon (years):	30		
11				
12				
13	**Year**	**2002**		
14	Potential gross income	$1,620,000		
15	Less: Vacancy and collection loss	(486,000)		
16	Effective gross income	1,134,000		
17	Less: Operating expenses	(216,000)		
18	Net operating income	918,000		
19	Less: Debt service	(1,510,826)		
20	Before-tax cash flow	(592,826)		
21	Less: Tax	(329,984)		
22	After-tax cash flow	($262,842)		
23				

One Year Model / One Year Simulation / Fig 2.7 / Figs 2.25 and 2.26 / Multiyear Model / Multiyr Scenario Comp / M months. Under these assumptions, the monthly payment will be $125,902, and they will pay $44,416 in principal and $1,466,410 in interest during the first year. The owners must also pay for the cost of operating the building, which includes taxes, insurance, maintenance, and certain utilities. They assume that the average operating cost per square foot will be $1.20, resulting in a total operating cost for the first year of $216,000.

The average rent per square foot is assumed to be $9, so if the building is fully occupied, the total potential gross income is $1.62 million per year. However, on average, 30 percent of the square footage is assumed to be vacant, resulting in a loss of $486,000. Thus, the estimated actual gross annual rental income is $1.134 million. Subtracting operating expenses and debt service (principal and interest), we find that the before-tax cash flow is –$592,826.

The net operating income from this investment (gross income minus operating expenses) is $918,000. Subtracting mortgage interest, depreciation (using the straight-line method) and adding other income (or subtracting, if we have a loss), we get a taxable income of –$1.113 million, with a tax of –$329,984 at a rate of 34 percent. Subtracting taxes from the before-tax cash flow, we get an ATCF of –$262,842.

To see how these calculations were made, define the following variables:

$$\begin{aligned} CSF &= \text{building cost per square foot} \\ SIZE &= \text{size of building in square feet} \\ RENT &= \text{average rent per square foot} \\ ERATE &= \text{assumed operating expense rate per square foot} \\ VRATE &= \text{assumed vacancy rate} \\ OINC &= \text{other income} \end{aligned}$$

and the function

$$TRATE(tinc)$$

is the tax rate for a firm whose taxable income is *tinc*. The other quantities are computed as shown in Figure 2.6.

2.3.2 Sensitivity Analysis for Office Building Model

If all of the assumptions involved with these calculations hold, then the after-tax cash flow will be exactly –$262,842. We can indeed count on some of the assumptions. Since the architect has already submitted final designs for the building, the size is fixed at 180,000 square feet. The percentage of the cost (85%) to be financed, interest rate (12%), and the term of the loan (360 months) are fixed by agreement with the financing institution. Other assumptions are certainly not definite since they depend on the condition of the local economy. For example, the average cost per square foot cannot be predicted exactly because the building has not been built. Unknown problems, changes in material and labor costs, building delays, and other

FIGURE 2.6
Summary of calculations for the office building model

Original cost of building	COST = CSF*SIZE
Potential gross income	PGI = RENT*SIZE
Vacancy and collection losses	VCL = VRATE*PGI
Effective gross income	EGI = PGI - VCL
Operating expense	OPEXP = ERATE*SIZE
Net operating income	NET = EGI - OPEXP
Debt service	DS = 12*PMT(0.85*COST, 12%, 360)
Before-tax cash flow	BTCF = NET - DS
Interest	INT = CUMINT(12%/12,12,0.85*COST, 1, 12,0)
Depreciation	DEP = SLN(COST, 0, 31.5)
Taxable income	TAXINC = NET - INT - DEP - OINCOME
Tax	TAX = TAXINC*TRATE(TAXINC)
After-tax cash flow	ATCF = BTCF - TAX

TABLE 2.1 Minimum, most likely, and maximum values of uncertain parameters in the office building model

Uncertain Parameter	Minimum	Most Likely	Maximum
Cost per square foot ($)	50	70	90
Average rent per square foot ($)	8	10	12
Average operating expense per square foot ($)	.6	1.0	1.4
Vacancy rate (%)	.2	.6	.6

factors may increase the cost. Alternatively, fortuitous circumstances may decrease the building costs. Similarly, the average rent per square foot, vacancy rate, and operating expenses depend on the local real estate market and economic conditions, and therefore they are uncertain. Table 2.1 shows the minimum, most likely, and maximum values for these variables.

We can be quite sure that once the building is finished and the first year's operation is completed, the ATCF will not be –$262,842. It might be close to this figure, or it might differ from it by a large amount. Perhaps the most important question for decision-making purposes is, "How much is the actual after-tax cash flow likely to differ from this amount?" Could it be as large as +$100,000, or +$200,000? Could the loss be as large as –$500,000? Methods to find answers to questions such as these come under the general heading of *sensitivity analysis.* Sensitivity analysis seeks to determine how sensitive model outputs are to variations in inputs. If outputs are highly sensitive, less certainty must be associated with the output, and the results must be considered to involve more risk.

2.3.3 What-If Calculations

One way to answer these questions is to do some "what-if" calculations. The results of some of these are given in Table 2.2. One can see from these calculations that if the most likely values of the uncertain variables are used, the after-tax cash flow is

TABLE 2.2 Results of some what-if calculations on the office building model

Version	Cost/Ft² ($)	Rent ($/Ft²)	Operating Expenses ($/Ft²)	Vacancy Rate (%)	After-Tax Cash Flow ($)
1	50	8	.60	.20	+188,654
2	70	10	1.00	.60	–337,340
3	90	12	1.40	.60	–483,175
4	50	12	.60	.20	+568,814
5	90	8	1.40	.60	–673.255

–$337,340. It could be as negative as –$673,255, or as large as +$568,814. What-if calculations are done by substituting alternative combinations of values of the inputs and recording the corresponding values of the outputs. For example, in row 1 of Table 2.2, the values of 50, 8, .60, and .20 were substituted for the cost per square foot, rent per square foot, operating expense per square foot, and vacancy rate, respectively. Then the ATCF was computed to be $188,654.

What-if calculations in a spreadsheet can be quite cumbersome because the process involves selecting combinations of values from lists, substituting them into specific cells, and recording the values of output cells after recalculation. Excel and virtually all other spreadsheets have built-in commands for forming what-if tables when one or two input parameters will be varied. Indeed, this is the command we used in the previous section to perform replications of the simulation. So if you want to substitute a list of values for cost per square foot and a list of values for operating expense per square foot, the spreadsheet can automate the process for you. However, most spreadsheet models involve more than two uncertain inputs, so these tools are not useful.

Table 2.1 has four parameters, three with three possible values and one with two values. The number of combinations of these values is $2 \cdot 3^4 = 54$, so to substitute all possible combinations would require 54 recalculations and produce 54 values of after-tax cash flow. Four is a modest number of uncertain parameters. Consider a model with eight uncertain parameters and three possible values of these parameters. The number of recalculations would be $3^8 = 6561$. This would clearly be a burdensome number of recalculations to do by hand and would require a program to be written so the recalculations could be done automatically. If the number of uncertain inputs is 10, then the recalculations would be slightly more than 59,000; and if the model has 15 uncertain inputs, then the number of recalculations for all combinations would be more than 14 million. The situation gets much worse as the number of uncertain variables and the number of possible values for each variable increase. The point of this discussion is that what-if calculations are usually computationally infeasible if a complete enumeration is done.

If complete enumeration of what-if calculations were computationally feasible, one would still have the problem of determining what to do with all of the resulting values of the outputs. For example, if we do enumerate all 54 combinations of the four inputs to the office building model and compute all 54 values of ATCF, what will we do with them? They do not constitute a random sample, so they are not amenable to statistical analysis. Their mean is not the mean of the possible values of the output. Their distribution does not estimate the distribution of a population. Suppose that the investors have decided that they can tolerate a loss of $400,000 the first year. It is clearly possible to have a negative after-tax cash flow that is more than –$400,000 because one of the what-if calculations resulted in a loss larger than this. But how likely is that? Is the probability 1 in 2, 1 in 4, 1 in 10, 1 in 100? What-if calculations cannot answer this question since the combination of values that are substituted do not include information about likelihood.

2.3.4 Simulation for Sensitivity Analysis

Simulation provides a means to overcome all of these difficulties. To simulate this model, we must provide more information about the uncertain values in it. We will represent these quantities as random variables and therefore must specify their probability distributions. (See Appendix for a review of random variables.) Suppose we assume that each of the four variables—building cost, average rent per square foot, average operating expense per square foot, and vacancy rate—has a triangular distribution between the largest and smallest possible values, with a peak at the most likely value. Figure 2.7 shows the probability density functions of these random variables. Note that the distributions of the first three quantities are symmetric, but

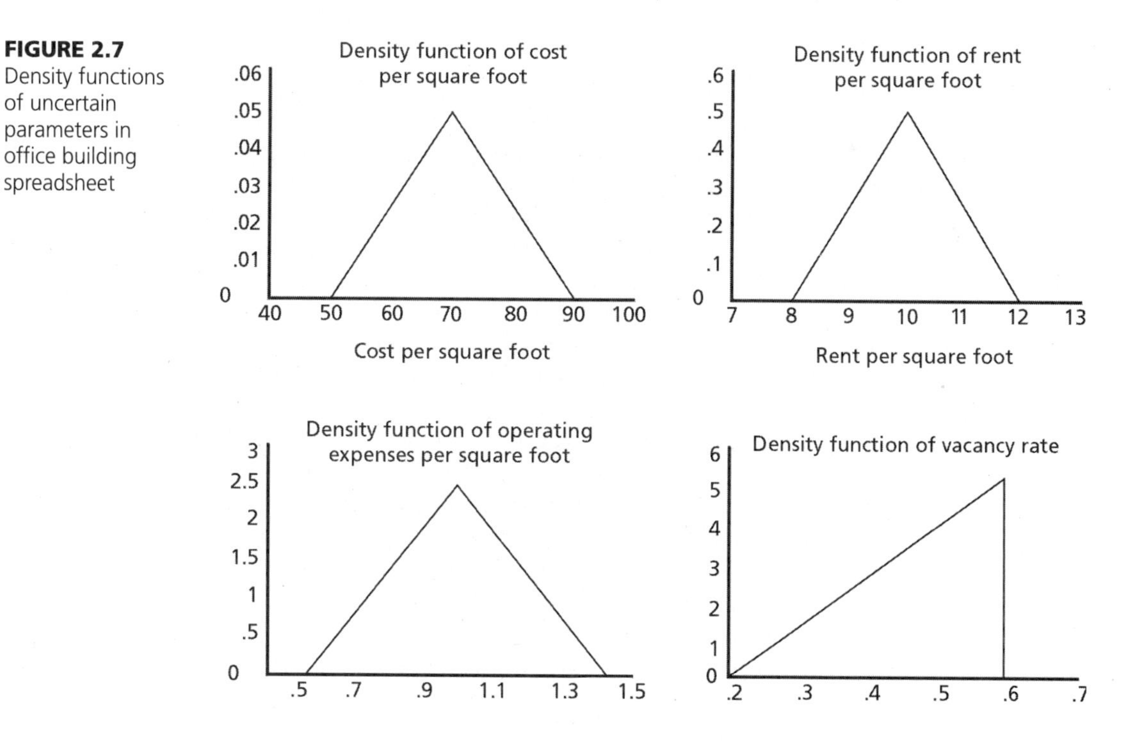

FIGURE 2.7 Density functions of uncertain parameters in office building spreadsheet

FIGURE 2.8

Histogram of after-tax cash flow

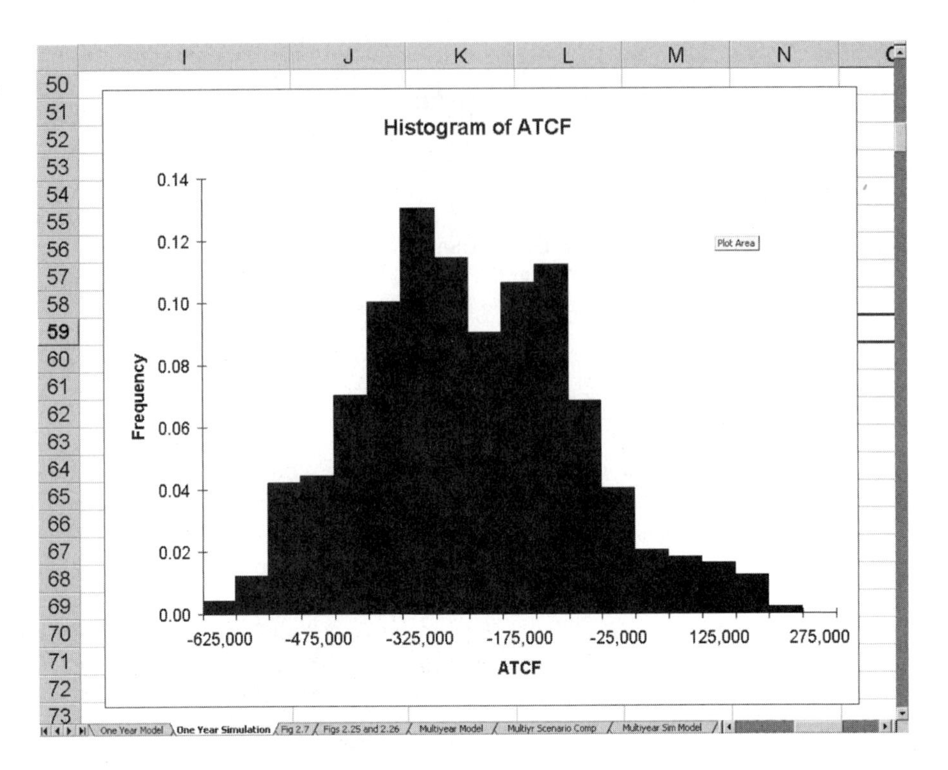

the vacancy rate is more likely to be close to the maximum value of .6 than the minimum value of .2. Running the simulation consists of selecting a random observation from each of these four distributions and then performing the calculations of the model using these values along with the other known quantities. This process is repeated many times independently, and the values of after-tax cash flow resulting from these replications are used to estimate the desired probabilities, mean, variance, or other parameters or to show the resulting distribution of the ATCF.

This simulation was performed using Microsoft Excel. In this case, 500 replications were performed, resulting in 500 independent observations of after-tax cash flow for the office building. Figure 2.8 shows the resulting histogram, and Figure 2.9 gives relevant sample statistics from these data. We can see that the variation in the four uncertain variables produces a great deal of variation in the ATCF. It is most likely to be between –\$500,000 and +\$50,000, but it could be as low as –\$650,000 or as high as +\$250,000. The 95 percent confidence interval for the expected value of after-tax cash flow is the range from –\$260,637 to –\$232,184. Looking back at Table 2.2, the after-tax cash flow computed using the most likely values of the uncertain variables was –\$337,340, which is not in this range. Indeed, none of our what-if calculations produced a value in this range. Thus, our simulation has given us information about ATCF that was not available from what-if analysis.

FIGURE 2.9
Simulation data for ATCF

	I	J	K	L
1	**Analysis of Simulation Data for ATCF**			
2				
3	No. of replications	500		
4	Sum	-1.2E+08		
5	Sum of squares	4.35E+13		
6	Mean	-246,410		
7	Variance	2.63E+10		
8	Standard deviation	162,308		
9	Minimum	-645,820		
10	Maximum	215,384		
11				
12	Confidence Interval for Mean of ATCF			
13	Upper limit	-232,184		
14	Lower limit	-260,637		
15				

One Year Model / One Year Simulation / Fig 2.7 / Figs 2.25 and 2.26 / Multiyear Model / Mul

A more careful look at the $n = 500$ observations shows that 86 were less than –\$400,000. Thus, we estimate the probability of having an after-tax cash flow of less than this amount to be $\hat{p} = 86/500 = .172$; a 95 percent confidence interval for this probability is

$$\hat{p} \pm 1.96 \sqrt{\frac{\hat{p}\,(1-\hat{p})}{n}} = .172 \pm 0.033$$

or from .139 to .205. Thus, the chances of having a bad investment are between 1 in 7 and 1 in 5. If we wish to more accurately estimate this probability, we can increase the number of replications.

Finally, looking more carefully at the sample cumulative distribution function (cdf) of these 500 observations, which is given in Figure 2.10, we see that, although the values range from –\$645,819 to +\$215,384, there is approximately a 90 percent probability that the after-tax cash flow will be between approximately –\$560,000 and +\$45,000, which is a much smaller range of values and provides much more accurate information for decision-making purposes.

The purpose of this example is to show how a simulation can provide useful information that can help decision makers evaluate the risk associated with a decision. A few what-if calculations provided a few possible values of the ATCF, but these values were not useful for estimating the expected value of after-tax cash flow, the probability this quantity would be less than a particular amount, or a range of most likely values. We will return to these topics in section 2.6.

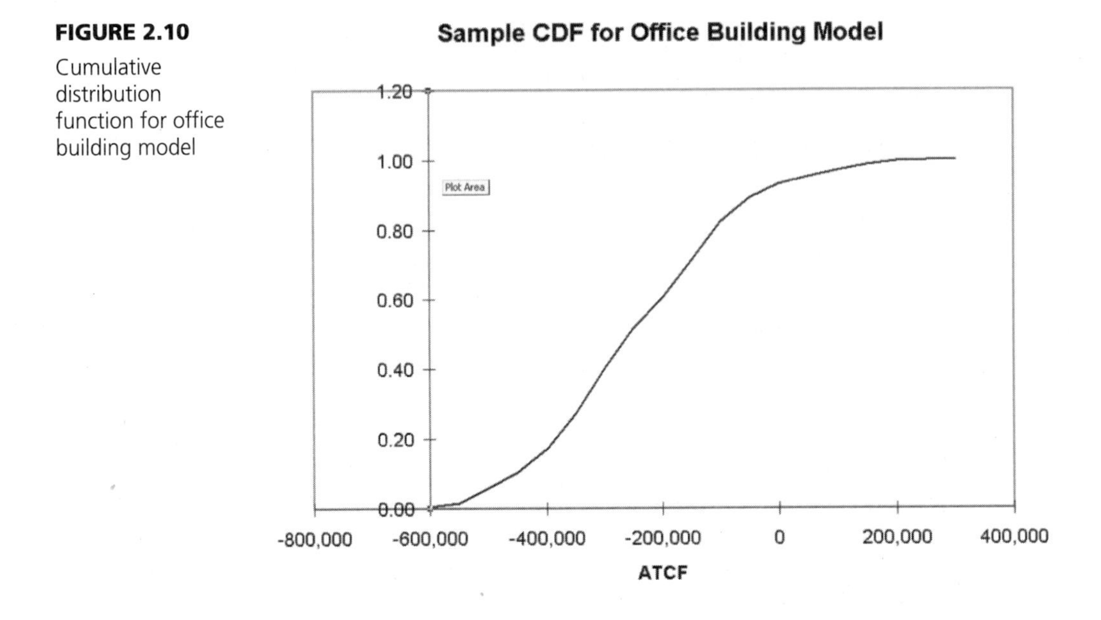

FIGURE 2.10
Cumulative distribution function for office building model

2.4 SIMULATION: SAMPLING ON THE COMPUTER

In this section, we will look at a simple simulation problem for which we already know the answer. Doing this will demonstrate two important concepts associated with simulation. First, it will show that the best we can do with a stochastic simulation is to compute an estimate of the parameter we seek to know. Second, if our model is a correct representation of the problem, the simulation will provide a reliable parameter estimate if the simulation is run long enough.

2.4.1 A Model to Estimate π

The problem that we will consider concerns computing a value for π. Elementary geometry tells us that the area of a circle of radius r is $A = \pi r^2$. It follows that if we have a circle of radius r and know its area, A, we can compute π from $\pi = A/r^2$. Unfortunately, it is difficult to compute the area, especially without knowing the value of π. Consider the following situation: Figure 2.11 shows a quarter-circle inscribed in a square. Suppose that the radius of the circle and each side of the square are 1 unit; so the area of the quarter-circle is 1/4 π, and the area of the square is 1. Now, if we randomly select a point (x, y) in the square, giving all points equal chance of being selected, the probability, p, that the point will also fall inside the quarter-circle is simply the ratio of the area of the quarter-circle to that of the square, or $p = \pi/4$.

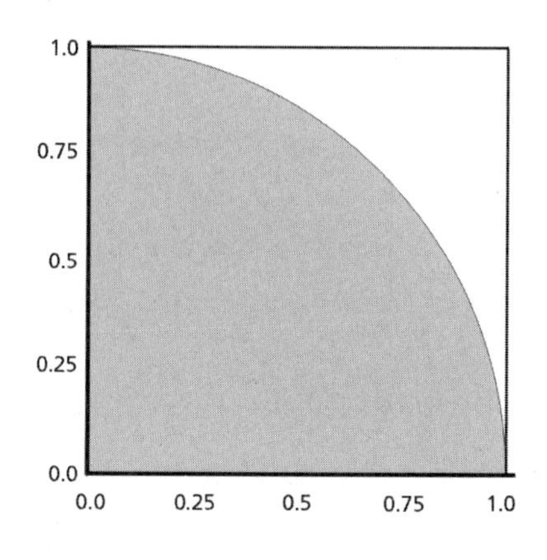

FIGURE 2.11
Quarter-circle inscribed in a square

Thus, we can estimate the value of π by estimating p and multiplying the resulting estimate by 4.

2.4.2 An Experiment to Estimate π

Suppose that we perform n replications of this experiment of randomly selecting a point (x, y) in the square and observing whether it falls in the quarter-circle. Let n_q denote the number of points that fall in the quarter-circle out of the n replications. Then, the estimate of p is

$$\hat{p} = \frac{n_q}{n}$$

Since the observations are independent, elementary statistics (see Appendix) provides the following 95 percent confidence interval for p:

$$\hat{p} \pm \sqrt{\frac{\hat{p}\,(1-\hat{p})}{n}}$$

Therefore, a 95 percent confidence interval for the value of π is:

$$4\hat{p} \pm 4(1.960)\sqrt{\frac{\hat{p}\,(1-\hat{p})}{n}}$$

The algorithm for estimating the value of π using simulation is summarized as follows:

1. Initialize n (the number of replications);
   Initialize n_q and *count* to 0.
2. REPEAT:

Generate X and Y independently from Uniform(0,1) distributions.

If $X^2 + Y^2 < 1$ THEN increment n_q by 1.

Increment *count* by 1.

UNTIL *count* = n.

3. Compute the estimate of π:

$$\hat{p} = \frac{n_q}{n}; \hat{\pi} = 4\hat{p}$$

Compute the confidence interval using the expression above.

2.4.3 Analysis of the Data

We implemented this algorithm and ran the resulting program for values of n ranging from 100 to 10,000,000. The results of these runs are given in Table 2.3. Note two important points about Table 2.3. First, we cannot compute an exact value of π using simulation as our tool. Instead, each entry in Table 2.3 is an estimate of π. If we run the simulation again using an independent stream of random numbers, then we will get different (but similar) entries in Table 2.3. This is true for any application of simulation to analyze a stochastic model. The simulation is a sampling experiment on the model, and, as such, it produces a sample of data that must be used to estimate the parameter in which we are interested.

Second, as the number of replications grows, the point estimate for π gets closer to the true value (3.14159265 . . .). This is shown graphically in Figure 2.12. Al-

TABLE 2.3 Estimates of π

					95% Confidence Interval		
n	n_q	Estimated p	Estimated π	Error	Upper	Lower	True π
1000	785	.78500	3.14000	.10185	3.24185	3.038148	3.14459
2000	1556	.77800	3.11200	.07286	3.18486	3.039144	3.14159
5000	3930	.78600	3.14400	.04547	3.18947	3.098527	3.14159
10,000	7850	.78500	3.14000	.03221	3.17221	3.107792	3.14159
20,000	15,628	.78140	3.12560	.02291	3.14851	3.102688	3.14159
50,000	39,314	.78628	3.14512	.01437	3.15949	3.130747	3.14159
100,000	78,483	.78483	3.13932	.01019	3.14951	3.129132	3.14159
200,000	157,071	.78536	3.14142	.00720	3.14862	3.134222	3.14159
500,000	392,569	.78514	3.14055	.00455	3.14511	3.135998	3.14159
1,000,000	784,982	.78498	3.13993	.00322	3.14315	3.136707	3.14159
2,000,000	1,570,258	.78513	3.14052	.00228	3.14279	3.138239	3.14159
5,000,000	3,925,810	.78516	3.14065	.00144	3.14209	3.139208	3.14159
10,000,000	7,852,765	.78528	3.14111	.00102	3.14212	3.140088	3.14159

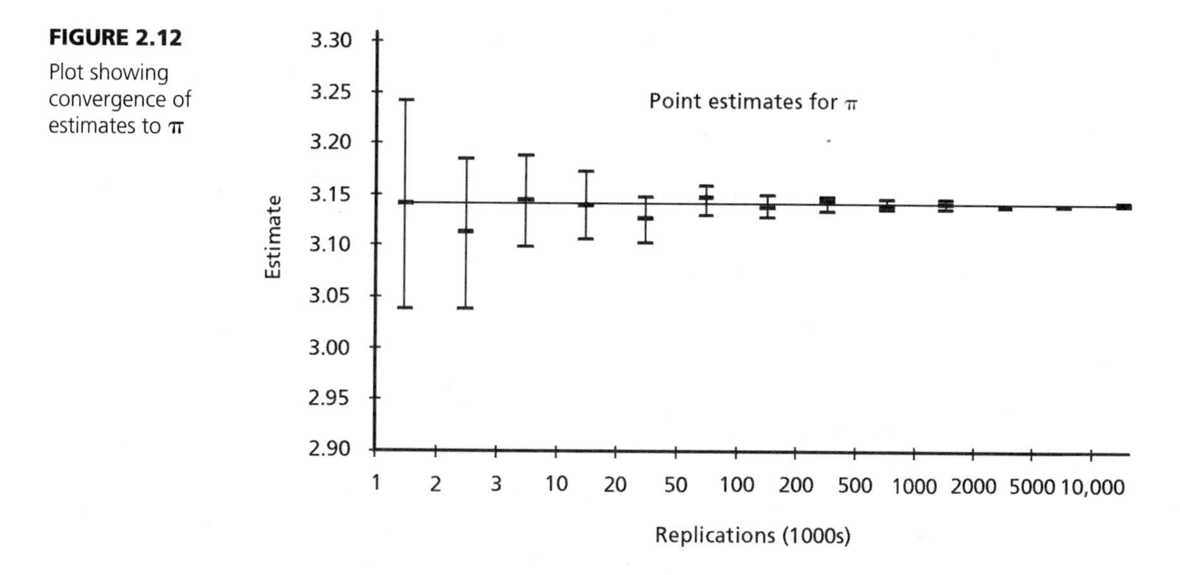

FIGURE 2.12 Plot showing convergence of estimates to π

though the last estimate we have, 3.14111, is accurate only to two decimal places, there appears no doubt that if we could perform an indefinitely large number of replications, the estimate would get arbitrarily close to the true value of π. This is true because the model we are using correctly relates the value of π to the area and radius of a circle. The simulation approach will give valid, accurate results if the model is correct and the number of replications is sufficiently large. Both conditions are important. The model we have presented in this section is quite simple; however, realistic models are usually complex and must be validated using techniques that we will discuss later in order to assure us that they faithfully represent the system we plan to study.

In Table 2.3, even with 10 million replications, we were able to accurately estimate π only to two decimal digits. Clearly, if we want more accurate estimates, we will have to greatly enlarge the sample size. Variance reduction techniques also are available to provide more accurate estimates with smaller sample sizes. Fortunately, when we analyze systems for decision-making purposes, estimates accurate to ±10 percent are often sufficient. Therefore, 100, 200, or 1000 replications may suffice.

2.5 SOME TECHNIQUES FOR GENERATING RANDOM VARIATES

Implementing this model for estimating π only required the generation of uniformly distributed random variates between 0.0 and 1.0. In the model discussed in the introduction to this chapter as well as most other models, it is necessary to generate

TABLE 2.4 Summary of methods for generating random variates from common distributions

Distribution	Parameters	Formula
Bernoulli	p	$X = \begin{cases} 1 \text{ if } U \leqslant p \\ 0 \text{ if } U > p \end{cases}$
Uniform	$a < b$	$X = a + (b - a)\,U$
Triangular	0, ½, 1	$X = \frac{1}{2}(U_1 + U_2)$
Symmetric triangular	$a < b$	$X = a + \frac{(b-a)}{2}(U_1 + U_2)$
Right triangular	$a < c$	$X = a + (c - a)\sqrt{U}$
Approximately normal	0, 1	$X = U_1 + U_2 + \ldots + U_{12} - 6$
Approximately normal	μ, σ	$X = \mu + \sigma(U_1 + U_2 + \ldots + U_{12} - 6)$
Exponential	μ	$X = -\mu \ln(\mathrm{U})$
Discrete uniform	$k, k+1, \ldots, k+m$	$X = k + \mathrm{int}[(m+1)U]$
Empirical	$a_1, a_2, \ldots, a_m$ $p_1, p_2, \ldots, p_m$ $a_1 < a_2 < \ldots < a_m$ $p_1 + p_2 + \ldots + p_m = 1$	$X = \begin{cases} a_1 \text{ if } U \leqslant p_1 \\ a_2 \text{ if } p_1 < U \leqslant p_1 + p_2 \\ a_2 \text{ if } p_1 + p_2 < U \leqslant p_1 + p_2 + p_3 \\ \vdots \\ a_m \text{ if } U > p_1 + p_2 + \ldots + p_{m-1} \end{cases}$

random variates from other distributions, both continuous and discrete. Here we present useful techniques for generating random variates from a few relatively simple distributions. We will then use these techniques in this and subsequent chapters. Other texts present the theory behind these techniques. See, for example, Banks (1998), Chapter 5; Banks et al. (2001), Chapter 8; Law and Kelton (2000), Chapter 8; Pidd (1992), Chapter 12, sections 5–7; and Fishman (1996), Chapter 3. All of these random variate generation techniques, which are summarized in Table 2.4, assume that you have a procedure or function that will generate a sequence of independent uniformly distributed random variates between 0 and 1. Such a function—called RAND() in Excel—is available in virtually all spreadsheet programs. We will assume that U represents a uniformly distributed random variate between 0 and 1; that is, U = RAND() in Excel.

2.5.1 Bernoulli Random Variates

A Bernoulli random variable, X, which has the value 1 with probability p and the value 0 with probability $1 - p$, can be used to randomly select between two alternatives. If $X = 1$, then alternative 1 is selected; otherwise ($X = 0$), alternative 2 is se-

FIGURE 2.13

Triangular density function with most likely value on extreme right

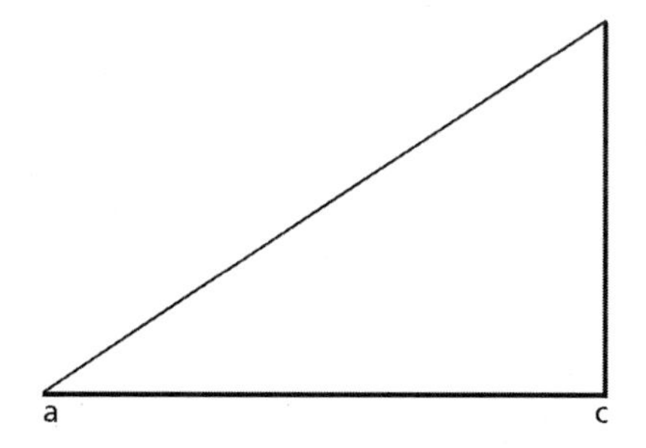

lected. It then follows that alternative 1 will be selected with probability p and alternative 2 with probability $1 - p$. To generate a Bernoulli random variate, let $X = 1$ if $U \leqslant p$, and $X = 0$ otherwise. If we want to give both alternatives an equal chance of being selected, we first generate the uniformly distributed random variate U and then select alternative 1 if $U \leqslant .5$ and alternative 2 otherwise.

2.5.2 Uniform Random Variates

A uniformly distributed random variate between a and b, where $b > a$, can be computed from $X = a + (b - a)U$. Thus, a uniformly distributed random variate between 0 and 10.0 is given by $X = 10.0U$, and a uniformly distributed random variate between 20.0 and 100.0 is given by $X = 20.0 + 80.0U$.

2.5.3 Triangular Random Variates

Probability theory tells us that if U_1 and U_2 are uniformly distributed between 0.0 and 1.0, then $(U_1 + U_2)/2$ has a symmetric triangular distribution between 0.0 and 1.0. If we want a random variate, X, to have a symmetric triangular distribution between a and b, X can be computed from

$$X = a + \frac{(b - a)}{2}(U_1 + U_2)$$

Sometimes we need to generate random variates from a nonsymmetric triangular distribution between a and c, where the most likely value is c, as shown in Figure 2.13. This can be done with $X = a + (c - a)\sqrt{U}$. This can be combined with the technique for randomly choosing between two alternatives to generate a random variate X from a nonsymmetric triangular distribution between a and b, with the most likely value at c, $a < c < b$, as shown in Figure 2.14. First, compute $p = (c - a)/(b - a)$, then generate two uniform random variates U_1 and U_2. If $U_1 \leqslant p$, then $X = a + (c - a)\sqrt{U_2}$; otherwise, $X = b - (b - c)\sqrt{U_2}$.

FIGURE 2.14
Asymmetric triangular density function

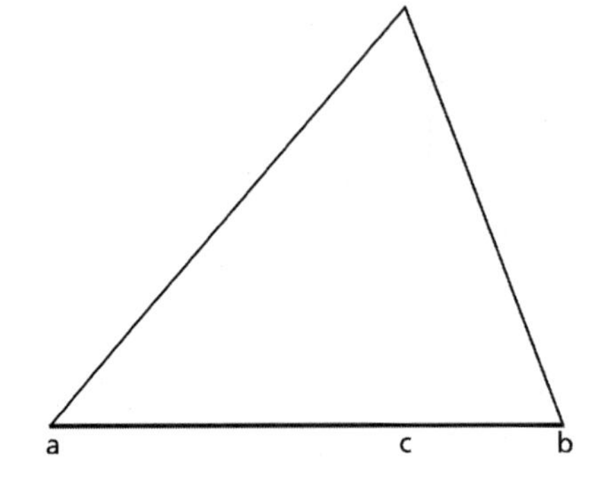

2.5.4 Normal Random Variates

The central limit theorem can be used to generate random variates that have approximately a normal distribution. Recall that the central limit theorem states that if Y_1, $Y_2, \ldots, Y_n$ are independent random variables drawn from the same infinite population having mean μ and variance σ^2, then if n is sufficiently large, the sample mean

$$\overline{Y} = \frac{1}{n}\sum_{i=1}^{n} Y_i$$

is approximately normally distributed with mean μ and variance σ^2/n. If Y_i is uniformly distributed between 0.0 and 1.0, then $\mu = 1/2$ and $\sigma^2 = 1/12$. Using $n = 12$, and assuming the central limit theorem holds, we get that

$$\frac{1}{12}\sum_{i=1}^{12} U_i$$

is approximately normally distributed with mean 1/2 and variance $1/(12)^2$. Thus,

$$X = \sum_{i=1}^{12} U_i - 6$$

is approximately normally distributed with mean 0.0 and variance 1.0, and therefore X has approximately a standard normal distribution. A random variate, Y, that is approximately normally distributed with mean μ and standard deviation σ can be generated from $Y = \mu + \sigma X$.

2.5.5 Exponential Random Variates

The exponential distribution is frequently used to represent the times between events and other random quantities, especially in reliability problems. If U is uniformly distributed between 0.0 and 1.0, then $X = -\ln(U)$ is exponentially distributed with mean 1.0, and $Y = \mu X$ is exponentially distributed with mean μ. Thus, to generate an exponential random variate with mean 20, we would compute $Y = -20 \ln(U)$.

2.5.6 Discrete Integer-Valued Random Variates

Suppose you want to randomly select one of the integers $k, k + 1, \ldots, k + m$ with equal probability. This is useful, for example, if you want to randomly select one of $m + 1$ alternatives with equal probability. To generate a random integer J from this range, let $J = k + \text{int}((m + 1)U)$, where int($x$) denotes the largest integer less than x. As an example, if you want to randomly select one of the values $1, 2, \ldots, 5$ with equal probability, then let $J = 1 + \text{int}(5U)$.

2.5.7 Other Discrete Random Variates

The preceding methods can all be implemented with a single formula in a spreadsheet cell. Sometimes a more involved method is needed that uses a range of cells. If you want to generate a random variate J having the value 1 with probability p_1, 2 with probability $p_2, \ldots, m$ with probability p_m, first compute the cumulative probabilities for this random variable: $F_1 = p_1$; $F_2 + p_1 + p_2 = F_1 + p_2$; $F_3 = p_1 + p_2 + p_3 = F_2 + p_3$; $F_j = p_1 + p_2 + \ldots + p_j = F_{j-1} + p_j$. Generate U from a uniform distribution between 0 and 1, and find the index J such that $F_{j-1} \leqslant U < F_j$. As an example, suppose we want to generate a random variate X having the value 1 with probability .25, 2 with probability .65, and 3 with probability .10. Then we would let $J = 1$ if $U < .25$, $J = 2$ if $.25 < J < .90$, and $J = 3$ otherwise.

This can be easily implemented in a spreadsheet using a *lookup table,* a two-dimensional table with at least two columns. The first column contains values that will be matched to an argument. The second column (and later columns if included) contains values that will be selected when a match is found. For most spreadsheets, the function that does the lookup is VLOOKUP(). Figure 2.15 contains an example. In this example, we want to sample the values 1, 2, 3, and 4 with probabilities .15, .20, .25, and .40, respectively. Column B in Figure 2.15 contains the probabilities. Column C contains the cumulative probabilities, which are computed from the probabilities by just adding the current probability to the previous cumulative probability—that is,

$$F_j = p_j + F_{j-1}$$

with $F_0 = 0$. Columns D and E contain the values to be sampled. Note that the value 1 is in the same row as the cumulative probability zero. Thus, the values are shifted up a row. This is done because of the way the VLOOKUP function works.

Now to sample the values in column D having the cumulative probabilities in column C, we apply the function

```
=VLOOKUP(RAND(),$C$4:$E$7,2)
```

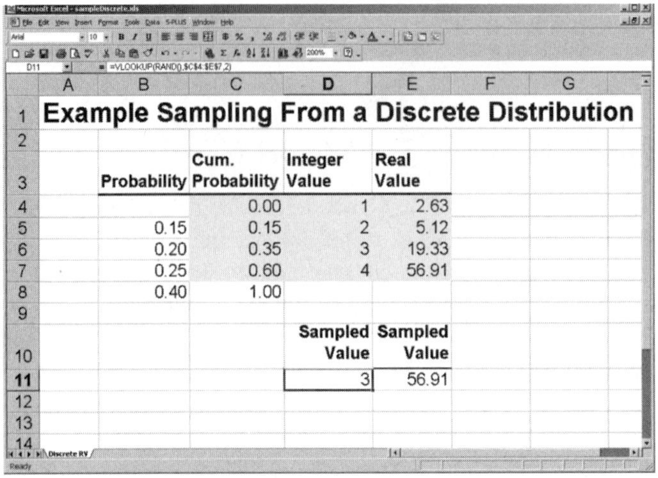

FIGURE 2.15
Use of a lookup table to sample from a discrete distribution

D11 =VLOOKUP(RAND(),$C$4:$E$7,2)

Example Sampling From a Discrete Distribution

	Probability	Cum. Probability	Integer Value	Real Value
4		0.00	1	2.63
5	0.15	0.15	2	5.12
6	0.20	0.35	3	19.33
7	0.25	0.60	4	56.91
8	0.40	1.00		

			Sampled Value	Sampled Value
11			3	56.91

This function has been placed in cell D11 and works as follows. First, a random value U, $0 < U < 1$, is sampled by RAND(). Then VLOOKUP attempts to find this value in the range C4..C7, which is the first column in the range that is the second parameter of the function—the lookup table. In Figure 2.15, the lookup table is shaded in gray. Normally, this value will not match a value in one of these cells, so VLOOKUP selects the cell so that U is at least the value in the cell but less than the value in the next cell. Thus, if U has the value .1736, VLOOKUP will select cell C5 because $.15 < .1736 < .35$. Once a value is found, VLOOKUP uses its last parameter to determine the column from which to select the return value. In this example, this parameter is 2, so the return value is selected from the second column in the table—that is, column D.

If we want to generate a random variate X that has the value a_1 with probability p_1, a_2 with probability p_2, and so on, we can first put the values $a_1, a_2, \ldots$, and so on in an array, **A.** Then generate a random variate J having the value 1 with probability p_1, 2 with probability p_2, and so on as just described, and let $X = a_J$ (i.e., X is the Jth element of the array **A**). This can be implemented in a spreadsheet using the same approach as we used to sample noninteger discrete values. Figure 2.15 has an example in column E and cell E11, where we sample the values in column E having probabilities in column B (shifted down one row). The approach is the same as for the integer values in column D, but instead of selecting values from column D, the function call selects values from column E. The contents of cell E11 are

```
=VLOOKUP(RAND(),$C$4:$E$7,3)
```

2.5.8 The Inverse Transform Method

One general way to produce observations from any distribution is the inverse transform method. See Fishman (1996) and Law and Kelton (2000) for background and more details. Suppose that $F(x)$ is the cumulative distribution function for the population we wish to sample. The inverse transform method operates by first sampling U from a uniform distribution between zero and 1. Then we apply the transform

$$X = F^{-1}(U)$$

That is, we find the value of X such that $F(X) = U$. This may seem abstract and difficult, but most spreadsheets provide handy functions that compute the inverse of many useful distributions.

As an example, suppose you want to sample from the gamma distribution with parameters 1.3 and 2.5. The function GAMMAINV(RAND(), 1.3, 2.5) will sample from this distribution by computing the inverse of the gamma distribution with parameters 1.3 and 2.5. Similar functions are provided in Excel, StarOffice Calc, and other spreadsheets for the T, normal, F, chi-square, beta, and lognormal distributions.

A note of caution: Many of these functions implement an iterative algorithm that searches for the correct value. This search can be time-consuming and thus greatly increase the time needed to run the replications. This is not usually a problem, but if it is, more efficient sampling methods often can be found.

We have illustrated the inverse transform method using continuous random variables, but it can also be applied to discrete random variables. The cumulative distribution function of a discrete random variable is a step function that increases by the probability assigned to each value at that value. Figure 2.16 is a plot of the cdf of a binomial distribution with $n = 4$ and $p = .3$. For discrete random variables, $F^{-1}(U)$ is the value X such that $F(X-1) < U \leqslant F(X)$. Using a spreadsheet, the technique is to create a lookup table containing the values of $F(x)$ and use VLOOKUP or HLOOKUP to find the value of X satisfying $F(X-1) < U \leqslant F(X)$.

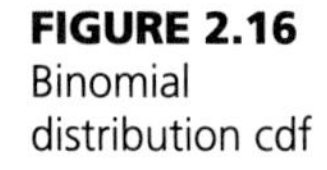

FIGURE 2.16
Binomial distribution cdf

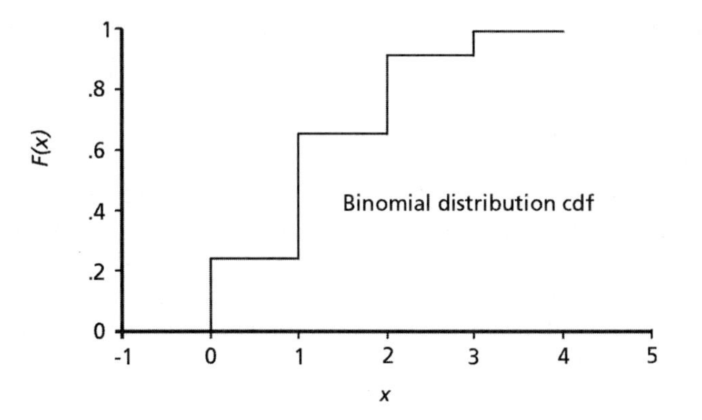

2.5.9 Special Considerations

The preceding methods can be used to generate a random variate from virtually any distribution, but frequently there are simpler and often faster ways to generate random variates from certain special distributions. As an example, consider the binomial distribution. A fact from probability theory that can be applied to generate observations from this distribution is that the sum of n independent Bernoulli random variables each with parameter p has a binomial distribution with parameters n and p. We have already discussed how to generate Bernoulli random variates, so to generate a binomial random variate, just generate n independent Bernoulli random variates and add them. We can shorten this process further with the following algorithm:

$X = 0$
REPEAT n times:
Generate U;
IF $U \leq p$, add 1 to X

This can be implemented in a spreadsheet by creating a column or row of n cells, each containing a sampled Bernoulli observation and summing the range of cells. Note, however, that to change the value of n, the spreadsheet structure must be altered—that is, the number of cells in the range must be changed.

2.6 EVALUATING DECISIONS: A ONE-PERIOD INVENTORY MODEL

The ultimate purpose of every model is to predict the likely effects of alternative decisions. This is true whether the analysis is done using mathematics, simulation, or a combination of both. In this section, we will look at a simple inventory model and show how simulation can be used to evaluate alternative ordering strategies.

2.6.1 The Inventory Model Revisited

Recall that in Chapter 1 we introduced a model for ordering soft drinks for a basketball game. Suppose you are responsible for deciding how many canisters to order for one game. One canister can serve 100 drinks. Let the demand be represented by a random variable, D, and assume we know that D has an exponential probability distribution with mean 5.0. The mean of D is expressed in canisters, or hundreds of drinks. Let s represent the number of canisters of soft drink that we order. Because

D is a random variable, we cannot, for a given value of s, predict whether D will be larger or smaller than s. If D is larger than s, then we will run out, and $D - s$ will be the amount of demand we were unable to fill. Suppose that each unit (100 drinks) of unmet demand costs us \$40. (We can consider this the cost associated with customer dissatisfaction and unrealized potential profits.) Thus, the cost associated with ordering s canisters of soft drink, when D is greater than s, is $40.0(D - s)$. Suppose that we order too much soft drink so that D is smaller than s. In this case, we will have $s - D$ canisters of soft drink left over. Suppose that the cost per unit of excess soft drink is \$10. (This is the cost associated with returning the soft drink to the bottler or otherwise disposing of it.) Thus, the cost associated with ordering s canisters, when D is less than s, is $10.0(s - D)$. We want to determine the value of s that minimizes the expected cost.

2.6.2 Optimal Order Quantity Using Simulation

To solve this problem using simulation, we first choose a value for s, and then set up an experiment in which the value of D is generated from an exponential distribution with mean 5.0. Finally, the cost associated with this particular value of s and D is computed. This experiment is replicated independently n times, producing n observations of the cost. A confidence interval for the mean cost is computed from these data. The whole process is then repeated for other values of s, producing a confidence interval for the mean cost for each order quantity, s. These estimates can be plotted and the plot used to locate the minimum value. The algorithm for the simulation is:

1. For $j = 1, 2, \ldots, n$, do:

   Generate D from the distribution of demand

   Compute: $$Y_j = \begin{cases} 10.0\,(s - D) \text{ if } D \leq s, \\ 40.0\,(D - s) \text{ if } D > s \end{cases}$$

   Accumulate the sum and sum of squares of the Y_j's.

2. Compute the sample mean of the Y_j's and a confidence interval for the mean.

The simulation was run for each value of $s = 5.0, 6.0, \ldots, 12.0$, using 4000 replications. Figure 2.17 gives the results of the simulation and presents them graphically. We can see that the minimum mean cost appears to be obtained when approximately 8.0 canisters of soft drink are ordered. It is possible to solve this problem mathematically and determine the exact value of s that minimizes the mean cost; see Hillier and Lieberman (1995) for a solution. The value of s that minimizes the mean cost is the solution to

$$F(s^*) = \frac{40.0}{40.0 + 10.0} = .80$$

FIGURE 2.17

Results of multiple runs for inventory model

Decision Variable:

Order Quantity	8

Simulation:

Trial	Demand (D)	Cost
1	6.69	52.25
2	0.17	313.12
3	16.00	1279.92
4	2.75	210.11
5	3.42	183.17
6	9.89	303.01
7	0.74	290.50
8	2.93	202.74
9	5.09	116.27
10	3.72	171.26
11	2.82	207.22
12	7.41	23.77
13	6.93	42.83
14	3.05	197.97
15	2.31	227.45
16	10.35	375.92
17	2.44	222.27
18	0.37	305.24
19	2.94	202.39
20	1.27	269.35
21	0.09	316.27
22	1.19	272.25
23	5.73	90.73
24	6.50	60.09
25	0.07	317.09
26	6.57	57.26
27	4.74	130.28
28	1.49	260.24
29	4.56	137.61

Data Analysis:

Trials	4000
Mean	313.05
Std. Dev.	421.00
Std. Error	6.66
Error	13.05
Lower c.l.	300.00
Upper c.l.	326.10
Min	0.28
Max	4506.50

Decision Analysis

	Mean	Error
Order Qty	313.05	13.05
5	372.7	18.55
6	331.68	15.60
7	336.625	15.50
8	313.083	13.29
9	318.169	11.10
10	329.693	10.80
11	350.284	9.87
12	375.489	9.62

where $F(x)$ is the cumulative distribution function of demand. Because the distribution of demand is exponential, $F(x) = 1 - e^{-x/5.0}$, and $s^* = 5 \ln(5) = 8.05$. We see, therefore, that the simulation gave us a valid and rather accurate answer to this problem.

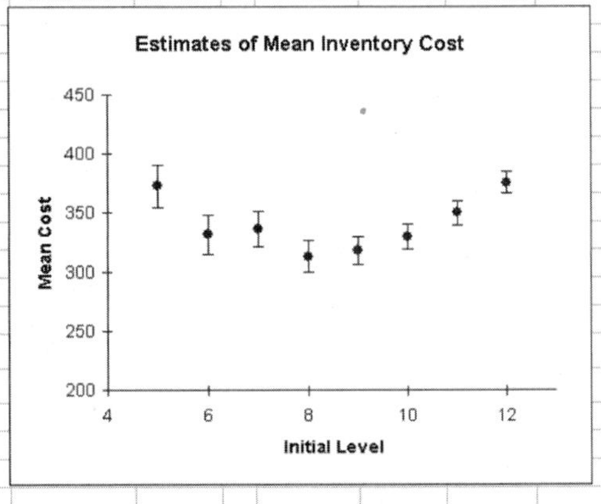

2.6.3 Other Cost Functions

The mathematical solution to this problem just given applies only when the cost functions are linear; that is, it applies only when the cost incurred for having an excess or shortage is proportional to the amount of excess or shortage (and, therefore, the marginal cost is constant). In realistic problems, the cost might not be linear; rather, the marginal cost might increase or decrease as the amount of excess or shortage increases. When this is the case, we can still use simulation to estimate the optimal order quantity even though the mathematical solution no longer applies.

FIGURE 2.18
Results of multiple runs for inventory model with nonlinear costs

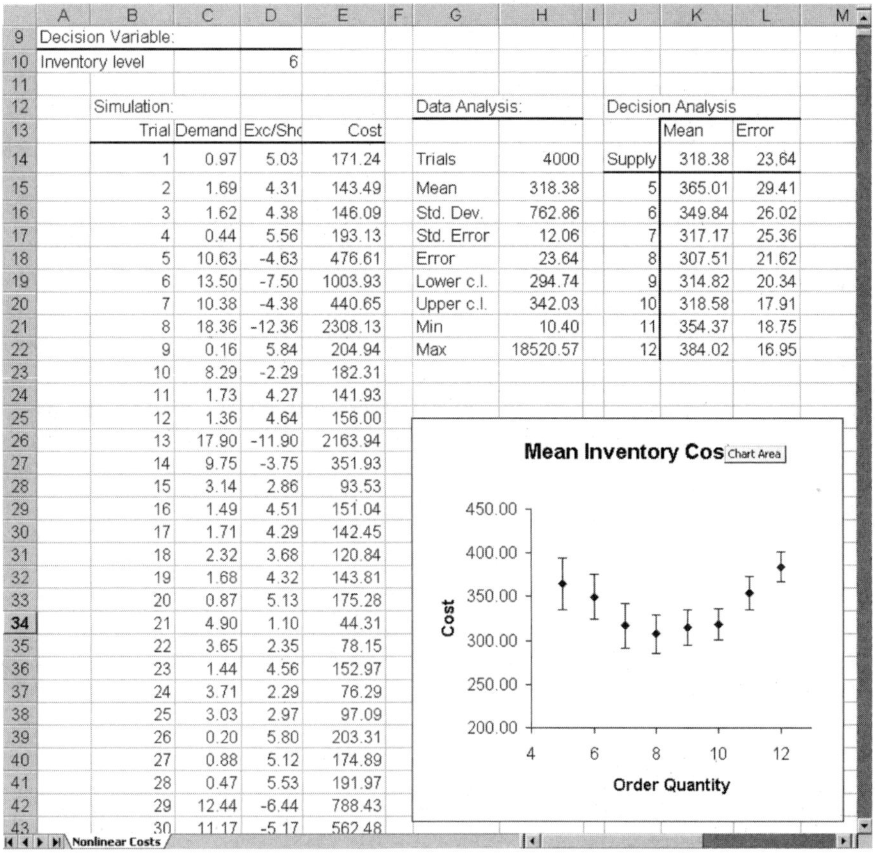

Decision Variable:

Inventory level	6

Simulation:

Trial	Demand	Exc/Shc	Cost
1	0.97	5.03	171.24
2	1.69	4.31	143.49
3	1.62	4.38	146.09
4	0.44	5.56	193.13
5	10.63	-4.63	476.61
6	13.50	-7.50	1003.93
7	10.38	-4.38	440.65
8	18.36	-12.36	2308.13
9	0.16	5.84	204.94
10	8.29	-2.29	182.31
11	1.73	4.27	141.93
12	1.36	4.64	156.00
13	17.90	-11.90	2163.94
14	9.75	-3.75	351.93
15	3.14	2.86	93.53
16	1.49	4.51	151.04
17	1.71	4.29	142.45
18	2.32	3.68	120.84
19	1.68	4.32	143.81
20	0.87	5.13	175.28
21	4.90	1.10	44.31
22	3.65	2.35	78.15
23	1.44	4.56	152.97
24	3.71	2.29	76.29
25	3.03	2.97	97.09
26	0.20	5.80	203.31
27	0.88	5.12	174.89
28	0.47	5.53	191.97
29	12.44	-6.44	788.43
30	11.17	-5.17	562.48

Data Analysis:

Trials	4000
Mean	318.38
Std. Dev.	762.86
Std. Error	12.06
Error	23.64
Lower c.l.	294.74
Upper c.l.	342.03
Min	10.40
Max	18520.57

Decision Analysis

	Mean	Error
Supply	318.38	23.64
5	365.01	29.41
6	349.84	26.02
7	317.17	25.36
8	307.51	21.62
9	314.82	20.34
10	318.58	17.91
11	354.37	18.75
12	384.02	16.95

Mean Inventory Cos Chart Area
450.00
400.00
350.00
300.00
250.00
200.00
Cost
4
6
8
10
12
Order Quantity

Nonlinear Costs

Suppose the cost for a shortage is $10.0 + 50.0S + 2.0S^2$, where $S = (D - s)$ is the shortage, and the cost for an excess is $20.0 + 20.0E + 11.0E^2$, where $E = (s - D)$ is the amount of excess. The analytical approach requires us to solve a difficult nonlinear equation. However, to find an approximate solution using simulation, we just change the cost computation in the simulation and repeat the simulation run. Figure 2.18 shows the results of doing this using inventory levels 5, 6, 7, . . . , 12, and 4000 replications. The minimum mean cost appears to be incurred when 9 canisters are ordered using this cost function. Note that the mean cost as a function of order quantity appears to have a relative minimum at $s = 7$, increases for $s = 8$, and decreases to a global minimum when $s = 9$. The confidence intervals overlap, so we cannot be sure of this relationship without doing additional replications. However, all of these estimates of mean cost are quite close, so the error involved with selecting the wrong order quantity is quite small.

2.6.4 Sensitivity Analysis

Usually, a simple answer such as "The optimal order quantity is seven canisters" is not sufficient for a decision maker to gain confidence in the model. Since many specifics in the model are determined by judgment, we usually want to determine how sensitive the conclusions are to these specifics by doing a sensitivity analysis. We have assumed that the demand is exponentially distributed with mean 5.0 canisters. The standard deviation of the exponential distribution is the same as the mean (see Appendix), so the standard deviation of demand is 5.0 also. Can the same conclusion concerning the optimal order quantity be made if the demand is normally distributed with mean 5.0 and standard deviation 5.0? We will assume that this demand distribution is truncated at 0, because we cannot experience negative demand. Figure 2.19 shows the results when the simulation is run using normally distributed demand. It seems evident that with this distribution of demand, ordering 7, 8, or 9 canisters is the best choice, and these choices are approximately equal. Figure 2.20 shows the results of running the simulation again with demand distributed as an Erlang random variable with two stages and mean 5.0. (The average of two indepen-

FIGURE 2.19
Results of simulation runs with normally distributed demand

Decision Variable:

Order Quantity	6

Simulation:

Trial	Demand	Cost
1	3.79	88.43
2	1.72	171.19
3	0.58	216.78
4	11.11	818.23
5	1.22	191.26
6	2.62	135.27
7	0.42	223.19
8	5.74	10.58
9	2.70	131.86
10	3.37	105.16
11	2.43	142.95
12	2.07	157.13
13	0.85	206.19
14	1.48	180.82
15	3.11	115.48
16	2.01	159.73
17	0.03	238.95
18	4.20	71.90
19	7.09	173.86
20	5.58	16.97
21	0.23	230.79
22	1.51	179.70
23	3.09	116.47
24	1.02	199.34
25	18.17	1946.49
26	10.17	667.13
27	1.79	168.53
28	2.78	128.91
29	6.36	57.94
30	0.56	217.64

Data Analysis:

Trials	4000
Mean	335.72
Std. Dev.	534.65
Std. Error	8.45
Error	16.57
Lower c.l.	319.15
Upper c.l.	352.29
Min	0.24
Max	6289.77

Decision Analysis:

	Mean	Error
Order Qty	335.72	16.57
5	361.21	17.94
6	347.422	17.02
7	321.078	14.59
8	330.141	14.23
9	327.399	13.64
10	324.964	10.39
11	349.103	10.22
12	366.335	8.47

Estimates of Mean Inventory Cost

Mean Cost: 200, 250, 300, 350, 400

Initial Level: 4, 6, 8, 10, 12

Normal Demand

FIGURE 2.20

Results of simulation runs with Erlang demand

Decision Variable:	
Order Quantity	8

Simulation:

Trial	Demand	Cost
1	1.73	250.86
2	9.54	247.18
3	4.83	126.82
4	5.53	98.87
5	8.66	105.54
6	2.00	239.98
7	4.44	142.31
8	12.70	752.79
9	5.76	89.68
10	7.91	3.44
11	1.13	274.75
12	6.61	55.65
13	4.64	134.46
14	23.67	2507.17
15	0.47	301.37
16	7.90	4.07
17	0.58	296.76
18	6.91	43.71
19	5.14	114.60
20	3.06	197.69
21	5.26	109.55
22	10.34	374.79
23	0.92	283.38
24	6.49	60.42
25	4.22	151.17
26	8.43	69.56
27	4.08	156.88
28	6.18	72.69
29	1.13	274.80

Data Analysis:

Trials	4000
Mean	228.83
Std. Dev.	262.77
Std. Error	4.15
Error	8.14
Lower c.l.	220.68
Upper c.l.	236.97
Min	0.01
Max	3936.50

Decision Analysis:

	Mean	Error
Order Qty	228.83	8.14
5	268.39	12.06
6	237.51	10.05
7	223.7	8.54
8	228.88	7.84
9	237.45	6.27
10	252.76	5.47
11	277.1	4.97
12	309.88	4.61

Estimates of Mean Inventory Cost

Mean Cost

Initial Level

Erlang Demand

dent exponential random variables has an Erlang distribution.) From this graph, 7 appears to be the optimal order quantity; however, we must remember that the numbers plotted in the graph are sample statistics, and therefore will vary somewhat from the true mean cost they are estimating. To further investigate whether 7 is truly the optimal order quantity, the simulation was rerun for 16,371 replications, using 7, 8 and 9 canisters as the order quantity. The results are in Figure 2.21. There is some evidence that 8 is the better order quantity; however, the mean cost associated with ordering 7 canisters appears to be only slightly higher than that for 8.

Suppose we decide to order eight canisters and continue to assume that the distribution of demand is exponential. The cost will be a random variable, because it depends on the demand, which is a random variable. The analytic model for determining the optimal order quantity does not provide much help in specifying the distribution of the cost that we will actually experience; however, we can collect the cost observations from the simulation and plot their histogram to see the distribution. Figure 2.22 shows this histogram. Remember that if you order eight canisters and the demand distribution is exponential, the actual cost incurred will be a single randomly selected observation from this distribution. One can see that this distribution has a

FIGURE 2.21
Results of a long simulation run with Erlang demand

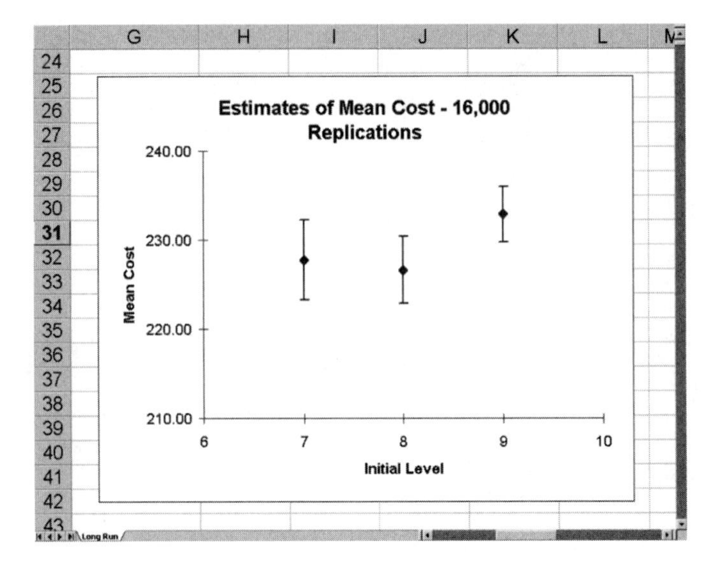

FIGURE 2.22
Histogram for actual inventory cost

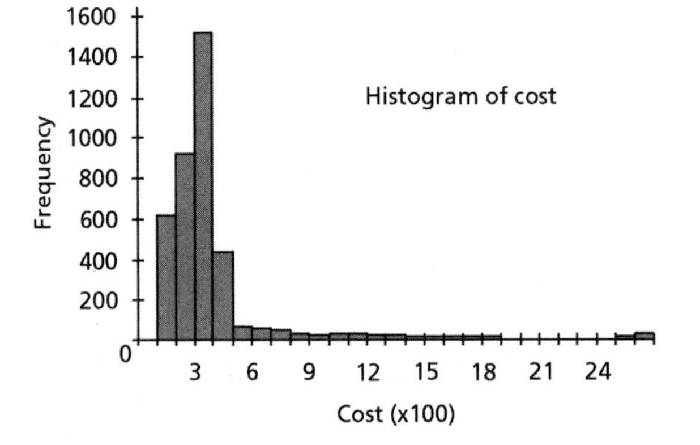

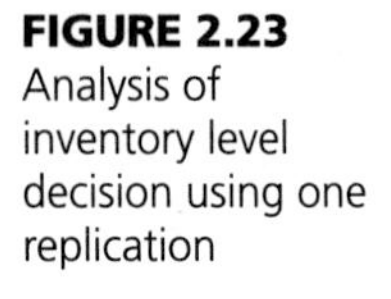

FIGURE 2.23
Analysis of inventory level decision using one replication

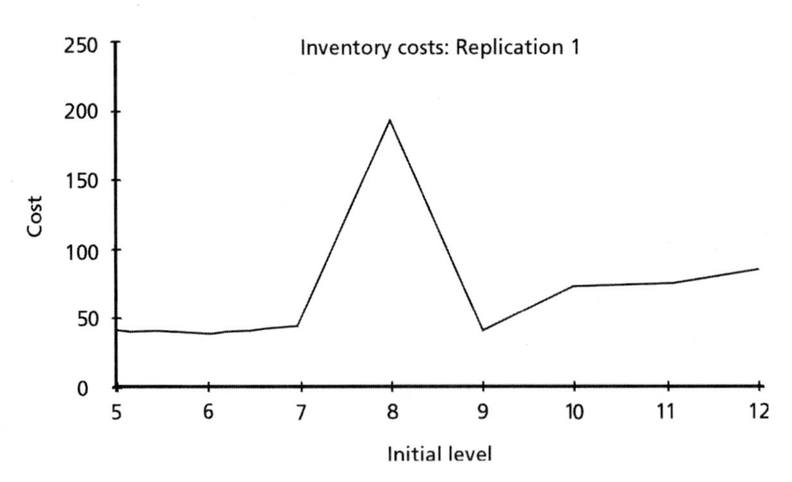

strange shape and certainly does not belong to any class of distributions that statisticians normally deal with. We can see that the most likely cost to incur is between \$300 and \$400; however, the chance that a cost in excess of \$500 is incurred is approximately 13 percent. This unusual shape is the result of the skewed shape of the exponential distribution and the particular cost function that was used.

2.7 DATA ANALYSIS FOR STATIC SIMULATIONS

In the simulations just discussed, we repeated the experiment several times and used statistical methods to estimate the mean from the data produced. If the model involves random variables, this *must* be done, because a single replication is normally too variable to allow reliable conclusions to be drawn. To illustrate, we ran the inventory simulation just discussed for a single independent replication using each initial inventory level of 5, 6, . . . , 12 canisters and then plotted the results in Figure 2.23. One can see that if these observations are accepted as the respective mean costs, then 6 canisters is the best initial inventory level and 8 canisters provide the largest cost. However, one might be suspicious of this conclusion because of the obvious variation in the cost observations. Each initial inventory level corresponds to a specific distribution for the resulting cost. The eight values plotted in Figure 2.23 are simply eight single observations randomly and independently sampled from each of these eight distributions or populations. Thus, on a single replication, any observation could exceed any other, regardless of the relationship of their means. Single replications were again run for initial inventories between 5 and 12 canisters, and this was repeated four times. Figure 2.24 shows the result graphically. In each plot, the costs look considerably different from those in Figure 2.23. Clearly, each individual replication has too much variation to allow reliable conclusions to be drawn.

What we actually would like to know is which of the eight distributions under consideration has the smallest *mean?* Of course, we can never answer this question with absolute certainty; however, by drawing a sufficiently large sample from each of the eight distributions (which is done by replicating the simulation using each of the eight initial inventory levels), we can use standard statistical methods to compute a confidence interval for each mean.

2.7.1 Performing Independent Replications

First, let's discuss how *independent* replications are generated. A more detailed explanation will be given in a later chapter; here we want to discuss the mechanics of ensuring that replications are independent. The basic mechanism for generating sample observations from given distributions (random variates) is the *random number generator.* A random number generator is a computer program that on each exe-

FIGURE 2.24 Repeat of Figure 2.23 four times

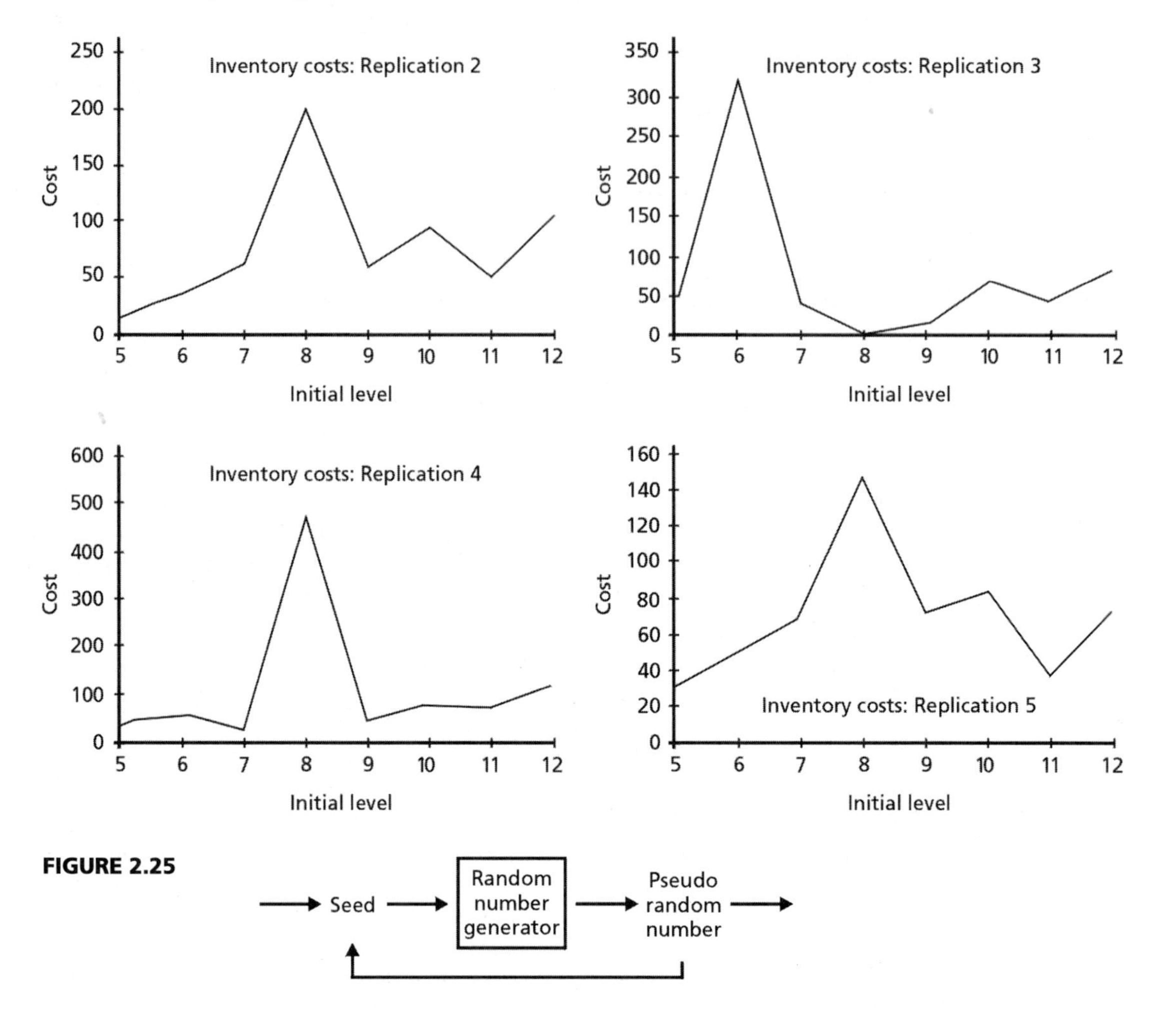

FIGURE 2.25

cution takes an integer, called the *seed,* as input and produces another integer as output. This output integer then serves two purposes: (1) It is divided by the largest possible integer that can be produced to return a uniformly distributed random variate between 0 and 1, which is then used to compute the next random variate; and (2) it also becomes the seed the next time the random number generator is executed. Figure 2.25 shows this relationship. Thus, each random number generated is a "transformation" of the previous one, and we can think of the sequence of random numbers as a "stream." A good random number generator produces a sequence of random numbers that behave as if they are mutually independent and uniformly distributed between 0 and 1. Thus, if replications use random numbers that come from distinct nonoverlapping segments of the random number stream, the output from the replications will be mutually independent because the random numbers that are input to the replications are mutually independent. For example, if the first replication uses the first 1531 random numbers in the stream, the second replication uses the next 2195 random numbers, the third replication uses the next 1299 random numbers, and so on, then the replications will produce sample statistics that are mutually independent. The important idea here is that none of the random numbers should be used in two or more replications.

The mechanics of ensuring that the replications are independent are deceptively simple. One needs only to make sure that the system is reinitialized to the same starting conditions at the beginning of each replication and that the initial seed for each replication is the final random number generated in the previous replication. If all replications are done sequentially in the same program (which is usually the case), then nothing explicit must be done because the next seed for the random number generator is always the previous random number generated. Therefore, if the system starts each replication with the same initial conditions, the replications will automatically be independent.

2.7.2 Estimating the Mean from Independent Replications

Suppose that each replication produces one observation, and let the observations be $X_1, X_2, \ldots, X_n$ for n replications (see Appendix for a more detailed discussion of statistical estimation methods). Let μ denote the mean and σ^2 denote the variance of each observation. The central limit theorem tells us that if σ^2 is finite, n is sufficiently large, and $X_1, X_2, \ldots, X_n$ are mutually independent, then the sample mean,

$$\overline{X} = \frac{1}{n} \sum_{i=1}^{n} X_i$$

has a distribution that is approximately normal with mean μ and variance σ^2/n. From this, one can compute a $100(1 - \alpha)$-percent confidence interval for μ using

$$\overline{X} \pm z_{1-\frac{\alpha}{2}} \frac{s}{\sqrt{n}}$$

where $z_{1-\alpha/2}$ is the $100(1 - \alpha/2)$ percentage point of the standard normal distribution and s is the sample standard deviation of $X_1, X_2, \ldots, X_n$.

2.8 SIMULATION WITH SPREADSHEETS: A MORE COMPLEX MODEL

Spreadsheets have opened a new world of modeling and problem analysis for managers and analysts. These programs allow the user to enter both numerical data (input parameters) relevant to a problem and formulas that specify the relationship between these values and other quantities of interest into the cells of an electronic table. When the input parameter values are changed, the values of other quantities depending on these parameters are automatically recalculated and displayed, allowing the modeler to play an unlimited number of what-if games with the spreadsheet. Most spreadsheet programs also have a large number of built-in functions to do various complex calculations. These include trigonometric and transcendental functions, financial functions to compute such things as present value and internal rate of return, statistical functions for the mean, standard deviation and other sample statistics, and other functions for various purposes. These functions enable the user to develop models with complex relationships between the variables. Many real spreadsheet models are large. They may have hundreds or thousands of cells. These models are frequently financial in nature, as in the case of the model presented in section 2.1. However, they can also involve production, inventory, personnel, marketing, economics, and any other type of problem where the relationships can be expressed as formulas that can be put into spreadsheet cells.

2.8.1 A More Realistic Real Estate Model

As we saw in section 2.3, simulation can provide a more reliable approach to risk analysis than what-if calculations. In this section, we will expand the model in section 2.3 to make it a more realistic model, and we will examine the steps necessary to perform a risk analysis on the model using simulation in a spreadsheet context. First, let's reexamine the model in Figure 2.5. This model computed ATCF at the end of one year, given the following input parameters:

1. building cost per square foot,
2. size of building in square feet,
3. average rent per square foot,
4. operating expense per square foot,
5. vacancy rate,
6. other income,
7. percentage of original building price mortgaged,
8. mortgage interest rate,

FIGURE 2.26 Multiyear office building cash-flow model

	A	B	C	D	E	F
1	CASH FLOW CALCULATION FOR PURCHASE OF AN OFFICE BUILDING - MULTIYEAR MODEL					
2						
3	Year	2002	2003	2004	2005	2006
4	Building cost per square foot:	$80.00				
5	Original cost of building:	$14,400,000				
6	Size of building (sq. ft):	180,000				
7	Avg. rent per square foot:	$9.00	$9.45	$9.92	$10.42	$10.94
8	Assumed Operating Exp./sq.ft.:	$1.20	$1.32	$1.45	$1.60	$1.76
9	Assumed vacancy rate:	0.30	0.26	0.22	0.18	0.14
10	Depreciation horizon (years):	30				
11						
12						
13	Year	2002	2003	2004	2005	2006
14	Potential gross income	$1,620,000	$1,701,000	$1,786,050	$1,875,353	$1,969,120
15	Less: Vacancy and collection loss	(486,000)	(442,260)	(392,931)	(337,563)	(275,677)
16	Effective gross income	1,134,000	1,258,740	1,393,119	1,537,789	1,693,443
17	Less: Operating expenses	(216,000)	(237,600)	(261,360)	(287,496)	(316,246)
18	Net operating income	918,000	1,021,140	1,131,759	1,250,293	1,377,198
19	Less: Debt service	(1,510,826)	(1,510,826)	(1,510,826)	(1,510,826)	(1,510,826)
20	Before-tax cash flow	(592,826)	(489,686)	(379,067)	(260,533)	(133,628)
21	Less: Tax	(329,984)	(293,001)	(253,232)	(210,499)	(164,611)
22	After-tax cash flow	($262,842)	($196,685)	($125,835)	($50,034)	$30,983
23						
24	PERFORMANCE MEASURE					
25	Risk-free Rate of Return	0.1				
26	Discount Factor	1.000	0.909	0.826	0.751	0.683
27	Contrib. to PV of ATCF	(262,842)	(178,805)	(103,996)	(37,591)	21,162
28	Present Value of ATCF	(562,072)				
29						

One Year Model / One Year Simulation / Fig 2.7 / Figs 2.25 and 2.26 / Multiyear Model / Multiyr Scenario Comp / Multiyear Sim Model

9. term of mortgage loan, and

10. tax rates.

Normally, a decision to build an office building such as this one would not be made based on a one-year analysis. Instead, the investors would evaluate the performance of the investment over a 5- or 10-year horizon. Figure 2.26 shows a portion of the same model after it has been modified to represent the investment over a five-year horizon. Columns 1 and 2 are the same as the corresponding columns in Figure 2.5. Columns 3 through 6 show the cash-flow calculations for the following four years. We are assuming that the rent, operating expense, vacancy rate, and other income will possibly change over time. The other parameters listed above are fixed when the building is finished. Thus, the columns for the final four years have new values for rent, operating expense, vacancy rate, and other income but not for the rest of the input parameters. The model assumes that values for rent increase by 5 percent per year, the values of operating expense increase by 10 percent per year, and the values of the vacancy rate decrease (linearly) by .04 per year. Cash-flow calculations are done for the following four years exactly as they are for the initial year. Tax calculations are also done as they are for the initial year, with the exception that we assume straight-line depreciation, and thus the yearly depreciation is the same as the initial year depreciation. The interest calculations for the final four years are based on the remaining balance at the end of the previous year and the remaining term of the loan. Otherwise, they are identical to the initial year calculations.

Because we are evaluating the after-tax cash flow for this investment over a five-year horizon, we need a way to summarize the sequence of yearly cash flows. The usual way to summarize a cash flow is through the present value, and this is what we choose to compute in this case. In Figure 2.26, the risk-free interest rate, i, is given as .1, and the present value of the sequence of five after-tax cash flows is

$$PV = \sum_{i=1}^{5} \alpha^{i-1} F_i = -\$562{,}072$$

where

$$\alpha = \frac{1}{(1+i)} = .909$$

and F_i is the after-tax cash flow in the ith year.

This spreadsheet model provides a realistic representation of the financial consequences over five years of building an office building. If all parameters are accurate, then the current value of the stream of yearly cash flows will be – $562,072. Of course, as in the case of the simplified model presented in section 2.3, several of the values are not known with certainty, so we will conduct a risk analysis on the model using simulation. Indeed, this can be done with any model implemented in a spreadsheet.

2.8.2 Risk Analysis for the Real Estate Model

First, we must determine which variables (input parameters) are uncertain and replace them with appropriate random variates. As before, we will assume that the building cost per square foot is unknown. Since the building is only built once, this cost is computed only for the first year. Other uncertain variables are, as discussed in section 2.3, the average rent per square foot, operating expense per square foot, and vacancy rate. These values are computed each year in the five-year horizon, so the total number of uncertain parameters is 20. If we were to do a series of what-if calculations on these 20 parameters, assuming three possible values for each, we would have to do about 10^{10} recalculations. Clearly, this is not feasible even if we can automate the process.

2.8.3 Fitting Distributions for Uncertain Variables

Each parameter must be replaced by a random variate whose distribution represents the relative likelihood for the parameter's possible values. In Chapter 3, we will discuss how to select a distribution to represent a given variable based on historical observations on that variate. Here we will take a nontechnical approach to this problem. Suppose that the building cost per square foot for each of 100 buildings similar

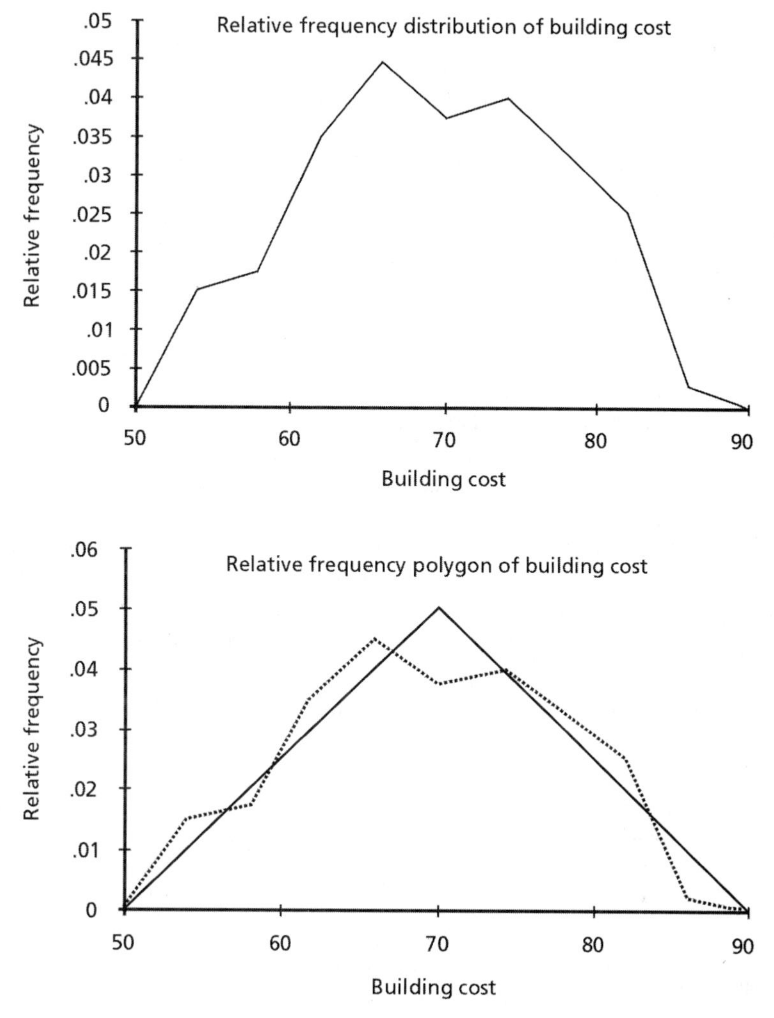

FIGURE 2.27 Empirical distribution of building cost

FIGURE 2.28 Empirical distribution with triangular distribution

to ours that were built during the past five years has the frequency distribution shown in Figure 2.27. We want to select a distribution that has a shape similar to that of Figure 2.27. There are, of course, many distributions that are similar to that figure's shape. Since the frequency distribution in this figure is based on only 100 observations, however, we can expect it to look fairly ragged, and thus normal, gamma, and triangular distributions might all appear to have the same shape. Figure 2.28 shows the same frequency distribution superposed with a triangular distribution. The two appear to have approximately the same shape, so we will use a triangular distribution to represent the building cost per square foot. Of course, if we were to try to fit several distributions, we might find that another distribution fits a little better. However, we will be content to use the triangular distribution because we do not have enough data to enable us to select confidently from among several distributions.

Once the distribution has been selected, we must compute the parameters of the distribution to use when generating random variates. The specific parameters that must be computed depend on the distribution. For the normal distribution, for example, we would use the sample mean and sample standard deviation to estimate the population mean and standard deviation, respectively. In the case of the triangular distribution, the three parameters are the minimum, maximum, and mode (most likely) values, and they can be estimated from the frequency distribution of the data. See Figure 2.28. Note that the minimum and maximum values in a finite sample are obviously biased estimates of the minimum and maximum values for the triangular distribution, because these parameters are the minimum and maximum values that could ever be observed, and, generally, an extremely large sample is required before values close to these would actually be observed. From the data, we see that the minimum and maximum building costs observed are 52.5 and 86.9, respectively. Thus, if the observations did come from a triangular distribution, the minimum building cost is less than 52.5 and the maximum cost is greater than 86.9. The most likely building cost appears to be approximately 70.0. Figure 2.28 shows that a triangular distribution with minimum, most likely, and maximum parameters of 50.0, 70.0, and 90.0 appears to fit these data rather well. Therefore, we will use these three values—50.0, 70.0, and 90.0—as the minimum, most likely, and maximum parameters of the triangular distribution that we will use to generate building costs.

The data we used to compute these parameters are one year old relative to our current model. Projections have been offered showing that building costs have risen by 4 percent over the year. Thus, we will increase the three parameters by 4 percent to adjust for this year's difference in time. Our final parameter estimates are 52.0, 72.80 and 93.60. If data are available for the other uncertain variables, we can go through the same process to select the form of the distribution that will be used to generate each variate and then compute estimates of the parameters for the selected distribution. Generally, data on office space rent, operating expenses, and vacancy rates are available and accurate for the near term. Of course, the accuracy of projections decreases as the length of the projections increases simply because longer time spans provide more opportunity for unexpected events to occur, which provides more variability in the outcomes. The spreadsheet in Figure 2.26 assumes that rent per square foot and operating expense increase each year at 5 percent and 10 percent rates, respectively, and the vacancy rate decreases by 4 percent each year. We will represent these variables using the same distributions as in section 2.3, but we will adjust their parameters so the means of the variables increase and decrease as in Figure 2.26. We will also assume that other income is a random variable with the same (approximately normal) distribution given in section 2.2, but we will not assume that the mean changes over the five years.

Table 2.5 shows the formulas used to compute the random variates in the 1995 column in the office building financing model shown in Figure 2.26. The formulas

TABLE 2.5 Formulas used to compute random variates in the five-year office building model

Variate	Formula
Building cost per sq. ft.	`IF(RAND(),0.4132, 55.64+15.08*RAND(), 92.14-21.42*RAND()`
Average rent per sq. ft.	`8 + 2*(RAND()+RAND())`
Average operating expense	`0.6 + 0.4*(RAND()+RAND())`
Vacancy rate	`0.2 + 0.4*SQRT(RAND())`
Other income	`50,000 + 141421*(RAND() + RAND() + . . . + RAND() - 3)`

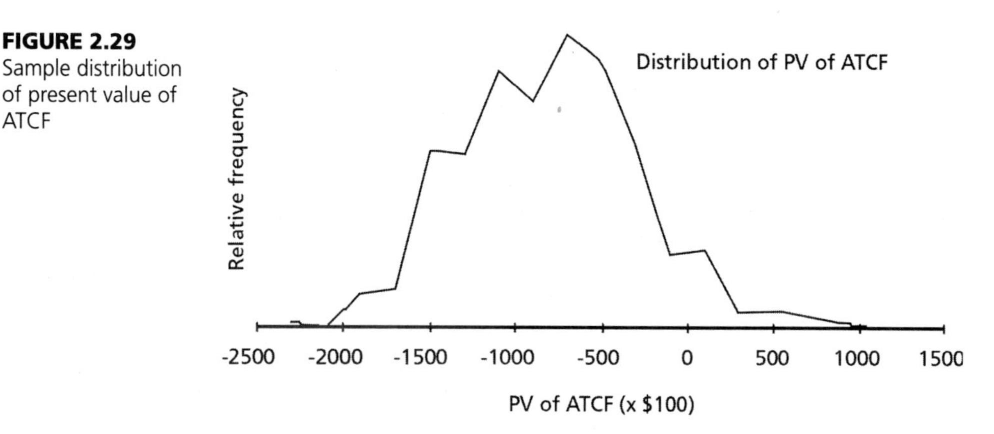

FIGURE 2.29 Sample distribution of present value of ATCF

for 1996 and later years are similar to those for 1995 except that the parameters are adjusted to reflect the percentage increases in the values.

2.8.4 Simulation Results

The results of 500 replications for the office building model are shown in Figures 2.29 and 2.30 and Table 2.6. Since the performance measure in which we are interested is the present value of a stream of five yearly cash flows, the results of this model are not directly comparable with those of the model in section 2.2. From Figures 2.29 and 2.30 and Table 2.6, we see that the expected PV of ATCF is –$806,094, and a 95 percent confidence interval for the mean is from –$850,512 to –$761,676. There is approximately a 1 in 4 chance that the PV of ATCF will be less than approximately –$1.2 million, and a 1 in 4 chance that it will be greater than –$500,000. The probability that the PV of ATCF will be positive is estimated to be .03, so this investment is unlikely to produce a positive cash flow.

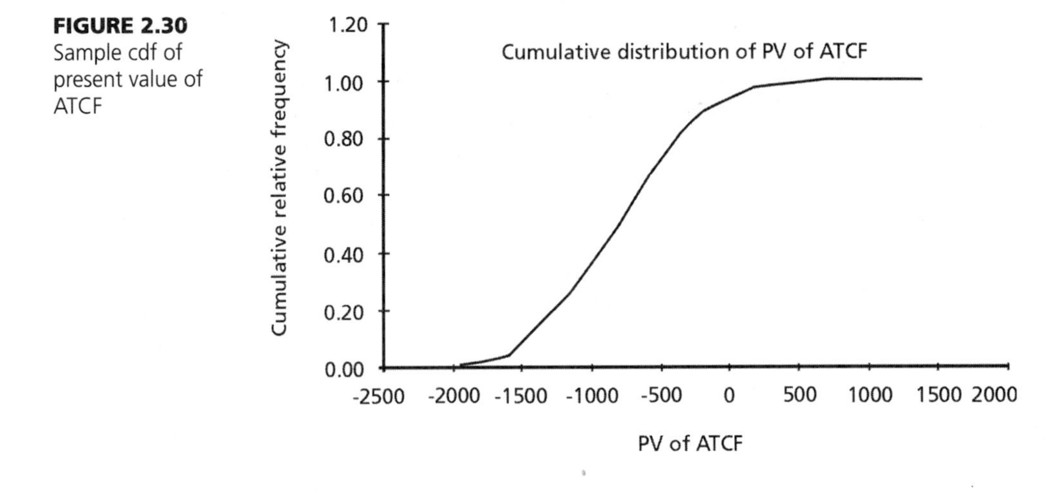

FIGURE 2.30 Sample cdf of present value of ATCF

2.8.5 Alternative Decisions to Improve Performance

There are several ways in which we can seek to improve the performance of this investment. Two of these are:

1. We can modify the building design. This will decrease the building cost, but it will also affect the rent and the vacancy rate. A less expensive building will bring lower rents or higher vacancy rates.
2. We can change the rent. If we decrease it, then we will decrease the vacancy rate, because lower rents will generate larger demand for office space in the building.

Table 2.7 provides projections of these parameters under different scenarios. For example, the mean building cost per square foot decreases in steps of \$2 from \$70 down to \$60. We assume that the whole distribution shifts when we change from one scenario to the next; for example, in the second scenario, the distribution of building cost is triangular from \$48 to \$88, with a maximum at \$68. The rent also shifts, so the distribution of rent in the second scenario is triangular from \$7.75 to \$11.75, with a maximum at \$9.75. We assume that the vacancy rate has the same triangular distribution with the lower limit fixed at .20 and the upper limit varying from scenario to scenario. In the second scenario, therefore, the distribution of vacancy rate is triangular from .20 to .50 with maximum at .50. The mean of this distribution is .467. Trends in the distribution of rent and mean vacancy were applied as before.

The simulation was run again for 500 replications for each scenario in Table 2.7, and the results are given in Table 2.7 and Figure 2.31. Note that we compute confidence intervals for the mean present value of ATCF for each scenario, because the observed present values are random variables and each scenario is replicated

TABLE 2.6 Sample cdf of present value of ATCF

Number of replications	500
Mean	–806,094
Variance	2.568E+11
Standard deviation	506,755
Minimum	–2,243,235
Maximum	817,186

Confidence interval for mean of ATCF	
Upper limit	–761,676
Lower limit	–850,512

Limits	Frequency	Relative Frequency	Cumulative Relative Frequency
–2.40 to –2.20	1	.0020	.0020
–2.20 to –2.00	0	.0000	.0020
–2.00 to –1.80	9	.0180	.0200
–1.80 to –1.60	10	.0200	.0400
–1.60 to –1.40	48	.0960	.1360
–1.40 to –1.20	47	.0940	.2300
–1.20 to –1.00	70	.1400	.3700
–1.00 to – .80	61	.1220	.4920
– .80 to – .60	80	.1600	.6520
– .60 to – .40	72	.1440	.7960
– .40 to – .20	49	.0980	.8940
– .20 to .00	20	.0400	.9340
.00 to .20	21	.0420	.9760
.20 to .40	4	.0080	.9840
.40 to .60	4	.0080	.9920
.60 to .80	3	.0060	.9980
.80 to 1.00	1	.0020	1.0000
1.00 to 1.20	0	.0000	1.0000
1.20 to 1.40	0	.0000	1.0000
	500	.9340	

TABLE 2.7 Outputs for additional simulation runs of office building model

Scenario	Mean Building Cost	Mean Rent	Vacancy Limit	Mean PV of ACTF	Upper Limit	Lower Limit
1	70	10.00	.600	–513,010	–462,263	–563,757
2	68	9.75	.500	–214,612	–163,505	–266,719
3	66	9.50	.395	16,236	216,503	115,970
4	64	8.75	.350	101,217	152,595	49,839
5	62	8.50	.350	88,142	138,736	37,547
6	60	8.00	.320	21,702	72,632	–29,227

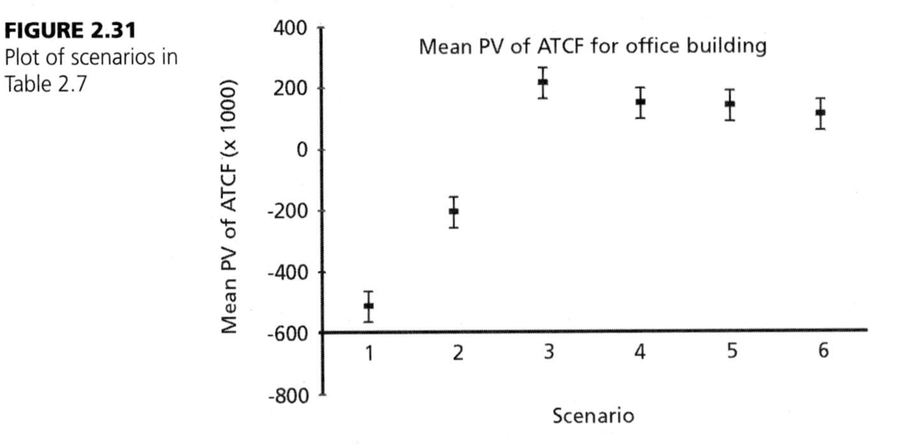

FIGURE 2.31
Plot of scenarios in Table 2.7

500 times. It appears from these runs that we can maximize the average of the ATCF's current value by aiming to build a less expensive building with mean building cost of $66 per square foot. Moreover, the model appears to be fairly sensitive to the building cost when it is more than $66, but less sensitive when it is less. This indicates that, according to the model, the market will accept a less expensive building design better than a more expensive one.

The example presented in this chapter is intended to demonstrate how to evaluate the risk associated with a spreadsheet model using simulation. We applied the techniques to an expanded version of the office building model presented in section 2.3; however, it is important to keep in mind that these techniques can be applied to any model implemented in a spreadsheet. We then used that spreadsheet to evaluate several decision alternatives. In evaluating these alternatives, we performed a crude optimization of the system by seeking to find the target building cost that maximizes the expected value of the present value of after-tax cash flow.

2.9 AN INSURANCE MODEL

A spreadsheet can be easily used to implement many models and run the simulation replications. For many others, this is difficult, inconvenient, or impossible. The structure of some of these models is such that a variable number of cells would be needed to represent the model, and an elaborate macro facility developed to perform each replication (instead of simply recalculating the model). This is the case with the model presented in this section. For others, the model is too large to fit into a spreadsheet or the recalculation time is long enough that the time to run the replications becomes excessive. Many corporate models as well as models of reliability, production, and inventory and other service systems fall in this category. Also, spreadsheets often do not work well with models where results must be computed iteratively.

2.9.1 A Model for Loss Ratio for an Insurance Agency

In this section, we will look at a model to compute the loss ratio for an agency selling automobile insurance. For any agency, some policy holders will experience losses, or claims, during the year for which the insurance company will be liable. At the end of the year, the insurance company uses the ratio of losses (total claims paid or pending) to premiums (total premium income for the particular group of policies) as a measure of how well the agency performed in choosing which motorists to insure. If the agency performed well, the loss ratio is likely to be small; however, if the agency performed poorly, it is more likely to be large. Clearly, over a long period of time the ratio of losses to premiums must be less than 1 for the insurance to be profitable. Some agencies will choose to insure classes of business that are unprofitable since it generates sales of other more profitable lines of insurance. Experience shows, however, that even good agencies will experience strongly fluctuating loss ratios from year to year. It is easy to see that there should be some fluctuation, because both the numerator (total claims for the year) and the denominator (total premiums for the year) are random variables. Because the loss ratio is computed only once each year, over an agent's career only perhaps 40 yearly loss ratios would be reported, and these would not be determined under the same conditions each year. Thus, historical data would be misleading and insufficient to give us insight into how to select risks to insure. We want to develop a model that will show us the distribution of the loss ratio and predict the probability that it will be greater than 1.

Our model assumes that the number of policies sold during the year is a normally distributed random variable with mean 100 and standard deviation 10. Because it must be an integer, we will just take the integer part. Each policy's premium is determined in the following manner. The premium is composed of two parts: basic liability and excess liability. Basic liability coverage is the minimum amount of liability insurance required by law. The premium for basic liability depends on many factors: the type and age of the vehicle, the driver's age and sex, and the location and distance the vehicle will be driven. A highly detailed model would consider the distribution of these factors, randomly select a category for each policy, and compute the premium of the policy from the premium tables using this category. Instead of taking all of these factors explicitly into account, we will assume that the premium for basic liability coverage is a uniformly distributed random variable between \$50 and \$200. Excess liability is the additional amount of insurance that some drivers choose to carry on their vehicles. We will assume that 40 percent of the customers purchase excess liability, and the model will select the specific customers who purchase it using a random trial for each customer, with the probability of selection being .4 and the probability the customer does not purchase additional coverage being .6. If a customer does select excess liability, then an additional 20 percent is added to the basic premium. The total premium for the year is the sum of the premiums for all customers.

Each policy may generate a claim, although most do not. If a claim is generated, then the amount of that claim, or loss, will be a random variable that is related to the type of vehicle, characteristics of the driver and passengers, the type of accident, and other factors. The relationship between these factors and the amount of the loss is quite complex, but generally the larger the premium, the larger we can expect the loss to be. Therefore, the loss will be modeled as an exponential random variable whose mean is proportional to the premium for that vehicle. The total losses for the agency are the sum of the losses for all vehicles insured. We use the term *losses* to denote the dollar amount of claims for which the insuring company is liable. Finally, the loss ratio is the ratio of total losses to total premiums. The simulation will examine the distribution of this ratio. To do this, we will do independent replications of this sampling experiment, saving the values generated for the loss ratio.

We can describe this model more explicitly using mathematical notation. Let S denote the number of policies sold during a year, and $P_1, P_2, \ldots, P_S$ denote the premiums paid by each of the customers. Then

$$\begin{aligned} P_i &= X_i \text{ with probability } .6 \\ &= 1.20\, X_i \text{ with probability } .4 \end{aligned}$$

where each $X_1, X_2, \ldots, X_S$ is a random variate having the distribution of the basic liability premium, which is uniformly distributed between \$50 and \$200. If TP represents the total premium, then

$$TP = \sum_{i=1}^{S} P_i$$

Let p_c denote the probability that each policy generates a claim during the year and r_c denote the constant of proportionality between the mean loss and the premium; that is, if the premium for the ith policy is P_i, then if the policy generates a claim the mean loss is $r_c P_i$. Let $C_1, C_2, \ldots, C_S$ be the losses generated from the policies. Then

$$\begin{aligned} C_i &= \mathrm{Y}_i, \text{ with probability } p_c \\ &= 0, \text{ with probability } 1 - p_c \end{aligned}$$

where Y_i is an exponential random variate with mean rcP_i. These exponential random variates are mutually independent. Total losses for all policies, TL, is

$$TL = \sum_{i=1}^{S} C_i$$

Finally, the loss ratio is the ratio of total claims, TL, to total premiums, TP:

$$R = \frac{TL}{TP}$$

We will use Monte Carlo simulation to examine the distribution of R. (This will be done for a number of different values of p_c and r_c chosen so their product is a constant. Note that $p_c r_c$ is proportional to the expected losses per policy.)

2.9.2 The Simulation

In practice, the parameters p_c and r_c can be estimated from agency data. The parameter p_c is the proportion of policies that produce one or more legitimate claims during a year, and r_c is the ratio of the losses to the premium, which can be estimated using regression. Thus, an agency can estimate these parameters using existing data and then examine the distribution of the loss ratio using the model just presented. Although the model used to compute the loss ratio is relatively simple, it is too complex to yield to mathematical analysis, leaving simulation as the only way to determine distributional properties of the loss ratios. Other parameters or characteristics of this model that also can be determined from agency data involve the distribution of the number of policies sold per year; the distribution of the premiums for those policies, given the agency's preference concerning the level of risk it is willing to assume in writing policies; and the distribution of the amount of claims per policy, given that one or more claims are generated.

The following is an algorithm for performing each replication that can be easily programmed in any programming language, provided that routines are available for generating uniform, normal, and exponential random variates:

1. Generate S from a normal distribution with mean 100 and standard deviation 10. Truncate S to be an integer.

2. $TP = 0$. $TL = 0$.

3. For $i = 1$ to S DO:

> **4.** Generate X_i from a uniform distribution between 50 and 200.
>
> **5.** Generate U from a uniform distribution between 0 and 1.
>
> **6.** If $U \leq .4$ then $P_i = 1.20\, X_i$,
> else $P_i = X_i$
>
> **7.** Generate U from a uniform distribution between 0 and 1.
>
> **8.** If $U \leq p_c$, then generate Y_i from an exponential distribution with mean $r_c P_i$ and let $C_i = Y_i$,
> else $C_i = 0.0$
>
> **9.** Add P_i to TP.
>
> **10.** Add C_i to TL.

11. $R = TL/TP$.

We can now see why this model is difficult to implement in a spreadsheet program. The premium and claims for each customer must be stored in a cell in the spreadsheet. Since the number of customers is a random variable, the number of cells required will vary from replication to replication. The necessary actions can be programmed using a macro facility that most spreadsheets have; however, this can be tedious and slow and requires expert knowledge. This model can be implemented more easily in a programming language such as Pascal, FORTRAN, C, and so on, or a mathematical package such as MATLAB, Mathematica, or MathCAD. This simulation was implemented using MathCAD.

The purpose of running this simulation is to explore how the distribution of the loss ratio changes with the two parameters r_c, the factor that determines the mean loss relative to the premiums, and p_c, the probability that a claim is generated. If p_c decreases, then the agents are being more selective (risk-averse) about the customers they are willing to insure. If r_c increases, then the agents are insuring more expensive automobiles or more potentially costly claims. Thus, if p_c is large and r_c is small, the customers are more likely to generate claims but the losses are not large on average. If p_c is small and r_c is large, then the customers are less likely to generate claims, but when they do, the losses are larger on average. We selected four scenarios to examine. All scenarios have a mean loss, relative to the premium, $r_c p_c$, of .8; however, they differ by how likely claims are to occur and the amounts of the losses. Table 2.8 gives the parameters of the four scenarios.

2.9.3 Simulation Results

Each simulation run consisted of 500 replications. Table 2.8 and Figure 2.32 show the results. Note that in all runs the mean claims ratio is .8. You can show this to be true mathematically. However, when r_c is 5 and p_c is .16, the claims ratios are grouped relatively close to .8. This can be seen in the plot of the distribution and in the fact that the standard deviation is relatively small (.30). The largest ratio in the 500 replications was 2.33, and the smallest was .17. On the other extreme, when p_c is .01 and r_c is 80, the distribution of the claims ratios is quite skewed and dispersed.

TABLE 2.8 Scenarios for loss ratio simulation

Scenario	1	2	3	4
r_c	5.0	10.0	20.0	80.0
p_c	.16	.08	.04	.01
Replications	500	500	500	500
Mean	.7983	.7731	.7735	.7828
Standard deviation	.3029	.4161	.6022	1.1380
Minimum	.1738	0.0	0.0	0.0
Maximum	2.3340	2.4610	3.5850	7.7870
Proportion > 1.0	.220	.234	.292	.290
Standard error of proportion	.018	.019	.020	.020

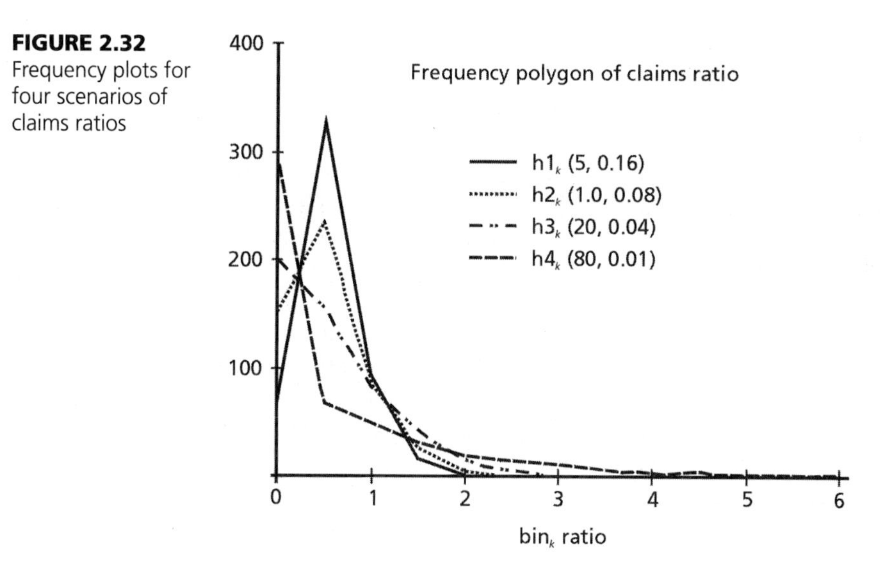

FIGURE 2.32 Frequency plots for four scenarios of claims ratios

Of the 500 replications, 145 produced no claims at all and thus a claims ratio of 0. On the other hand, the largest claims ratio was almost 7.8, and 11 of the 500 replications produced a claims ratio of 4.0 or larger. For the other two runs, which had values of p_c and r_c between these extremes, the mean was close to .8, and the variation was between these two extremes.

Another way to look at the results of these runs is to consider the proportion of replications in each run where the claims ratio was larger than 1. These proportions are also summarized in Table 2.8. When p_c is large and r_c is small, the ratios are bunched around .8, which is less than 1.0, and approximately one-fifth (22 percent) had claims ratios larger than 1.0. On the other hand, 29 percent of the replications in run 4 had a claims ratio larger than 1.0. Thus, even though the average losses for the collective of risks being insured, as measured by the expected claims ratio, are identical for the four runs, the actual shapes of the distributions of losses vary considerably for the four scenarios, but the proportion of risks with losses more than 1.0 is relatively stable. To the agency's management, this says that if the performance criterion involves having a loss ratio greater than 1.0, then the changes involved in the four scenarios provide little control.

2.10 SUMMARY

We have looked at several *static* simulations in this chapter. A static simulation is one in which we do not have to deal with a time dimension in the observations produced by the simulation. This may occur because the model does not have a time dimension, as in the case of four models in this chapter: the profit model of section 2.2, the office building model for one year in section 2.3, the inventory model in section 2.6, and the insurance loss ratio model in section 2.9. The model in section 2.8 involved a five-year time span, but we summarized the performance of the investment over that span using the present value, and we collected one observation to represent the entire five years. In all of these cases, the data collected from the simulation consisted of a sequence of independent observations from the same conceptual population. As a result, we can use traditional statistical methods to estimate parameters, such as the mean, variance, median, and so on, of this population.

Any stochastic model can be analyzed using simulation. If the model can be described, then it can be analyzed using simulation. Models that are implemented in an electronic spreadsheet, such as the office building investment model discussed in sections 2.3 and 2.8, are especially suited to analysis using simulation. The procedure here is to replace uncertain variables in the model with random variates having a distribution appropriate to the particular variables; we then observe how the performance measure of concern changes as we recalculate the spreadsheet numerous times. Each time the spreadsheet is recalculated, a new observation is produced using the newly generated random variates for the uncertain variables. We saw in section 2.8 that these data can be used to evaluate the risk associated with various decisions. *Risk* refers to the possible variation from unpredictable or unknown circumstances, as well as to the possibility of undesirable outcomes. For many spreadsheet models, evaluation of the outcome resulting from the most likely scenario is a straightforward process; however, evaluation of the outcome's *distribution* induced by the variation in uncertain variables is usually difficult or impossible to do using an analytical or mathematical approach. Thus, a simulation analysis is an easy and useful way to determine how likely a particular decision is to produce an undesirable outcome. It can also be used to explore relationships between decision variables and the outcomes of the decisions.

Some models are best simulated using the facilities of a general-purpose computer programming language (Pascal, FORTRAN, C, Java). The structure of the insurance loss ratio model in section 2.9 was such that it could not easily be implemented in a spreadsheet. The inventory model in section 2.6 required a large number of replications. Since recalculations are slow relative to running a compiled program, the time required do this using a spreadsheet would be prohibitive. In addition, it would have been difficult or impossible to store this many observations in a spreadsheet.

Some relatively simple techniques for generating random variates from various discrete and continuous distributions were introduced. These techniques can be used in any situation where a random variate generator is available to generate uniformly distributed random variates between 0 and 1, and they are especially useful if you are using a spreadsheet to do the simulation.

References

Banks, J. Ed. 1998. *Handbook of simulation.* New York: Wiley.

Banks, J., J. S. Carson, B. L. Nelson, and D. M. Nicol. 2001. *Discrete-event system simulation.* Upper Saddle River, New Jersey: Prentice Hall.

Evans, J. R., and D. Olson. 1998. *Introduction to simulation and risk analysis.* Upper Saddle River, New Jersey: Prentice Hall.

Fishman, G. S. 1996. *Monte Carlo concepts, algorithms and applications.* New York: Springer.

Hillier, F. S., and G. J. Lieberman. 1995. *Introduction to operations research.* 6th ed. New York: McGraw-Hill.

Law, A. M., and W. D. Kelton. 2000. *Simulation modeling and analysis.* Boston: McGraw-Hill.

Pidd, M. 1992. *Computer simulation in management science.* Chichester, England: Wiley.

Winston, W. L. 1996. *Simulation modeling using @RISK.* Belmont, California: Duxbury.

Problems

2.1 Create a spreadsheet to implement the furniture sale model in section 2.2. Run 200 replications of the model. From the observations produced, estimate mean net profit given that 3000 chairs are ordered. Compare your numbers with those in section 2.2. Why are your numbers different?

2.2 In the office building model in section 2.3, assume the interest rate is uniformly distributed between 7.0 percent and 14.0 percent. Run the simulation again to observe the sensitivity of ATCF to variations in interest rate along with the other uncertain parameters. How do your results change?

2.3 If you performed a sensitivity analysis of the office building model in section 2.3 using the values in Table 2.9 for uncertain parameters, how many recalculations would you need to perform?

2.4 The mean of the uniform distribution between zero and 1 is $\mu = .5$. Estimate this value with a 95 percent confidence interval using samples of 100, 200, 400, 800, 1600, 3200, and 6400. Plot the confidence intervals and show graphically that the estimates converge to .5.

2.5 Generate 200 independent observations from a normal distribution with mean 10 and standard deviation 2 using the approximate method of section 2.5.4. Compute a 95 percent confidence interval for the mean of the distribution. Plot the histogram of the data and compare it with a normal density with mean 10 and standard deviation 2.

TABLE 2.9

Parameter	Values
Building cost per square foot	60, 62, 64, 66, 68, 70, 75, 80, 90
Rent per square foot	8, 9, 10, 11, 12, 15
Operating expense per square foot	.90, .95, 1.00, 1.05, 1.10, 1.20
Vacancy rate	.1, .2, .3, .4, .5, .6, .7
Other income (in \$1000)	50, 100, 150, 200

2.6 Generate 200 independent observations from a normal distribution with mean 10 and standard deviation 2 using the inverse transform method. Compute a 95 percent confidence interval for the mean of the distribution. Plot the histogram of the data and compare it with a normal density with mean 10 and standard deviation 2.

2.7 Sample 500 observations from the chi-square distribution with 5 degrees of freedom using the inverse transform method. Plot the sample cumulative distribution function and compare it with the cdf of the population.

2.8 If $X_1, X_2, \ldots, X_6$ are observations having a standard normal distribution, then

$$Y = \sum_{i=1}^{6} X_i^2$$

has a chi-square distribution with 5 degrees of freedom. Use this fact to generate 500 observations from a chi-square distribution with 5 degrees of freedom. Use the inverse transform method to sample from the standard normal distribution. Do you think this method is more or less efficient than the method in problem 2.7?

2.9 Sample 200 independent observations from the binomial distribution with $n = 6$ and $p = .4$. Compute a 95 percent confidence interval for the mean and compare it with the true mean. Plot the sample cdf along with the population cdf. Do they appear similar?

2.10 Use the inverse transform method in section 2.5. to sample 100 observations from a Poisson distribution with mean 1.0. What approximation is needed to do this?

2.11 Sam and Mary are salespeople. Sam's weekly sales, X (in thousands of dollars), are uniformly distributed between 1.0 and 5.0. Mary's weekly sales, Y, have a symmetric triangular distribution between 1.0 and 5.0. Each week Sam and Mary compare their performance by computing the ratio X/Y. If $X/Y < .3$, then Sam buys dinner. If $X/Y > .7$, then Mary buys dinner. Otherwise, they split the check. Use a simulation to estimate how often Sam buys dinner and how often Mary buys dinner.

2.12 In a store, 40 percent of customers make a single purchase. This activity requires a time that has an exponential distribution with mean 2.0 minutes. The other 60 percent of customers ask for information before making a purchase. This two-stage process requires time that has an Erlang distribution with mean 4.0 minutes and two stages. Use Bernoulli, exponential, and Erlang random variates to generate a sample of shopping times for 200 customers. Plot the histogram of these observations.

2.13 The actuaries for an insurance company represent the total losses experienced during a week as follows: Let N represent the number of policies that experience a loss during the week, and $Z_1, Z_2, \ldots$ be independent random variables that have the same distribution representing the amounts of each loss. Then the total losses for the week are

$$X = Z_1 + Z_2 + \ldots + Z_N$$

Using probability theory, one can show that if N and $Z_1, Z_2, \ldots$ are statistically independent, the expected value of X is:

$$E(X) = E(N)\,E(Z_1)$$

Thus, under these assumptions, it is easy to compute the mean loss during a week; however, it is not so easy to compute the probability distribution of the total loss.

a. Let N have a discrete uniform distribution over the values 100 to 200, and let each Z_1 have a triangular distribution between \$100 and \$500, with maximum density at \$150. Develop a simulation of this model and run the simulation for 500 independent replications.

b. Using the data from the simulation, compute the 90 percent confidence interval for the mean weekly loss. Compare the confidence interval to the value computed from the expression for E(X). The mean of each Z_1 is \$250, and the mean for N can be easily computed using the symmetry of its distribution.

c. Plot the histogram of the 500 observations from part a. Estimate the probability that the loss will be greater than \$40,000. What is the probability it will be less than \$35,000?

2.14 It is known that if the times between events are independent and have an exponential distribution with mean $1/\lambda$ minutes, then the number of events that occur during one minute will have a Poisson distribution with mean λ events. This fact can be used to generate Poisson random variates. Generate exponential random variates and add them until they just exceed 1.0. Suppose M is the number of variates that add to a value greater than 1.0, but $M - 1$ variates sum to a value less than 1.0. Then, let X, the Poisson variate, be $M - 1$.

a. Implement this method for generating Poisson random variates in a spreadsheet. You will probably find it useful to put the exponential variates, their counts, and their sums in a table and use the spreadsheet's table lookup functions. You will also probably find it necessary to limit the range of values the Poisson variate can assume in order to make the method fit into a table. To determine how large to make the table, use the Poisson tables for the largest value of λ you plan to use.

b. Use the implementation in part a to generate 500 observations from the Poisson distribution with mean 2.0, and plot the histogram of these observations. Does it look like that of a Poisson distribution? If it does not, check your answer to part a.

c. Use the 500 observations generated in part b to compute confidence intervals for the mean and variance. The mean and variance of the Poisson distribution are both λ, so these confidence intervals should be close to the value 2.0. If they are not, generate another sample and calculate new confidence intervals. If the new confidence intervals also are not close to 2.0, recheck your answer to part a.

2.15 A financial management company offers a tax-sheltered annuity called a *fixed* account. The following is a model for this investment. Each quarter (three months), a flat interest rate is paid on all money on deposit at the start of the quarter. Initially, the annual rate is 10.0 percent. The board of directors meets at the end of each quarter to decide the interest rate for the next quarter. The board's decision will depend on many uncertain factors, including the health of the economy and the bond market, previous and anticipated expenses, and others. We will assume that it has a probability of .25 of decreasing the interest rate by .5 percent, a probability of .5 of leaving it unchanged, and a probability of .25 of increasing it .5 percent. However, the directors will never decrease the interest rate below 4 percent annually or increase it above 16 percent. If the rate is 4.0 percent, they will leave it the same with probability .75 and increase it .5 percent with probability .25; if it is 16.0 percent, then they will leave it the same with probability .75 and decrease it .5 percent with probability .25. All interest is compounded quarterly. To compute quarterly interest rates, divide the annual rate by 4. Assume that $600 will be invested each quarter.

a. Implement this model to simulate the investment over a period of five years (20 quarters). The observation we want to collect from each replication is the value of the investment at the end of that period. Assume $200 is invested each month.

b. Run this simulation for 200 replications. Use the data gathered to estimate the expected value and variance of the value of the investment at the end of five years. The mean can be used to evaluate the performance of the investment, and the variance can be used to evaluate the associated risk.

c. Use the data obtained in part b to plot the cumulative distribution function of the value at the end of five years. Use this plot to select the 10th and 90th percentiles of the distribution of value after that period. These can be used to evaluate the risk associated with the investment.

d. Modify the model to remove the limits on the interest rate, rerun the simulation, and recompute estimates of the parameters in parts b and c. How do these compare to the estimates computed previously?

2.16 The following is a model for a stock mutual fund. Monthly changes in the fund index, which is the price of one unit of the fund, are independent random variables. We will assume that the initial index is 1.0 and the monthly changes are uniformly distributed between –.008 and +.025. Assume that you plan to invest $200 in the fund each month. The number of units you can buy each month is 200, divided by the current index value. You accumulate units as they are purchased. To determine the current value of your holdings, just multiply the number of units by the current index value.

a. Implement this model to simulate the investment over a period of five years (20 quarters). The observation that we want to collect from each replication is the value of your holdings at the end of that time.

b. Run this simulation for 200 replications. Use the data gathered to estimate the expected value and variance of the value of the investment at the end of five years. The mean can be used to evaluate the performance of the investment, and the variance can be used to evaluate the associated risk.

c. Use the data obtained in part b to plot the cumulative distribution function of the value at the end of five years. Use this plot to select the 10th and 90th percentiles of the distribution of value after that time. These can be used to evaluate the risk associated with the investment.

d. Modify the model so the monthly change in the index has a triangular distribution between –.008 and +.025, with maximum density at 0.0. Run the simulation again and recompute estimates of the parameters in parts b and c. How do these compare to the estimates computed previously?

2.17 Consider the following model for the reliability of a computer system installed at a major bank. The system has a single power source that supplies two independently operating central processing units (CPUs). Each CPU is connected to each of four input–output (I/O) channels. Under normal circumstances, both CPUs and all four I/O channels are used; however, the system can still be operational (though at a diminished capacity) if only one CPU and only one I/O channel are operating along with the power. Thus, the system fails as soon as either the power unit fails, both CPUs have failed, or all four I/O channels have failed.

Each unit fails independently of the others. The times to failure are random variables with the following distributions:

Power	Normal with mean 400 hours and standard deviation 100 hours
CPUs	Uniform between 250 and 750 hours
I/Os	Exponential with mean 400 hours

Note: These are the distributions of time to failure for each individual unit, not the collection of units of each type.

a. First specify the model for computing the time to failure for the entire system from each component's time to failure. The time at which the CPUs fail is the maximum of the two CPU failure times, and the failure time for the entire system is the minimum of the failure times for each of the three components.

b. Implement this model in a spreadsheet and generate 400 observations on failure times.

c. Use the observations generated in part c to compute 90 percent confidence intervals for the mean time to failure and the probability that the time to failure is less than 350 hours.

d. Use the observations in part c to produce a plot of the probability the system is operating versus time (the reliability function). Use this plot to select a maintenance time such that the probability the system goes down before this time is no larger than .02.

SFSU Bookstore
San Francisco State University
1650 Holloway Avenue
San Francisco, CA 94132
(415)338-2092

3296 001 009 1 279

780534381592 01000
APPLIED SIMULATION MODELING 122.35tx
Sub Total 122.35
California Sales Tax Total 10.40

TOTAL 132.75**

astercard# ************2348 132.75
uthorization Code: 834385
ef: 702918505353

:50 01/29/07 Sale
EXT RETURN POLICY FOR FULL REFUND ARE
FOR WINTER 2007 DEC 11TH - JAN 9TH
FOR SPRING 2007 DEC 11TH -FEB 9TH

OTHING EXCHANGE/RETURN WITHIN 30 DAYS
ECTRONICS MUST BE UNOPENED FOR RETURN
NO RETURN ON COMPUTER HARDWARE
) REFUNDS ON STUDY-AIDS & SALE ITEMS
SOFTWARE MUST BE UNOPENED FOR RETURN

L REFUNDS/EXCHANGES REQUIRE RECEIPT
www.sfsubookstore.com

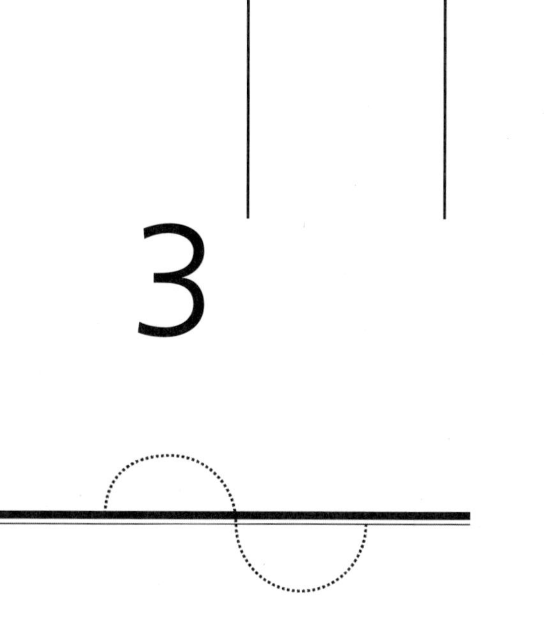

3

Financial Models and @RISK

The material in this chapter serves two purposes: (1) to introduce the reader to some financial models and show how they can be analyzed using simulation and (2) to present a spreadsheet add-in, @RISK, and show how it can be used to facilitate building, running, and analyzing simulations in Excel. Many spreadsheet simulations involve financial models such as option pricing or evaluation of investment choices. The models in this chapter are all in this category. First, we examine briefly the basic notions of derivative securities and then show how to estimate the price of a European stock option using simulation. The notion of modeling a process over time is introduced in the context of the price changes for a stock over time. Using this model, we estimate the price of an Asian option. In many financial models, multivariate observations must be sampled from a population with a specific correlation structure. We develop a model for a portfolio of stocks with correlated prices and show how to evaluate the return for such a portfolio and compare the return to a portfolio with independent asset values. Finally, we discuss the basic concepts of distribution fitting and show how a tool, BestFit (part of @RISK), can be used to find the distribution that best fits your data.

3.1 INTRODUCTION

Financial models deal with money or value directly. They usually involve the management of investments, especially predicting the value of one or more assets at some time in the future. Financial models have assumed great importance in the day-to-day activities as well as strategic-planning functions of many financial institutions. Many organizations, especially financial institutions such as investment banks, use financial models to decide how to allocate financial resources and price securities. Because the future is uncertain, most financial models include random components to capture this uncertainty. Simulation is therefore a natural tool for evaluating financial models and, in some cases, is the only tool available for this. In this chapter, we will introduce some fairly standard financial models and show how to analyze these models using simulation. It is not our purpose to teach financial modeling in this chapter; that is a much bigger job than we can accomplish in the limited space we have. See Hull (1997), Benninga (1998), or Luenberger (1998) for a detailed exposition of the financial concepts used in this chapter. Our purpose here is to provide an understanding of how simulation can be used to get the desired information from any stochastic financial model.

The models in this chapter are similar to some of the models in Chapter 2, especially those in sections 2.3, 2.8, and 2.9. However, we will extend those models in three ways: First, we will introduce the geometric random walk that is used to model the price of an asset at a time in the future. Second, we will use a new tool, @RISK, to set up and run the simulations. Third, we will introduce *dynamic* models that observe a stochastic variate over *time*. In Chapter 4, we will continue with our discussion of dynamic models.

@RISK is an add-in that can be used to perform simulations in Microsoft Excel. This software tool is designed to facilitate the activities of simulation, and it provides many useful output reports and graphs. Because @RISK is an add-in, all of the original spreadsheet's features are still available. @RISK helps with the following tasks:

- sampling from various distributions by providing additional functions for random variate generation,
- defining the simulation experiments by specifying cells to observe and the number of replications to perform,
- running the simulation and collecting data, and
- analyzing the output data and displaying the results graphically.

@RISK must be installed before it can be used. Directions for installing @RISK are included on the CD accompanying this book. The installation program creates a program group—Palisade Decision Tools—for @RISK and, once installed, @RISK can be run either by selecting @RISK for Excel from the Palisade Decision

Tools program group under the Start menu or by double-clicking the @RISK for Excel icon if you have it installed on the desktop. When @RISK runs, it actually runs Excel and installs itself as an add-in. (You can also install the @RISK add-in manually in Excel, using the same procedure as for any manual installation of an add-in.)

@RISK is a large and complex software product with extensive facilities for designing, running, and analyzing simulations in Excel. We cannot present all of the features of @RISK in this chapter, but we will show how an add-in package such as @RISK can enhance the capabilities of the spreadsheet and facilitate the building and use of simulations. The reader is encouraged to explore the other facilities of @RISK not covered in this chapter. Information about them can be found in the documentation on the accompanying CD-ROM.

3.2 A MODEL FOR THE PRICE OF A STOCK

In this section, we will define a model for the price of a stock at a future point in time and use @RISK to determine the distribution of the future stock price. This model is the well-known *geometric random walk* (Hull, 1997, Chapter 10). In this model, the log of the percentage increase in the stock price between now and t units of time in the future has a normal distribution. Unless noted otherwise, we will assume that the unit of time is a year. In particular, define the following parameters: μ, called the *drift parameter,* and σ^2, called the *volatility.* Let Z denote a standard normal random variable—that is, a random variable having a normal distribution with mean 0 and standard deviation 1. Then the geometric random walk model for stock prices is

$$\ln\left(\frac{P(t)}{P(0)}\right) = \left(\mu - \frac{\sigma^2}{2}\right)t + \sigma\sqrt{t}\,Z \tag{3.1}$$

In this expression, $P(t)$ is the price of the stock t time units in the future, and $P(0)$ is the price of the stock now. The expression on the right is just a normal random variable having mean

$$\left(\mu - \frac{\sigma^2}{2}\right)t$$

and variance σ^2t. A more useful version of this expression can be obtained with a little algebra:

$$P(t) = P(0)\,e^{\left(\mu - \frac{\sigma^2}{2}\right)t + \sigma\sqrt{t}Z} \tag{3.2}$$

Thus, given the current price $P(0)$, the drift and volatility parameters μ and σ^2 and the time t, we can sample the price t time units later by sampling a standard normal

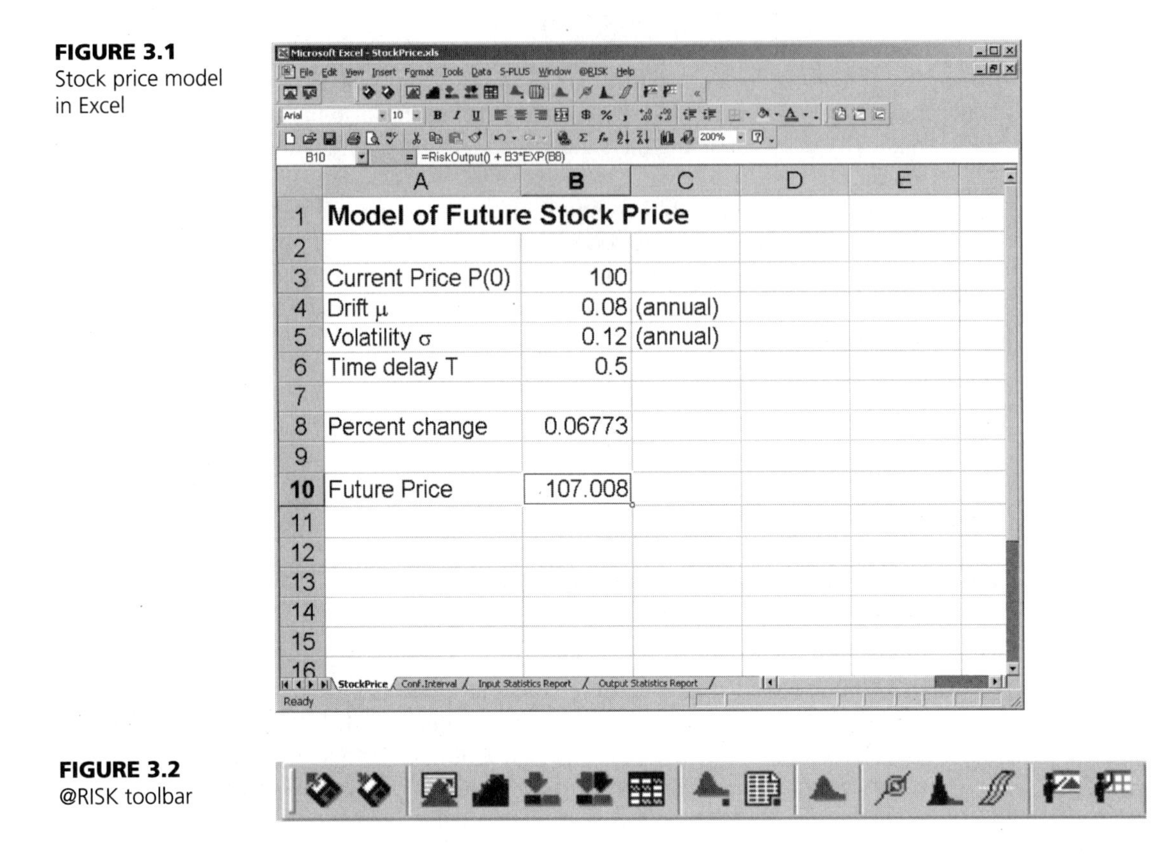

FIGURE 3.1
Stock price model in Excel

FIGURE 3.2
@RISK toolbar

random variate, Z, and plugging it into this expression. Figure 3.1 shows a worksheet containing this model.

First, notice that @RISK has added a new toolbar containing @RISK commands just below the main menu bar. Your toolbars may appear a little different from those in Figure 3.1, depending on how Excel is configured. The @RISK toolbar is shown in Figure 3.2. We will use the buttons on this toolbar to set up and run the simulation.

In the worksheet labeled **Stock Price,** cells A3 to A6, A8, and A10 contain labels; cells B3 to B6 contain the values of the parameters for this model. Cell B4 contains μ, and cell B5 contains σ. The price will be computed for a point in time six months in the future. Because the unit of time is years and six months is one-half year, cell B6 contains .5. In this spreadsheet, we implemented equation (3.1) in cell B8, using the formula

```
=(B4-B5^2)*B6+B5*SQRT(B6)*RiskNormal(0,1)
```

Here we have used the @RISK function `RiskNormal(0,1)` to sample an observation from a normal distribution with mean 0 and standard deviation 1. Thus,

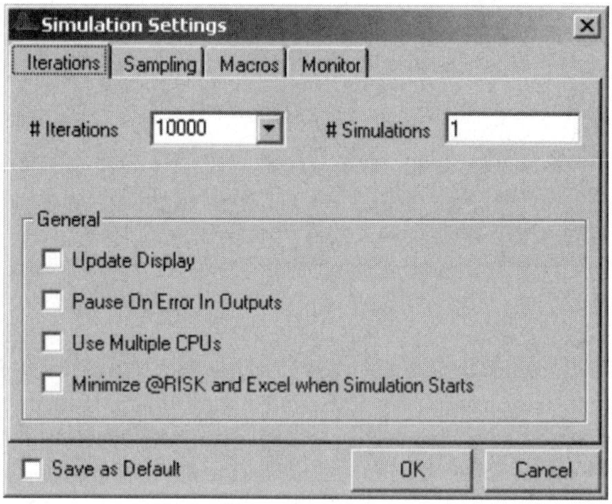

Add Output button | Data Window button | Detailed Statistics button

Graph button | Run Simulation button | Simulation Settings button

FIGURE 3.3
Simulation Settings dialog

Simulation Settings

Iterations | Sampling | Macros | Monitor

Iterations 10000 # Simulations 1

General

Update Display

Pause On Error In Outputs

Use Multiple CPUs

Minimize @RISK and Excel when Simulation Starts

Save as Default OK Cancel

`RiskNormal(0,1)` substitutes for Z in (3.1). The first parameter of the function `RiskNormal` is the mean of the normal distribution, and the second parameter is the standard deviation. Because the right-hand side of (3.1) is a normal random variable with mean

$$\left(\mu - \frac{\sigma^2}{2}\right)t$$

and variance $\sigma^2 t$, we could also have used the following formula in cell B9:

```
=RiskNormal(((B4-B5^2)*B6,B5*SQRT(B6))
```

In cell B10, the formula `=B3*EXP(B8)` computes the price in six months. @RISK includes a large number of functions for sampling from various distributions, including all common distributions such as beta, binomial, chi-square, exponential, gamma, lognormal, normal, Poisson, and Weibull, as well as many uncommon distributions such as extreme value, negative binomial, Pareto, and Pearson types 5 and 6. The @RISK help menu lists all available distributions.

Now, let's change some settings in @RISK to make it behave the way we want it to. Click the **Simulation Settings** button in the @RISK toolbar. The label **Simulation Settings** will appear if the cursor is held over the icon for a few seconds. A dialog similar to that in Figure 3.3 will appear. Click on the tab labeled **Sampling** to get a dialog that looks like that in Figure 3.4.

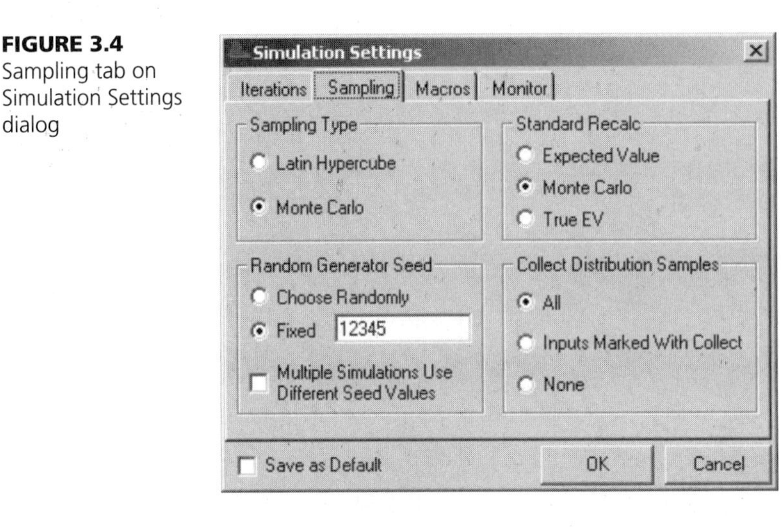

FIGURE 3.4
Sampling tab on Simulation Settings dialog

Now select the option **Monte Carlo** under **Sampling Type.** @RISK offers two types of sampling. *Latin Hypercube sampling* is a form of stratified sampling that is designed to distribute the sample more evenly over the space of possible samples; in some cases, it can produce estimates that are more precise. *Monte Carlo sampling* simply samples each observation independently from the specified distributions. This is the type of independent, identically distributed observations we generated in Chapter 2. On the bottom right of this dialog is a section box labeled **Collect Distribution Samples.** If **All** is checked, observations will be saved on each random variate generated during each replication. This is handy for examining and checking the distribution of the simulation's variates to see if their shape is what you expect.

Selecting **Monte Carlo** in the dialog under **Standard Recalc** controls what is displayed in each cell containing a formula that samples from a distribution. In our example, the only cell concerned here is B8, because it contains a formula with the function `RiskNormal()`. We chose **Monte Carlo** in order to display the actual sampled values in B8. If **Expected Value** is selected, the mean of the distribution for the cell will be displayed except for those cases when discrete random variables are sampled. In those cases, the model usually expects an integer value to be computed and used in subsequent computations. For example, a cell might contain the number of policies sold by an insurance company; subsequent computations might sample the amounts of claims for each of the policies. Thus, it would not make sense to compute 53.296 as the number of policies. If **Expected Value** is selected, the integer nearest to the mean value is displayed. Selecting **True EV** causes the actual expected value to be displayed and used in subsequent calculations. On the right side of the dialog is a section labeled **Random Generator Seed.** The options in this section let you control how the random number generator is initialized. For the current example, we initialized the seed with the fixed value 12345 so that the simulation runs will always produce the same output. If this is not important, as is the case in many experiments, the option **Choose Randomly** could be selected. The checkbox **Save As Default** is a handy way to keep these settings so they do not have to be reset for each simulation.

Before we can actually run the simulation, we must tell @RISK two additional things: (1) what cells to observe and record as the replications are performed (i.e., which cells contain the output data), and (2) how many replications to perform. @RISK uses the term *outputs* to refer to the cells that are observed and recorded during the simulation. In fact, the model of a simulation from @RISK's perspective is that random variates are *inputs* and the desired observations are *outputs.* To specify the outputs, first select the cell or cells containing the output data and then click the **Add Output** button in the @RISK toolbar to direct @RISK to add the selected cells to the list of outputs. We will only record the future price of the stock, which is in cell B10. So click cell B10 to select it and then click the **Add Output** button in the @RISK toolbar to add this cell to the outputs. @RISK provides a dialog that allows you to give the variable (i.e., output cell values) a name and suggests a name based on the contents of neighboring cells. In a model with multiple outputs, we would repeat this process until all outputs have been selected.

Now we are ready to specify the number of replications and run the simulation. If it is not still displayed, click on the **Simulation Settings** icon again to display the **Simulation Settings** dialog. Select the **Iterations** tab. In the edit box labeled **# Iterations,** you will see the default value of 100. Let's change this to 1000. @RISK uses the term *iterations* for what we called *replications* in Chapter 2. Leave the value of **# Simulations** equal to 1. This parameter is used when a model is simulated multiple times with different parameters. Click **OK** to close this dialog. To actually perform the replications, simply click on the **Run Simulation** button. If everything has been done correctly up to this point, you should see a window that looks similar to Figure 3.5.

FIGURE 3.5 @RISK output windows

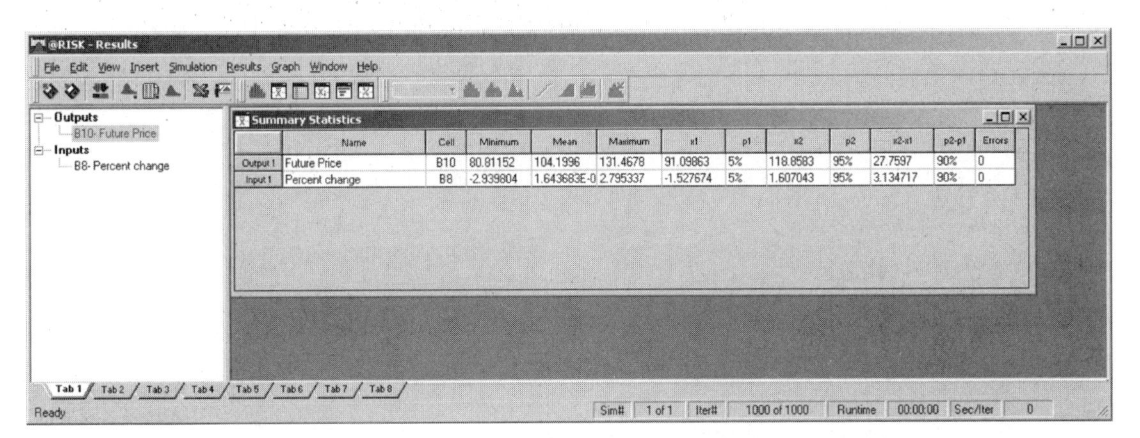

FIGURE 3.6
Summary Statistics window

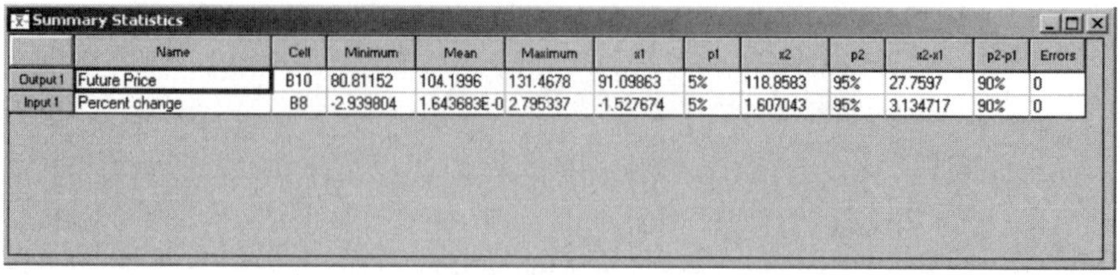

Summary Statistics

	Name	Cell	Minimum	Mean	Maximum	x1	p1	x2	p2	x2-x1	p2-p1	Errors
Output 1	Future Price	B10	80.81152	104.1996	131.4678	91.09863	5%	118.8583	95%	27.7597	90%	0
Input 1	Percent change	B8	-2.939804	1.643683E-0	2.795337	-1.527674	5%	1.607043	95%	3.134717	90%	0

The main @RISK window has two panes: A navigation pane on the left shows the outputs and inputs, and the pane on the right contains windows with the simulation results. Because we specified that distribution samples were to be collected, the left pane shows both the output (**Future Price**) and the input, which is a standard normal random variable used in cell B8. Although the input is labeled **Percent change,** the data are just the values of the `RiskNormal` function.

Figure 3.6 shows the **Summary** window. We can obtain some information by examining the summary statistics output. We see that the sample mean price of the stock six months in the future is $104.20, giving a 4.20 percent increase over the starting price of $100. This window also shows that the minimum value observed is $80.81 and the maximum value observed is $131.47. The minimum and maximum statistics are not especially useful for decision making because they depend on the sample size. If we make the sample larger, the minimum and maximum observed values will become more extreme. A more useful range might be the range between the 5th and 95th percentiles. This estimates a range of values that the future price will have a 90 percent probability of being within. These statistics are also provided. The 5th sample percentile is $91.10 and the 95th percentile is $118.86, so there is approximately a 90 percent probability that the price of the stock will be between $91.10 and $118.86 six months hence. As the sample size increases, these sample percentiles will become more precise.

Clicking on the **Detailed Statistics** button produces a window with more summary statistics, including many percentiles. The output data can also be accessed in the **Output** window by clicking the **Data Window** button. These data also can be copied to the spreadsheet or another application for further analysis.

Finally, we can compute a confidence interval for the mean value of the future price of the stock. Recall that the confidence interval is given by

$$\overline{X} \pm t_{\alpha/2,\, n-1} \frac{s}{\sqrt{n}} \qquad \textbf{(3.3)}$$

where $\overline{X}$ is the sample mean, s is the sample standard deviation, n is the number of observations (i.e., the number of replications or iterations), and

$$t_{\alpha/2,\, n-1}$$

is the $1 - \alpha/2$ quantile of the Student's t-distribution with $n - 1$ degrees of freedom. We can find the sample mean and sample standard deviation in the **Detailed Statistics** window. Click the **Detailed Statistics** button to open this window. Figure

FIGURE 3.7
Detailed Statistics window

Detailed Statistics

Name	Future Price	Percent change	
Description	Output	RiskNormal(0,1)	
Cell	B10	B8	
Minimum	80.81152	-2.939804	
Maximum	131.4678	2.795337	
Mean	104.1996	1.643683E-02	
Std Deviation	8.556371	0.9630286	
Variance	73.21148	0.9274241	
Skewness	0.3058059	8.987407E-02	
Kurtosis	2.872773	2.766953	
Errors Calculated	0	0	
Mode	107.0383	0.3726032	
5% Perc	91.09863	-1.527674	
10% Perc	93.63867	-1.203575	
15% Perc	95.43805	-0.9792585	
20% Perc	96.90353	-0.7996703	
25% Perc	98.11192	-0.6536186	
30% Perc	99.06828	-0.5392971	
35% Perc	100.0858	-0.4188708	
40% Perc	101.2423	-0.2834692	
45% Perc	102.376	-0.1522348	
50% Perc	103.7469	4.523452E-03	
55% Perc	104.7335	0.116072	
60% Perc	105.9522	0.2524143	

3.7 shows a portion of the **Detailed Statistics** window. The simplest approach is to copy the output statistics to an Excel worksheet and then do the calculations. In the @RISK window, select the menu item **Results,** then select **Report Settings. . . .** You will see the dialog in Figure 3.8. This dialog has a large number of options that can be selected to meet your needs. For our purposes, we will select **Detailed Statistics** and **Active Workbook** as shown in Figure 3.8. Notice that you can place the data, graphs, and a selection of other outputs into Excel; at the top of this dialog you have the option to put these outputs into Excel automatically. Click **Generate Report Now.** You should now see your Excel model window with two new worksheets labeled **Input Statistics Report** and **Output Statistics Report.**

Figure 3.9 shows the worksheet in Excel to which the **Output Statistics Report** has been copied. Now the mean and standard deviation can be copied to another worksheet where the confidence interval can be computed. The result is shown in Figure 3.10, which shows that with a sample size of 1000 (i.e., 1000 replications), the confidence interval for the expected price of the stock in six months is the range 103.67 to 104.73. This is a fairly narrow range, but, depending on your needs, it can be reduced further by increasing the number of replications.

FIGURE 3.8
Reports to Worksheet dialog

@RISK Reports

At the End of Each @RISK Simulation
- [x] Show Interactive @RISK Results Window
- [x] Generate Excel Reports Selected Below

Excel Reports
- [ ] Quick Summary
- [ ] Quick Output Report
- [ ] Simulation Summary
- [x] Detailed Statistics
- [ ] Output Data
- [ ] Summary Graphs
- [ ] Input Data
- [ ] Output Graphs
- [ ] Sensitivities
- [ ] Input Graphs
- [ ] Scenarios
- [ ] Tornado Graphs
- [ ] Template Sheet myTemplate

Place Excel Reports in
- (•) New Workbook
- ( ) Active Workbook

Graph Format
- (•) Metafile
- ( ) Native Excel

Generate Reports Now
- [ ] Save as Default

OK Cancel

FIGURE 3.9
Copying simulation results to Excel

@RISK Output Details Report

Output Statistics

Outputs	Future Price
Simulation	1
Statistics / Cell	$B$10
Minimum	80.81151581
Maximum	131.4678497
Mean	104.1995724
Standard Deviation	8.556370582
Variance	73.21147754
Skewness	0.30580591
Kurtosis	2.872773307
Number of Errors	0
Mode	107.0382919
5%	91.09863281
10%	93.63867188

Simulation results can also be graphed in several formats to show the distributions of the outputs. We can graph the distribution of the future price of the stock by first selecting the output **Future Price** in the **Results** window and then clicking on the **Graph** button in the @RISK window and selecting the type of graph we want. We will select **Histogram** to produce a histogram like that in Figure 3.11.

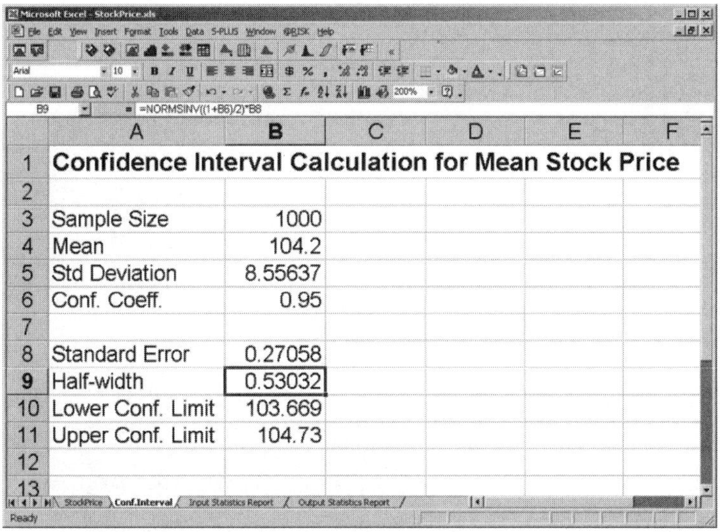

FIGURE 3.10
Confidence interval calculation for stock price

Microsoft Excel - StockPrice.xls

B9 =NORMSINV((1+B6)/2)*B8

	A	B	C	D	E	F
1	Confidence Interval Calculation for Mean Stock Price					
2						
3	Sample Size	1000				
4	Mean	104.2				
5	Std Deviation	8.55637				
6	Conf. Coeff.	0.95				
7						
8	Standard Error	0.27058				
9	Half-width	0.53032				
10	Lower Conf. Limit	103.669				
11	Upper Conf. Limit	104.73				
12						
13						

StockPrice / Conf.Interval / Input Statistics Report / Output Statistics Report

Ready

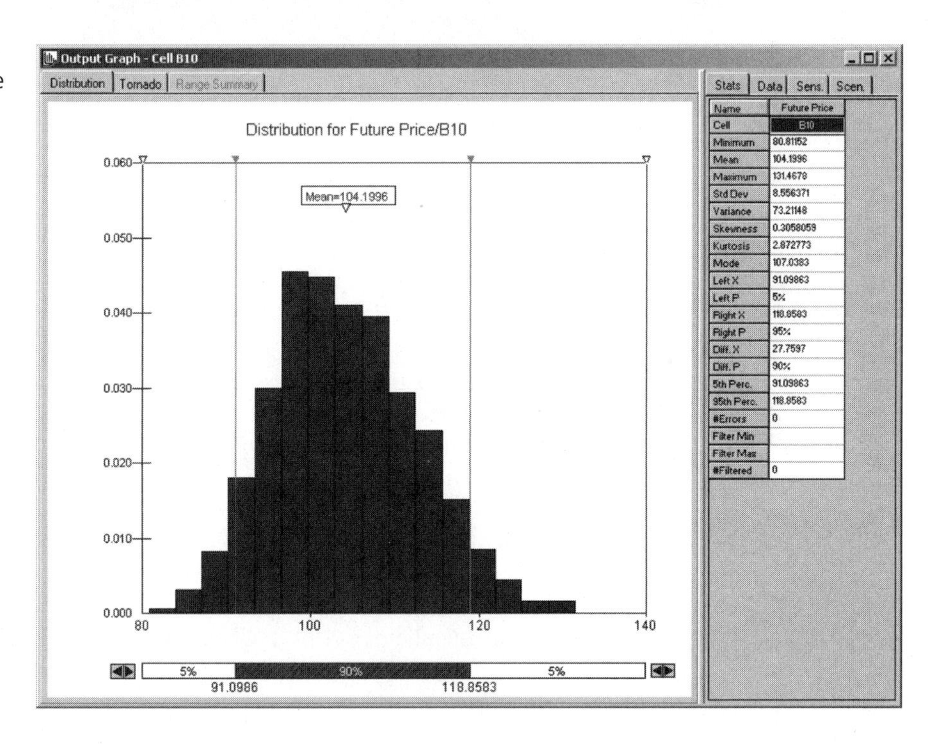

FIGURE 3.11
Histogram of future prices for stock

The type and format of the graph can be changed by right-clicking on the mouse when the mouse pointer is on the graph. For example, by selecting **Ascending Cumulative-Solid** from the **Graph Type** menu, we can get a sample cumulative distribution function like the one in Figure 3.12.

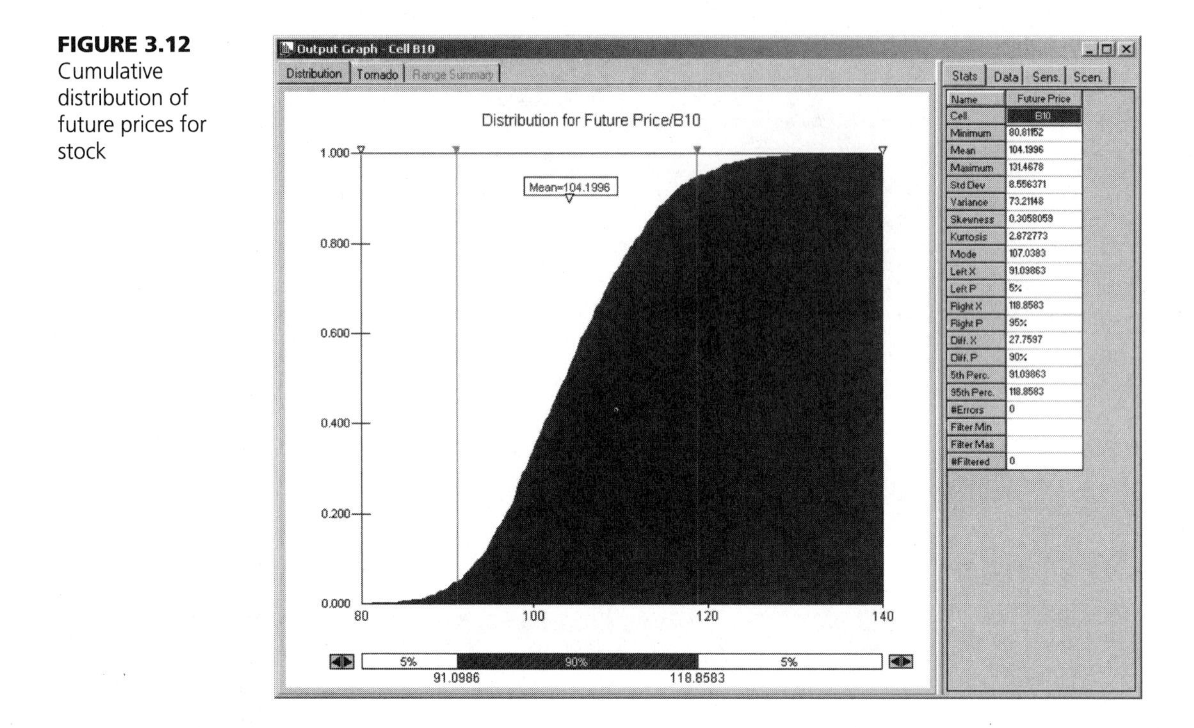

FIGURE 3.12 Cumulative distribution of future prices for stock

3.3 OPTIONS, FUTURES, AND SIMULATION

Options and futures are examples of *derivative securities* or *derivatives,* which are also called *contingent contracts.* Derivatives are securities whose value is derived from the value of an underlying asset. Usually, the price or value of the underlying asset is currently known but subject to considerable uncertainty in the future. Simulation is a useful tool to estimate the price of derivatives and evaluate their effectiveness for controlling the variation in a portfolio of securities. The primary focus of this book is neither financial modeling nor derivatives, so we are not able to provide an in-depth discussion of these topics. See Hull (1997), Chapter 6; Benninga (1998), Chapters 10–13; or Luenberger (1998), Chapters 10–12 for more information on these topics.

3.3.1 Options and Futures

What are options and futures? Let's start with futures. A *future* is an agreement to buy a particular asset at a specified price at a specified time in the future. Futures are sold for many commodities such as lumber, soybeans, and copper, and they are useful for business planning when a firm knows that an asset will be needed at some future point. A future involves the obligation to purchase the asset. An *option,* on the other hand, provides the right but not the obligation to buy or sell an asset at a specified price sometime in the future. Because options involve the right but not the obligation to buy, they are more flexible than futures and are often used as a hedge against adverse events. Options are sold on many exchanges for many types of assets, especially stocks. The concept of an option as a contingent contract has also been useful in pricing insurance and other contingent obligations.

There are two basic types of options. A *call option,* or just *call,* involves the right to buy the asset for a given price, and a *put option,* or just *put,* involves the right to sell the asset at a specified price. The price involved is called the *strike price* or *exercise price,* and the date on or before which the holder can buy or sell the asset is called the *expiration date.* A *European option* is one that must be exercised on the expiration date. An *American option* can be exercised on or before the expiration date. This terminology has nothing to do with where the option originates. Indeed, most options sold on exchanges in the United States and Europe are American options. European options are simpler to analyze and price than American options, and arguments can be made that American options can be priced as if they were European options.

Let's look at a European call option for a stock with expiration date 90 days and strike price \$86.00. Suppose that, on the expiration date, the price of the stock is \$88.36. The holder of the call will buy the stock for \$86.00 and immediately sell it for the current price \$88.36, realizing a profit of \$88.36 – \$86.00 = \$2.36 on each share. In fact, if we let S be the strike price, T be the time until expiration, and P_T be the price at expiration, then the return from the call option at expiration will be

$$R_T = \begin{cases} 0 & \text{if } P_T \leq S \\ P_T - S & \text{if } P_T > S \end{cases} \tag{3.4}$$

The present value of this cash flow is

$$R_0 = R_T e^{-rT} \tag{3.5}$$

if r is the risk-free rate. Because P_T is stochastic, R_0 is also stochastic.

3.3.2 Estimating the Price of a Call Option Using Simulation

Financial theory says that the fair price of the call option is the expected value of R_0. To compute $E(R_0)$, we must know the distribution of P_T. If we assume certain distributions, it is possible to compute the expected value using probability calculus. However, with simulation we can estimate the expected value by repeatedly sampling P_T, applying equation (3.4), computing the mean of these observations, and discounting the mean to the present time. These computations can be done using

FIGURE 3.13 Option pricing worksheet

	A	B	C	D	E	F	G
1	**Computation of Option Price for European Call Option**						
2							
3	Current Price P(0)	100			Replications	1000	100000
4	Drift μ	0.08	(annual)		Mean C. F.	3.26468	3.40369
5	Volatility σ	0.1	(annual)		Std. Dev. of C. F.	4.54897	4.69779
6	Expiration Time T	0.5			Std. Error	0.14385	0.01486
7	Risk-free Rate r	0.059			Lcl	2.98274	3.37458
8	Exercise Price S	103			Ucl	3.54662	3.43281
9							
10	Price at Expiration	116.311					
11	Cash Flow at Exp	13.3112					
12	Discounted C.F.	12.9243					
13							
14							
15							

StockPrice / Option / Put Option / Hedging /

any appropriate distribution for P_T. Figure 3.13 shows a worksheet in which this sampling is done using @RISK.

This worksheet is quite similar to Figure 3.1. Cells B3 through B6 are the same in both worksheets. Cell B7 contains the risk-free rate used to discount the cash flow at the expiration time, and cell B8 contains the exercise price, which is $103.00 for this option. Cell B10 implements (3.2), B11 implements (3.4), and B12 implements the present value of the cash flow, given in (3.5). @RISK is used to replicate cell B12 and estimate the mean of the value sampled in this cell. The results of simulations using 1000 and 100,000 replications are in cells E3:G8. When 1000 replications are run, the 95 percent confidence interval for the mean—that is, the fair value for the option—is $2.98 to $3.54, a confidence interval with width $.56. Because this is a rather large interval, it is not really useful. When the sample size is increased to 100,000, the confidence interval width narrows to $.05, which is much more reasonable. Note, however, that the length of time needed to run the simulation with 100,000 replications is 100 times that needed for 1000 replications. To price a large number of options can require a powerful computer to do in a reasonable length of time.

A put option works similarly to a call option except that one has a positive cash flow only if the price of the stock goes down. The holder has the right to sell the asset at a specified exercise price. If that price is higher than the current price of the asset, the holder would buy the asset at market and sell it at the exercise price, making a profit of the difference. Thus, the discounted cash flow for a put option is

$$R_T = \begin{cases} 0 & \text{if } P_T > S \\ e^{-rT}(S - P_T) & \text{if } P_T < S \end{cases} \tag{3.6}$$

Because the fair value for a put option is the expected value of R_T, a put option can be priced in exactly the same way as a call option, except that the entry in cell B12 implements equation (3.6).

3.3.3 Hedging Using Put Options

Options are useful for managing risk. Financial analysts define risk in several ways. Usually, it is defined as the variance or standard deviation of the return on investment. Thus, risk is a measure of the *uncertainty* in the return. We can also define risk to be the probability of an unfavorable outcome. For example, if you own a stock portfolio for a year, the risk could be defined as the probability that the value of the portfolio will decrease over the year. When buying a stock, you are most interested in controlling the risk that the price of the stock will fall too low—that is, the "downside risk." If you buy a put option that allows you to sell the stock at a given price, then the loss due to dropping prices can be limited. However, the option has a cost that cannot be ignored. To evaluate the effectiveness of a put option as a hedge to control this downside risk, we can set up a simulation to sample the value of a portfolio consisting of both the stock and a put option on the stock, and compare the distribution of the return on the portfolio to the distribution of the return on the stock alone.

Figure 3.14 shows a worksheet in which portfolio returns are simulated with and without hedging using put options. The basic parameters—current price, drift, volatility, and so on—are contained in cells B3 through B8. In addition, cell B10 contains the price of a put option for the stock. Because the put option must be paid for, its cost must be included in the computation of the rate of return. Cell B11 contains the price of the stock at expiration, which is computed using equation (3.2). At expiration, there will be two cash flows: the cash flow from the put option, given by

$$R_T = \begin{cases} 0 & \text{if } P_T \geqslant S \\ S - P_T & \text{if } P_T < S \end{cases} \tag{3.7}$$

and the cash flow from the gain on the value of the stock, given by $P_T - P_0$. The sum of these two expressions is the value in cell B12. Cell B13 uses the risk-free rate to

FIGURE 3.14 Worksheet to evaluate hedging using put options

	A	B	C	D	E	F
1	Demonstration of Hedging using Put Options					
2		With Hedging	No Hedging		No Hedging	Hedging
3	Current Price P(0)	100		Replications	10000	10000
4	Drift μ (annual)	0.08		Mean	0.0404	0.0406
5	Volatility σ (annual)	0.1		Std. Dev.	0.0725	0.0635
6	Expiration Time T	0.5		Std. Error	0.001	0.001
7	Risk-free Rate r	0.059		Lcl	0.039	0.039
8	Exercise Price S	97		Ucl	0.042	0.042
9						
10	Price of put option	0.57				
11	Price at Expiration	101.410				
12	Cash Flow at Exp	1.410	1.410			
13	Discounted C.F.	1.369	1.369			
14	Return on Portfolio	0.799	1.369			
15	Percentage Return	0.80%	1.37%			
16						

StockPrice / Option / Put Option / Hedging

FIGURE 3.15 @RISK output for evaluating hedging with put options

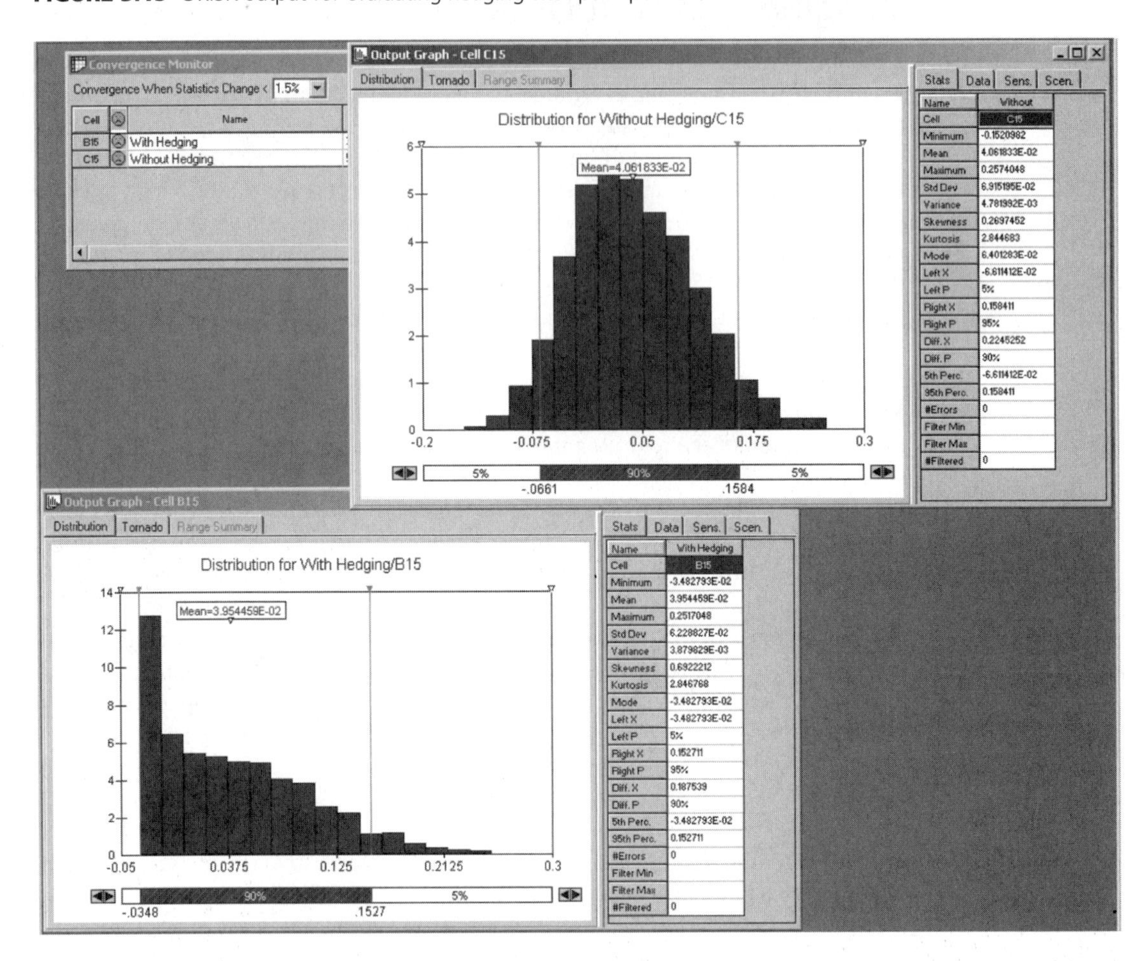

discount the cash flow in B12 to the present, and cell B14 subtracts the cost of the option from the value in B13 to get the return on the portfolio. Cell B15 expresses the return as a percentage by dividing B14 by the current price. It is cell B15 that we want to observe in our simulation.

The computation of rate of return without a put option is straightforward. In cell C12, we compute the gain (or loss) on the stock $P_T - P_0$; in cell C13, this value is discounted to the present. The return on the stock is just the discounted gain or loss, so C13 is copied to C14; in cell C15, the return is divided by P_0 to compute the percentage return.

In the worksheet in Figure 3.14, Cell B15 contains the rate of return when a put option is bought with the stock, and cell C15 has the rate of return for the stock alone. These two cells are the desired outputs for the simulation. We then simulate the model for 10,000 replications. The @RISK **Graph** windows for the two scenarios are shown in Figure 3.15.

The left window shows the distribution of return for the stock alone. We see that it is almost a normal distribution. Actually, the distribution is lognormal and should be skewed to the right slightly. The right window shows the distribution of return for the portfolio that consists of the stock and a put option. Since the put option limits the loss on the stock to \$3 if the price drops below \$97, we see that the distribution of return on the right has a spike at –3.48 percent, whereas the distribution on the left without the hedge has a long tail on the left.

We can also use the simulation output to compute a confidence interval for the mean return on the investments. Means and standard deviations for the outputs were copied from the @RISK **Statistics** window to cells E4:F5 on the worksheet in Figure 3.14. The number of replications was placed in cells E3 and F3, and the confidence interval computations were done in cells E6:F8 as before. Note that the mean returns on the two investments do not differ significantly. Thus, we have demonstrated that the use of a put option as a hedge is effective to improve the distribution of return without sacrificing the expected return.

3.4 DYNAMIC FINANCIAL MODELS OF STOCK PRICES

Dynamic models are those that observe a variable of interest over time. In financial modeling, these types of models are important because managers usually wish to control some financial process over time. For example, an insurance company might want to model its cash position to optimally allocate excess cash to investments. As another example, we might want to compute a fair price for an *Asian option* on a stock. The cash flow for such an option depends on the average price of the stock over the period of the option. Recall that the cash flow from a European option only depends on the price of the stock at the expiration time, so the stock price must only be sampled at one point in time—the expiration time. For an Asian option, the price of the stock must be sampled each day so the average can be computed.

3.4.1 A Worksheet to Sample Stock Prices over Time

Before we model the cash flow for an Asian option, we will first set up a worksheet to model the price of the stock over a period of time. Such a worksheet is shown in Figure 3.16. The geometric random walk model, which we used in (3.1), can be used to model the stock price over time. The price of the stock after a delay of δt can be computed from

$$P(t + \delta t) = P(t)\, e^{\left(\mu - \frac{\sigma^2}{2}\right)\delta t + \sigma\sqrt{\delta t}\,Z} \tag{3.8}$$

FIGURE 3.16 Model of stock price over time

	A	B	C	D	E
1	**Dynamic Model of Future Stock Price**				
2					
3	Current Price P(0)	100		Week	Price
4	Drift μ (annual)	0.08		0	100
5	Volatility σ (annual)	0.12		1	100.643
6				2	100.346
7	Time increment	1	week	3	99.5113
8				4	100.544
9				5	99.33
10				6	102.198
11				7	102.173
12				8	102.869
13				9	103.883
14				10	104.421
15				11	104.287
16				12	105.458
17				13	104.237
18				14	106.01
19				15	106.235
20				16	104.829
21				17	101.947
22				18	101.283

Price Series / Stock Price / Asian Option

Thus, we can start with a current price $P(t)$, then use (3.8) to sample $P(t + \delta t)$.

Now substitute $t + \delta t$ for t in (3.8) to have an expression to sample $P(t + 2\delta t)$. Continuing in this way, we can sample P at discrete points in time t, $t + \delta t$, $t + 2\delta t$, ...

To create the worksheet in Figure 3.16, enter the current price, drift (μ) and volatility (σ) in cells B3:B5 and the labels shown in cells A3:A5 and A7. In this worksheet, we want to sample the price of the stock at the end of each week for 26 weeks. We will let the increment be one week, so put "1" in cell B7 and put "week" in cell B8 to remind us that the unit of time is a week for sampling $P(t)$. To compute the series of prices, start by creating a column containing the week numbers in column D. We will not actually use these numbers to sample the price, but they are handy to keep track of which week we are following. In cell E4, put the formula `=B3` to set the starting price in cell E4 to the same value in cell B3. In cell E5, put the formula

```
=E4*EXP(($B$4-$B$5^2/2)*$B$7/52+$B$5
   *SQRT($B$7/52)*RiskNormal(0,1))
```

This formula implements equation (3.8). $P(t)$ is in cell E4, and δt, in units of years, is computed from \$B\$7/52. In this formula, all cell references are absolute except for E4, so when we copy it to cell E6, E4 becomes E5 in the formula and everything else is unchanged. This is exactly what we want. (Why?) Now copy this formula to cells E6:E30 to create 26 weekly price observations.

FIGURE 3.17
Plot of price series for stock

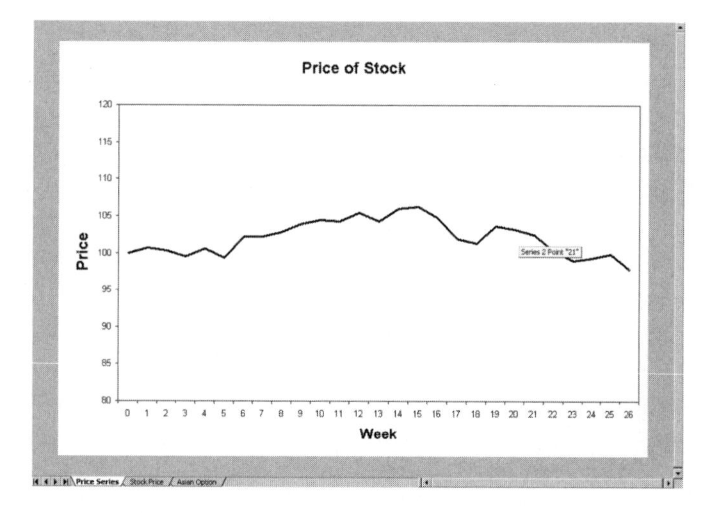

To see what this geometric random walk looks like, we can plot the prices over time as shown in Figure 3.17. You can also get a feel for how this series can vary by pressing the recalculate key (**F9**) several times. (Be sure you have set the @RISK **Standard Recalc** setting under the **Sampling** tab of the **Simulation Settings** dialog to **Monte Carlo** so the new values will be displayed after each recalculation.)

Now that we have created a spreadsheet to sample the series of prices for the stock, we can simulate the series using @RISK. Select cells E4:E30, which contain the series of prices, and then right-click and select **Add Output** from the menu to add these cells to the outputs. Click the **@RISK Settings** button and make sure the number of replications is set under the **Iterations** tab—we used 1000—and that **Monte Carlo** is selected as the sampling type and **Standard Recalc** option under the **Sampling** tab. Now click the **Run Simulation** button on the @RISK toolbar to run the simulation.

After the simulation runs, you will see output similar to Figure 3.18. Because we have specified 27 output cells, the simulation output will have statistics on 27 output variables. The first variable is just the price at the start of the time period, which is a constant, so it will not vary over the replications. The remaining 26 output variables will vary from replication to replication, and data and sample statistics for these variables are available in the @RISK window. We can use this output to estimate the mean price at the end of any week in the period, or we can plot the distribution of the prices.

We will plot the distribution of stock price at the end of both 13 and 26 weeks and also show a summary plot of all 27 stock prices. To plot the stock price distributions at the two points in time, first select the output in the @RISK **Results** window, then click the **Graph** button in the @RISK window. The output for the end of

FIGURE 3.18 @RISK window for dynamic simulation of stock price series

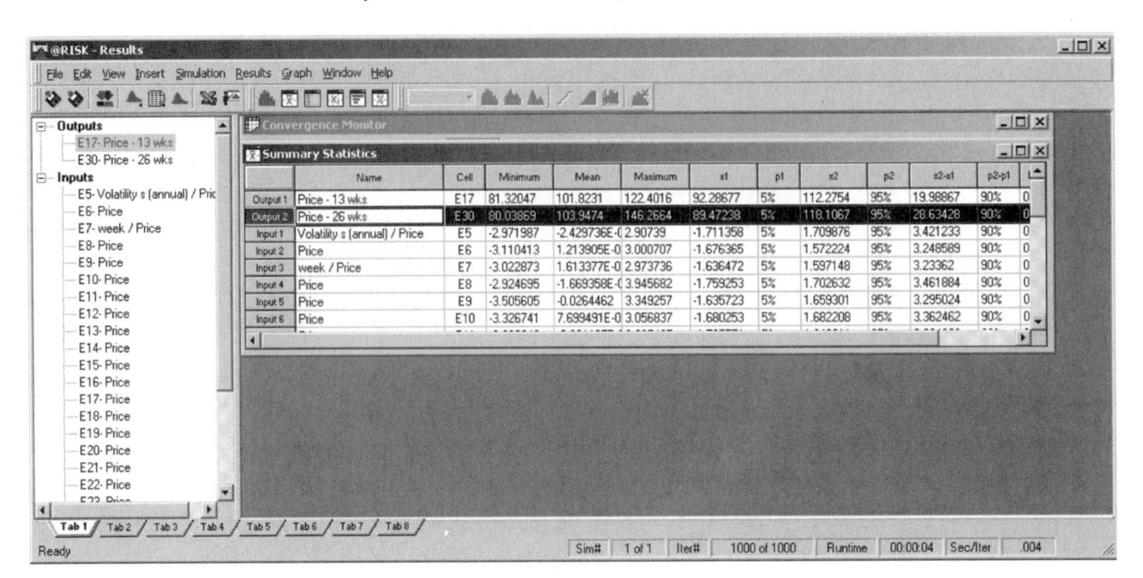

the 13th week is in cell E17, and the output at the end of the 26th week is in cell E30. The two sample distributions are shown in Figure 3.19. The graphs are quite similar, but the means and standard deviations are larger in the bottom plot after 26 weeks than the top plot after 13 weeks.

We can further investigate the change in the distribution of the stock price over the 26-week period. @RISK provides the **Summary Statistics** graph button on the toolbar just for this. To plot the summary, first select the range of cells you want to summarize. In our case, it is the range from E5 to E30. Then click the **Summary Statistics** button. The summary graph is shown in Figure 3.20. Although much detail is lost here, the plot shows the sample mean, the points one standard deviation above and below the mean, and the upper and lower 95th percentiles. This shows how the mean price grows slowly over time—and that the variation in price grows more rapidly over time.

3.4.2 Estimating the Price of an Asian Option

To price an Asian option, we need to set up another worksheet that computes the current value of the cash flow at expiration of the option. The only difference between estimating the value for an Asian option and estimating the value for a European option is in the way the cash flow is computed. The cash flow for the former depends on the *average* price over the term of the option, whereas the cash flow for the latter depends only on the price of the stock at expiration. Figure 3.21 shows

FIGURE 3.19 Graphs of stock price after 13 and 26 weeks

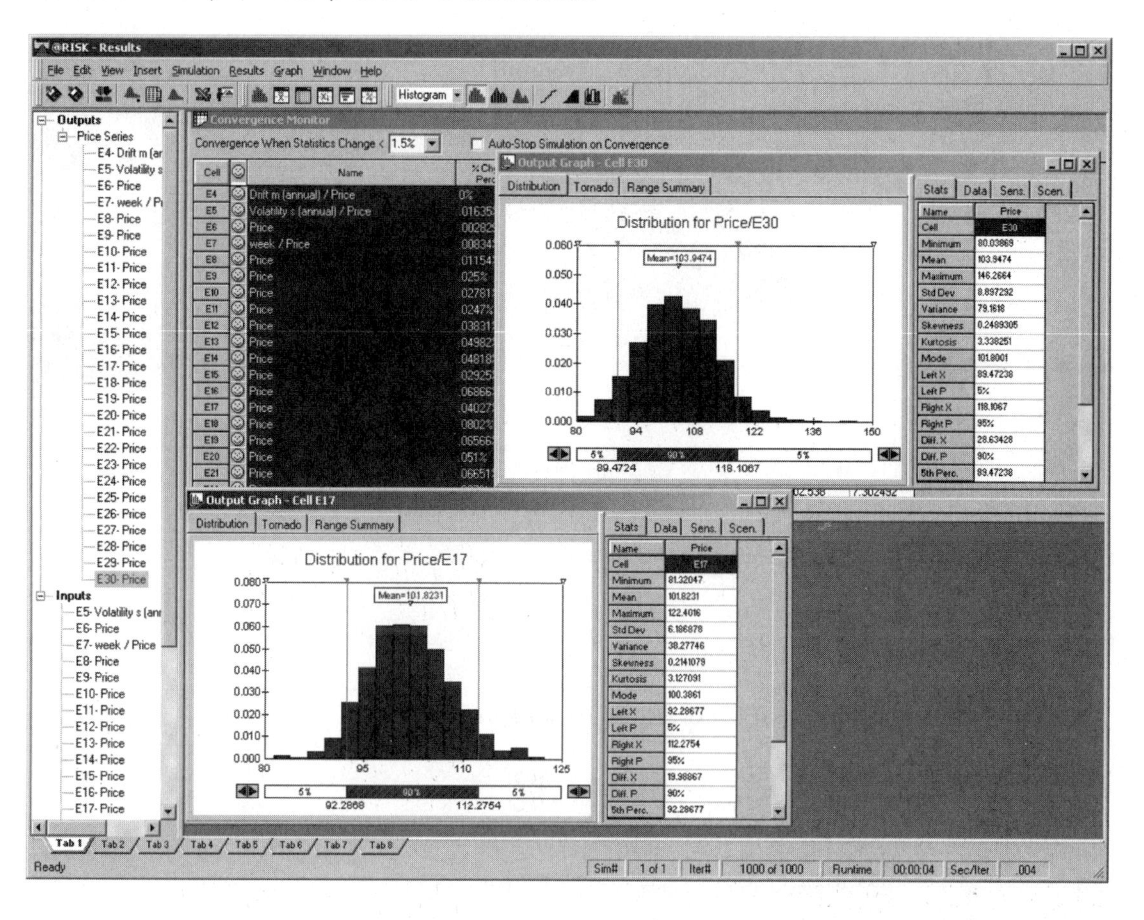

such a worksheet. Cells B3, B4, and B5 contain the strike price, expiration time, and risk-free rate, respectively. The expiration time was actually computed in this case with the formula

```
='Stock Price'!D30/52
```

which is the ratio of the last week number in the price series to the number of weeks in a year. Cell B7 has the formula

```
=AVERAGE('Stock Price'!E4:E30)
```

which computes the average of the 27 values of stock price on the previous worksheet. Cell B8 calculates the cash flow at expiration using the formula

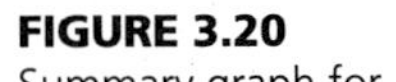

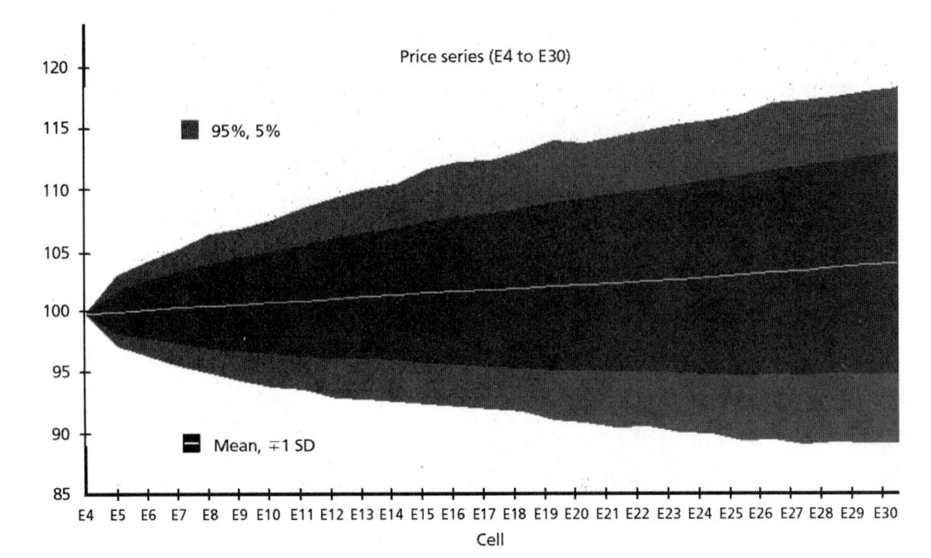

FIGURE 3.20
Summary graph for stock price series

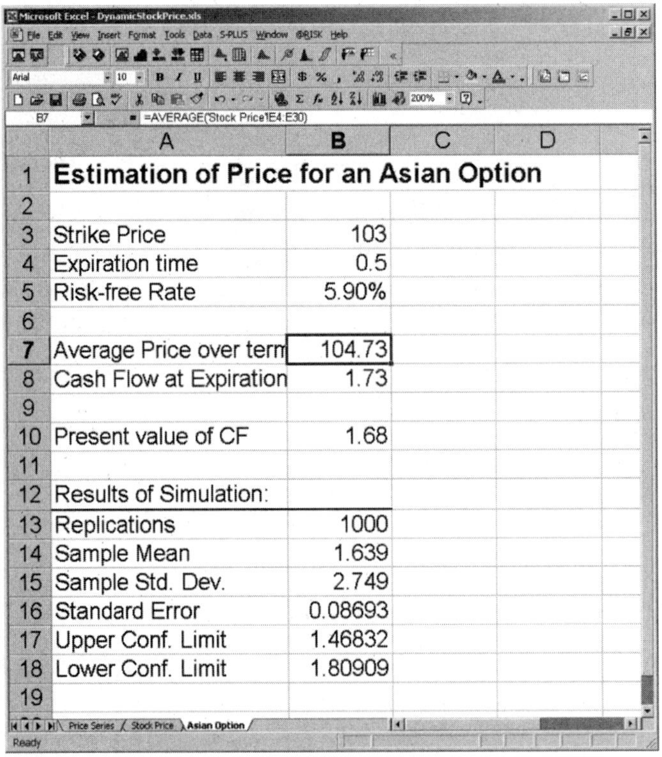

FIGURE 3.21
Worksheet for pricing an Asian option

Microsoft Excel - DynamicStockPrice.xls

B7 =AVERAGE('Stock Price'!E4:E30)

	A	B	C	D
1	**Estimation of Price for an Asian Option**			
2				
3	Strike Price	103		
4	Expiration time	0.5		
5	Risk-free Rate	5.90%		
6				
7	Average Price over term	104.73		
8	Cash Flow at Expiration	1.73		
9				
10	Present value of CF	1.68		
11				
12	Results of Simulation:			
13	Replications	1000		
14	Sample Mean	1.639		
15	Sample Std. Dev.	2.749		
16	Standard Error	0.08693		
17	Upper Conf. Limit	1.46832		
18	Lower Conf. Limit	1.80909		
19				

Price Series / Stock Price / Asian Option

Ready

Display Outputs and Inputs button

Summary Statistics button

```
=IF(B7-B3>0,B7-B3,0)
```

and in cell B10, the present value of cell B8 is computed with the formula

```
=B8*EXP(-B5*B4)
```

Before running the simulation, we want to remove all outputs from the previous simulation. Click the **Display Outputs and Inputs** button on the Excel window and select all outputs in the left-hand pane. Then right-click and select **Remove Functions** to remove them. Now select cell B10 as the new output and click the **Add Selected Cells . . .** button to add this cell as an output. We are now ready to run the simulation. Set the number of iterations to 1000 in the **Simulation Settings** window and click the **Run Simulation** button.

3.5 CORRELATED ASSET VALUES

In financial models, asset values or other variables are often correlated, and the model must capture this statistical behavior. Appendix A discusses the notion of correlation and the statistical tools that are available to estimate correlation. Variables that are correlated tend to move together but not in a completely predictable manner. For example, suppose that the prices of two stocks, General Electric (GE) and Home Depot (HD), are positively correlated. To be more precise, assume that the daily price *changes* in these two stocks are positively correlated. Then, if the price of GE increases more than average, it is likely that the HD price also increases more than average; similarly, if the price of GE decreases more than average it is likely that the price of Home Depot also decreases more than average. Usually, when asset values are positively correlated, it is because they tend to respond in the same way to market circumstances.

3.5.1 Estimating the Correlation Matrix

A client has a portfolio consisting of the six stocks shown in Table 3.1. For this model, the unit time increment is one week. The drift and volatility components of the price model and the correlation matrix for changes in price can be computed from historical data. Figure 3.22 shows a worksheet containing the logarithms of the weekly price changes for the six stocks. The sample mean of these data for each stock is an estimator of the drift parameter, μ, and the sample standard deviation is the estimator of the volatility, σ. These statistics can be computed using the `Average()` and `StDev()` functions in Excel. To sample the price changes for all six stocks from an appropriate multivariate distribution, we must also know the correlation matrix for the six random variables. Using Excel, this can be estimated from

TABLE 3.1 Holdings in stock portfolio

Company	Symbol	Shares	Price on January 11, 1999
General Electric	GE	200	133.625
Home Depot	HD	500	75.625
Intel	INTC	250	77.625
Coca Cola	KO	350	58.25
Pfizer	PFE	500	39.50
Wal Mart	WMT	400	57.25

FIGURE 3.22 Logs of weekly price changes for six stocks in portfolio

	A	B	C	D	E	F	G
1	**Data are differences of logs of weekly prices**						
2							
3							
4	Week of	GE	HD	INTC	KO	PFE	WMT
5	1-Nov-99	-0.04916	0.016367	0.000807	-0.02141	-0.04017	-0.00334
6	25-Oct-99	0.07567	0.046441	0.053036	0.065669	-0.04766	-0.01323
7	18-Oct-99	0.081869	0.054163	0.035517	0.101096	0.121167	0.116005
8	11-Oct-99	-0.07488	-0.08397	-0.06569	-0.0607	-0.06877	-0.08599
9	4-Oct-99	0.060938	0.068573	0.009958	0.078375	0.056972	0.136438
10	27-Sep-99	-0.00239	0.045024	-0.00975	-0.04118	0.071149	0.038238
11	20-Sep-99	-0.01681	-0.01029	-0.11182	-0.05471	-0.04742	-0.00268
12	13-Sep-99	0.006793	-0.00834	-0.03198	-0.01771	-0.02205	-0.02147
13	6-Sep-99	0.02361	0.063801	-0.02193	-0.04001	-0.0523	0.002625
14	30-Aug-99	-0.00134	-0.00621	0.0733	-0.07853	0.016052	0.018567
15	23-Aug-99	0.032709	0.018728	0.037595	0.053788	0.044672	0
16	16-Aug-99	0.045907	-0.00891	0.002349	-0.01686	0.068268	0.040989
17	9-Aug-99	0.006987	0.100551	0.108325	0.009449	0.052056	0.082836
18	2-Aug-99	-0.01852	-0.10743	0.036876	-0.02191	-0.03143	-0.02395
19	26-Jul-99	-0.05522	-0.01265	0.072297	-0.02748	-0.0255	-0.06167
20	19-Jul-99	-0.0315	-0.07452	-0.04475	-0.04323	-0.08446	-0.06593
21	12-Jul-99	0.014832	0.056313	0.013121	0.021381	0.021722	0.001304

Simulation / Sample Statistics / All Data

the data in Figure 3.22 using the **Correlation** tool from the **Data Analysis** selection on the **Tools** menu. This computation produces a lower triangular matrix whose (i, j)th element is the sample correlation between the ith and jth variables. Figure 3.23 shows the result of using this tool. To be precise, these values are the sample correlations between the logs of the weekly price changes for each pair of stocks. Because a variable is perfectly correlated with itself, the diagonal values are always going to be 1. From the figure, we see that the estimated correlation between GE and HD is .6327.

The **Correlation** tool only produces half of the full correlation matrix because the upper triangular portion is symmetric with the lower triangular portion. However, @RISK's function for sampling correlated observations requires the full correlation matrix as input. Thus, we must assemble the full correlation matrix from

FIGURE 3.23
Sample correlations between six log stock price changes

Correlation matrix produced by Tools -> Data Analysis -> Correlation

	GE	HD	INTC	KO	PFE	WMT
GE	1					
HD	0.63273	1				
INTC	0.31943	0.27582	1			
KO	0.53689	0.38835	0.13853	1		
PFE	0.58423	0.49987	0.4082	0.27063	1	
WMT	0.6518	0.68422	0.24117	0.39495	0.59471	1

FIGURE 3.24
Full correlation matrix for six stock price changes

Full correlation matrix produced by Paste Special with (Skip Blanks, Transpose)

	GE	HD	INTC	KO	PFE	WMT
GE	1	0.63273	0.31943	0.53689	0.58423	0.6518
HD	0.63273	1	0.27582	0.38835	0.49987	0.68422
INTC	0.31943	0.27582	1	0.13853	0.4082	0.24117
KO	0.53689	0.38835	0.13853	1	0.27063	0.39495
PFE	0.58423	0.49987	0.4082	0.27063	1	0.59471
WMT	0.6518	0.68422	0.24117	0.39495	0.59471	1

the portion in Figure 3.23 before we can ask @RISK to sample the price changes. To get the full correlation matrix, we need to copy the lower triangular portion, transpose the rows and columns, and paste the copied portion in the empty upper triangular cells in Figure 3.23. The result is shown in Figure 3.24. This is easily done using a copy-and-paste operation. First, copy the range from B4 to G9. Then, keeping this range marked, right-click and select **Paste Special.** At the bottom of the **Paste Special** dialog, click **Skip Blanks** and **Transpose.** This will cause the entire matrix to be transposed (rows and columns exchanged), and when the values are pasted, blanks will not replace nonblank cells. Once the correlation matrix in Figure 3.24 is created, we are ready to build the worksheet to sample the correlated price changes.

3.5.2 Sampling Correlated Variates

The worksheet that simulates this portfolio is primarily a collection of columns in which each stock price is sampled as in the earlier price models. Once the prices have been sampled, the value of the entire portfolio is computed by multiplying each stock price by the holdings for that stock and summing the results. What is

FIGURE 3.25
Parameter portion of the portfolio model

	A	B	C	D	E	F	G	H
1	**Simulation of Portfolio Value**							
2								
3								
4		*GE*	*HD*	*INTC*	*KO*	*PFE*	*WMT*	
5	Starting Prices							
6	1-Nov-99	133.63	75.625	77.625	58.25	39.5	57.25	
7								
8	Price Model Parameters							
9	Component	1	2	3	4	5	6	
10	μ	0.00685	0.00970	0.00668	0.00036	0.00469	0.01120	
11	σ	0.04011	0.05058	0.05282	0.04711	0.05135	0.04888	
12	Number of Share	200	500	250	350	500	400	
13								
14	Time (weeks)	52						
15								

Simulation / Sample Statistics / All Data /

FIGURE 3.26
Computations to sample prices

	A	B	C	D	E	F	G	H	I
16	Correlation Matrix:								
17									
18		*GE*	*HD*	*INTC*	*KO*	*PFE*	*WMT*		
19	GE	1	0.6327	0.3194	0.5369	0.5842	0.6518		
20	HD	0.6327	1	0.2758	0.3883	0.4999	0.6842		
21	INTC	0.3194	0.2758	1	0.1385	0.4082	0.2412		
22	KO	0.5369	0.3883	0.1385	1	0.2706	0.3950		
23	PFE	0.5842	0.4999	0.4082	0.2706	1	0.5947		
24	WMT	0.6518	0.6842	0.2412	0.3950	0.5947	1		
25									
26	Simulation using correlations							Entire Portfolio	
27								with Correlation	
28	Exponent	0.2147	0.4342	0.2330	-0.2600	0.4208	0.4815		
29	Simulated Price	165.63	116.74	98.00	44.91	60.17	92.66		
30	Simulated Value	33126	58371	24499	15720	30084	37064		198,863
31									
32	Value on 11/1/9	26725	37813	19406	20388	19750	22900		146,981
33									
34	Growth	6401	20558	5093	-4667	10334	14164		51,881
35									
36	Return (percent)	23.95%	54.37%	26.24%	-22.89%	52.32%	61.85%		35.30%

Simulation / Sample Statistics / All Data /

new here is the way the price changes are sampled to include correlation with the other price changes. Figure 3.25 shows a portion of the worksheet with the input parameters and the column setup.

The actual computations to sample the prices are shown in Figure 3.26. The only difference between this model and those used earlier are the formulas used to sample price changes. In this model, the price change was sampled using

```
=+RiskNormal((B$10-B$11^2/2)*$B$14,B$11
*SQRT($B$14),RiskCorrmat($B$19:$G$24,B$9))
```

FIGURE 3.27 Portfolio simulation assuming uncorrelated stock prices

	A	B	C	D	E	F	G	H	I
38									
39	Simulation assuming independence:							Entire Portfolio	
40								with Independence	
41	Exponent	0.4792	0.5786	0.3575	0.2725	-0.0735	0.7090		
42	Simulated Price	215.78	134.89	110.99	76.49	36.70	116.33		
43	Simulated Value	43157	67443	27747	26773	18350	46534		230,003
44									
45	Value on 11/1/9	26725	37813	19406	20388	19750	22900		146,981
46									
47	Growth	16432	29630	8341	6386	-1400	23634		83,022
48									
49	Return (percent)	61.48%	78.36%	42.98%	31.32%	-7.09%	103.20%		56.48%
50									

Simulation / Sample Statistics / All Data /

In Figure 3.26, an additional optional function, `RiskName` ("Exponent/GE"), is included to identify the input. This formula is composed of two parts: The call to function `RiskNormal` is supplemented by a call to function `RiskCorrmat`. The first parameter, $B$19:$G$24, is the correlation matrix, and the second parameter, B$9, is the index of the row and column in the correlation matrix that represents this variable (GE). Note that the row is anchored with a "$," but the column is not, so once the formula is entered in column B, it can be copied to columns C through G. The result is that cell B28 contains the log of the growth in price for GE over the one-year time horizon; when cells B28:B36 are copied to the corresponding rows in columns C through G, samples of correlated observations for the logs of the growth of the stocks will be in cells B28 through G28, and these observations will be used to compute the values of each component in the portfolio in cells B30 through G30. The total value of the portfolio is computed in cell I30 as the sum of these values. The percentage growth is computed in cell I36 from the growth in cell I34. Each time the worksheet is recalculated, a new value for the percentage growth is sampled.

The portfolio can also be simulated with the assumption that the stock price changes are not correlated. By simulating the portfolio both with and without correlation, we can determine the effect of correlation among the stocks on the performance of the portfolio. The range A28:I36 was copied and pasted starting at A41. Then the formulas in cells B41:G41 were edited to remove the call to `RiskCorrmat`. Without this call, the values in this range are sampled independently. All other computations are identical. The result is shown in Figure 3.27.

In the simulation, 10,000 replications (iterations) were run. The outputs were cells I34, I36, I47, and I49, which are the growth in the portfolio in dollars and the percentage growth for both the (real) case where the stock prices are correlated and the (assumed) case where the stock prices are independent. Figure 3.28 shows a portion of the output. The first two columns contain the results for the correlated stocks, and the last two columns contain results assuming independence. We can see that the *mean* return is roughly the same in both cases, but the *variation* in the return

FIGURE 3.28
Output from simulation of a portfolio of six correlated stock prices

Detailed Statistics

Name	Growth Correlated	Return (percent) with Cor	Growth Uncorrelated	Return (percent) without	E
Description	Output	Output	Output	Output	F
Cell	I34	I36	I47	I49	B
Minimum	-65370.57	-0.4447545	-32840.17	-0.223431	-0
Maximum	423786.9	2.883272	293889.9	1.999506	1.
Mean	68560.03	0.4664543	68455.03	0.4657399	0.
Std Deviation	58679.17	0.399229	34336.62	0.2336122	0.
Variance	3.443245E+09	0.1593838	1.179003E+09	5.457468E-02	8.
Skewness	0.8515319	0.8515319	0.6084594	0.6084594	-6
Kurtosis	4.281928	4.281928	3.817948	3.817948	2.
Errors Calculated	0	0	0	0	0
Mode	92670.41	0.2985385	81652.52	0.6363603	0.
5% Perc	-12290.69	-8.362082E-02	18291.79	0.1244498	-0
10% Perc	999.5212	6.800331E-03	27100.28	0.1843792	-6
15% Perc	10476.82	7.127997E-02	33768.08	0.2297442	1.
20% Perc	18789.11	0.1278334	39606.06	0.2694634	6.
25% Perc	26521.59	0.180442	44519.79	0.3028944	0.
30% Perc	33861.76	0.2303815	48843.5	0.3323111	0.
35% Perc	40463.09	0.2752942	53013.57	0.3606826	0.
40% Perc	46980.05	0.3196329	56911.34	0.3872014	0.
45% Perc	53783.6	0.3659215	60960.95	0.4147532	0.
50% Perc	61137.93	0.4159573	65214.32	0.4436914	0.
55% Perc	68062.35	0.4630683	69350.25	0.4718306	0.
60% Perc	75508.07	0.5137259	74219.31	0.5049577	0.
65% Perc	83829.1	0.5703387	78762.28	0.5358661	0.
70% Perc	92653.09	0.6303736	83714.8	0.5695611	0.
75% Perc	102252.9	0.6956864	89102.48	0.6062167	0.

is considerably less for the case where the stock prices are independent. This conclusion is further supported by viewing the histograms of the returns, shown in Figure 3.29. The histogram on the top plots percentage return with correlation; the histogram on the bottom plots percentage return without correlation. Although the histograms look rather similar, observe the scales. The scale on the top is larger than that on the bottom, and the histogram on the top has a longer tail on the right side. We can conclude that the portfolio with correlation is riskier than a similar portfolio without correlation. This explains why funds such as sector funds that invest in securities representing the same sector of the economy tend to be riskier than balanced funds, which invest in a broader selection of securities.

3.6 FITTING A DISTRIBUTION TO DATA

In equation (3.1), we assumed the random variable Z, which represents the log of the proportional change in a stock price, has a normal distribution. A body of theory in mathematical finance predicts that the log of price changes for stocks and other pub-

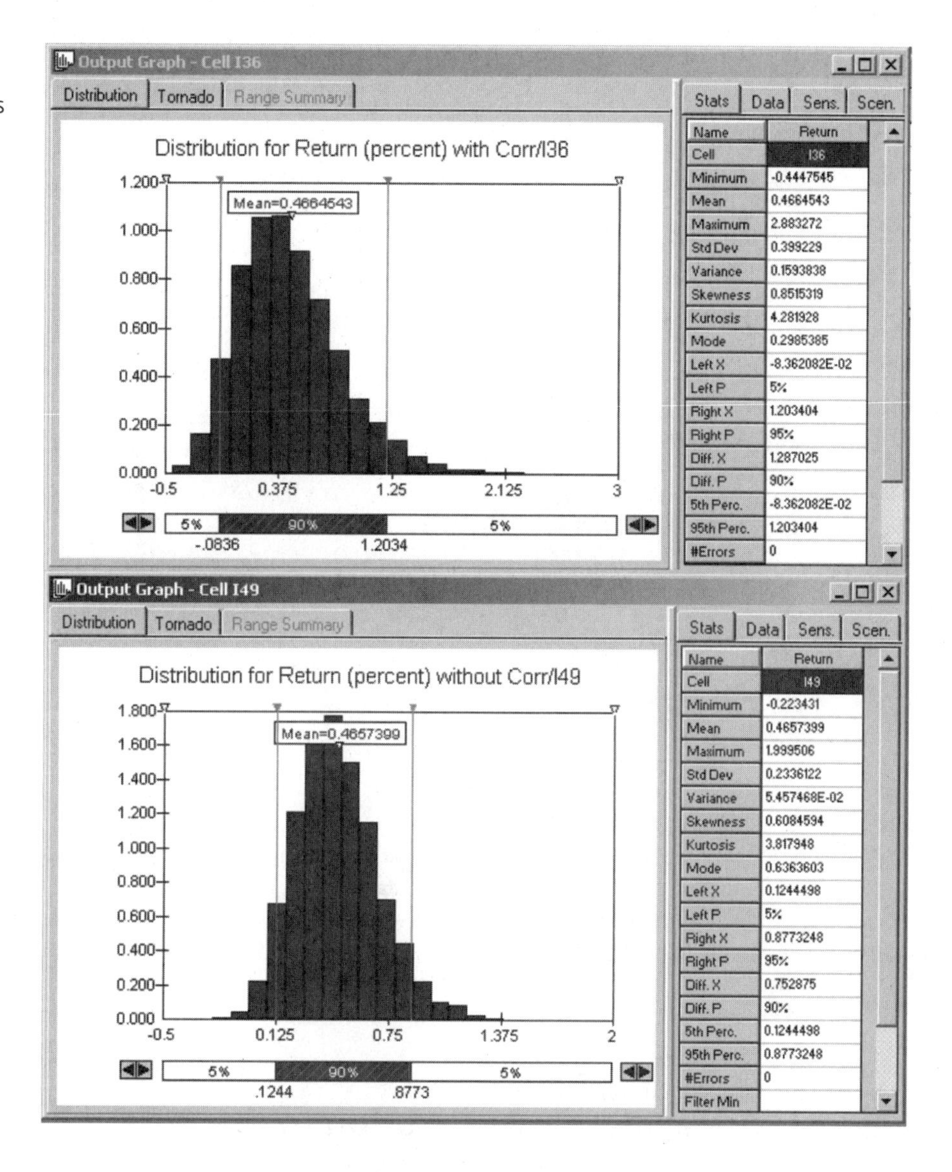

FIGURE 3.29 Comparison of percentage returns with and without correlation

licly traded assets should be normally distributed. Rather than just accept this theory, we would like to actually test this assumption—that is, we would like to collect data on the logs of price changes for a stock and test the hypothesis that our data follow a normal distribution. If our data do not provide strong evidence to the contrary, then we will accept this hypothesis and sample Z from a normal distribution.

In other cases, we will have data that are sampled from a population with an unknown distribution. For example, we might have observations on after-tax earnings for a sample of Internet companies. If our simulation model uses this variable,

FIGURE 3.30
Differences of logs of daily closing prices for Daimler-Chrysler

	A	H	I	J
1	Date	Close	ln	Change
2	22-Oct-99	48.1875	3.8751	-0.0263
3	21-Oct-99	46.9375	3.8488	-0.0148
4	20-Oct-99	46.25	3.8341	-0.0288
5	19-Oct-99	44.9375	3.8053	-0.0282
6	18-Oct-99	43.6875	3.7771	-0.0364
7	15-Oct-99	42.125	3.7406	0.0421
8	14-Oct-99	43.9375	3.7828	-0.0028
9	13-Oct-99	43.8125	3.7799	0.0378
10	12-Oct-99	45.5	3.8177	0.0082
11	11-Oct-99	45.875	3.8259	0.0149
12	8-Oct-99	46.5625	3.8408	-0.0149
13	7-Oct-99	45.875	3.8259	0.0108
14	6-Oct-99	46.375	3.8368	-0.0259
15	5-Oct-99	45.1875	3.8108	-0.0111
16	4-Oct-99	44.6875	3.7997	-0.0327
17	1-Oct-99	43.25	3.7670	0.0172
18	30-Sep-99	44	3.7842	-0.0303
19	29-Sep-99	42.6875	3.7539	0.0246

Fit Distributions to Data button

we must sample it from some distribution. But we do not know which distribution to use, so we will need to search for an appropriate distribution. The way this is done is by fitting several distributions to the data and selecting the distribution that fits the best. Vincent (1998) and Law and Kelton (2000), Chapter 6, have more detailed discussions of distribution fitting.

3.6.1 Using BestFit to Fit a Specific Distribution to the Data

@RISK includes a distribution-fitting module that can be used to accomplish these tasks. This module, BestFit in the DecisionTools Suite, is available as a stand-alone program but is also integrated into versions of @RISK 4.0 and later. It fits several distributions to a set of observations and displays the results of these fittings both numerically and graphically. We will demonstrate the use of BestFit through two examples. Figure 3.30 shows some observations for the difference of the log of the daily closing prices of Daimler-Chrysler stock. Column H contains historical stock prices. The figures in column I were computed by taking the natural log of the corresponding value in column H, and then the figures in column J were computed as

@RISK - Fit Excel Data
Fit Tab Name
Fit2
Excel Data Range
[c.xls]c!$J$2:$J$253
Type Of Data
Sampled Values
Density Curve
Cumulative Curve
Domain
Continuous
Discrete (Integral Domain)
Filtering Options
No Filtering
Filter Data Outside Range: Xmin
to Xmax
Filter Data that Falls
1
Standard Deviations Beyond the Mean
Automatically Refit When Input Data Changes
OK
Cancel

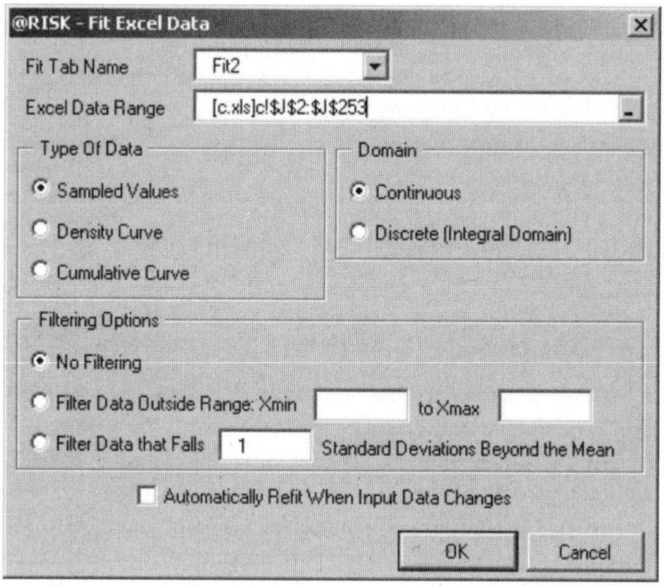

FIGURE 3.31 Distribution-fitting dialog

FIGURE 3.32 Distribution-fitting output window

Fit Results
Fitted Distributions
Weibull
Normal
BetaGeneral
Logistic
Triang
ExtValue
Uniform
Expon
7 Invalid Fits
Rank by Chi-Sq
Comparison
Difference
P-P
Q-Q
Normal(-0.0019112, 0.025404)
5.0%
90.0%
5.0%
-0.0437
0.0399
Stats
GOF

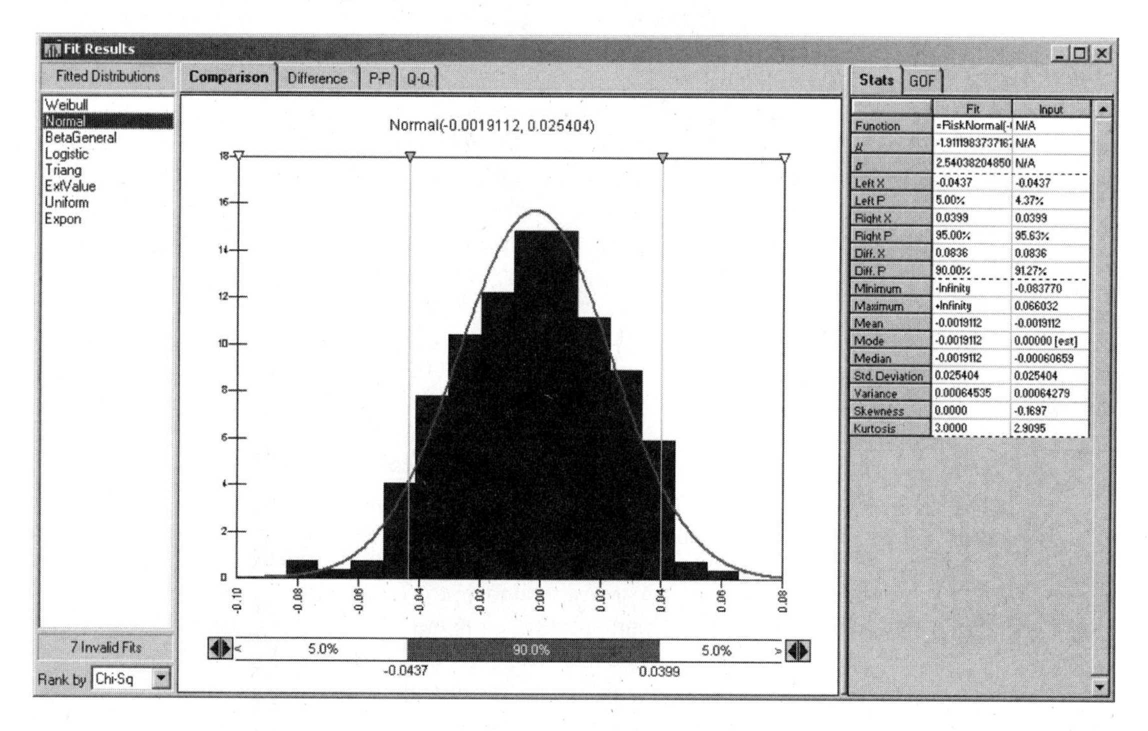

the difference of successive values in column I. For example, the entry in cell I2 is =LN(H2), and the entry in cell J2 is =I3-I2.

The entire dataset consists of 252 observations in column J. We want to see if these observations are consistent with a sample from a normally distributed population. The procedure is as follows:

1. Select the range containing the data to be fitted. We selected the range I2:I253.
2. Click the **Fit Distributions to Data** button. You will get a dialog like that in Figure 3.31.
3. You can edit the value in the text box **Fit Tab Name** if you wish. Otherwise, click the **OK** button.
4. You will be presented with the **Fit Results** window as in Figure 3.32.

This window contains considerable useful information. The left pane shows the distributions @RISK attempted to fit to the data, the middle pane shows a graphical summary of the fit, and the right pane contains the numerical results of the fitting procedure. First, look at the left pane, which is titled **Fitted Distributions.** If you click on one of the distributions listed, the center and right panes will change to show the graph and numerical results for fitting that distribution. Since we are only interested now in testing to see if a normal distribution fits our data, we only care about the results for the normal distribution fit. The center pane shows a graph of the best-fitting normal density function overlaid on the histogram of the data. A good fit is indicated by a density function that fairly closely follows the histogram. In this case, the fit appears to be rather good. On other tabs of this pane, you can find alternative graphical displays. The **Difference** tab shows the absolute difference between the density function and the histogram. If the fit is good, this plot will jump up and down but should not show any clear trends. The last two tabs show probability-probability (**P-P**) and quantile-quantile (**Q-Q**) plots comparing the fitted and sample distributions. If the fit is good, then these plots should be close to a straight line; if the fit is poor, they will have noticeable curvature, usually at one or both ends of the plot. Banks et al. (2001) discuss q-q plots in section 9.2.3, and Law and Kelton (2000) provide a detailed discussion of both p-p and q-q plots in section 6.6.1.

The process of fitting a distribution consists of two distinct steps for each specific family of distributions to be fit: First, compute estimates of the parameters for the candidate distribution from the data, then compute the test statistic for the goodness-of-fit test. For example, in fitting a normal distribution, we will first estimate the mean and variance, which are the two parameters of the normal distribution, using the sample mean and sample variance from the data. Once these parameters have been estimated, we can compute the test statistic, which measures how much the observed distribution differs from the best-fitting normal distribution. For all candidate distributions, BestFit computes the maximum likelihood estimates of all parameters. The details of this computation are explained in the @RISK manual and in most general statistics texts (see Hogg and Craig, 1995, Chapter 6).

The right pane contains the numerical results of the computations to fit the distribution. There are two tabs: **Stats** and **GOF.** The **Stats** tab provides the parameter estimates for the candidate distribution, and the **GOF** (goodness of fit) tab has the test statistics for the goodness-of-fit tests. Let's look at the **GOF** tab first. The three columns labeled **Chi-sq, A-D,** and **K-S** refer to the three goodness-of-fit tests that can be performed: chi-squared, Anderson-Darling, and Kolmogorov-Smirnov. The chi-squared test applies to any distribution. The Anderson-Darling test is used to test the goodness of fit for normal, logistic, and certain other distributions, and the Kolmogorov-Smirnov test is applied to tests of fit for certain continuous distributions, including the normal and exponential distributions. In the cases where they apply, the Kolmogorov-Smirnov and Anderson-Darling tests are more powerful and therefore preferred. See Law and Kelton (2000; section 6.6.2) and Banks et al. (2001; section 9.4) for discussions of these statistical tests. The values of the test statistics are given in the first row of this table. The larger the test statistic value, the greater the evidence of lack of fit. The second row contains the p-values for the goodness-of-fit tests. In all hypothesis testing, small p-values indicate strong evidence to reject the null hypothesis. In tests of goodness of fit, a small p-value indicates a lack of fit because the null hypothesis is that the distribution fits the data. Generally, if a p-value is less than .10, you should be concerned that the distribution does not fit. If the p-value is greater than .15, it is reasonable to conclude that the fit is adequate. From Figure 3.32, we can conclude that the data fit a normal distribution quite well; all p-values are well above .15.

The **Stats** tab in the right panel contains a lot of sample statistics for the input data as well as the fitted distribution. The two columns **Fit** and **Input** refer to the fitted distribution and the input data, respectively. The first row gives the formula for the @RISK function to generate observations from the fitted distribution. This formula contains the numerical value(s) of the parameter(s); however, good spreadsheet practice provides that the parameter values be put in clearly labeled cells and the cells be referenced in the formula. The next group of rows contain the estimates of the parameter values for the candidate distribution. In the case of the normal distribution, these are the mean and standard deviation. Although the rows are labeled μ and σ, the values given are not the true parameter values but *estimates* of them. If the same table were produced using another sample of data from the same population, the parameter estimates would have *different* values because they are computed from a sample. However, the conclusions of the goodness-of-fit tests would likely be the same. The rows after the parameter values contain numerical measures that can be used to compare the input data and the fitted distribution. For example, the 5th and 95th percentiles of the fitted distribution are given. In Figure 3.32, these are (Left X) –.0437 and (Right X) .0399. We can then compare the cumulative probabilities—.05 and .95—for these points with the empirical probabilities for the same points from the input data, which are .0437 and .9563. Since these values are quite close to .05 and .95, we can conclude that the distribution fits well at this point in the tails.

FIGURE 3.33
Data for effective average interest rates by state

	A	B	C	D	E
1	**Effective average mortgage rates in 1999 by state.**				
2					
3	Alabama	7.43			
4	Alaska	7.73			
5	Arizona	7.38			
6	Arkansas	7.42			
7	California	6.75			
8	Colorado	7.30			
9	Connecticut	7.16			
10	Delaware	7.37			
11	District of Columbia	7.22			
12	Florida	7.29			
13	Georgia	7.35			
14	Hawaii	7.27			
15	Idaho	7.27			
16	Illinois	7.20			
17	Indiana	7.46			
18	Iowa	7.43			
19	Kansas	7.08			
20	Kentucky	7.40			

Sheet1

3.6.2 Finding the Distribution with the Best Fit

Sometimes we have data for a stochastic variate and are faced with the problem of selecting a distribution that best represents the variate. There is no theoretical reason why one distribution should fit better than another, so we would just like to fit several candidate distributions and select the one that best fits the data as long as the fit is sufficiently good. As an example, Figure 3.33 contains effective average mortgage-interest-rate data for each state and the District of Columbia for 1999. "Effective" interest rate includes the fees and other expenses associated with obtaining the mortgage. These data were obtained from the Federal Housing Finance Board Web site at <http://www.fhfb.gov/>. Imagine that your model uses an interest rate variate and these data constitute samples from the appropriate distribution.

First, we note that interest rates are measured on a continuous scale, so only continuous distributions would be candidates to fit these data. Using @RISK, the procedure to fit the data is the same as before: First, select the range of cells containing the data to which we want to fit a distribution. In this case, the data are in cells B3 through B53. Click the **Fit Distributions to Data** button in the @RISK toolbar. In the dialog, the tab on the @RISK sheet and range containing the data should be set correctly. Be sure **Sampled Values** is selected under **Type of Data** and **Continuous** is selected under **Domain.** Since we want to use all of the data, no filtering options are selected either. Finally, click OK. @RISK will show the distribution fitting window as in Figure 3.34. The fitted distributions are listed in order of best fit to worst fit according to the criterion at the bottom of the left pane. We see immediately that the Logistic distribution provides the best fit to this data. You can select **K-S** in the **Rank By** select box at the bottom of the **Fitted Distributions**

FIGURE 3.34 Output window for fitting a distribution to interest rate data

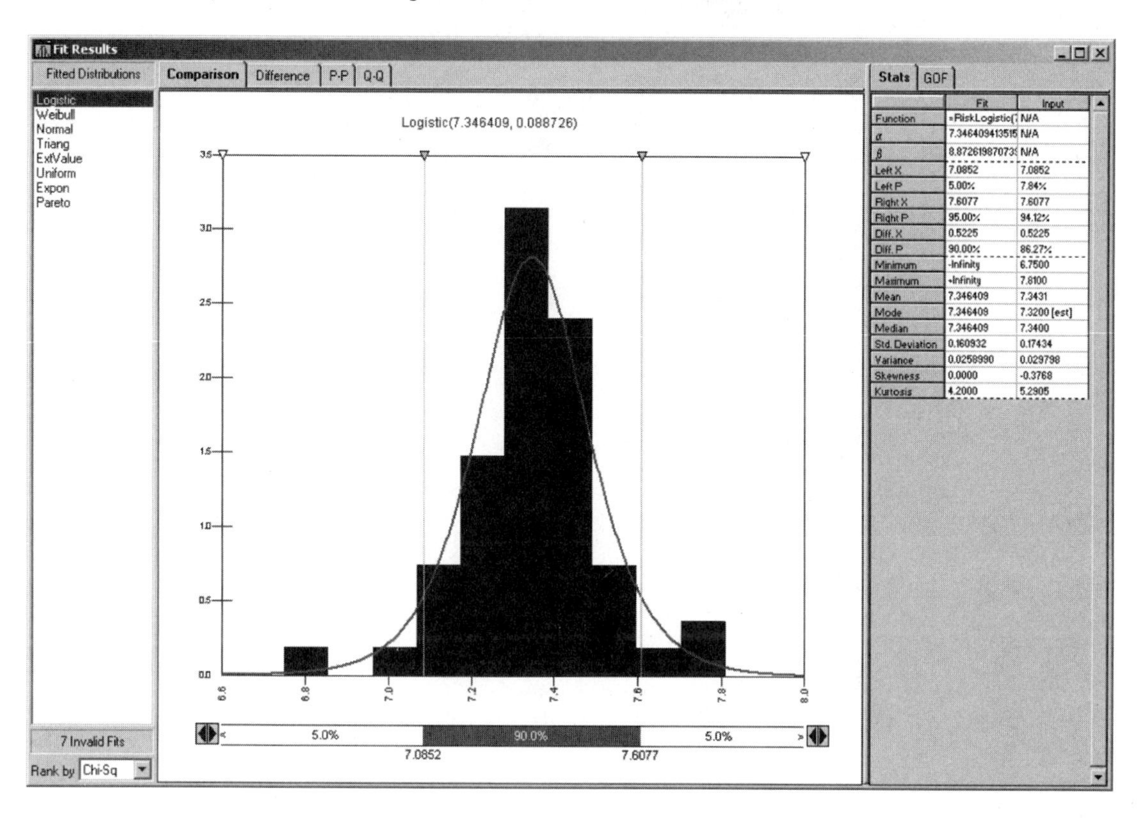

panel to see that the order remains essentially the same if the Kolmogorov-Smirnov test is selected for the goodness of fit.

Next, examine the goodness-of-fit tab in the right pane. Figure 3.35 shows the output. By examining the p-values, we see that the fit for the logistic distribution is good. This means that there is no substantial evidence of lack of fit, because all p-values are greater than .1. The goodness-of-fit tabs for the other distributions show that the logistic distribution is the only distribution that fits these data. A more accurate statement is that there is substantial evidence of lack of fit for the other distributions.

Other graphical evidence of the fit of the logistic distribution to the data is given by the middle pane. The **Comparison** tab shows that the density function follows the histogram of the data reasonably well. The **Difference** tab shows that the difference between the histogram and the fitted distribution appears to be a random function and without obvious patterns. The best evidence of fit is given by the **P-P** and **Q-Q** tabs. Figure 3.36 shows the **P-P** tab. Here we can see that the probability plot of the data is quite close to a straight line. This plot has more variation than the

FIGURE 3.35
Goodness-of-fit statistics for interest rate data fitted to the logistic distribution

	Chi-Sq	A-D	K-S
Test Value	3.5294	0.3714	0.0844
P Value	0.8969	> 0.25	> 0.1
Rank	1	1	1
C.Val @ 0.75	5.0706	N/A	N/A
C.Val @ 0.5	7.3441	N/A	N/A
C.Val @ 0.25	10.2189	0.4239	N/A
C.Val @ 0.15	12.0271	N/A	N/A
C.Val @ 0.1	13.3616	0.6568	0.0991
C.Val @ 0.05	15.5073	0.7652	0.1080
C.Val @ 0.025	17.5345	0.9016	0.1141
C.Val @ 0.01	20.0902	0.9016	0.1233
C.Val @ 0.005	21.9550	1.0051	N/A
C.Val @ 0.001	26.1245	N/A	N/A
# Bins	9	N/A	N/A
Bin #1 Min	-Infinity	N/A	N/A
Bin #1 Max	7.1619085	N/A	N/A
Bin #1 Input	6	N/A	N/A
Bin #1 Fit	5.67	N/A	N/A
Bin #2 Min	7.1619085	N/A	N/A
Bin #2 Max	7.2352565	N/A	N/A
Bin #2 Input	3	N/A	N/A
Bin #2 Fit	5.67	N/A	N/A
Bin #3 Min	7.2352565	N/A	N/A
Bin #3 Max	7.2849091	N/A	N/A
Bin #3 Input	7	N/A	N/A
Bin #3 Fit	5.67	N/A	N/A
Bin #4 Min	7.2849091	N/A	N/A
Bin #4 Max	7.3266107	N/A	N/A

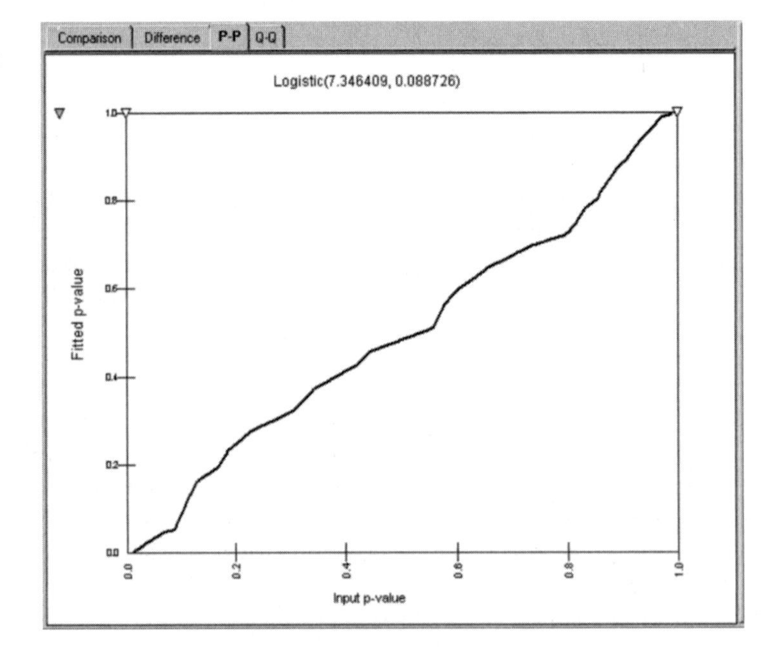

FIGURE 3.36
P-P plot for logistic distribution and interest rate data: P–P

plot of the Daimler-Chrysler data in Figure 3.30 because there are fewer observations, but it clearly lacks curvature or other evidence of nonlinear behavior.

Now that we have determined that the logistic distribution fits the data, we can use the @RISK function `=RiskLogistic(7.35,.089)` to sample values for this variate. Note that it is not necessary to carry all of the digits for the parameter estimates. @RISK does not give standard errors for these estimates, but it is reasonable to round them off to two or three significant digits. In this case, we rounded them off to two significant digits.

Take some time to explore the distribution fitting tools of @RISK.

3.7 SUMMARY

This chapter has presented several financial models and analyzed them using the facilities of @RISK, a large and complex add-in for Excel that facilitates building, running, and analyzing simulation models. Useful @RISK facilities include functions for sampling from a variety of distributions, output tables with a large number of statistics, and graphical representations of the output data. In this chapter, we were only able to look at a small subset of these facilities, but they were helpful to speed the process of simulation model building. We suggest that the reader take the time to examine the other capabilities of @RISK presented in the software's documentation.

The first model considered in this chapter was the *geometric random walk model* for the price of a stock. We used this model to evaluate the distribution of a stock price in the future. A *European call option* involves the right, but not the obligation, to buy shares of a stock for a fixed price, called the *strike price,* at or before a fixed time in the future (the *expiration date).* Using this notion, we examined how simulation can be used to estimate a fair price for a stock option. Other methods are available for computing option prices. Simulation has the advantage that the same methodology can be applied when the distribution of the future price is not lognormal. The results of the simulation were presented using both output tables and graphs from @RISK.

A *European put option* provides the right, but not the obligation, to sell shares of a stock at or before an expiration date in the future for a specific strike price. Put options have value only if the price of the stock drops below the strike price. Thus, put options can be used as a hedge against the event that the price of the stock falls more than a desirable amount. We developed a simulation to demonstrate this situation and used the simulation to show that a portfolio consisting of a number of shares of a stock and an equal number of put options for the stock can provide almost equivalent returns but at a lower risk.

The payout for some options depends on the entire series of prices over a fixed period of time. This is the case for an *Asian option,* which has a payout based on the average price of the underlying stock. To simulate such a model, we must first model the price process for the stock. We showed how to implement the geometric random walk model to sample the price of a stock at specific points in time. Using this model and @RISK, we simulated the price of a stock and examined its distribution. Once this model was set up, the procedure for pricing an Asian option was the same as that for a European option except that the payout was changed to be the average value of the stock.

Financial models, especially those that involve a portfolio of securities, include variates that are correlated. Using stock price data from six well-known companies, we estimated the correlation matrix for the logs of their price changes and used this estimated correlation matrix to sample their price changes from a multivariate lognormal distribution. With this sampling procedure, we could examine the distribution of the value of the portfolio at a future time. By comparing this distribution to that of an equivalent portfolio but without correlation among the price changes, we were able to show that the risk in a portfolio of positively correlated stocks is greater than that of a portfolio of independent assets.

If a simulation model is going to truly represent a real system, then the distributions sampled in the model must represent the populations in the real system. We can ensure this by collecting data from the real populations and fitting a theoretical distribution to those data. @RISK includes a tool, BestFit, that performs this work. Given a sample of data, BestFit will estimate the parameters of each of several candidate distributions and compute test statistics for three tests of goodness of fit: chi-square, Anderson-Darling, and Kolmogorov-Smirnov. This tool can be used to examine whether a specific distribution fits a set of data as well as examine the goodness of fit for several distributions in order to select the distribution that fits the best. We examined both of these approaches using price data for a stock and mortgage rate data for 50 states.

References

Banks, J., J. S. Carson III, B. L. Nelson, and D. M. Nicol. 2001. *Discrete-event system simulation.* Upper Saddle River, New Jersey: Prentice Hall.

Benninga, S. 1998. *Financial modeling*. Cambridge, Massachusetts: MIT Press.

Hogg, R. V., and A. T. Craig. 1995. *Introduction to mathematical statistics*. Upper Saddle River, New Jersey: Prentice Hall.

Hull, J. C. 1997. *Options, futures, and other derivatives.* Upper Saddle River, New Jersey: Prentice Hall.

Law, A. M., and W. D. Kelton. 2000. *Simulation modeling and analysis*. Boston: McGraw-Hill.

Luenberger, D. G. 1998. *Investment science.* New York: Oxford University Press.

Vincent, S. 1998. Input data analysis. In *Handbook of simulation,* ed. J. Banks. New York: Wiley.

Problems

3.1 In the geometric random walk model in section 3.2, sample $P(t)$ assuming that the log of the price change has a triangular distribution. In the model (3.2), let Z have a symmetric triangular distribution between $-\sqrt{3}$ and $\sqrt{3}$. Use the appropriate @RISK distribution function to sample Z.

3.2 Calculate the price of a put option using the methods of section 3.3 and the following parameters: $\mu = .08$, $\sigma = .10$, $t = .5$, $P(0) = 100$, $r = .059$, and $S = 97$. Use 1000 replications. Compute a 95 percent confidence interval for the price. Increase the sample size to 10,000 replications and rerun the simulation.

3.3 Evaluate the following policy for hedging. If the price of the stock increases more than 3 percent in the first six months, buy a call option for the next six months. Otherwise, buy a put option for the next six months. Use the parameters in section 3.3 and assume the cost of the call option is $3.40. Set the exercise price of the put option at $97 and assume the price of the put option is $2.35.

3.4 Consider the following retirement plan. You invest $1000 per year in a tax-sheltered savings plan for 30 years. Assume the money is invested in an aggressive growth stock fund that averages 20 percent growth per year. Assume volatility is $\sigma = .30$. Use the geometric random walk model for the value of the fund. Show the distribution of the fund value at the end of 30 years (at retirement). Compute a 90 percent tolerance interval for the fund value after 30 years.

3.5 Consider the following retirement plan, which is more conservative than the plan in problem 3.4. You invest $1000 per year in a tax-sheltered savings plan for 30 years. Half of the investment each year is placed in the aggressive growth fund with the parameters in problem 3.4, and half is placed in a bond fund that pays a fixed return of 5 percent per year. Show the distribution of the fund value at the end of 30 years (at retirement). Compute a 90 percent tolerance interval for the fund value after 30 years.

3.6 Develop a model of the value of a retirement plan that combines those of the two previous problems. The money is placed into funds according to the schedule in Table 3.2.

Table 3.2

Years	Aggressive Growth (%)	Bond (%)
1–10	100	0
11–20	50	50
21–30	0	100

Show the distribution of the fund value at the end of 30 years (at retirement). Compute a 90 percent tolerance interval for the fund value after 30 years. Compare this distribution with the results of the two previous problems.

3.7 A fund has a value of $1,000,000. Half of the fund is invested in a stock portfolio whose value follows the geometric random walk model of section 3.2, and half is invested in bonds paying 5 percent per year. You plan to withdraw $120,000 per year at the end of each year. Your plan is to withdraw $60,000 from the stock portfolio and $60,000 from the bond portion. If either portion does not have $60,000, then the portion will be emptied and the remaining amount withdrawn from the other portion until both parts are depleted. Show the distribution of the length of time until the fund is fully depleted.

3.8 Repeat problem 3.7 but change the withdrawal policy to the following. Withdraw the money from the bond fund until it is depleted, then withdraw the money from the stock portfolio. Compare the distribution of the life of the funds with that obtained from problem 3.7.

3.9 Consider an exotic option whose value is determined by the 75th percentile of the price over the duration of the option. Using the parameters in section 3.3, estimate the price of this exotic option.

3.10 File interest2000.xls contains a sample of interest rates for bonds issued in 2000. Use the distribution-fitting procedures of @RISK to find an appropriate distribution to use to sample observations for this variate.

3.11 File mortgages.xls contains observations on the number of mortgages sold each day by a small mortgage company. Because mortgages are an integer-valued variate, it is appropriate to fit a discrete distribution to these data. Use the distribution-fitting procedures of @RISK to find a discrete distribution that best describes these data.

3.12 Test the normal random variate generator function in @RISK, RiskNormal, by first using this function to generate a sample of 500 observations from a normal distribution with mean 0 and standard deviation 1, and then fitting a normal distribution to this sample of observations using the techniques in section 3.6.1.

3.13 Repeat problem 3.12 but use only the first 50 observations in the dataset. Comparing the results of distribution fitting in this problem with that in problem 3.12, what conclusions can you draw concerning the number of observations in the dataset?

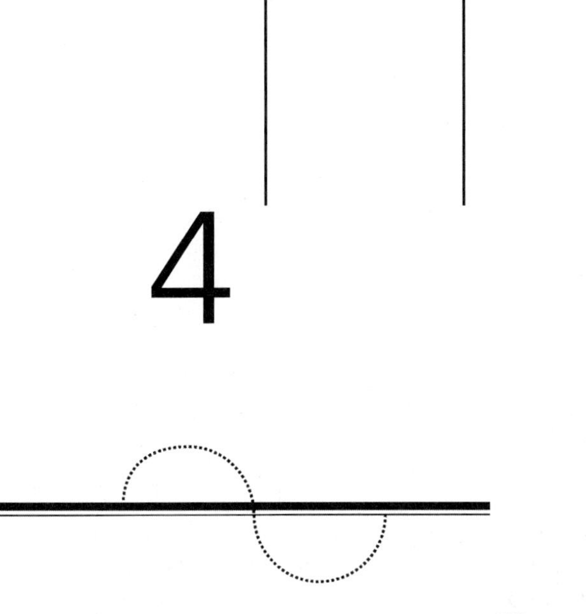

4

Dynamic Simulations

In this chapter we introduce the reader to the concept of a dynamic simulation in which we observe the system as it operates over time. The key concept here is that a model involves time and must be sampled over time if we are to obtain the information we seek from it. The time advance mechanism used here is the *fixed time advance;* the models are sufficiently simple that a more sophisticated worldview involving entities, attributes, sets, and so on will not be required; and the models can be programmed "from scratch" using a spreadsheet or a general-purpose programming language. The entity–attribute–set worldview will be presented in Chapter 5.

The example models are designed to increase in complexity so that the final models illustrate the kinds of more complex systems that can be simulated without utilizing the advanced modeling facilities discussed in Chapter 5. The models may be transient or stationary. If the model is transient, analysis of data requires the use of replications, which brings us back to the analysis methods presented in Chapter 2. We mention transient analysis in this chapter but concentrate on estimating stationary performance measures. Data analysis for stationary performance measures involves batch means and regenerative analysis.

4.1 INTRODUCTION

Chapters 2 and 3 presented the concept of a static simulation as a model observed at a single point in time. Dynamic models describe the behavior of a system over time. Such models are of interest to managers and engineers to describe various aspects of marketing, production, inventory, personnel, economic, financial, and other systems and processes. The remainder of this book is devoted to techniques for representing and sampling dynamic models. The models in this chapter are the simplest kind of dynamic models and can be implemented using a general-purpose language or a spreadsheet. Later chapters will examine more complex models from several areas of business and engineering and present more complex techniques for representing the models.

There are two basic mechanisms for advancing time in a dynamic simulation: fixed time advance and variable time advance. As the name implies, the *fixed time advance* mechanism moves time by a fixed amount when the system time changes. The *variable time advance* mechanism moves time by an amount that is not fixed but can vary each occasion that time is advanced. The variable time advance mechanism is the subject of Chapter 5; this chapter will focus on models using the fixed time advance mechanism.

Suppose the amount by which the time is increased is Δt. We will call the time between t and $t + \Delta t$ a *period.* If Δt is one day, then the period of the simulation is one day; if Δt is four seconds, the period of the simulation is four seconds. Frequently, time will be measured in discrete units (seconds, minutes, days, months, etc.), or time will just be a mechanism to maintain the order of the observations collected. When this is the case, we may use n to denote the current "time," and Δt will be equal to 1.

The logic of a model using the fixed time advance mechanism can be summarized in the following algorithm:

> Initialize time to 0. Initialize all other variables.
> WHILE *end condition* is not met DO:
>   Compute system state changes during period.
>   Record data for period.
>   Increment time by Δt.

We assume that time starts at 0 and is incremented by an amount Δt when it is updated. There must be an *end condition* that specifies when to stop the simulation, otherwise the simulation will run forever or until it causes a computer error. For example, the simulation might run until 1000 observations are collected or until time exceeds 1000 units. In each iteration of the simulation, a formula or algorithm is used to specify how the variables that describe the system state (i.e., the state variables) change during the period. This transformation is the essence of the model. Once the model's new state is computed, data relating to that state can be collected if required and the next period begun if the simulation does not stop at the end of this period.

In this chapter, we will illustrate fixed time advance models using several examples from business and engineering, and we will discuss how to collect and analyze data from these simulations.

4.2 WAITING TIMES IN A SINGLE-SERVER QUEUEING SYSTEM

Queueing systems have three components: (1) a stream of arriving customers, jobs, or other entities; (2) a mechanism to provide some sort of service or processing to the arriving entities; and (3) one or more facilities for entities to wait until service can be provided when demand for service exceeds the capacity to provide service. Familiar examples include lines of people at checkout counters in stores, lines of passengers waiting to board buses or planes, lines of vehicles at toll booths, and lines of planes waiting to take off on a particular runway. Other examples that may not be so familiar but are just as important involve jobs waiting to be processed by a computer (including instructions from an automated teller machine), messages and calls waiting to be processed by a telephone switch, orders waiting to be filled by a mail-order supplier, items waiting in inventory, patients waiting to use an operating room in a hospital, and operating room equipment waiting to be used by patients in the hospital. Queueing systems are an everyday part of our lives and will become more important as society seeks to more efficiently use limited resources and businesses seek to serve customers' needs at lower cost. In the coming chapters, we will explore how to model and simulate various queueing systems at the level of detail required by the problem at hand. Here, however, we will look at some simple models that can be represented in a simple way and yet are surprisingly useful.

4.2.1 Lindley's Formula

Consider a queueing system in which there is a single server that serves arriving customers one at a time in the same order as their arrival—that is, first come, first served. Service times are independent random variables with identical distributions, and successive interarrival times are also independent random variables with identical distributions, which may be different from the distribution of service times. Let $X_1, X_2, \ldots$ represent the successive service times and $Y_1, Y_2, \ldots$ represent the successive interarrival times. If $1/\lambda$ is the mean of each Y_i, and $1/\mu$ is the mean of each X_i, then λ is called the arrival rate (because $1/\lambda$ is the mean interarrival time), and μ is called the service rate (because $1/\mu$ is the mean service time). Intuitively, λ is the mean number of arrivals per minute, so $1/\lambda$ is the mean number of minutes between arrivals if the time unit is minutes.

Let W_n represent the waiting time in the queue—that is, the time between the moment of arrival and the beginning of service—for the nth customer to enter the system. A simple model for the successive waiting times is given by the following expression due to Lindley (1952):

$$W_{n+1} = \max(0, W_n + X_n - Y_n) \quad \textbf{(4.1)}$$

This relationship says that the time the $(n + 1)$st customer must wait is the time the nth customer waits, plus the nth customer's service time, X_n (because that customer is in front of the nth customer), less the time between the arrivals of the nth and $(n + 1)$st customers, Y_n. If $W_n + X_n - Y_n$ is less than 0, then the $(n + 1)$st customer arrived after the nth customer finished service, and thus the waiting time for the $(n + 1)$st customer is 0, because his service can begin immediately.

4.2.2 Taxonomy of Queueing Systems

If the distributions of interarrival times and service times are exponential (with different means), then the queueing system is known as the "M/M/1" queue. This notation, which is credited to M. G. Kendall, uses "M" to denote the exponential distribution. "M" actually refers to "Markov" because the exponential distribution has the "lack-of-memory" property that gives the arrival process or service process the Markov property. See Cox and Smith (1961), Ross (1993), Taylor and Karlin (1998), or Hillier and Lieberman (1995) for a discussion of this topic. The first letter in the notation *a/s/n* denotes the *interarrival time distribution,* the second denotes the *service time distribution,* and the number (n) denotes the number of servers in the system. The notation "M/G/1" refers to a system with exponential interarrival times (M), any general service time distribution (G), and one server; the notation GI/M/1 refers to a system with any interarrival time distribution (that is, general independent or GI), exponential service times (M), and one server; and GI/G/1 (also G/G/1) refers to a system with any interarrival time distribution, any service time distribution, and one server. Since the expression given in equation (4.1) does not specify the interarrival time or service time distributions, it represents all of these systems. This expression is the model that forms the basis for the simulation of the single-server queue in the next section.

Queueing theorists have shown that if the arrival rate is less than the service rate ($\lambda < \mu$), then the system will be stable in the sense that the number of customers waiting in the queue will not grow without bound. Instead, the number in the queue will return to zero at randomly selected times as the server completes processing all customers that have arrived up to that point. In a stable queueing system, the average capacity to provide service over a long period of time, μ, is greater than the average demand for service over a long period of time, λ.

When analyzing stationary queueing systems, one frequently wants to determine some characteristic of the waiting time for an arbitrary customer who arrives at the system after it has been operating for a long time. For example, one might want to know the mean waiting time or the probability that a randomly selected customer must wait longer than a specified amount of time. Queueing theorists have used probability theory to derive expressions for these quantities in certain special cases. These include the M/M/1 queue (exponential interarrival and service times), the M/G/1 queue (exponential interarrival times, general independent service times), the GI/M/1 queue (general independent interarrival times, exponential service times) and queues having Erlang interarrival or service time distributions (or both). These systems can be analyzed because they have the Markov property, which allows them to be decomposed in a certain sense, and this decomposition permits expressions for the mean waiting time to be derived. Cox and Smith (1961), Hillier and Lieberman (1995), Ross (1993), and Taylor and Karlin (1998) present analytical solutions for these and several other simple queueing models. We will discuss the Markov property in section 3.5. If neither the interarrival times nor the service times have the Markov property, then such a derivation is not possible, and expressions for the mean waiting time are not available. For example, if the service times are uniformly distributed between 3 and 5 minutes, and the interarrival times have a symmetric triangular distribution between 3 and 7 minutes, we have no way to compute the mean waiting time, even though we know that the system is stable because the mean service time (4.0) is less than the mean interarrival time (5.0). In this case, simulation is the only technique available to determine the mean waiting time.

The model given in equation (4.1) will allow us to generate successive waiting time observations by sampling service times ($X_1, X_2, \ldots$) and interarrival times ($Y_1, Y_2, \ldots$), and combining them according to equation (4.1). This can be done using a spreadsheet such as Microsoft Excel or a general-purpose language such as C, Pascal, FORTRAN, or Java. We will discuss how to do these calculations using Excel.

4.2.3 A Spreadsheet Simulation of M/M/1 Queue Waiting Times

Figure 4.1 contains a portion of a spreadsheet simulating waiting times in the M/M/1 queue. At the top of the spreadsheet, the mean interarrival time and mean service time are given in cells C4 and C5, respectively. Column B contains the indices of the waiting times generated: 1, 2, ... Column C contains the waiting times. The sequence of waiting times is started by setting $W_0 = 0$, indicating that the 0th customer arrives to find the system empty and idle and can therefore start service immediately without having to wait. Column D contains the service times. These random variates are sampled from an exponential distribution with the mean given in cell C4. The formula in each of these cells is `-$C$4*LN(RAND())`. Recall from Chapter 2 that this is the formula for sampling a random variate from the exponential distribution. Column E contains the interarrival times, computed from the formula `-$C$53*LN(RAND())`. Column F computes the value C*n* + D*n* – E*n*, where *n* denotes the current row. Finally, except for the first numerical entry, which is zero,

FIGURE 4.1
Spreadsheet for simulating a M/M/1 queue

Simulation of an M/M/1 Queue

Mean service time:	0.7
Mean interarrival time:	1.0

Customer number (n)	Waiting Time (W_n)	Service Time (X_n)	Interarrival Time (Y_n)	$W_n+X_n-Y_n$	Busy/Idle
0	0	0.4328	0.3852	0.0476	0
1	0.0476	0.1770	0.0454	0.1792	1
2	0.1792	1.3494	0.1230	1.4056	1
3	1.4056	0.4659	1.6279	0.2436	1
4	0.2436	0.3996	1.0828	-0.4395	1
5	0.0000	1.9593	2.8404	-0.8811	0
6	0.0000	0.1485	1.7174	-1.5689	0
7	0.0000	0.5877	0.2410	0.3467	0
8	0.3467	1.8964	1.6122	0.6310	1
9	0.6310	1.9285	0.0878	2.4717	1
10	2.4717	0.1926	1.5142	1.1500	1

MM1 Model / MM1TRANS / IND WT TIME 1 / IND WT TIME 2 / IND WT TIME 3 / Transient Plot /

column C contains the formula `IF(Fn – 1 < 0, 0, Fn – 1)`; that is, if the entry in column F, row $n - 1$, is less than zero, then the value is zero; otherwise the value is what is in column F, row $n - 1$. Once the first two rows of the table are created in this way, the remaining rows can be copied from the second row. Note that in this simulation, we are producing the observations by copying a formula rather than using a data table. This can make a huge spreadsheet, but it is a simple way to do the calculations in this case.

4.3 CHARACTERISTICS OF DATA FROM DYNAMIC SIMULATIONS

In Chapter 2, we discussed collection and analysis of data from static simulations. In this case, the output data were produced by multiple independent replications of the simulation experiment. Because the observations are independent and identically distributed, we could apply the classical statistical techniques that are taught in any introductory statistics course to compute confidence intervals. In a dynamic simulation, observations are produced by the model as the system moves through time. These observations are usually not independent, and they usually have other characteristics that make estimation of parameters more difficult than in the case of static simulations. In this section, we will look at data from a dynamic simulation and discuss those characteristics that have an important bearing on estimation of parameters. In the next section, we will present the batch means method for computing a confidence interval for the mean.

The data we will use are waiting times from a simulation of the M/M/1 queue. This is a specific and particularly simple system; however, it is important to understand that the characteristics of the waiting time data from this system are present to a greater or lesser extent in the data from all simulations of dynamic systems. The waiting times were generated using arrival rate $\lambda = .7$ and service rate $\mu = 1.0$. This system is "moderately" congested. The mean waiting time (in queue) for a stable M/M/1 queue is known (Hillier and Lieberman, 1995), so we can actually compare parameter estimates from the simulation to the known mean waiting time. In particular, the mean waiting time is given by

$$W_q = \frac{\lambda}{\mu(\mu - \lambda)} \tag{4.2}$$

and the probability that the server is busy and therefore the customer must wait on arrival is $\rho = \lambda/\mu$. The symbol ρ represents traffic intensity and is a measure of the amount of congestion in the system since it measures the proportion of the time the server is busy.

This system is stable because the arrival rate is less than the service rate, and the server is busy approximately 70 percent of the time over a long period. Thus, if we let the system run for a long time, it will exhibit stationary characteristics, and we can estimate the stationary mean waiting time. This is the expected waiting time that an arbitrary customer would experience if she arrived after the system had been operating for a very long time.

Twenty independent replications of the model presented in Figure 4.1, consisting of 500 observations each, were run. In each replication, the initial waiting time was set to zero, indicating that the system started in the empty-and-idle state. The "jagged" plot in Figure 4.2 shows the average over the 20 replications of each observation in the series—that is, the first point plotted in the series is the average, over 20 replications, of the waiting time of the first customer, which is always zero. The second point is the average of the waiting times of the second customer in each replication, the third point is the average of the waiting times of the third customer, and so on. This plot still has a lot of variation, so it was smoothed by applying a five-term moving average. Each point in a five-term moving average is the average of five adjacent observations. The purpose of smoothing the average waiting times is to allow the trend to be more easily observed.

4.3.1 The Initial Transient Period

From Figure 4.2, we can observe that there is a "warm-up period" during which the mean waiting time appears to increase to a value close to 2.0 and then settle back to a value close to 1.5. Using $\lambda = .7$ and $\mu = 1.0$, we can compute that the true stationary mean waiting time is 1.633. Thus, we see that in the first 100 observations, there is much evidence that the waiting times have settled into their stationary behavior

FIGURE 4.2 Results of 20 replications of queueing system

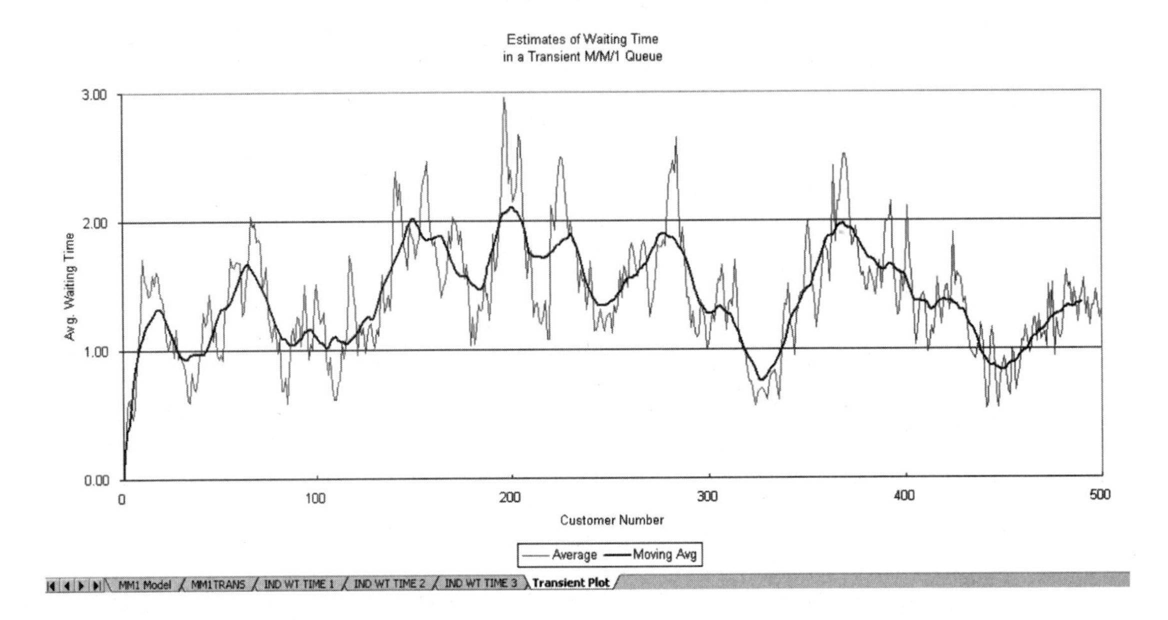

because the plot of mean waiting times increases from zero to a relatively stable value close to the mean. If we look further, we see that the plot of average waiting times meanders up and down, moving above the mean of 1.633 but spending most of its time below that value. We can conclude that during the first 500 observations, the waiting times probably do begin to exhibit stationary behavior, but it is difficult to detect on this plot.

The actual stationary distribution of the waiting time is known for the M/M/1 queue. (See Cox and Smith, 1961; Hillier and Lieberman, 1995; Ross, 1993; or Taylor and Karlin, 1998.) This distribution is

$$\begin{aligned} \Pr\{W \leq w\} &= 1 - \rho && \text{if } w = 0 \\ &= 1 - \rho\, e^{-(\mu - \lambda)w} && \text{if } w > 0 \end{aligned} \tag{4.3}$$

In other words, the probability that the waiting time will be zero is $1 - \rho$, which is also the probability that the system is empty on arrival and the arriving customer therefore does not have to wait. If the system is not empty on the arrival of a new customer, then his or her waiting time is an exponentially distributed random variable with mean $1/(\mu - \lambda)$. If the *first* waiting time is chosen from this distribution, then *all* subsequent waiting times will be from the stationary distribution. Thus, by generating the *first* waiting time from the stationary distribution, we can produce a sequence of waiting times, all of which are from the stationary distribution. This

FIGURE 4.3 Twenty replications of stationary single-server queue

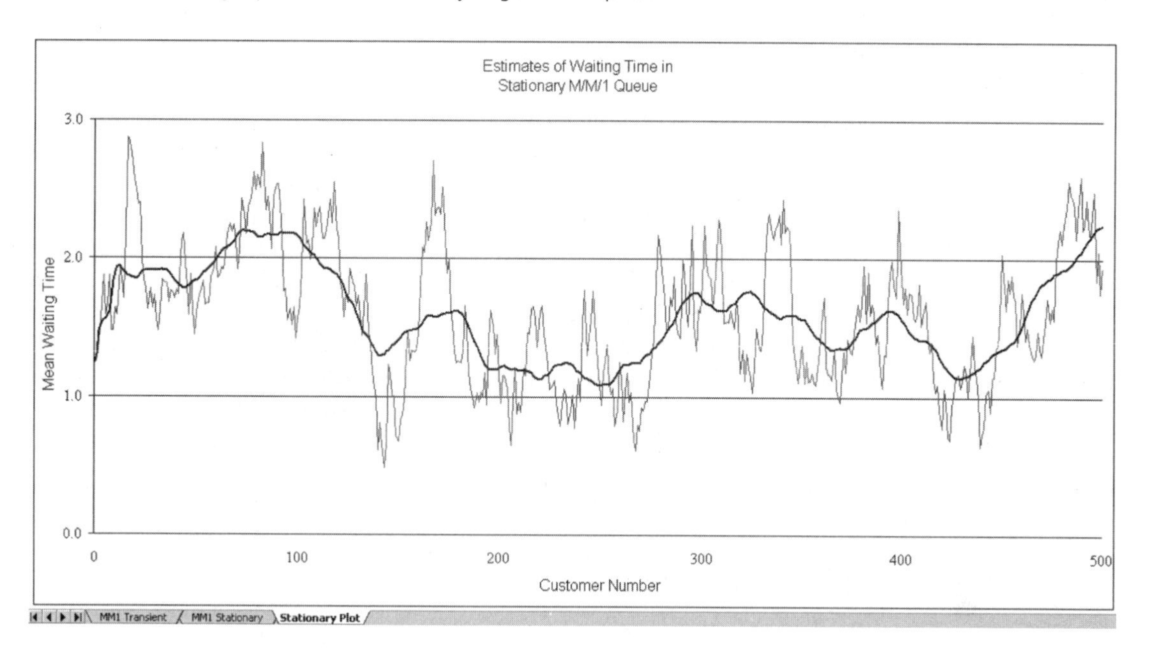

was done and is displayed in Figure 4.3. Twenty independent replications were run with 500 waiting times in each series. Then the series were averaged as before: The first observations in each series were averaged to form the first point in Figure 4.3; the second observation in each series were averaged; and so on. The resulting averages and the series after applying a five-term moving average also are shown in Figure 4.3. You can see that this series has the same general appearance as much of Figure 4.2, allowing us to conclude that the system that started empty and idle in Figure 4.2 was probably fairly close to its stationary behavior within the first 100 observations. However, we still cannot be sure exactly where the initial conditions cease to have a significant effect on the waiting times—that is, where the initial transient portion ends.

The initial transient behavior is an important characteristic of observations from a dynamic system. Every simulation of a dynamic system *must* begin in a fixed state. The initial state *cannot* be selected from the stationary distribution because the stationary distribution is not known for those systems that we wish to simulate. Indeed, if the stationary distribution were known, there would be no reason to simulate the system because we could compute the stationary mean from the stationary distribution using numerical integration! Since the initial state cannot be selected from the stationary distribution, *every* simulation of a dynamic system must go through an initial transient period before exhibiting the behavior of the stationary distribution.

4.3.2 Autocorrelated Observations

Another characteristic has an important influence on our ability to estimate the mean. The waiting times are not independent but are strongly related to one another. In statistical terms, we say that they are *autocorrelated.* If a particular customer's waiting time is small, then the next customer's waiting time is likely to be small also. On the other hand, if a customer's waiting time is large, then the next customer's waiting time is likely to be large also. This property shows up on the plot in Figure 4.3 through the characteristic meandering of the waiting time averages. Instead of "bouncing around" above and below the stationary mean, the waiting times will "make excursions," staying above the mean, then dipping below the mean for periods of time. The practical effect of this characteristic is that ordinary statistical methods cannot be naively applied to compute a reliable confidence interval for the mean, and when appropriate techniques are used a *great deal* of data are frequently needed to get a precise estimate of the mean.

4.4 BATCH MEANS TO ESTIMATE THE MEAN FROM STATIONARY DATA

We have seen that two problems present difficulties in estimating the stationary mean using data from a stable dynamic simulation: the initial transient period and autocorrelated observations. In this section, we will present a simple and useful method to compute a confidence interval for the mean using a single series of data from a dynamic simulation: the batch means method.

First, let's deal with the problem of the initial transient period. Because the initial observations are not drawn from the stationary distribution, we have little choice other than to discard them and only use observations that are generated after the system's behavior is close to the stationary behavior. We thus must determine how many observations to discard at the start of the series. A number of techniques have been proposed to formally choose a truncation point (Law, 1983). All of the techniques are fairly difficult to implement. Therefore, we will use a relatively simple graphical approach due to Welch (1981). As in the previous section, we will replicate the system many times, compute the mean of the series of observations from the replications, and smooth this series of means. Then by observing the pattern of this smoothed curve, we can visually select a point at which the stationary behavior seems to dominate. In the plot of Figure 4.2, one could truncate at approximately 100 observations. Even though the plot spends much of the time below the theoretical mean of 1.633 (information that is not generally available), the behavior has stabilized by this time and thus this looks like a good truncation point. We will not be too concerned about this point because, as we will later argue, the batch means method is fairly robust when it comes to errors in choosing the truncation point.

4.4.1 Batch Means Computation

Now that the initial transient observations have been removed, we assume that all remaining observations are sampled from the system operating in steady state. The next problem is to choose a method to compute a confidence interval for the mean. Several methods have been developed for this situation (Law, 1983; Alexopoulos and Seila, 1998). The simplest method, called the *batch means method* (Conway, 1963; Fishman, 1979; Law and Kelton, 1984), groups observations into several batches, computes the sample mean of the observations in each batch, and computes the confidence interval from these batch means using traditional statistical methods that apply to independent observations. As a simple example, suppose that our data, after removing the initial transient observations, consist of the following 12 observations:

$$1.2, 3.3, 2.6, 5.1, 4.4, .6, 0.0, 1.3, 1.5, 3.7, 3.5, 2.4$$

(Of course, in an actual application we would generate thousands of observations; this small dataset is just used to illustrate the computational procedure.) Suppose that we group the observations into three batches of four observations each. Then, the batches and their means would be

Batch 1	1.2, 3.3, 2.6, 5.1	Mean	$\overline{X}_1 = 3.05$
Batch 2	4.4, .6, 0.0, 1.3	Mean	$\overline{X}_2 = 1.58$
Batch 3	1.5, 3.7, 3.5, 2.4	Mean	$\overline{X}_3 = 2.78$

Now we will treat the three batch means as if they are a sequence of independent observations from the same population whose mean we want to estimate. The sample mean of the batch means is $\overline{\overline{X}} = 2.47$, and the standard deviation is

$$s_{\overline{X}} = .78$$

Thus, a 95 percent confidence interval for the mean is

$$\begin{aligned}\overline{\overline{X}} \pm t_{.025,2} \frac{s_{\overline{X}}}{\sqrt{3}} &= 2.47 \pm 4.303 \frac{.78}{\sqrt{3}} \\ &= 2.47 \pm 1.95 \\ &= (.52, 4.42)\end{aligned}$$

Recall that $t_{\alpha,k}$ is the $100(1 - \alpha)$-percentage point on the Student's t-distribution with k degrees of freedom. A review of basic statistics is presented in Appendix A.

Estimating the mean of a stationary sequence of observations produced by a dynamic simulation is a much more difficult problem than estimating the mean using independent observations, which we discussed in Chapter 3. If the amount of data is too small or the method is not applied appropriately, the confidence coefficient of the resulting confidence interval will be smaller than the presumed value. For example, if a 95 percent confidence interval is computed, we presume that the coverage probability is .95, when in fact it could be considerably less if the method is applied with too few observations or not applied correctly. The actual 95 percent confidence interval would necessarily be larger. This tricks us into thinking that our estimate of the mean is more precise than it really is.

4.4.2 Some Guidelines to Applying the Batch Means Method

Studies (Law and Kelton, 1984) have shown that the batch means method is highly competitive with other methods in terms of the accuracy and reliability of confidence intervals if it is applied correctly. When applying the batch means method using a fixed total number of observations, we must decide how many batches to form and therefore how many observations each batch will have. Theoretical considerations show that, if the batch size is sufficiently large, the batch means from separate batches will be approximately uncorrelated, even though the individual observations are autocorrelated. A central limit theorem similar to the one for independent observations applies to many processes of observations from dynamic simulations. If this is also the case, then the batch means will be approximately normally distributed when the batch size is sufficiently large, and this implies that they will be approximately independent because normally distributed random variables that are uncorrelated are also independent. Thus, our objective in applying the batch means method is to select the batch size large enough that the batch means are approximately uncorrelated—but small enough that the maximum number of batches is formed. As a rule of thumb, one should normally trade batch size for number of batches. For example, if we have generated 2000 observations, then we would generally be better off forming 10 batches of 200 observations than 20 batches of 100 observations. We usually want the number of batches to be at least 10 so that the percentiles of the t-distribution will be relatively close to those of the normal distribution, which is the smallest they can be. We also want the size of each batch to be at least 100 under most circumstances. Experience has shown that for many moderately congested systems this is the minimum number of observations required to give any assurance that the batch means are approximately uncorrelated or approximately normally distributed. Using these values, one should still consider the confidence interval computed to be a "rough" estimate.

An obvious alternative to the batch means approach is to run several independent replications, compute a sample mean from each, and use these sample means to compute a confidence interval for the mean. This approach has the appealing property that, because each replication starts from the same initial condition and is run independently of the others, we are guaranteed that the sample means are mutually independent, not just approximately uncorrelated. Any consideration of whether the batches are large enough is moot. However, this approach has two critical shortcomings: First, we must allow the simulation to run through the initial transient period in *each replication.* The batch means method requires this to be done only once—at the beginning of the run. Obviously, using independent replications is more wasteful of computation time because all data in the initial transient period are discarded. Second, if the length of the initial transient period is misjudged too small, then some or perhaps all of the observations in each replication will be collected before the system has reached steady state. This will cause the sample mean in every replication to be biased. In the batch means method, however, the second batch comes after the first, and the third after the second, and so on, so any bias in the batch means will reduce and likely be eliminated in successive batches because they are farther from the start of the run. Indeed, since batch sizes must be large, it is likely that only the first batch mean has significant bias. Thus, the batch means method is robust with respect to errors in determining the length of the initial transient period.

A useful graphical method to judge the adequacy of the initial transient period is to plot the batch means as a time series. If the first batch mean is either smaller or larger than all of the others, then there is evidence that the transient period is too short. The transient period can then be lengthened by eliminating the first batch from the sample. This procedure should be applied until the sequence of batch means appears to be stationary.

4.5 SIMULATING DISCRETE-TIME MARKOV CHAINS

Discrete-time Markov chains are a group of models that have been studied extensively and used to represent certain aspects of many systems in government, business, engineering, social science, biology, medicine, and other areas. A presentation of the mathematical theory of Markov chains is beyond the scope of this textbook, but the interested reader is referred to the several excellent textbooks on the subject (Kemeny and Snell, 1960; Ross, 1993; Taylor and Karlin, 1998). Although the parameters of Markov chains can frequently be computed analytically, it is often easier and more instructive to analyze these processes using simulation. Other processes that are similar to Markov chains but do not meet all of their assumptions can also be analyzed using simulation but not with the analytical methods that apply only to Markov chains. We will see in this section how to simulate Markov chains and estimate transient and stationary parameters using the data generated. First, we consider an example.

4.5.1 A Markov Chain Inventory Model

The following is a simple model of the inventory of computers in a small computer store: The number of computers in the inventory may be between zero and four. Daily demand for computers is a (discrete) random variable with the distribution given in Table 4.1.

If there are enough computers in the inventory to satisfy the demand, they are removed and sold. If demand is greater than the number of computers available, all

TABLE 4.1 Daily demand distribution for computers

Demand	Probability
0	.30
1	.30
2	.25
3	.15

computers available are sold and the remaining sales are lost. Inventory is replenished according to the following rule: If the number of computers is less than two, enough computers are ordered to refill the inventory to four; otherwise, no computers are ordered. Thus, if one computer is left at the end of the day, three computers are ordered; if none are left, four are ordered. Computers ordered at the end of a day are delivered at the start of the next day and are available to meet the next day's demand.

We are interested in the number of computers in the inventory at the start of each day. If we let X_n represent the inventory level at the start of day n and D_n represent the demand on day n, then the number of computers in the inventory at the end of the day is given by $R_n = \max(0, X_n - D_n)$. D_n is a random variable with the probability distribution given in Table 4.1. Letting A_n represent the number of computers ordered at the end of day n, the inventory at the start of day $n + 1$ is given by:

$$X_{n+1} = R_n + A_n \tag{4.4}$$

We can express A_n as

$$\begin{aligned} A_n &= 0 && \text{if } R_n > 1 \\ &= 4 - R_n && \text{if } R_n \leq 1 \end{aligned} \tag{4.5}$$

Figure 4.4 shows a sample path of X_n. Notice how X_n "cycles" among the values 2, 3, and 4. The values in Figure 4.4 were generated by setting $X_0 = 4$, and generating $X_1, X_2, \ldots$ using equations (4.4) and (4.5).

A careful examination of these relationships shows that X_{n+1} depends on the previous values $X_n, X_{n-1}, \ldots$ only through the value of X_n; that is, given the value of X_n, we can sample the values of $X_{n+1}, X_{n+2}, \ldots$ without knowing any previous value of the process. This property is called the *Markov property,* and a stochastic process that has this property is said to be a *Markov process.* The future behavior of a Markov process depends on the entire past history of the process only through the current value of the process. A process that has this property is said to be *memoryless.* The model of waiting times in a single-server queue presented in section 4.1 is also a Markov process because W_{n+1} depends on $W_n, W_{n-1}, \ldots$ only through W_n.

The *state space* for a process is defined to be the set of all possible values that the process can assume. In the inventory example, the state space consists of the values {2, 3, 4}; in the queue waiting time example, the state space consists of all nonnegative real numbers $[0, \infty)$. As these examples illustrate, the state space can be

FIGURE 4.4 Sample from Markov chain inventory model

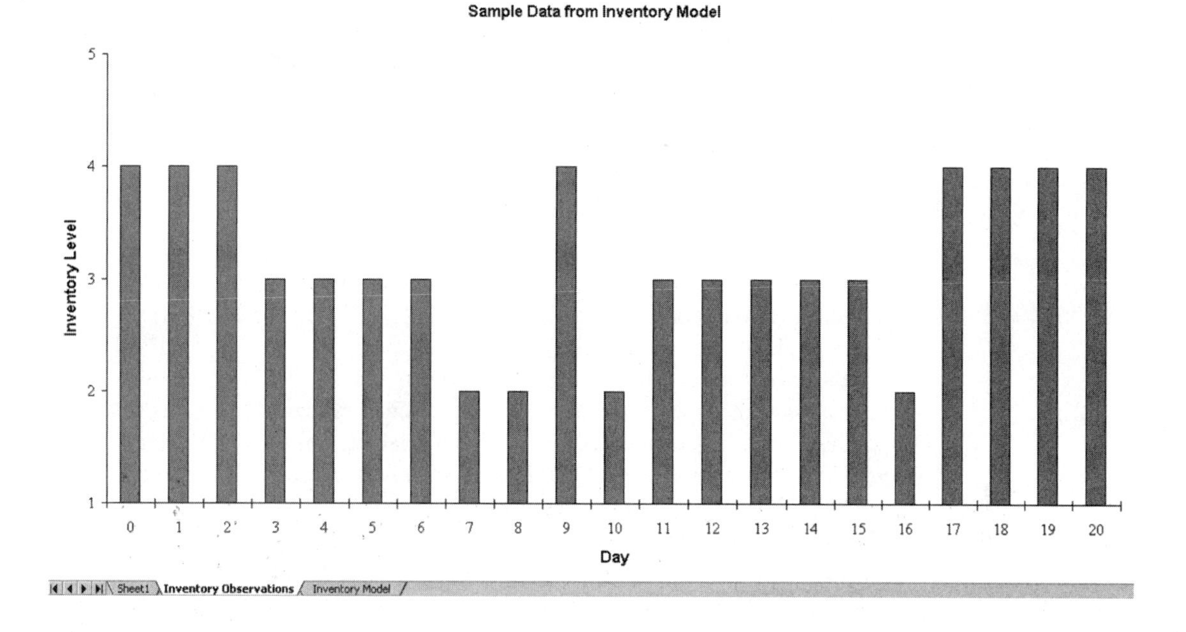

discrete as in the inventory example or continuous as in the waiting time example. If a process is a Markov process and, in addition, the state space is discrete, the process is called a *Markov chain.* Note that the waiting time process is not a Markov chain because the state space is continuous.

Suppose that $X_1, X_2, \ldots$ represents a Markov chain. The values in the state space may be finite or infinite in number. The state space is finite in our computer inventory model because it consists of only three values $\{2, 3, 4\}$. A second model will illustrate a Markov chain with an infinite state space.

4.5.2 A Markov Chain Queueing Model

A batting machine provides five minutes' operation each time it is activated. The rules allow each batter only one five-minute turn on the machine. If the machine is busy, then other batters will wait until it is available, and we assume that the waiting room is unlimited (as is the batters' patience!). Batter arrivals are independent and form a Poisson process: The number of batters to arrive in a five-minute period is a Poisson random variable with mean 5λ, where λ is the arrival rate per minute. Let X_n denote the number of batters waiting when the nth batter finishes batting. We will show that $\{X_1, X_2, \ldots\}$ is a Markov chain. Because there is no limit on the number of batters that can wait, the state space consists of the nonnegative integers $\{0, 1, 2, \ldots\}$ and is therefore infinite. Let A_n denote the number of batters to arrive during the $(n + 1)$st five-minute turn. We can see that if $X_n = 0$, then the machine will sit idle until the next batter arrives. Then on arrival the next batter begins her turn immediately, and X_{n+1} is just the number of arrivals during this turn, which is A_n. However, if $X_n > 0$, X_{n+1} is X_n, less the one batter that begins batting, plus the number of arrivals during her turn, A_n. We can express this succinctly as

$$X_{n+1} = \begin{cases} A_n & \text{if } X_n = 0 \\ X_n - 1 + A_n & \text{if } X_n > 0 \end{cases} \tag{4.6}$$

This relationship is completely determined if X_n is known, because the terms on the right-hand side do not depend explicitly on $X_{n-1}, X_{n-2}, \ldots$. Therefore, $\{X_1, X_2, \ldots\}$ is a Markov chain.

In the inventory and queueing models, the states are numerically valued, but states may have other values also. For example, let X_n represent the status of a loan in month *n;* it can have one of the following four values: up-to-date, in arrears, paid, or defaulted. We can use *U, A, P,* and *D,* respectively, to represent these values.

A Markov chain can be described by its *transition probability matrix.* This matrix specifies the probability the process moves to state *j* given it is currently in state *i,* for all possible values of *i* and *j;* that is, if P_{ij} is the entry in this matrix corresponding to row i, column j, then P_{ij} is the probability that $X_{n+1} = j$, given that $X_n = i$. A little calculation will show that the transition probability matrix for the inventory example is

$$\mathbf{P} = \begin{array}{c|ccc} & 2 & 3 & 4 \\ \hline 2 & .30 & 0 & .70 \\ 3 & .30 & .30 & .40 \\ 4 & .25 & .30 & .45 \end{array} \tag{4.7}$$

Note that the sum of the probabilities in each row is 1. This is true of all transition probability matrices.

For example, suppose $X_n = 3$. Then X_{n+1} will equal 2 only if exactly one computer is sold; the probability that this happens is .30. Therefore, the probability of a transition from state 3 to state 2 is .30, as shown in the (3,2) entry of this transition probability matrix. Although the transition probability matrix is usually necessary for mathematical analysis of Markov chains, it may or may not be used explicitly in the model when simulation is used for such analysis. Some Markov chains cannot easily be described using formulas such as (4.4) through (4.6) and therefore must be described through the transition probability matrix.

Now, consider using this transition probability matrix to simulate the Markov chain. Suppose that the process starts in state 4—that is, $X_0 = 4$. The next observation is sampled using the last row of **P**—the probabilities $\{.25, .30, .45\}$. Thus, we would sample a uniform random variate U and

$$X_1 = \begin{cases} 2 & \text{if } 0 < U \leq .25 \\ 3 & \text{if } .25 < U \leq .55 \\ 4 & \text{if } .55 < U \leq 1.00 \end{cases}$$

Suppose that $U = .40$, so $X_1 = 3$. Then X_2 is sampled using the second row of **P**—that is, we would sample a new uniform random variate U and

$$X_2 = \begin{cases} 2 & \text{if } 0 < \mathrm{U} \leq .30 \\ 3 & \text{if } .30 < \mathrm{U} \leq .60 \\ 4 & \text{if } .60 < \mathrm{U} \leq 1.00 \end{cases}$$

This sampling process continues until all desired observations are collected.

In general, the transition probability matrix is used to sample a Markov chain process as follows: Suppose that the states of the Markov chain are $\{1, 2, \ldots k\}$. X_0, the initial state, is given at the start of the simulation. Suppose $X_0 = i$. The probability distribution of X_1 is then given by the ith row of the transition probability matrix $P_{i,1}, P_{i,2}, \ldots, P_{i,k}$, where k is the number of states in the state space. Thus, if $X_0 = i$, we sample X_1 from the set of values $\{1, 2, \ldots k\}$ using the probability distribution $P_{i,1}, P_{i,2}, \ldots, P_{i,k}$. Next, we sample X_2 from the values $\{1, 2, \ldots k\}$ using the probability distribution $P_{X_1,1}, P_{X_1,2}, \ldots, P_{X_1,k}$. In general, we sample X_{n+1} from the values $\{1, 2, \ldots k\}$ using the probability distribution $P_{X_n,1}, \mathrm{P}_{X_n,2}, \ldots, \mathrm{P}_{X_n,k}$ in the X_nth row of the transition probability matrix. This sampling continues recursively until an appropriate end condition is met to stop the simulation.

Each state in a Markov chain can be classified as *transient* or *recurrent.* A transient state is a value such that the Markov chain $\{X_i, i = 1, 2, \ldots\}$ may take on that value a finite number of times but eventually will reach the point where that value can never be assumed again. In the example of a loan, the states U (up-to-date) and A (in arrears) are both transient because eventually the loan must be either paid or defaulted. If X_n represents the number of customers in a queue with infinite capacity and the arrival rate exceeds the service rate, then the value of X_n will tend to grow without bound. In this case, all states are transient.

Recurrent states are values that the process can assume infinitely often. All states (2, 3, and 4) in the inventory example are recurrent. The process will continue visiting each state indefinitely. In the second example where X_n represents the number of customers in a single-server queue, if the arrival rate is less than the service rate, then the system will be stationary and all states are recurrent. This means, for example, that given the system is empty when the nth customer leaves ($X_n = 0$), the system will again be empty sometime in the future—that is, state 0 will definitely be visited again.

4.5.3 A Markov Chain Reliability Model

If the number of states in the state space is infinite, it is possible for all states to be transient; however, if the total number of states is finite, then the Markov chain must have at least one recurrent state. For example, when X_n represents the number of customers in a nonstationary queue with infinite capacity, this number can grow without bound so that all states are transient. When the number of states is finite, on the other hand, there must be at least one state that the process continues to visit.

A third example will illustrate this point. A piece of production machinery is purchased with a one-month warranty. At the end of each month, the machine can be in one of three states of operation: fully functional (FF), partially functional (PF), or nonfunctional or failed (NF). A fully functional machine can perform all tasks for which it was designed; a partially functional machine can perform only some of the tasks but is useful for some operations; a failed machine cannot be used and must be repaired or discarded. If the machine fails during the warranty period, it is replaced by a new machine and has a new one-month warranty. If the machine becomes partially functional during the warranty period, it may be replaced by a new machine with a new warranty or it may continue to be used and the warranty expires. If the machine remains fully functional for the first month, then it is kept. The warranty expires at the end of the first month if the machine is not replaced and cannot be renewed. Because the machine may fail and be replaced during a given month, it is possible to start the month under warranty and also start the next month under warranty, because the machine was replaced during the month. We will consider six states for the machine that include all combinations of being under warranty (W) and not under warranty ($\overline{\text{W}}$), and the three statuses FF, PF, and NF. The following matrix gives the transition probabilities for this Markov chain:

$\mathbf{P} =$

	$\overline{\text{W}}$FF	$\overline{\text{W}}$PF	$\overline{\text{W}}$NF	WFF	WPF	WNF
$\overline{\text{W}}$FF	.80	.15	.05	0	0	0
$\overline{\text{W}}$PF	.50	.40	.10	0	0	0
$\overline{\text{W}}$NF	1.0	0	0	0	0	0
WFF	.60	.16	.04	.15	.04	.01
WPF	.10	.50	.10	.22	.06	.02
WNF	0	0	0	.70	.25	.05

Thus, the probability the machine is nonfunctional at the end of a month, given it was fully functional and not in warranty at the start of the month—that is, the transition from state $\overline{\text{W}}$FF to state $\overline{\text{W}}$NF—is .05, and the probability it remains fully functional is .80. One can see that the three states WFF, WPF, and WNF are transient because once the process enters one of the three states $\overline{\text{W}}$FF, $\overline{\text{W}}$PF, or $\overline{\text{W}}$NF, transitions back to any of the three states WFF, WPF, and WNF are impossible because the transition probabilities are all zero. Intuitively, this is so because once the warranty expires, it is gone and cannot be renewed. Because the states WFF, WPF, and WNF denote a machine in warranty, once the process leaves these states, it cannot return.

4.5.4 Performance Measures for Markov Chains

The important distinction between transient and recurrent states concerns the parameters or performance measures that we want to estimate when simulating the Markov chain. Consider a Markov chain with a finite number of states. If the process starts in a transient state, it may make a transition to a recurrent state in the next period, or it may make a transition to another transient state. However, at some (random) period, it will make a transition to a recurrent state. After this point, the process will never again assume a transient state. Thus, the history of a Markov chain with both transient and recurrent states can be divided into two phases: the initial periods that are spent in transient states, and the remainder of the periods, which are spent in recurrent states. For the transient phase, which has a finite (but random) duration, the quantities of interest are:

1. The expected number of periods before the process enters a recurrent state, given $X_0 = k$. For example, if the production machine starts out fully functional and under warranty (X_0 = WFF), what is the average number of months it will remain under warranty before the warranty expires?
2. The probability of entering state j first when the process exits the transient states, given it starts in transient state k. For example, if the machine starts out fully functional and under warranty (X_0 = WFF), what is the probability it will be nonfunctional at the end of the first month during which the warranty expires ($X_N = \overline{\text{W}}$NF), where N is the month when the process makes a transition from a transient state to a recurrent state?

We can associate a cost, C_j, per period with each state j (transient and recurrent). Then a relevant performance measure for a process that begins in a transient state is:

3. The mean cost incurred before leaving the transient states, given $X_0 = k$. For example, if the machine starts out fully functional and under warranty, what is the expected cost incurred before the warranty expires?

Each performance measure relating to a transient Markov chain depends on the initial state of the process and will have different values for different initial states.

For the recurrent phase, the theory of Markov chains provides that every recurrent Markov chain with a finite number of states has a stationary distribution that is independent of the initial value of the process. The interpretation of this statement is that if we start the process and let it run for a very large number of periods, then select an arbitrary period and observe the state, or value, of the process, that observation is a random variable whose distribution does not depend on the initial state of the process when it started. Note that this is true in the asymptotic sense as the number of periods becomes infinitely large; however, some studies demonstrate that the effect of the initial state is virtually removed after only a few, perhaps three or four, periods. Nevertheless, like any stationary process, given the initial state of the process, there is a segment, usually called a *transient period,* during which the initial value does assert an important influence on the probability distribution of the process, followed by a stationary segment during which the distribution of the process is not significantly affected by the initial state. The transient segment while the process is in the recurrent states is distinct from the periods when the process is in the transient states.

The performance measures usually desired for a stationary Markov chain include:

4. The stationary probability that the process is in a particular state. For example, if the production machine runs for many months, what is the probability it will be fully functional at the end of an arbitrary month?
5. Given the process is in state j, the mean number of periods before returning to state j. For example, given that the machine is nonfunctional at the end of a particular month, what is the expected number of months before the machine will again be observed to be nonfunctional? This would be called the *mean time between failures* (mtbf).
6. The mean cost per period. For example, what is the mean cost per month to operate the production machine?

The first and second performance measures are actually closely related. The theory of Markov chains provides that if p_j is the probability that the process is in state j, then $1/p_j$ is the mean number of periods between entrance into state j, which is called the *mean recurrence time* (see Ross, 1993). It is easy to see how this is true. Suppose the process runs for n periods and during that time spends N_j periods in state j. Then, p_j is the proportion of periods spent in state j, so $p_j \cong N_j/n$. But the mean recurrence time is approximately n/N_j because this ratio is the total number of periods divided by the total number of entrances into state j, or the average number of periods between entrances into state j.

4.6 THE REGENERATIVE METHOD FOR ESTIMATING THE MEAN

In this section we will examine another method for estimating the mean of a stationary sequence of observations that applies to Markov processes and certain other output observation processes produced by some simulations. This method, which is called the *regenerative method,* was first introduced by Fishman (1973, 1974) and Crane and Iglehart (1975). It does not always produce better confidence intervals than the batch means method, but it exploits the structure of the process when it can be applied and solves the problem of determining when the initial transient portion ends. First, we must provide some definitions.

4.6.1 Regenerative Processes

Suppose that $X_1, X_2, \ldots$ denotes a sequence of observations produced by a dynamic simulation. This process is said to be *regenerative* if there is a sequence of indices T_1, $T_2, \ldots$, which are called *regeneration points,* such that the following conditions hold:

1. The portion of the process after each regeneration point $X_{T_i}, X_{T_i+1}, X_{T_i+2}, \ldots$ is probabilistically independent of the portion before that regeneration point: $\ldots, X_{T_i-2}, X_{T_i-1}$.
2. The laws governing the evolution of the portion of the process after each regeneration point are identical to the laws governing the portion after any other regeneration point.

The effect of the first condition is to allow the process $X_1, X_2, \ldots$ to be divided into a sequence of independent cycles of observations. The first cycle consists of X_{T_1}, $X_{T_1+1}, \ldots, X_{T_2-1}$, the second cycle consists of $X_{T_2}, X_{T_2+1}, \ldots, X_{T_3-1}$, and so forth. If the initial observation is not recorded at the start of a regeneration cycle, then the start of the observation sequence will have a portion of an incomplete cycle. The second condition guarantees that all cycles will have identical probability distributions. Thus, although the cycles will not contain identical observations, they will be probabilistic replicas of one another.

These two conditions mean that at the points $T_1, T_2, \ldots$, something happens in the system that causes the past history of the system behavior to no longer have any influence on the behavior of the system. For example, consider the first model presented in this chapter—the single-server queue. Each time a customer arrives to find the queue empty and the server idle, a regeneration occurs. At this point, an interarrival time has just ended and a service time is beginning. Because the interarrival times and service times are mutually independent, the times until all future events (arrivals and service completions) are independent of the times since all previous events. Moreover, the server has no way of knowing whether this is the first time or the thousandth time that a customer arrives to find it idle. The system responds to this circumstance exactly as it would any other time the customer finds the server idle. Thus, T_i is the index of the ith customer to arrive and find the system empty and the server idle. Figure 4.5 presents a sequence of waiting time observations from this model. Because the initial waiting time was set equal to zero, the simulation began with the system in the empty-and-idle state, and the first observation began a cycle. Each waiting time having a value 0 denotes the start of a new cycle; therefore, in this particular replication, cycles begin with customers 0, 1, 6, 8, 14, 15, ... Note that cycles vary in length.

As a second example, all Markov chains are regenerative. Select any specific state (value) for the process. Each time the process enters that state, the future evolution becomes independent of the past. Consider the computer inventory model in section 4.5, for example. Each time the inventory has exactly four computers in it, a regeneration occurs, and the process evolves from that point in exactly the same manner as it does any other time the inventory has exactly four computers.

FIGURE 4.5 Waiting times from a regenerative M/M/1 queue

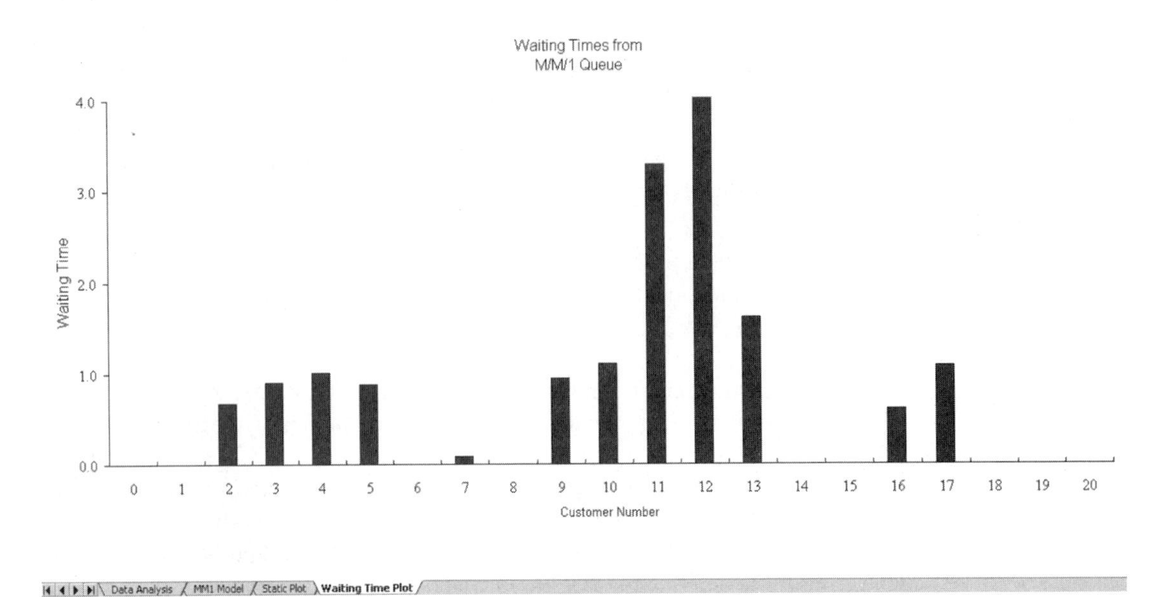

To determine if a process is regenerative when a particular condition exists, consider the following two questions:

1. When this condition is present, can the future of the process be simulated without knowledge of the past values?
2. When this condition is present, are all random variates generated to compute the future values of the process independent of all random variates used to determine the past values?

If both of these questions can be answered in the affirmative, then the condition denotes a regeneration. For Markov chains, one can see that the Markov property guarantees that the answers to these questions will be "yes" when the condition is such that the process enters a specific state.

4.6.2 The Regenerative Method for Estimating the Mean

All of the properties of a regenerative process are contained within each cycle. The regenerative method for estimating the mean treats the cycle as the basic unit of data, and it uses a collection of cycles to compute a confidence interval for the mean. Assume the sequence of observations $X_1, X_2, \ldots$ is stationary and regenerative, and let μ represent the mean of $X_1, X_2, \ldots$. If the process is regenerative and the mean length of a cycle is finite, then $X_1, X_2, \ldots$ can be shown to be stationary also. For an arbitrary cycle, let Y denote the sum of all observations and N the number of observations in the cycle. Then μ is related to Y and N by

$$\mu = \frac{E(Y)}{E(N)} \tag{4.8}$$

Thus, the mean value of the process is the ratio of the mean sum of the observations within a cycle to the mean number of observations in the cycle, and we are faced with the problem of computing an estimate of this ratio. The data from a regenerative process can be summarized as n pairs of sums: $(Y_1, N_1), (Y_2, N_2), \ldots, (Y_n, N_n)$, with each pair consisting of the sum of observations within a cycle and the number of observations within the cycle. The regenerative estimator for μ is the ratio

$$\bar{\mu} = \frac{\bar{Y}}{\bar{N}} \tag{4.9}$$

where

$$\bar{Y} = \frac{1}{n}\sum_{j=1}^{n} Y_j$$

and

$$\bar{N} = \frac{1}{n}\sum_{j=1}^{n} N_j$$

Notice that $\bar{\mu}$ is just the sample mean of all observations collected, because eliminating the $1/n$ term in the numerator and denominator leaves the sum of all observations in the numerator and the total number of observations in the denominator. However, keep in mind that the number of observations is a *random variable* and not fixed as when the estimator is the ordinary sample mean. Thus, both the numerator and denominator in the estimator $\bar{\mu}$ are random variables, and $\bar{\mu}$ is said to be a *ratio estimator*.

If the number of cycles, n, is large enough, then

$$Z = \frac{\bar{\mu} - \mu}{\sigma/(E(N)\sqrt{n})} \tag{4.10}$$

has approximately a standard normal distribution. In this expression, $\sigma^2 = E(Y - \mu N)^2$. Let s^2 denote an estimate of σ^2. Then, the approximate $100(1 - \alpha)$-percent confidence interval for μ is

$$\bar{\mu} \pm Z_{\alpha/2} \frac{s}{\bar{N}\sqrt{n}} \tag{4.11}$$

where $Z_{\alpha/2}$ is the $100(1 - \alpha)$ percentage point of the standard normal distribution. To implement this, we only need to specify how to compute s. The following description explains the entire process, including how to compute s:

1. If possible, start the simulation in a regenerative state; otherwise, run the simulation until a regeneration occurs and discard any data produced.
2. Run the simulation for each cycle k; accumulate Y_k and N_k during the cycle.
3. After each cycle, increment the following sums:

$$\sum_{j=1}^{k} Y_j, \quad \sum_{j=1}^{k} N_j, \quad \sum_{j=1}^{k} Y_j^2, \quad \sum_{j=1}^{k} N_j^2, \quad \text{and} \quad \sum_{j=1}^{k} Y_j N_j$$

4. Repeat the previous two steps until n cycles of data have been collected.
5. Compute

$$s_{11} = \frac{1}{n-1}\left[\sum_{j=1}^{n} Y_j^2 - \frac{1}{n}\left(\sum_{j=1}^{n} Y_j\right)^2\right] \tag{4.12}$$

$$s_{22} = \frac{1}{n-1}\left[\sum_{j=1}^{n} N_j^2 - \frac{1}{n}\left(\sum_{j=1}^{n} N_j\right)^2\right]$$

$$s_{12} = \frac{1}{n-1}\left[\sum_{j=1}^{n} Y_j N_j - \frac{1}{n}\left(\sum_{j=1}^{n} Y_j\right)\left[\sum_{j=1}^{n} N_j\right]\right]$$

6. Compute an estimate of s using

$$s^2 = s_{11} - 2\,\bar{\mu}\,s_{12} + \bar{\mu}^2\,s_{22} \tag{4.13}$$

7. Compute the confidence interval using equation (3.2).

Table 4.2 gives a portion of data from a simulation of the computer inventory model. This model assumed \$1.00 holding cost per computer per day and \$2.00 ordering cost per order. We will use the full level (4) as the regenerative state, although any of the other two states could also be used. In Table 4.2, an asterisk denotes the start of each cycle. The actual run consisted of 500 cycles, and the summary data for the entire run is given at the bottom of Table 4.2. From these data, we can compute that $\overline{Y} = 2801/500 = \$5.60$ and $\overline{N} = 1020/500 = 2.04$, so the point estimate for the mean cost per period is $\bar{\mu} = 5.60/2.04 = \$2.75$. Applying (3.2), we compute $s_{11} = 9.671$, $s_{12} = 3.621$, and $s_{22} = 1.501$. So, from (3.2), $s^2 = 1.221$ and $s = 1.105$. The confidence interval then is \$2.75 ± .05, or from \$2.70 to \$2.79, rounded to the nearest two decimal points. Mathematical analysis of this Markov chain (see Ross, 1993) allows us to compute that $\mu = \$2.74$, so the point estimate just computed is quite accurate and the confidence interval does include the true mean.

Of course, in those real-world models for which simulation will be used to estimate parameters of interest, this sort of validation cannot be done because the true

TABLE 4.2 Sample path for Markov chain

Period (n)	State (X_n)	Inventory Level	Holding Cost	Order Cost	Total Cost
0*	3	4	4	0	4
1*	3	4	4	0	4
2*	3	4	4	0	4
3	2	3	3	0	3
4	2	3	3	0	3
5	2	3	3	0	3
6	2	3	3	0	3
7	1	2	2	0	2
8	1	2	2	2	4
9*	3	4	4	0	4
10	1	2	2	2	4
11	2	3	3	0	3
12	2	3	3	0	3
13	2	3	3	0	3
14	2	3	3	0	3
15	2	3	3	0	3
16	1	2	2	2	4
17*	3	4	4	0	4
18*	3	4	4	0	4
19*	3	4	4	0	4
20*	3	4	4	0	4

$$n = 500 \qquad \sum_{j=1}^{k} Y_j = 2801 \qquad \sum_{j=1}^{k} N_j = 1020 \qquad \sum_{j=1}^{k} Y_j^2 = 20{,}517$$

$$\sum_{j=1}^{k} N_j^2 = 2832 \qquad \text{and} \qquad \sum_{j=1}^{k} Y_j N_j = 7521$$

value of the parameter is unknown. In these two cases, we have computed the true mean and shown that the estimates computed were close to it in order to demonstrate the validity of the simulation and the estimation techniques, as well as develop confidence in the techniques. These estimates could have been close to the true mean just by chance. A better demonstration of the validity of the method would involve repeating the entire experiment of running the simulation and computing a 95 percent confidence interval 100 times, and then showing that the proportion of confidence intervals that include the true mean is reasonably close to 95 percent. However, this topic is beyond the scope of the current discussion.

4.7 AN ADVANCED QUEUEING MODEL

The relationship $W_{n+1} = \max(0, W_n + X_n - Y_n)$ applies under more general circumstances than those presented in section 4.2. This expression for waiting times will remain valid, and the Markov property will continue to hold if the service times, X_n, and the interarrival times, Y_n, depend on the most recent waiting time, W_n. In this section, we will develop and analyze one extension to this model in which the service times depend on the previous waiting time. Other extensions will be suggested in the problems at the end of the chapter.

Suppose the server can work at either of two speeds: slow and fast. If a waiting time, W_n, is greater than T_0, then the next service time will be chosen from the distribution for fast services; if the waiting time is less than or equal to T_0, then the next service time will be chosen from the distribution for slow services. Therefore, the system compensates for congestion by speeding up and slowing down, depending on the length of wait the customers have. Service times for the slow speed have a symmetric triangular distribution between 0 and 1.4 minutes with mean and most likely value .7 minutes. For the fast speed, service times have a symmetric triangular distribution between 0 and .70 minutes, with mean and most likely value .35 minutes. Thus, the fast speed is twice as fast as the slow speed. Interarrival times are uniformly distributed between .2 and 1.8 minutes, so the mean interarrival time is 1.0 minute. Because the mean service time is either .7 or .35 minutes, depending on which speed the server is working, it is clear that the system is stable, and the waiting times have a stationary distribution. The waiting time process is also regenerative, with regenerations occurring each time a customer arrives to find the system empty and idle (and therefore has a waiting time of 0). This model differs from the models presented in sections 4.2 and 4.6 in three important respects:

1. Service times have a triangular distribution rather than an exponential distribution.
2. The interarrival times have a uniform distribution rather than an exponential distribution.
3. The server switches between the slow service time distribution and the fast distribution, depending on the value of the previous waiting time.

Since none of the distributions is exponential, it is not possible to analytically derive the mean waiting time for this model. Although this is a simple model, simulation is the only tool available to determine the mean waiting time.

Our objective in analyzing this model is to determine the most appropriate value for the threshold. We are willing to let customers wait .2 minutes on average, but longer mean waiting time is undesirable. Therefore, we will run the simulation using values of the threshold between 0 and 2.2 minutes. For each run, we will produce 4000 cycles of data and compute a confidence interval for the mean waiting time. Using these results, we will then seek to find the largest value of the threshold that will keep the confidence interval below .2 minutes. The resulting confidence inter-

TABLE 4.3 Confidence intervals (95%) for mean waiting time as a function of service time threshold

Threshold	Lower Limit	Point Estimate	Upper Limit
0.00	.0141	.0162	.0183
.10	.0979	.1041	.1103
.20	.1118	.1183	.1249
.30	.1170	.1235	.1301
.40	.1305	.1378	.1451
.50	.1363	.1440	.1517
.60	.1489	.1581	.1673
.70	.1514	.1613	.1711
.80	.1619	.1722	.1826
.90	.1701	.1816	.1931
1.00	.1787	.1912	.2037
1.10	.1909	.2048	.2188
1.20	.1774	.1909	.2044
1.30	.1925	.2062	.2200
1.40	.1994	.2156	.2319
1.50	.2055	.2228	.2401
1.60	.2104	.2299	.2494
1.70	.2175	.2362	.2550
1.80	.2270	.2454	.2639
1.90	.2273	.2480	.2686
2.00	.2222	.2427	.2632
2.10	.2120	.2305	.2491
2.20	.2269	.2483	.2697

vals are given in Table 4.3 and plotted in Figure 4.6. This figure clearly shows that mean waiting time is an increasing function of the threshold; if the threshold is approximately 1.0, then the mean waiting time is most likely to be less than .2 minutes.

4.8 A MARKETING MODEL

Having examined several relatively simple models, we now want to conclude this chapter with a somewhat more complex model to illustrate the process of developing and using more realistic models for decision making. Many discrete-time dynamic models that are developed in the business sector fall under the general label of *corporate planning models.* This is a broad categorization, but it mainly denotes models that attempt to relate sales, employment, environmental variables, and other

FIGURE 4.6 Plot of mean waiting time estimate versus threshold

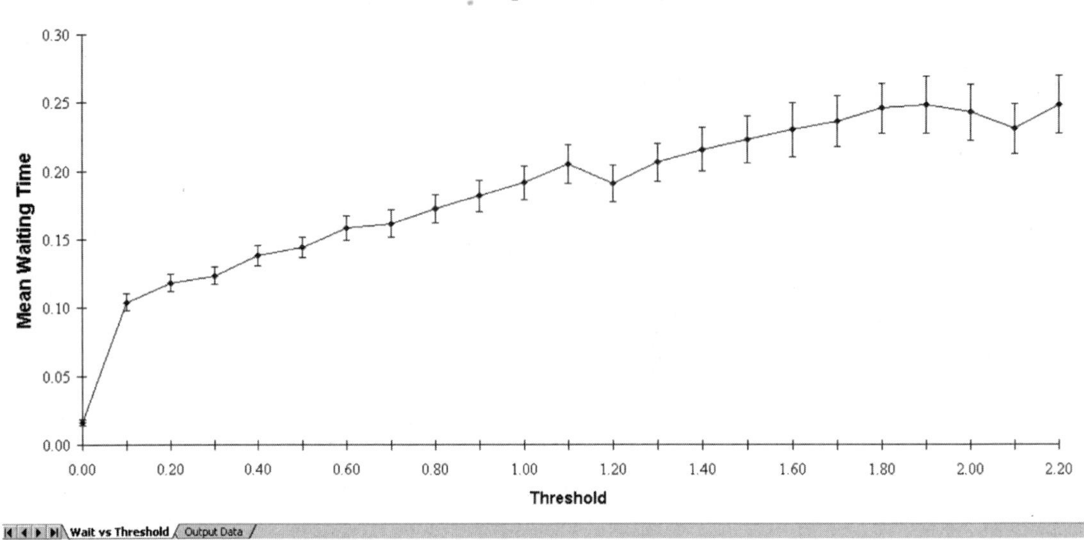

factors to profit or net income for a business. See Naylor (1979) for an early discussion of corporate planning models. These models are frequently highly complex, and special languages have even been created for such models (Gray, 1987; Simplan Systems, 1982). The model that we will consider in this section is not so large or complex, but it has many of the same characteristics as more involved models.

The model that we will consider attempts to relate market share for a product to the level of advertising expenditures. Market share is defined to be the ratio of sales of our product (in dollars) to the total sales of all products in the particular market. We assume that the *potential* market is a constant size, M. However, the total market (defined to be the *actual* total sales) varies over time. Our time unit will be a week, and the model will predict market share for each week in the planning period. Let T_n denote the total sales in week n, and X_n denote sales of our product in week n. Then our market share in week n is $S_n = X_n/T_n$. Both X_n and T_n can be expected to vary from week to week in response to our advertising expenditures and those of our competitors. In addition, we can expect the variation to be random from factors such as economic activity, weather, and political events, among others that cannot be predicted.

Define the following variables: Let A denote our advertising expenditures per week and B denote those of our competitors. Let α, β, λ, and φ denote parameters. Then the following two expressions specify a deterministic model for market share:

$$T_n = \beta T_{n-1} + [\mathrm{M} - \beta T_{n-1}]\, e^{-\left[\frac{\lambda}{A+B}\right]} \quad \textbf{(4.14)}$$

$$X_n = \alpha X_{n-1} + [T_{n-1} - \alpha X_{n-1}]\, e^{-\left[\varphi \frac{B}{A}\right]} \quad \textbf{(4.15)}$$

The first expression divides the market into two portions: One part is unaffected by advertising, βT_{n-1}, and one part does respond to advertising, $\mathrm{M} - \beta T_{n-1}$. The parameter β measures market volatility for the industry as a whole. If β is 1.0, then the market is constant from week to week, regardless of the amount of advertising. If β is close to 0, then the market will be small unless the amount of advertising is large. The term

$$e^{-\left[\frac{\lambda}{A+B}\right]}$$

specifies the effect of advertising on the market as a whole. If $A + B$ is large, this term will be close to 1 and T_n will be close to T_{n-1}. If $A + B$ is small, this term will be small and T_n will be close to βT_{n-1}. The parameter λ establishes how effective total advertising is on the entire market. A smaller value of λ indicates more effective advertising.

The second equation determines our product's portion of the total market. The first term, αX_{n-1}, is that part of our product's market that is insensitive to advertising; the second term, $T_{n-1} - \alpha X_{n-1}$, is the part that is responsive to advertising. If α is small, then our market is volatile and we must advertise heavily to retain it. If α is large, then our market is stable and advertising is of less use. The effect of advertising is contained in the exponential term

$$e^{-\left[\varphi \frac{B}{A}\right]}$$

If A is large relative to B, then B/A is small and this term is close to 1; if A is small relative to B, then B/A is large and this term is close to zero. The parameter φ denotes the effectiveness of our competitors' advertising relative to ours. Larger values of φ mean our competitors' advertising is more effective.

Equations (4.14) and (4.15) do not contain any random variables, and therefore do not include the random variation that we referred to earlier. In reality, there are several possible sources of randomness; however, we will represent the randomness by replacing λ and φ with uniformly distributed random variables, L_n and K_n, over the ranges $\lambda \pm \omega_\lambda$ and $\varphi \pm \omega_\varphi$, respectively. Thus, the model represents the actual effectiveness of advertising as a random variable. We will assume that all L_n and K_n are mutually independent. The model can now be described as:

$$T_n = \beta T_{n-1} + [\mathrm{M} - \beta T_{n-1}]\, e^{-\left[\frac{L_n}{A+B}\right]} \quad \textbf{(4.16)}$$

$$X_n = \alpha X_{n-1} + [T_{n-1} - \alpha X_{n-1}]\, e^{-\left[K_n \frac{B}{A}\right]} \quad \textbf{(4.17)}$$

where

$$L_n \sim \mathcal{U}(\lambda - \omega_\lambda, \lambda + \omega_\lambda) \quad \textbf{(4.18)}$$

TABLE 4.4 Parameter values for two scenarios

Parameter	Scenario 1	Scenario 2
α	.750	.750
β	.950	.950
φ	.552	.552
λ	1.0 x 107	1.0 x 107
ω_λ	2.0 x 106	2.0 x 106
ω_φ	.092	.138
A	\$634,176	\$761,011
B	\$3,536,148	\$3,536,148
T_0	1.0	1.0
X_0	.05	.05

and

$$K_n \sim \mathcal{U}(\varphi - \omega_\varphi, \varphi + \omega_\varphi) \qquad \textbf{(4.19)}$$

The notation $\sim \mathcal{U}(a, b)$ is read "has a uniform distribution between a and b."

We will analyze two scenarios using this model. The first represents the current environment. For this scenario, we will assume the parameter values given in the middle column of Table 4.4.

If we put the values for the first eight parameters (α, β, φ, λ, ω_λ, ω_φ, A, and B) into equations (4.14) and (4.15), and assume sales are constant each period, then the market share will be stable at 25 percent. These parameter values represent a relatively stable total market ($\beta = .95$) but a more volatile market for our product ($\alpha = .75$). The value of φ indicates that our advertising effectiveness is somewhat greater than our competitors'. We can compute that if $B/A = 1.256$, then

$$e^{-\left[\varphi \frac{B}{A}\right]} = .5$$

Thus our advertising can be less than half of the total, but we will get half of the available market. The value of λ is large because it is approximately the same order of magnitude as the total advertising expenditure ($A + B$). If $\lambda = .693(A + B)$, then half of the available market, M, will respond to the advertising. The values of ω_λ and ω_φ in scenario 1 are 20 percent and 16.67 percent of their respective parameters. We assume that T_n and X_n are measured in millions of dollars, so their initial values are \$1.0 million and \$50,000, respectively.

In the second scenario, we plan to change our advertising media mix. This will increase our advertising expenditures by 20 percent, and it will increase the variation in advertising effectiveness, ω_φ, by 50 percent to .138. We want to determine how much this new level of advertising increases our market share, if any. The parameters for this scenario are contained in the last column of Table 4.4.

The model is expressed in equations (4.16) through (4.19). To run the simulation for N weeks, we must implement the following pseudocode:

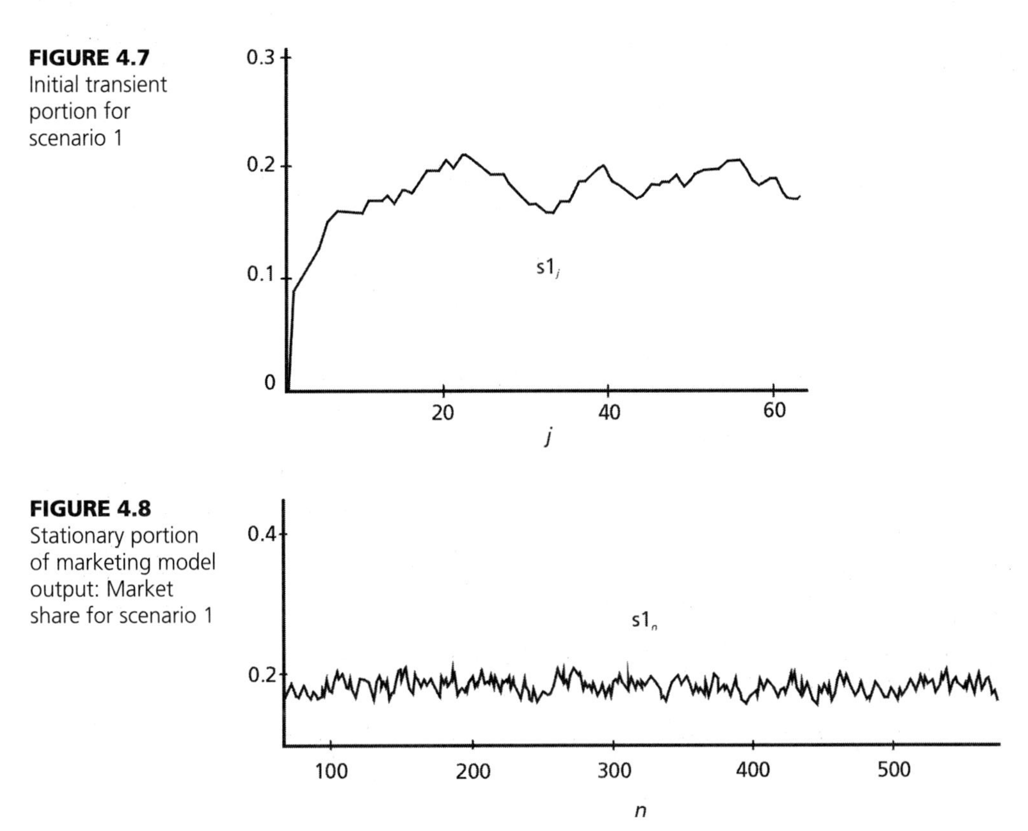

FIGURE 4.7 Initial transient portion for scenario 1

FIGURE 4.8 Stationary portion of marketing model output: Market share for scenario 1

1. Initialize parameters α, β, λ, φ, ω_λ, ω_φ, A, and B.
2. Initialize T_0 to 1.0 and X_0 to .05.
3. For $n = 1$ to N, apply (4.16) and (4.19) to compute T_n and X_n from T_{n-1} and X_{n-1}.

This algorithm can be implemented using a spreadsheet, but for longer runs it is most easily implemented using a general-purpose programming language or mathematical software such as Mathematica, MATLAB, or MathCAD. Regardless of the software used, a trusted random number generator should be employed to generate the uniform random variates.

The model was run for $N = 576$ weeks. Figure 4.7 shows the predicted market share for the first 64 weeks. The initial transient portion, where the market share increases from its initial value of .05 to its steady state value close to .19, is clearly visible in this plot. The first 64 observations were omitted from the data analysis in order to eliminate data from the initial transient portion. The remaining 512 observations, which are plotted in Figure 4.8, appear to be stationary. These observations were grouped into eight batches of 64 observations per batch and the batch means

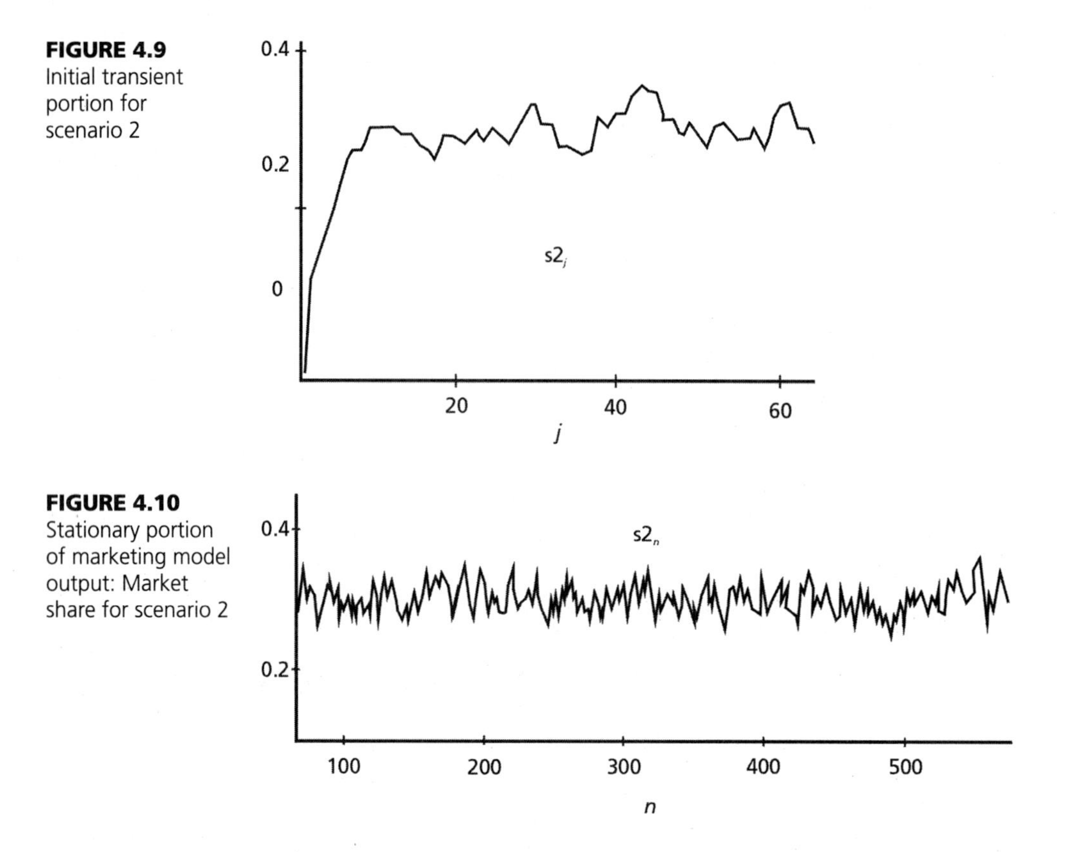

FIGURE 4.9 Initial transient portion for scenario 2

FIGURE 4.10 Stationary portion of marketing model output: Market share for scenario 2

method was used to compute a 95 percent confidence interval for the mean market share. This confidence interval for the mean market share is .1822 ± .0038. From Table 4.4, we see that our advertising budget, *A*, is 15.2 percent of the total, $A + B$, but our market share is approximately 18.2 percent. Therefore, our advertising effectiveness is somewhat better than our competitors' because our market share is larger than our share of the total advertising budget.

The second run also consisted of 576 weeks. The initial transient portion is also evident in the first 64 observations, which are shown in Figure 4.9. The remaining 512 observations, which are shown in Figure 4.10, appear stationary and therefore were used to compute a 95 percent confidence interval for the mean using eight batches of 64 observations each. The confidence interval is .303 ± .008. Under this scenario, our advertising expenditures, *A*, increased to 17.7 percent of the total, but our market share increased to 30.3 percent. This indicates that the new advertising mix is much more effective than the previous mix, and therefore the decision to adopt the new mix appears to be a good one.

4.9 SUMMARY

This chapter has introduced the concept of a dynamic simulation. The key idea in dynamic simulations is that the models involve time and must be observed over time to obtain the information we seek. The fixed time advance mechanism was introduced as a method to represent the movement of the model through time. This mechanism goes through a series of iterations, moving time a fixed amount and updating the state of the system in each iteration. Several example models were introduced. Simple queueing models based on the relationship $W_{n+1} = \max(0, W_n + X_n - Y_n)$ were used to show how mean waiting time and other parameters could be estimated. Markov chain models were used to illustrate an important group of models (and their extensions to non-Markov models) that can be analyzed using the methodology in this chapter. Examples from inventory and reliability provided illustrations. Finally, a model for market share was developed and analyzed to show the process of implementing a more complex model and using it to evaluate two decision alternatives. All of these models were sufficiently simple that they could be programmed from scratch using Microsoft Excel or a general-purpose programming language.

We examined the data produced by dynamic models and saw that observations from all dynamic models have two characteristics that make estimation of the mean and other parameters difficult: a transient portion at the start of the simulation run and dependence (or correlation) among the observations, even after the influence of starting conditions has dissipated. Some models may have a transient portion that lasts indefinitely. Such models are never stationary, and techniques to estimate stationary parameters do not apply to them because they do not possess stationary parameters. An example of such a model is a queueing system in which the arrival rate is greater than the service rate. In such a system, the number of customers in the queue will always increase and never stabilize; therefore, this system does not have a stationary mean waiting time. It may nevertheless be useful to estimate certain parameters associated with nonstationary systems, such as the mean number of customers waiting after the system has been operating for five hours or the length of time until the queue length exceeds 10 customers. These parameters must be estimated using independent replications and the data analysis methods presented in Chapter 2 for independent, identically distributed observations.

Two methods were presented for estimating the mean of a stationary sequence of observations such as those obtained after deleting transient observations from the output of a stable dynamic system. The first method, which can be applied to any stationary system, was the batch means method. This method partitions the observations into a number of equal-sized batches of adjacent observations, computes the mean for each batch, and then treats these batch means as if they were a sequence of independent, identically distributed observations to compute a confidence interval for the mean. In establishing the size of batches and number of batches, empirical evidence indicates that one should use batches that are as large as possible, as long as the number of batches is greater than 10. The second method, the regenerative method, applies only to output data that are regenerative. A regenerative process can be divided into a sequence of independent and identically distributed cycles of data. These cycles are of random length, and each cycle becomes the basic unit of data for the simulation. This method uses the ratio of the average of the sum of observations in a cycle to the average number of observations in a cycle as the estimator of the mean. The confidence interval can then be computed using summary statistics for the sums and numbers of observations in the cycles. The regenerative method applies to all Markov chain models and many queueing models.

References

Alexopoulos, C., and A. F. Seila. 1998. Output data analysis. In *Handbook of simulation,* ed. J. Banks. New York: Wiley.

Conway, R. W. 1963. Some tactical problems in digital simulation. *Management Science* 10(1):47–61.

Cox, D. R., and Smith, W. L. 1961. *Queues.* London: Methuen & Co.

Crane, M. A., and Iglehart, D. L. 1975. Simulating stable stochastic systems, III: Regenerative processes and discrete-event simulation. *Operations Research* 23(1):33–45.

Fishman, G. S. 1979. *Principles of discrete event simulation.* New York: Wiley & Sons.

———. 1974. Estimation in multiserver queueing simulations. *Operations Research* 22:72–78.

———. 1973. Statistical analysis for queueing simulations. *Management Science* 20:363–369.

Gray, P. 1987. *Guide to IFPS.* 2d ed. New York: McGraw-Hill.

Hillier, F. S., and Lieberman, G. J. 1995. *Introduction to operations research.* 6th ed. New York: McGraw-Hill.

Kemeny, J. G., and Snell, J. L. 1960. *Finite Markov chains.* Princeton, New Jersey: Van Nostrand Reinhold.

Law, A. M. 1983. Statistical analysis of simulation output. *Operations Research* 31(6):983–1029.

Law, A. M., and Kelton, W. D. 1984. Confidence intervals for steady-state simulation, I: A survey of fixed sample size procedures. *Operations Research* 32:1221–1239.

Lindley, D. V. 1952. The theory of queues with a single server. In *Proceedings of the Cambridge Philosophical Society* 48:277–289.

Naylor, T. H. 1979. *Corporate planning models.* Reading, Massachusetts: Addison-Wesley.

Ross, S. M. 1993. *Introduction to probability models.* 5th ed. Boston: Academic Press.

Simplan Systems, Inc. 1982. *Corporate planning and modeling with SIMPLAN.* 2d ed. Reading, Massachusetts: Addison-Wesley.

Taylor, H. M., and Karlin, S. 1998. *An introduction to stochastic modeling.* 3d ed. San Diego: Academic Press.

Welch, P. 1981. *On the problem of the initial transient in steady-state simulation.* Yorktown Heights, New York: IBM Watson Research Center.

Problems

4.1 The M/D/1 queue has exponential interarrival times and constant service times. Using the model $W_{n+1} = \max(0, W_n + X_n - Y_n)$, simulate the M/D/1 queue with arrival rate 1.0 and service times equal to .7. Use eight batches of 64 observations per batch and omit the first 64 observations in the run as the initial transient. Compare the mean waiting time for this system with that of the M/M/1 queue in section 4.6. If the confidence interval is too large to draw a conclusion, then increase your sample size by a factor of two or four until your estimate is precise enough. What conclusion can you make concerning the mean waiting time in an M/D/1 queue with traffic intensity .7 versus an M/M/1 queue with the same traffic intensity?

4.2 Simulate an M/G/1 queueing system with service times that are uniformly distributed between 0.0 and 2.0 minutes and arrival rate .6. Using 16 batches of 256 observations per batch, compute a 95 percent confidence interval for the mean waiting time. The actual mean waiting time for this system is .875. Does your confidence interval include this value?

4.3 Using the model $W_{n+1} = \max(0, W_n + X_n - Y_n)$, where X_n is the nth service time and Y_n is the time between the nth and $(n + 1)$th arrivals, simulate a queueing system where the arrivals depend on the amount of congestion in the queue. Suppose that the service times are exponentially distributed with mean 1.0 minute, and the interarrival times are exponentially distributed with mean equal to 1.2 minutes if the previous waiting time (W_n) is less than a threshold of 1 minute, and 2.0 minutes if the previous waiting time is greater than the threshold. Compute a 95 percent confidence interval for the mean waiting time for this model.

4.4 Using the same model as in the previous problem, but allowing the threshold to vary between 0.0 and 3.0 minutes, compute a series of confidence intervals for the mean waiting time. Use these confidence intervals to plot the mean waiting time as a function of the threshold value. If you wish mean waiting time to be less than .3 minutes, what threshold value should you use?

TABLE 4.5 Batch data for inventory model

Batch 1	3, 4, 4, 5, 4
Batch 2	5, 6, 4, 5, 6
Batch 3	4, 6, 6, 4, 5
Batch 4	4, 5, 4, 4, 6

4.5 Table 4.5 contains data for four batches with five observations per batch from a stationary simulation of 20 weeks of an inventory model. Compute a 95 percent confidence interval for the mean cost per week.

4.6 Refer to problem 4.1. Use the batch means method to compute confidence intervals for the mean waiting time using 64 batches of 32 observations, 32 batches of 64 observations, 16 batches of 128 observations, and 8 batches of 256 observations. Which confidence interval was shortest? Which confidence interval(s) included the actual mean of .8167? Can you draw any conclusions about the trade-off between batch size and number of batches?

4.7 Consider the computer inventory problem in section 4.5. Run this simulation assuming the demand for computers per day has a Poisson distribution with mean 2.0, maximum inventory level 8, and reorder points 1, 2, 3, 4, 5, 6, and 7. For each reorder point value, compute a 95 percent confidence interval for the mean cost per day. Plot these confidence intervals versus the reorder point. Which reorder point provides the minimum cost per day?

4.8 For the computer inventory problem in section 4.5, suppose the demand depends on the number of computers in inventory. (We can assume that the sales people market their computers more aggressively when they have more computers.) In particular, suppose the demand has a Poisson distribution with mean equal to $X_n/2$, half the number of computers in inventory. Compute a 90 percent confidence interval for the mean cost per day, using maximum inventory level 8 and reorder point 3.

4.9 Simulate the transient portion of the warranty model presented in section 4.5. Assume the machine is initially fully functional and under warranty. Run 100 independent replications, and on each replication, collect the following data: (a) the number of months until the warranty expires, and (b) the state of the machine in the first month the machine is not under warranty. Use these data to estimate (a) the mean number of months the machine will be under warranty, (b) the probability the machine will be under warranty more than one month, and (c) the probability the machine will be fully functional the first month the warranty is expired.

4.10 Simulate the stationary portion of the warranty model presented in section 4.5. Assume that initially the machine is fully functional. Note that only the first three rows and columns of the transition probability matrix are needed to do this simulation. Use the fully functional state as a regeneration state and collect data on 200 cycles. Estimate the probability the machine is fully functional by setting Y_n equal to 1 if the machine is fully functional in month n and equal to 0 otherwise. Then the probability that the machine is fully functional is the mean of Y_n. Associate the costs \$20.00, \$50.00, and \$200.00 with the states $\overline{\text{W}}$FF, $\overline{\text{W}}$PF, and $\overline{\text{W}}$NF, respectively. Run the simulation for 200 cycles, collect appropriate data, and estimate the mean cost per month.

4.11 Suppose that in the batting cage problem in section 4.5, there is room for at most five customers to wait. Use mean interarrival time six minutes, or arrival rate 1/6 customer per minute. Run the simulation and collect data on the number of customers in the queue. Use these data to compute a 95 percent confidence interval for the mean number of customers waiting and the probability the queue is full.

4.12 In the batting cage problem in section 4.5, run the simulation using arrival rate 1/6 and maximum queue lengths of 2, 3, . . . , 10. For each queue limit, compute a point estimate and 95 percent confidence interval for the probability the queue is full. Plot these point estimates and confidence intervals to show the relationship between the likelihood the queue is full and the limit on the queue size. Does this plot correspond to your intuition? If you want the probability that the queue is full to be no larger than .10, how much waiting room must you provide?

4.13 In the marketing model presented in section 4.8, replace equations (4.15) and (4.16) with

$$T_n = \beta T_{n-1} + \frac{M - \beta T_{n-1}}{1 + \frac{L_n}{(A+B)^2}} \qquad \textbf{(4.20)}$$

$$X_n = \alpha X_{n-1} + \frac{T_{n-1} - \alpha X_{n-1}}{1 + K_n \left(\frac{B}{A}\right)^2} \qquad \textbf{(4.21)}$$

Reevaluate scenarios 1 and 2 using this model for the market share and the parameters in Table 4.4.

4.14 The following is a simple corporate planning model. Define the variables:

S_n = sales in month n,
A_n = advertising expenses in month n,
C_n = cost of goods sold in month n,
E_n = total expenses in month n,
N_n = net income in month n,
T_n = taxes in month n, and
X_n = profit in month n.

Then, $S_n = K_1 A_{n-1} + K_2 A_{n-2} + K_3 A_{n-3} + \alpha(S_{n-1} - \mu) + \epsilon_n$, where $K_1 = 1.03$, $K_2 = .43$, $K_3 = .16$, $\alpha = .33$, $\mu = 1.25$, and ϵ_n is a sequence of independent normally distributed random variables with mean 0 and variance $\sigma^2\epsilon = .05$. The units for S_n are millions of dollars. The other variables are determined by:

$$C_n = D_n S_n$$

where D_n are independent random variables uniformly distributed between .42 and .48;

$$A_n = .23 S_n$$

$$E_n = A_n + B_n$$

where B_n are independent random variables uniformly distributed between .05 and .11;

$$N_n = S_n - E_n - C_n$$

and

$$X_n = N_n - T_n$$

where $T_n = f(N_n)$, and $f(\cdot)$ is the tax function given by

$f(x) = 0$ if $x < 10{,}000$
$= .10\,(x - 10{,}000)$ if $10{,}000 \leqslant x < 40{,}000$
$= 3000 + .20\,(x - 40{,}000)$ if $40{,}000 \leqslant x < 80{,}000$
$= 11{,}000 + .40\,(x - 80{,}000)$ if $80{,}000 \leqslant x < 200{,}000$
$= 41{,}000 + .60\,(x - 200{,}000)$ if $x \geqslant 200{,}000$

We have starting conditions $S_0 = 2.35$, $A_0 = A_{-1} = A_{-2} = .54$. Implement this model and generate 100 months of data (X_n) from the model. Plot these values and determine a cutoff point for the initial transient. Then generate a large sample and estimate the mean monthly profit using the batch means method.

4.15 The following is a model of a lumber yard. The yard is able to hold a maximum of 1 million board feet of lumber. Initially it has 250,000 board feet. Demand for lumber in week n will be denoted D_n, and D_n is computed from D_{n-1} according to the formula

$$D_n = \mu + \alpha(D_{n-1} - \mu) + \epsilon_n$$

where $\alpha = .45$, $\mu = 100{,}000$, and ϵ_n denotes a sequence of independent normally distributed random variables with mean zero and standard deviation 120,000. After weekly demand is determined, the amount sold per week is given by the following policy:

- If the inventory level is greater than the week's demand, then the entire amount of demand is sold.
- If the inventory level is between 50 percent and 100 percent of the week's demand, then all inventory is sold.
- If the inventory level is less than 50 percent of demand, then none is sold.
- Any demand not met is lost; none is carried over to the next week.

Each week, a certain amount of spoilage and deterioration occurs due to insects and rot. The amount of rot is uniformly distributed between 5.0 and 10.0 percent of the amount of lumber in the yard and is computed before demand is met.

The following control limit policy is used to replenish the inventory at the end of each week:

- If the amount of lumber, after meeting demand, is less than 250,000 board feet, then enough is bought to bring the inventory up to the full 1 million board feet. Lumber is delivered immediately, so lumber ordered can be used to meet next week's demand.

Define the following costs: C_h is the cost per week per thousand board feet of storing the lumber; C_o is the cost incurred each time an order is placed and is independent of the amount of lumber ordered; and C_l is the cost per thousand board feet for lost sales when all of the demand cannot be met. The total cost over N weeks is the sum of C_h times the sum of the weekly inventory levels, C_o times the number of orders placed, and C_l times the total amount of lost sales.

Use this model to estimate the following quantities:

a. mean total cost per week,

b. mean inventory level,

c. mean proportion of demand met, and

d. the probability that 95 percent of demand can be met each week.

Use the batch means method to compute confidence intervals for these parameters.

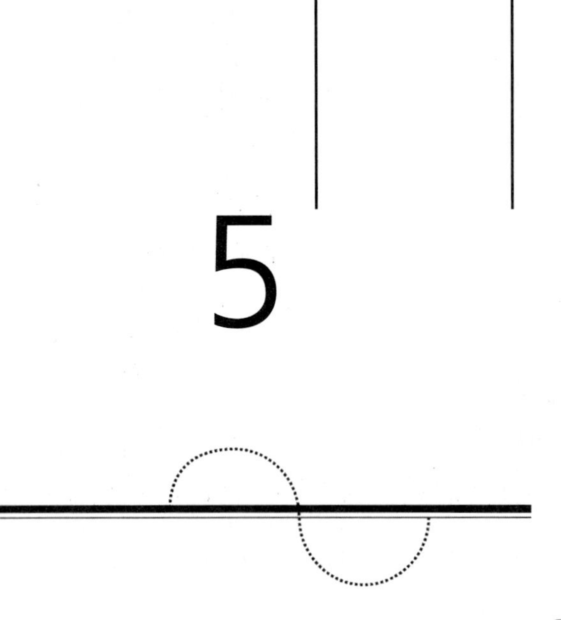

5

System Modeling Concepts for Discrete-Event Simulation

This chapter concentrates on the concepts and data structures required to implement the event scheduling "worldview," as well as those needed to implement the entity–attribute–set–resource system structure representation, which we will call the "modeling framework." First, the concept of a *discrete-event system* will be presented, along with the ideas of events, event notices, and the event list. This will be illustrated using a simple single-server queue where the system state is denoted by the number of customers in the queue and the status of the server—idle or busy. Event parameters will be introduced and illustrated using a two-server queue. The mean number of customers in the queue will be estimated from the output data. Then entities, attributes, and lists will be introduced. A simple message router model will be used to demonstrate the use of entities, attributes, and lists without events. Finally, these concepts will be combined with the event scheduling approach and illustrated with the single-server queue. For these systems, we will estimate the mean number in queue and the mean waiting time.

These models are carefully selected to illustrate the capabilities of the modeling method, however. *The objective here is to show the reader exactly how a discrete-event simulation operates.* To achieve this objective, we will go carefully through some hand simulations. In this chapter, we will concentrate heavily on simple queueing systems. Additional models in the following chapters will further reinforce and extend the material of this chapter. The presentations of the models will include analysis of data and estimation of parameters using replications or batch means; however, the emphasis will be on model development and representation.

5.1 INTRODUCTION

In Chapter 4, we examined dynamic models—that is, models that evolve over time. These models had a certain simplicity that allowed us to use such devices as equations and transition probability matrices to describe changes in numerical variables such as inventory level and waiting time as the system evolved over time. Models like these are useful, but many other important systems do not permit a representation using this paradigm. For example, consider the system of checkout lanes at a supermarket. Suppose there are six lanes and the checkout process has the following characteristics:

1. None of the times involved in the checkout process, such as interarrival times and service times, has an exponential distribution.
2. Customers can pay by cash, check, credit card, or debit card.
3. The length of time a customer spends checking out depends on both the number of items purchased and the method of payment.
4. Two express lanes are reserved for customers with 12 or fewer items and who pay by cash only.
5. Waiting customers will jockey to a shorter line when their line is at least two customers longer than the shorter line.

Such a model is impossible to represent as a Markov chain or similar model.

To use simulation to analyze the supermarket model and other models of complex systems, we need a more powerful modeling framework or "language" that can describe these models. Our ultimate goal is to be able to represent models of complex systems such as manufacturing plants, communication networks, transportation systems (railroad networks, rapid transit systems, harbor traffic, etc.), and other systems that involve service, queueing, and processing. These systems are frequently quite large and complex, but they have certain characteristics in common and can be described using a common language. In this chapter we will develop a general modeling framework that can be used to describe many operational systems. We will then extend this framework further in the next chapter and utilize it later when we examine some special types of models.

5.1.1 Static and Dynamic Model Descriptions

Imagine the supermarket checkout system mentioned earlier. A complete description of this system must involve two aspects: (1) what components are involved, and (2) *how* they interact. If a photograph of the system were taken, it would show the components that constitute the system as well as their characteristics (size, color, capacity, speed, etc.). The *static model description* provides a list of all components in the system along with all relevant information about them and describes the relationships among these components. For example, the supermarket checkout system consists of six checkout lanes. Each lane is either an express lane or a regular lane, and each lane has a queue that holds waiting customers. Each customer has two characteristics: (1) a number of items and (2) a method of payment, which can be by cash, check, credit card, or debit card. In section 5.6, we will present a framework that can be used to describe virtually every system of interest.

The static model description models the components that make up the system, but it does not tell *how* they interact. The *dynamic model description* provides a set of rules telling how the components interact as time advances. For example, the dynamic model description for the supermarket checkout might specify that the time between customer arrivals is uniformly distributed between 10 seconds and 60 seconds, and an arriving customer selects the checkout lane containing the smallest number of customers and waits in the line until all customers in front of her have finished being served before service begins for her—that is, service order is first-come, first-served for each lane. It would also describe the distribution of service times and their relationship to the number of items purchased and the method of payment, as well as the conditions under which a customer jockeys to another queue. Section 5.2 introduces the event scheduling approach to describe model dynamics—that is, how the system changes over time.

5.2 EVENTS AND EVENT SEQUENCING

The *state* of a system model is defined to be a collection of variables containing all information necessary to operate the model and record relevant changes in it over time. Any change in any state variable represents a change in the system, and any change in the system is reflected by a change in some state variable(s). For example, in a simple queueing system with a single server and a single queue, in which all customers are identical, we could represent the state of the system using two variables, (Q,S), where Q is the number of customers in the queue and S represents

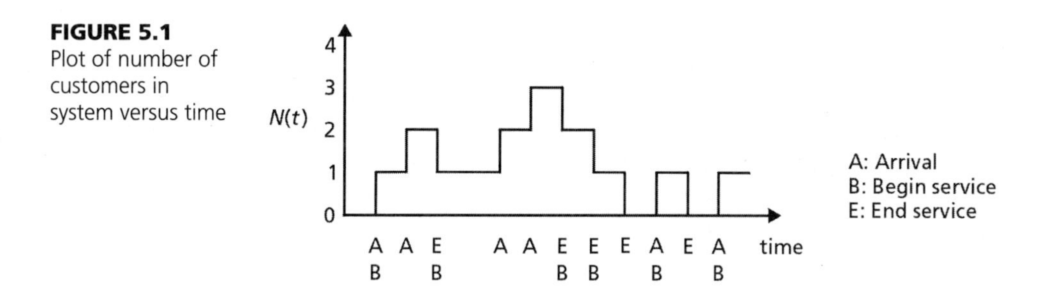

FIGURE 5.1 Plot of number of customers in system versus time

the status of the server: busy or idle. However, a large collection of variables is required to represent the system state in realistic systems.

5.2.1 Events and State Changes

In a *discrete-event dynamic system,* the system state changes only at discrete points in time, which mark the occurrence of an *event.* The system state remains constant between these points. For example, in the supermarket checkout system (as well as other queueing systems) the system state changes the instant a customer arrives, the instant a customer begins service, and the instant a customer completes service and departs. If the model were more detailed and the system state included the cumulative number of items that have been checked, then the completion of scanning an individual item would also represent the occurrence of an event, and the system state would also change at each of these instances.

An *event* is defined to be any occurrence that causes an instantaneous change in the system state. In any system, there will be many event occurrences; however, there is usually only a small number of distinct *types* of events. For example, in a simple queueing system, there are only three distinct types of events: (1) arrival of a customer, (2) beginning of service, and (3) completion of service for a customer.* However, in a simulation run, each type of event will occur many times, as there will be many distinct customers to arrive, begin service, and complete service. Figure 5.1 shows a typical plot of the total number of customers in a queueing system, $N(t)$, and identifies the instances of the arrival, beginning of service, and service-completion events.

It is important to understand that the definition of events depends on how the system state is defined and what changes in the system are captured in the model. For example, most queueing system models are not concerned with the small amount of idle time a server may experience between customers. If this time were important, then the completion of service for a customer and the beginning of service for the next customer would have to be distinct events with a (possibly ran-

*Later we will show how this system can be represented with only two distinct types of events: arrival of a customer and beginning of service.

dom) time delay between them. If the model allowed for server breakdowns or vacations, the server would need to have an attribute that denotes whether or not it is available, and at least two additional events would be required: (1) one to start a breakdown or vacation period and (2) another to end it. If an airport model includes weather effects, there would be one or more variables representing weather conditions, along with one or more types of events denoting changes in the weather. The weather variables in the system state would also be used in other events to determine how the system—specifically, entering and exiting airplanes—responds in some or all events. If the airport model does not include weather effects, the system state would not need these weather variables, and there would be no need for events denoting changes in the weather. Thus, the determination of which variables will be included in the system state and which events will be included in the model depends on the level of detail to be included and the purpose for which the model is developed.

Each type of event has an associated set of instructions, called an *event routine,* that specifies exactly how the system state changes when an event of the given type occurs. For example, when an *arrival* event occurs in a queueing system, a set of rules for an *arrival* event is applied to determine exactly how the new arrival interacts with the system to change the system state. This set of rules might say, "Place the arriving customer last in the queue; then if there is an idle server, start service for the arriving customer using that server."

5.2.2 Event Scheduling

In all of the dynamic models in Chapter 4, time was incremented by the same fixed amount between each state change; that is, the simulation operated by repeatedly incrementing time a fixed amount, then updating the system state by computing the next value of the process. Basically, the only state variable for the system was the value of the process. To implement the dynamics of a discrete-event simulation, a mechanism must be provided to increment time by a *variable* amount between events. The simulation must also be able to sequence events according to the time of occurrence and apply the associated event routine to make the changes to the system state. This is accomplished by creating an *event notice* for each occurrence when its future time is determined. At least two items of information must be provided on this event notice: (1) the actual time the event will occur, and (2) the type of event that is scheduled to occur (whether the event is an arrival event or an end of service event, for example). In addition, any other information that must be known to the event routine when it is activated is recorded with the event notice. These event notices are stored in a list, called the *(future) event list,* which is ordered by time of occurrence. Thus, the event notice with the earliest time of occurrence—that is, denoting the next event to occur—will be at the front of the event list. The process of creating an event notice, recording the necessary information about the event on the event notice, and placing it in the event list is called *scheduling* the event. At any point in the simulation, the next event to occur is the one represented by the event notice at the front of the event list—that is, the event notice with the earliest time of occurrence.

The *system clock* is a special system variable that holds the current system time. The simulation progresses through time in the following way: The first event notice on the event list is selected. The system clock is set to the event time on this event notice, then the event routine for this type of event is executed to produce the appropriate changes in the system state for this type of event. Finally, the event notice is discarded. This process repeats until the event list is empty or some other signal is given to stop the simulation. In a discrete-event simulation, the routine that implements this process of repeatedly selecting the first event notice, updating the system clock, executing the appropriate event routine, and discarding the event notice is called the *timing routine.* The timing routine is the heart of a discrete-event simulation. Once the system is initialized, the simulation runs by turning complete control over to the timing routine.*

5.2.3 Discrete-Event Simulation Model Development

The essence of modeling discrete-event dynamic systems is to determine:

1. what variables or components are needed to adequately represent the system state,
2. what events are needed to represent the system changes, and
3. what state changes occur within each event, including the details of which events are scheduled and when are they scheduled.

Unfortunately, these three parts of the model specification cannot be established one at a time. For example, it is necessary to know what (state) changes occur during an event in order to determine what variables or other components must be included in the system state. Rather, the process of specifying the model is usually an iterative one: Establish a preliminary set of state variables and a preliminary set of events, then start writing the event routines. When additional information is needed, modify the variables or components in the system state to accommodate this information, and then look again at the event routines.

It is important to understand that there is no one *unique* model that represents a given system. In Chapter 4, we saw a model of a single-server queue; in this chapter, we will see two additional representations of this same system. Multiple repre-

---

*The routine that initializes the system prior to execution of all other routines may be thought of as an event routine for an *initialization event.* In this case, the initialization event is executed only once at the start of the simulation run, and control is turned over to the timing routine before the system is initialized.

sentations can be used for the same model. To some extent, the representation used is a matter of taste. However, another consideration is the ease of implementing the representation. The particular computer language used to implement the model also may dictate the representation.

5.2.4 A Generic Simulation Language

In the following sections we will reexamine Chapter 4's single-server queueing system. First, however, we will show how to model this system using events and a simple set of state variables. To express our models, we will develop a simple *generic simulation programming language* (GSL) that can be used to describe the system components and the changes that occur in the event routines. This is not a language that is implemented on a computer. Instead, it is simply a vehicle to describe the logic of the discrete-event simulation models we will be discussing and to help you understand the operation of these models.

Several computer languages are available for programming discrete-event simulations. The most popular of these are SIMAN, GPSS, SIMSCRIPT II.5, and SLAM II. In addition, libraries of simulation routines are available for general-purpose programming languages—such as C, Java, Pascal, and FORTRAN—that allow discrete-event simulations to be programmed. Each program implements all of the concepts and structures presented in this chapter in its own unique way. Each also has its own structure, syntax, and additional special capabilities. What is important to us is that the language can implement the concepts we talk about in this chapter. To avoid focusing on one or two languages and excluding the rest, we will use our idealized simulation language to express the *algorithms* for our discrete-event simulation examples. This language has the basic capabilities that any simulation language should have. In particular, we will assume that it has the following general and event scheduling features.

General Programming Language Features

Constants

Variables

Arrays of variables

Arithmetic and logical operators

Assignment statement: $A := B + C/D$

Commands or procedures to read in and write out constant and variable values

Looping statements: *While . . . Do, Repeat . . . Until* and *For i:=k to n Do . . .*

Conditional statements: *If . . . Then . . . Else*

Procedures with parameters

Functions with parameters

Event Definition and Scheduling Facilities

Event list

System clock: *NOW*

Timing routine: *simulate*

Event routines

Commands to:

- Schedule an event at a given time: *schedule* event-type at *time* T
- Find an event notice in the event list: *find* event
- Cancel a scheduled event: *cancel* event

A second reason for introducing the GSL is to be able to focus on simulation concepts without becoming involved with learning the syntax of a computer language. With GSL, we will not be concerned with such details as data types, for example. We will assume the data type represented by a variable or parameter is appropriate for the values we intend the variable or parameter to store.

Of course, to gain skill and experience with these ideas, one must actually develop and implement models. To do this, you do have to learn to use one of the real simulation packages and implement the model. In a later chapter we will present Arena, which is a simulation software package designed to do just this. We cannot possibly discuss all simulation languages and software packages, but if you understand the concepts presented in this chapter, learning to use them is relatively straightforward.

5.3 EXAMPLE: A SINGLE-SERVER QUEUE

The queueing system consists of a single server and a single queue. Service times are independent random variables that have an exponential distribution with mean $1/\mu$. Customers arrive one at a time with independent, exponentially distributed interarrival times with mean $1/\lambda$. (The interarrival and service times are assumed to be exponential just to be specific; we can easily sample these values from any other distribution.) Recall that λ is the arrival rate, or mean number of arrivals per minute, and μ is the service rate, or mean number of services per minute when the server is busy. We assume that the arrival rate is less than the service rate, so the system has a stationary distribution, and we are interested in estimating the mean number of customers in the queue when the system is operating in steady state.

5.3.1 State Variables for the Single-Server Queue

Because all customers are identical, we do not need to track them individually. Instead, we can just maintain a count of the number of customers in the queue. We also need to know whether the server is idle or busy. If the customer in service is not in the queue, it is possible to have an empty queue and the server to be busy. In this case, an arriving customer must be placed in the queue. Thus, the two necessary state variables are the number of customers in the queue, Q, and the status of the server, S. Q can assume any nonnegative integer value 0, 1, 2, . . . , and S can assume one of the values *idle* or *busy.* Of course, we could use the value 0 to represent idle and 1 to represent busy, but any two distinct values will do; we will just call these values *idle* and *busy.*

5.3.2 Event Routines for the Single-Server Queue

This system has three types of events: *arrival, begin service*, and *end service.* The event routines for these three events are shown in Figure 5.2.

In the GSL, the term *NOW* is the variable that contains the current system time—that is, the system clock, whose value is the time on the current event notice. The terms *interarrival time* and *service time* refer to routines or functions that return randomly sampled values from the interarrival time and service time distributions, respectively. The statement "Schedule *event-type* at time *T*" means to place an event notice with event time *T* and event type *event-type* on the event list. Thus, "Schedule *begin service* at time *NOW*" puts an event notice with time *NOW* and event type *begin service* on the event list. Because *NOW* always contains the current event time, an event scheduled at time *NOW* will be at the front of the event list and will execute immediately after the current event with no time lapse.* "Schedule *end service* at time *NOW* + service time" puts an *end service* event notice with time *service time* later than *NOW* on the event list.

Before the timing routine can take control of the simulation, the system must be initialized. We will assume that the system is initialized in the empty-and-idle state—that is, there are no customers in the queue and the server is idle. The initialization routine is shown in Figure 5.3.

The main program for this simulation, then, is as follows:

```
Initialize
Simulate
```

---

*Some models can schedule multiple events at the same time and require a priority system to ensure that they execute in the appropriate order. For example, if arrivals and service completions occur only at discrete times, we want service completions to have a higher priority and occur before an arrival so the server will be idle and ready to begin service for the new arrival. We will not deal with these details at this time.

FIGURE 5.2
Event routines for single-server queue

```
Event routine arrival

Schedule arrival event at time NOW + interarrival time
Q := Q + 1
IF S := idle THEN schedule begin service event at time NOW

Event routine begin service

Q := Q - 1
S := busy
Schedule end service event at time NOW + service time

Event routine end service

S := idle
If Q > 0
   THEN schedule begin service event at time NOW
```

FIGURE 5.3
Intialization routine

```
Routine Initialize

Q := 0
S := idle
Schedule arrival event at time NOW
```

5.3.3 A Detailed Description of Event Execution and State Changes

We will now examine carefully what happens when this simulation runs. The procedure `Simulate` is the timing routine that moves the model through time. Figure 5.4 shows the two state variables, *Q* and *S,* the clock (*NOW*), and the empty event list at the start of the simulation (before the initialization routine is called). An empty box represents an uninitialized value. After the initialization routine has been exe-

FIGURE 5.4
State variables and event list for queue before initialization

State Variables

Q	*S*	*NOW*

Event List

Number	Time	Event Type

FIGURE 5.5
State variables and event list for queue after initialization

State Variables

Q	*S*	*NOW*
0	*idle*	0.0

Event List

Number	*Time*	*Event Type*
→ 1	0.0	Arrival

cuted, we have the conditions in Figure 5.5. At this point, the event list holds one event notice, denoting an *arrival* event to occur at the start of the simulation. The arrow in the first event notice indicates that this is the current event.

After initialization, control is turned over to the timing routine, `Simulate`, which executes the following algorithm:

```
Simulate
WHILE the event list is not empty and stop signal has not been
     received DO:
   Update the clock to the time on the first event notice.
   Execute the event routine for the type of event on the first
   event notice.
Discard the first event notice.
```

The timing routines in some languages remove the first event notice from the event list before updating the clock and executing the event routine; others will leave it on the event list and remove it when it is discarded. Whether the event notice remains on the event list is irrelevant, but no event should be scheduled before the current event if its event notice remains on the list.

When the timing routine takes control, there is one event notice on the event list, as Figure 5.5 shows. The event time, 0.0, is assigned to the system clock. The clock initially had the value 0.0; therefore, this particular assignment does not change the current time, and *NOW* has the value 0.0. Since the current event is an *arrival* event, the *arrival* event routine is executed (refer to Figure 5.2). The first action in the *arrival* event routine is to schedule the next *arrival* event after an interarrival time delay. Table 5.1 contains lists of randomly sampled values from the interarrival time and service time distributions. We will select values from each list when the simulation calls for an interarrival time or a service time. The first interarrival time is .5, so this event is scheduled to occur at time *NOW* + .5, or .5. Next

TABLE 5.1 Randomly sampled observations for interarrival and service times for the single-server queue model

Interarrival times	.5	1.4	.1	1.7	.8	.1	.1	2.6	3.5	.2
Service times	1.0	.2	.8	.8	.4	.5	.4	2.6	.5	.0

FIGURE 5.6
State variables and event list for queue after first *arrival* event

System State

Q
1

S
idle

NOW
0.0

Event List

Number	*Time*	*Event Type*
→ 3	0.0	*Begin service*
2	.5	*Arrival*

we increase the value of *Q*, which represents the number of customers in the queue, by 1. Finally, we look at the server status variable, which has the value `idle`, indicating the server is not currently working. In this case, we schedule a *begin service* event at time *NOW*. The event notice for the first *arrival* event is discarded, and we have now completed one iteration of the loop in the timing routine. The system state at the end of the first *arrival* event looks like Figure 5.6.

The first event notice on the event list now is for the *begin service* event for the first customer. This event occurs at time 0.0, so the clock remains at time 0.0, and the *begin service* event routine is executed. In this routine, we first remove the customer from the queue by decreasing *Q* by 1. Then the state variable *S* is set to `busy`, indicating that the server is now working. Lastly, the *end service* event is scheduled at time equal to the service time later than *NOW*. The random variates in Table 5.1 indicate that the first service time is 1.0, so the time at which the first *end service* event occurs is *NOW* + 1.0, or 1.0. The event notice for the first *begin service* event is discarded, and the system state looks like Figure 5.7.

Now the first event notice on the event list is one for the *arrival* event at time .5. After the clock is assigned the event time, .5, the *arrival* event routine is executed. In this event routine, the next *arrival* event is scheduled at time *NOW* + interarrival time, which is .5 + 1.4 = 1.9. We increment *Q* by 1 and then check the server's status. The server's status is `busy`, so the arriving customer cannot begin service.

FIGURE 5.7
State variables and event list for queue after first *begin service* event

System State

Q
0

S
busy

NOW
0.0

Event List

Number	*Time*	*Event Type*
→ 2	.5	*Arrival*
4	1.0	*End service*

FIGURE 5.8
State variables and event list for queue after second *arrival* event

System State

Q	*S*	*NOW*
1	*busy*	.5

Event List

Number	*Time*	*Event Type*
→ 4	1.0	*End service*
5	1.9	*Arrival*

Figure 5.8 shows the system state at the end of the second *arrival* event after discarding the event notice for this event.

At this point, the first event notice on the event list is one for an *end service* event. The clock is set equal to the time of this event, 1.0, and the *end service* event routine is executed to make the state changes involved with the first customer's completing service. We first set the server's status to `S := idle`. Then, looking at Figure 5.8, we see that *Q* is 1, so we can schedule a *begin service* event at time *NOW*. Figure 5.9 shows the system state after discarding the *end service* event notice.

The clock is updated to 1.0, indicating no change, and the *begin service* event removes the next customer from the queue ($Q := Q - 1$) and sets the server's status to `S := busy`. An *end service* event is scheduled at time 1.2 (1.0 + .2), and Figure 5.10 shows the system state after discarding the event notice for the *begin service* event.

This process continues until the timing routine breaks out of its loop. Figure 5.11 shows a summary of the operations of this system for the first 20 events. If the event routines are executed as given and no signal is sent to the timing routine to quit, then the process will go on forever (or until the computer is turned off!). Because each *arrival* event schedules the next *arrival* event, the event list always contains exactly one *arrival* event. (What we actually mean is that exactly one event notice for an *arrival* event is always in the event list, but because each event notice is associated with exactly one event occurrence, this way of stating it provides no

FIGURE 5.9
State variables and event list for queue after first *end service* event

System State

Q	*S*	*NOW*
0	*busy*	1.0

Event List

Number	Time	Event Type
→ 6	1.0	*Begin service*
5	1.9	*Arrival*

FIGURE 5.10
State variables and event list for queue after second *begin service* event

System State

Q	S	NOW
0	*busy*	1.0

Event List

Number	Time	Event Type
→ 7	1.2	*End service*
5	1.9	*Arrival*

ambiguity.) Thus, the event list will never be empty. There are various ways to stop the simulation. For example, if we wanted to make the simulation stop as soon as the first event on or after time 10.0 is executed, then we could replace the statement:

```
Schedule arrival event at time NOW + interarrival time.
```

with the statement:

```
IF NOW < 10.0
THEN Schedule arrival event at time NOW + interarrival time.
```

Once the clock equals or exceeds time 10.0, the next *arrival* event will not be scheduled. Then if the server is busy, *begin service* and *end service* events will be scheduled and later executed for each of Q customers in the queue, Q will be decremented to 0, and the server will become *idle.* If the server is idle, then no more events will occur because the next *arrival* and *begin service* events can be scheduled only by an *arrival* event in this case. Thus, interrupting the sequence of *arrival* events will cause the event list to empty and halt the simulation.

5.3.4 Review: Event Scheduling

Let's review the event scheduling process. For the event scheduling time advance mechanism to work properly, two things must be done: First, at least one event must be scheduled when the system is initialized. If no events are scheduled, then the event list will be empty and the timing routine will cease without doing anything. Second, at least some events must schedule other events. If only the initialization routine schedules events, then when the timing routine executes, these event routines will be executed and the simulation will come to an end without accomplishing anything. The process of running a discrete-event simulation can be thought of as a form of "bootstrapping," because instances of some events are initially scheduled and then these events schedule additional instances of themselves, so the system can run indefinitely.

Note also that because each *arrival* event schedules the next *arrival* event at time *NOW* + interarrival time, the time *between* arrival events is guaranteed equal to

FIGURE 5.11
Summary of 20 events in M/M/1 queue model

```
           Simulation of an M/M/1 Queueing System:
                   Arrival Rate     0.700
                   Service Rate     1.000

 Time      Event                    Action(s)
====================================================================
                      Run trace for 20 events.
                      System initialized to have 0 customers.
                      Schedule Arrival event at time            0.000
0.000   Arrival       ------------ Start event ------------
                      Schedule Arrival event at time            1.787
                      Add customer to queue. Length is 1
                      Server status is idle.
                      Schedule Begin Serv event at time         0.000
0.000   Begin Serv    ------------ Start event ------------
                      Take customer from queue. Length is 0
                      Server is now busy.
                      Schedule End Serv event at time           0.124
0.124   End Serv      ------------ Start event ------------
                      Queue length is 0. Server is idle.
1.787   Arrival       ------------ Start event ------------
                      Schedule Arrival event at time            2.602
                      Add customer to queue. Length is 1
                      Server status is idle.
                      Schedule Begin Serv event at time         1.787
1.787   Begin Serv    ------------ Start event ------------
                      Take customer from queue. Length is 0
                      Server is now busy.
                      Schedule End Serv event at time           4.492
2.602   Arrival       ------------ Start event ------------
                      Schedule Arrival event at time            3.708
                      Add customer to queue. Length is 1
                      Server status is busy.
3.708   Arrival       ------------ Start event ------------
                      Schedule Arrival event at time            4.625
                      Add customer to queue. Length is 2
                      Server status is busy.
4.492   End Serv      ------------ Start event ------------
                      Queue length is 2. Server is idle.
                      Schedule Begin Serv event at time         4.492
4.492   Begin Serv    ------------ Start event ------------
                      Take customer from queue. Length is 1
                      Server is now busy.
                      Schedule End Serv event at time           8.460
4.625   Arrival       ------------ Start event ------------
                      Schedule Arrival event at time            5.002
                      Add customer to queue. Length is 2
                      Server status is busy.
5.002   Arrival       ------------ Start event ------------
                      Schedule Arrival event at time            7.444
                      Add customer to queue. Length is 3
                      Server status is busy.
7.444   Arrival       ------------ Start event ------------
                      Schedule Arrival event at time            8.003
```

FIGURE 5.11 (continued)

```
  Time      Event                        Action(s)
==================================================================
                    Add customer to queue. Length is 4
                    Server status is busy.
 8.003   Arrival    ------------ Start event ------------
                    Schedule Arrival event at time              9.250
                    Add customer to queue. Length is 5
                    Server status is busy.
 8.460   End Serv   ------------ Start event ------------
                    Queue length is 5. Server is idle.
                    Schedule Begin Serv event at time           8.460
 8.460   Begin Serv ------------ Start event ------------
                    Take customer from queue. Length is 4
                    Server is now busy.
                    Schedule End Serv event at time             8.557
 8.557   End Serv   ------------ Start event ------------
                    Queue length is 4. Server is idle.
                    Schedule Begin Serv event at time           8.557
 8.557   Begin Serv ------------ Start event ------------
                    Take customer from queue. Length is 3
                    Server is now busy.
                    Schedule End Serv event at time             8.715
 8.715   End Serv   ------------ Start event ------------
                    Queue length is 3. Server is idle.
                    Schedule Begin Serv event at time           8.715
 8.715   Begin Serv ------------ Start event ------------
                    Take customer from queue. Length is 2
                    Server is now busy.
                    Schedule End Serv event at time             9.125
 9.125   End Serv   ------------ Start event ------------
                    Queue length is 2. Server is idle.
                    Schedule Begin Serv event at time           9.125
```

the interarrival time. If the sequence of interarrival times is a sequence of independent and identically distributed (iid) random variables, then this guarantees that times between *arrival* events are independent and identically distributed. In any discrete-event simulation, the way to ensure that the time between two events has a particular value or a value sampled from a particular distribution is to let the first event schedule the second at time *NOW* + the interevent time.

5.3.5 Estimating the Mean Number in the Queue

We stated earlier that the purpose of this simulation is to estimate the mean number of customers in the queue when the system is in steady state. The number of customers in the queue, or queue length, at any point in time is contained in the system variable Q. Thus, we must use this variable to collect the data that will be used to estimate the steady state mean. Because Q changes over time, we will use the nota-

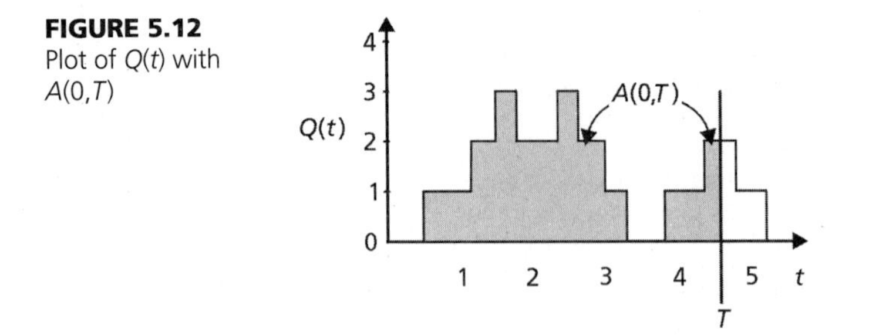

FIGURE 5.12
Plot of $Q(t)$ with $A(0,T)$

tion $Q(t)$ to denote the queue length at time t. Figure 5.12 shows a plot of $Q(t)$ for a portion of the simulation run. Because this is a discrete-event dynamic system, $Q(t)$ remains constant except when events occur and it changes to a new value. The parameter we wish to estimate is the average number of customers in the queue. This can be expressed as the average area under the graph of $Q(t)$:

$$\mu = \lim_{T \to \infty} \frac{A(0,T)}{T}$$

where $A(0,T)$ is the area under the function $Q(t)$ between 0 and T^*—that is, μ is the limit of the time average of $Q(t)$ over the interval from 0 to T as T approaches infinity. Because the process is stationary, we could begin computing the time average at any point $\tau > 0$. Just as when we estimated the mean waiting time in Chapter 4 for this same system, the initial observations will not be representative of the steady state because the system here is started in a fixed state, $Q = 0$. To remove this initialization bias, a portion of the data at the start of the run must be omitted from the statistical analysis. For now, we will assume that the data beyond some point $t = t_s$ is free from initialization bias. The point estimate for μ, computed from T time units of data, is

$$\hat{\mu}_T = \frac{A(t_s, t_s + T)}{T}$$

The expression $A(t_s, t_s + T)$ just represents the area under the function $Q(t)$ between t_s and $t_s + T$. Since $Q(t)$ is constant except where it jumps up or down by 1, this area is the sum of areas of rectangles (see Figure 5.12). Note that each point t where $Q(t)$ changes is an event occurrence. We could express the area as

$$A(t_s, t_s + T) = \sum_{k=1}^{K} Q_k (t_k - t_{k-1})$$

---

*This area can be represented as the integral $A(0,T) = \int_0^T Q(t)\, dt$.

where Q_k represents the number in queue between the $(k-1)$st and kth events, and t_k is the time of the kth event. Events are numbered beginning at $t = t_s$. To accumulate the area under $Q(t)$, the time of the previous event, t_{prev}, must be saved each time an event occurs. Then we compute $A = Q\,(NOW - t_{\text{prev}})$ and add it to the cumulative area A_c under $Q(t)$. The point estimate, then, is

$$\hat{\mu}_T = \frac{A_C}{T}$$

The batch means method that was introduced in Chapter 4 can be used to compute a confidence interval for the mean waiting time. The implementation in Chapter 4 was described for discrete observations. When the observations are made over continuous time, batches must be defined in terms of time segments. Let the length of each batch (in time units rather than number of observations) be L and the number of batches be B. Then the first batch mean is

$$\overline{X}_1 = \frac{A(t_s, t_s + L)}{L}$$

the second batch mean is

$$\overline{X}_2 = \frac{A(t_s + L, t_s + 2L)}{L}$$

and the last batch mean is

$$\overline{X}_B = \frac{A(t_s + (B-1)\,L, t_s + BL)}{L}$$

The total run length is $T = BL$. One can easily show that the average of the batch means

$$\overline{\overline{X}} = \frac{1}{B}\sum_{k=0}^{B} \overline{X}_k$$

is just the time average over the entire length from t_s to $t_s + T$.

To compute the variance of $\overline{X}$ as before, we treat $\overline{X}_1, \overline{X}_2, \ldots, \overline{X}_B$ as a sequence of independent, identically distributed observations and compute the usual sample variance:

$$s^2 - \frac{1}{B-1}\sum_{i=1}^{B} (\overline{X}_i - \overline{\overline{X}})^2$$

Then the confidence interval is computed using the t-distribution:

$$\overline{\overline{X}} \pm t_{B-1,\,\alpha/2} \frac{s}{\sqrt{B}}$$

where $t_{B-1,\alpha/2}$ is the $\alpha/2$ upper percentage point of the Student's t-distribution with $B - 1$ degrees of freedom. For the confidence interval to be valid, the batch size, L, must be large enough that the batch means are approximately independent.

One detail remains to be resolved: How does one make sure the data collection starts at time t_s and that each batch then ends L time units thereafter? The answer is that in a discrete-event simulation, if an action is to be taken at a given point in time or sequence of times, then an event must be defined to do whatever actions are to be taken—in this case to record data and reinitialize variables used to collect the data. The event must then be scheduled at the appointed time or times. In this case, we would define an event, *data collection,* and schedule it initially to occur at time t_s. Then the *data-collection* event would schedule itself to occur after a delay of L time units. On each occurrence of the *data-collection* event except the first, the time average for the batch would be computed and the variable used to collect the area under the function $Q(t)$ reinitialized to 0. The time average would not be computed the first time the event occurs because it is scheduled first at the end of the transient period and the data are to be discarded at this point. Therefore, the event routine must record the time average for the batch only after the first batch has been completed.

The event routines with the data-collection activities added are shown in Figure 5.13. One new event has been added: the *data-collection* event. This event occurs at the end and at the start of each batch because the end of one batch is the start of the next batch. Seven new variables have also been added. L contains the length of each batch, B contains the number of batches in the simulation run, and *TRANS* contains the length of the initial transient portion of data that will be omitted. All three values are assumed to be specified as external to the simulation run. *AREA* contains the cumulative area under the $Q(t)$ function, and C contains the cumulative batch count. Finally, *OLDT* contains the time of the immediately previous event, and *BM* is an array with at least B elements containing all of the batch mean values.

A couple of changes in these event routines, as compared to the first version in Figure 5.2, are worth noting. First, the initialization routine initializes both the batch count, C, and the cumulative area to zero, then schedules the first *data-collection* event at the end of the transient period. When the first *data-collection* event occurs, a batch has not ended. Rather, the transient period has ended and we must prepare for the start of the first batch. In the *data-collection* event routine, the cumulative area is first updated and the time of the previous event, *OLDT,* is updated to the present time, *NOW.* Note that this is also done at the start of all of the other event routines to accumulate the area under the $Q(t)$ function. It must be done in the *data-collection* event routine because the final bit of area must be accumulated since the end of a batch does not normally coincide with other events. Next in the *data-collection* event, the batch mean is computed when $C > 0$. If $C = 0$, the initial transient portion has just ended and we do not want to compute the batch mean.

FIGURE 5.13
Event routines for single-server queue with data collection

```
Initialize
  Q := 0
  S := idle
  Schedule arrival event at time NOW
  C := 0
  AREA := 0.0
  Schedule data-collection event at time NOW + TRANS

Event routine data collection
  AREA := AREA + Q * (NOW - OLDT)
  OLDT := NOW
  IF C > 0 THEN BM[C] := AREA/L
  IF C < B THEN
    AREA := 0
    C := C + 1
    Schedule data-collection event at time NOW + L
  ELSE stop simulation

Event routine arrival
  AREA := AREA + Q * (NOW - OLDT)
  OLDT := NOW
  Schedule arrival event at time NOW + interarrival time
  Q := Q + 1
  IF S := idle THEN schedule the begin service event at time
    NOW

Event routine begin service
  AREA := AREA + Q * (NOW - OLDT)
  OLDT := NOW
  Q := Q - 1
  S := busy
  Schedule end service event at time NOW + service time

Event routine end service
  AREA := AREA + Q * (NOW - OLDT)
  OLDT := NOW
  S := idle
  IF Q > 0 THEN schedule begin service event at time NOW
```

The *data-collection* event is also used to determine when to stop the simulation. This is accomplished by counting the batches and terminating the simulation when the batch count, C, is at least B, the total number of batches in the simulation run. If C is less than B, we prepare for another batch by resetting *AREA* to 0, incrementing C by one, and scheduling another *data-collection* event at the end of the next batch. If C is at least B, then a routine, *stop simulation,* is called to end the

FIGURE 5.14
Results of simulation run for single-server queue

```
Simulation of an M/M/1 Queueing System:
    Arrival Rate:              .700
    Service Rate:             1.000
System initialized to have 2 customers.

Final r.n. seed:    612501395

Results of simulation run:
Batch means method applied to queue lengths.

                   Number of batches:      10
                        Batch length:     500.000
                   Initial transient:      50.000
                    Grand batch mean:       1.532
               Variance of batch means:      .357
     Standard deviation of batch means:      .597
         Standard error of batch means:      .189
```

simulation. The simulation can be stopped by simply emptying the event list of all event notices or by sending a signal to the timing routine to break out of its loop, depending on the specific logic in the timing routine. We assume the routine *stop simulation* produces some action that causes the timing routine to cease executing.

Figure 5.14 shows the results of running the simulation using an arrival rate of .7 customers per minute and a service rate of 1.0 customers per minute. An initial transient period of 50 minutes was used, and the run consisted of 10 batches of 500 minutes each. The overall mean number in the queue, or sample mean of the batch means, was $\overline{\overline{X}} = 1.532$, and the standard deviation of the batch means was $s = .597$. A 90 percent confidence interval for the stationary mean number in the queue is

$$1.532 \pm 1.833\left(\frac{.597}{\sqrt{10}}\right)$$

or 1.532 ± .346, or from 1.186 to 1.878. In Chapter 4, we saw that the true mean number of customers in the queue is 1.633, which is included in this confidence interval. Of course, in simulations of real-world systems, the true mean would not be known. We mention it here only to demonstrate that this simulation does produce reliable estimates. Note also that even though the sample size included 5000 minutes of simulation time and involved approximately 3500 customers, the confidence interval is still rather large. Generally, because of the correlation between observations, extremely large sample sizes are required to produce precise confidence intervals for the mean or other parameters of interest. It has been observed that it is more difficult to estimate parameters in the M/M/1 queue than in most systems because the distributions of interarrival times and service times are highly skewed, and there is no control mechanism in the system to limit extreme behavior. In more complex systems with control mechanisms, it is reasonable to expect more precise estimates of parameters.

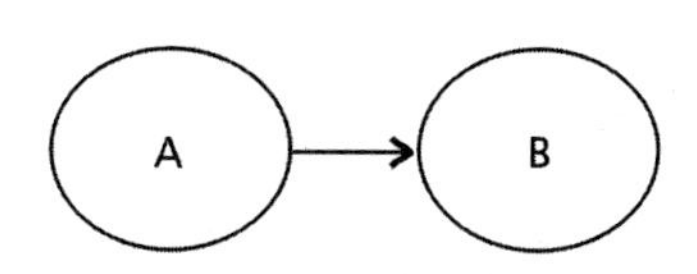

FIGURE 5.15
A simple event graph with two events

5.4 EVENT GRAPHS

In the previous example and all other discrete-event simulations, the simulation proceeds by having some events schedule other events. The relationships among various events can be displayed graphically using an *event graph.* Event graphs were first introduced by Schruben (1983). Schruben (1995) and Sargent (1988) have a more detailed discussion of event graph modeling. Event graphs use nodes to represent each type of event; directed arcs, or lines, between these nodes represent scheduling relationships between the events. Each arc points from the node representing the scheduling event to the node representing the scheduled event. Consider the graph in Figure 5.15 with two event nodes, A and B, and one directed arc from node A to node B. We interpret this graph to mean, "When event A occurs, it will schedule event B to occur."

Figure 5.16 is an event graph for the queueing system in section 5.3. In this event graph, we treat the initialization as an event that is scheduled once at the start of the simulation. Thus, there are four nodes for the four (types of) events: *initialization, arrival, begin service,* and *end service.* The arcs show that the *initialization* event schedules an *arrival* event, an *arrival* event schedules an additional *arrival* event and a *begin service* event, a *begin service* event schedules an *end service* event, and an *end service* event schedules a *begin service* event (for the next customer to be served). These relationships can also be seen by examining the event routines for this system in Figure 5.2 in section 5.3.

When one event schedules another, we must specify how much time elapses between the scheduling event and the scheduled event. On the event graph, this is shown by associating a number, variable, or expression with the scheduling arc between the two events. Figure 5.17 shows the event graph for the single-server queueing system

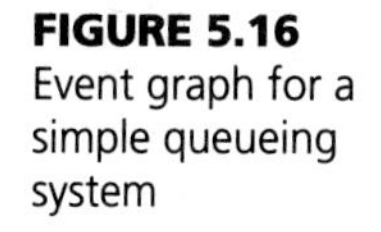

FIGURE 5.16
Event graph for a simple queueing system

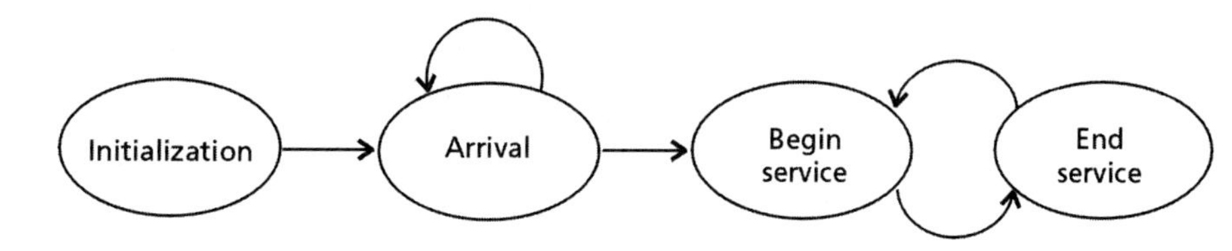

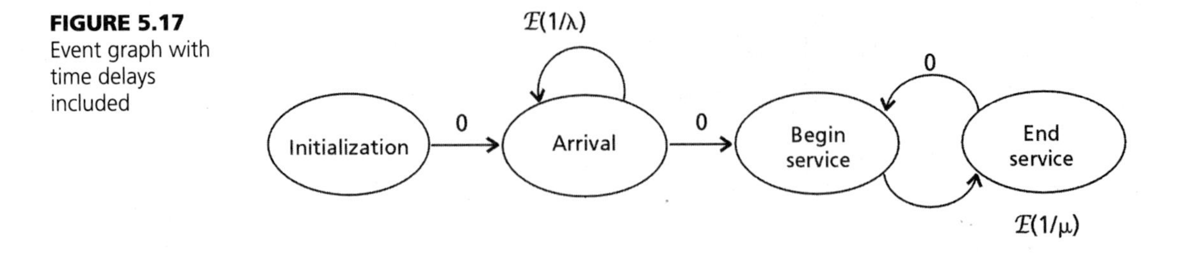

FIGURE 5.17 Event graph with time delays included

with the time delays between scheduling events and scheduled events. The time delays between the *arrival* and *begin service* events and between the *end service* and *begin service* events are zero, but the time delay between the *begin service* event and the *end service* event is the service time (which has an exponential distribution with mean $1/\mu$ in this case), and the delay between an *arrival* event and the next *arrival* event is the interarrival time, which is exponential with mean $1/\lambda$ in this case.

5.4.1 Simplifying an Event Graph

Because the delay between a *begin service* event and each event that schedules it is always zero, the *begin service* event always occurs at the same time as the scheduling event, either the *arrival* event or the *end service* event. This can be seen by observing that all arcs pointing to the *begin service* event node have zero times on them. Any event node that has all zero delays represents an event that always occurs simultaneously with another event. If this is the case, then the event structure of the model can be simplified by eliminating that event as a separate scheduled event. We can think of that event routine as being a subroutine of the event routines for the events that occur simultaneously with it—that is, we can combine the actions of the event routine with those of each scheduling event.

Figure 5.18 is the event graph for the single-server queue with the *begin service* event node removed. This model is equivalent to the model in Figure 5.17 but has one fewer event. The new event graph is derived from the original graph by replacing the *arrival* → *begin service* → *end service* arc sequence with the *arrival* → *end service* single arc, and the *end service* → *begin service* → *end service* arc sequence with the *end service* → *end service* arc. The time associated with each new arc is the

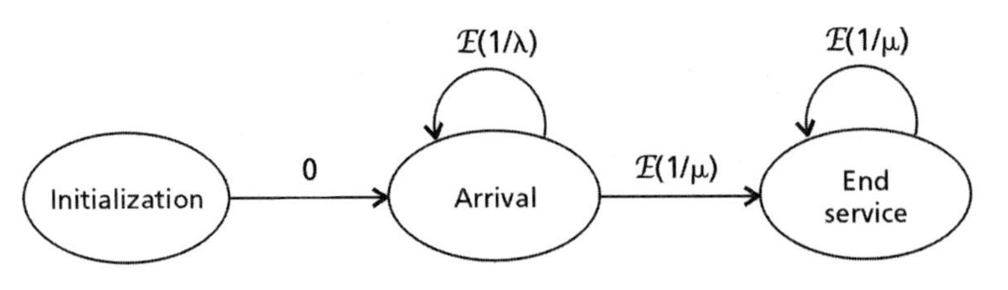

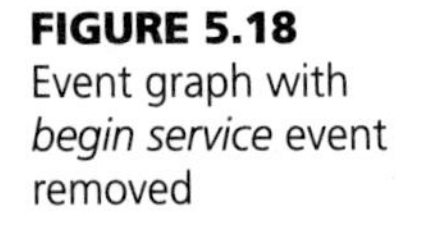

FIGURE 5.18 Event graph with *begin service* event removed

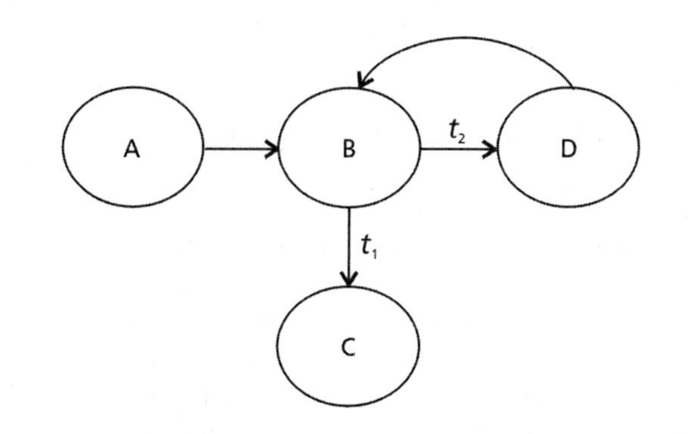

FIGURE 5.19
A sample event graph with unnecessary event B

service time. Since the delay between the *arrival* and *begin service* events is zero, and the delay between the *begin service* and *end service* events is the service time, the delay between the *arrival* and *end service* events is the sum of these two—the service time. A similar argument shows that the delay between an *end service* event and the next *end service* event is also just the service time.

A node—that is, an event—can be removed if every arc that terminates at the node has zero time associated with it. To remove the node, we replace every pair of arcs such that (1) the first arc in the pair is directed from an initial node toward the node to be removed and (2) the second arc is directed from the node to be removed to another node, with a single arc having the same initial node and terminating node and the time associated with the arc exiting the removed node. For example, consider the graph in Figure 5.19. Because event node B has zero time associated with all arcs directed toward it, the node can be removed. To remove node B, first form a graph with only nodes A, C, and D. Then replace the pair of arcs A → B → C with the arc A → C, having time t_1. Replace the pair of arcs A → B → D with an arc having time t_2. Finally, replace the pair D → B → D with the arc D → D having time t_2. The final graph is shown in Figure 5.20. To simplify an event graph to the maximum extent, this process is repeated for each node that can be removed.

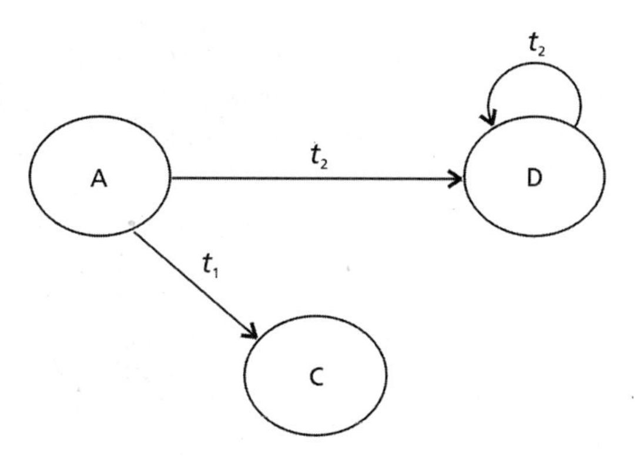

FIGURE 5.20
The event graph in Figure 5.18 with event B removed

5.4.2 Conditional Events

When an *arrival* event occurs, the next *arrival* event is always scheduled. However, the next *begin service* event is scheduled only if the server is idle. The scheduling of an *arrival* event by the previous *arrival* event is unconditional—it always occurs. The scheduling of a *begin service* event by an *arrival* event is conditional—it occurs only if a specified condition, S := *idle,* exists when the *arrival* event occurs. To specify the model correctly using an event graph, we must include the information on the conditions that must exist for the scheduling to take place. This is done by placing a line through the arc and putting the condition above or below the line. Figure 5.21 shows the original event graph for the single-server queueing system with all conditions included. This representation shows the scheduling relationships, time delays between events, and conditions that govern event scheduling.

5.4.3 Event Routines and State Changes

The model is now fully specified if we provide the state changes for each event. Referring back to the event routines, we see that, omitting the *data-collection* activities, the state change for the *arrival* event is $Q := Q + 1$. The other actions in the *arrival* event routine are captured in the event graph structure. Similarly, the state changes for the *begin service* event are $Q := Q - 1$; S := *busy*, and the state changes for the *end service* event are S := *idle.* The state changes for the initialization event are $Q := 0$ and S := *idle.* Figure 5.22 shows the completed event graph with event routines included. This representation of the single-server queueing system is entirely equivalent to the event routine representation given in Figure 5.2. Study both representations to be sure that all information in one is also in the other.

5.5 EVENT PARAMETERS

Sometimes information must be passed from a scheduling event to a scheduled event. For example, suppose that the queueing system in the previous section has two servers instead of one, and suppose that server 1 serves at a rate of one cus-

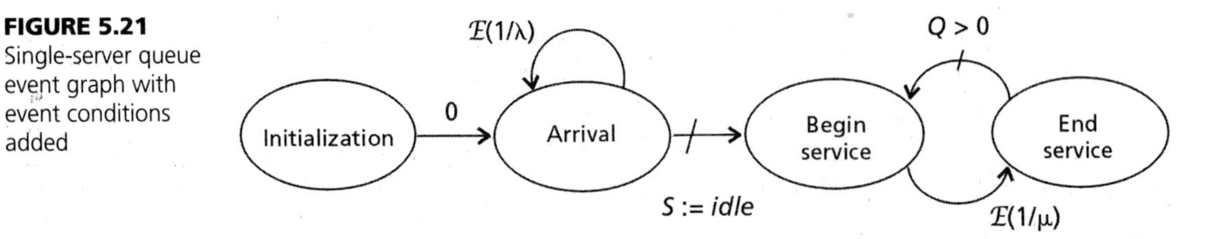

FIGURE 5.21 Single-server queue event graph with event conditions added

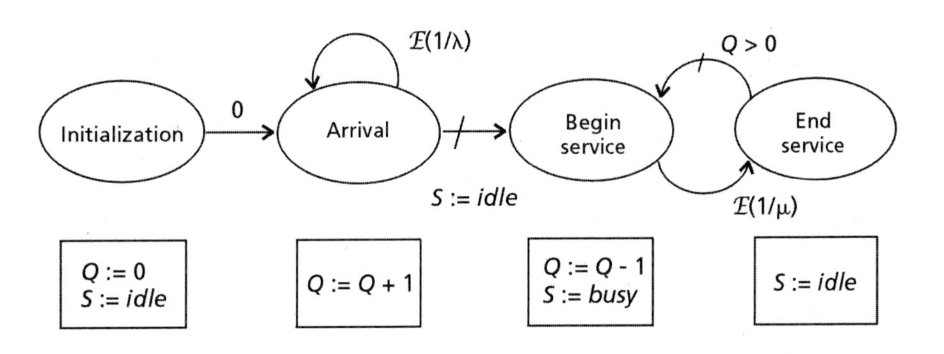

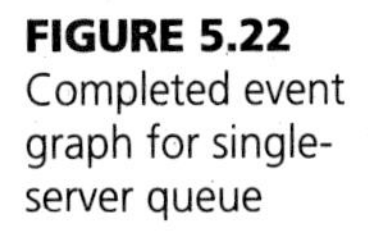

FIGURE 5.22 Completed event graph for single-server queue

tomer per minute but server 2 has a service rate of 1/2 customer per minute. In this case, the servers are not identical and each server must have a separate status variable. Then the status for all servers in the system would be represented by an array of status variables—say, $S[1]$ and $S[2]$—where each variable could have either of the values *busy* or *idle.* When a *begin service* or an *end service* event occurs, the event routine must know which server is involved so the appropriate status variable can be changed and the appropriate service rate can be applied to compute the time of the *end service* event. The three event routines in Figure 5.23 model a two-server queueing system in which each server has its own service time distribution.

5.5.1 Passing Event Parameters to Event Routines

These event routines differ from those of the single-server case in two important ways. First, the *arrival* event routine must search for an idle server. It first checks server 1 to see if it is idle and can be engaged to begin service. If the first server is busy, then the second server is checked. Second, both the *begin service* and the *end service* events have a single parameter, j, that identifies the index of the server involved in the event occurrence. This is denoted by the notation *begin service(j)* and *end service(j)*. It is important to understand that a parameter defined for an event routine is just a storage location or variable for a value to be passed to the event routine. We will call this the *formal parameter.* Thus, j is the formal parameter for the *begin service* and *end service* routines. When the event is actually scheduled, an *actual parameter* value is supplied. When either of these events is scheduled, the actual parameter value is stored with the event notice so that it will be available to the event routine when it is executed. This is the technique of passing a parameter by value that is used by many programming languages. In this case, the event notice is used to store parameter values because the event routine is not executed immediately. We can think of this as assigning the actual parameter value to the formal parameter variable. Thus, in the *arrival* event, if the first server is idle, the statement "schedule the *begin service*(1) event at time NOW" stores the value 1 on the event notice and assigns this value to the formal parameter, j, of the *begin service* event routine when this event occurs, thus passing the value 1 as the event parameter.

FIGURE 5.23
Spreadsheet for office building cash-flow analysis

```
Routine initialize
  Q := 0
  S[1] := idle
  S[2] := idle
  Schedule arrival at time NOW

Event routine arrival
  Schedule arrival event at time NOW + interarrival time
  Q := Q + 1
  IF S[1] := idle
    THEN schedule begin service(1) event at time NOW
    ELSE IF S[2] := idle
      THEN schedule begin service(2) event at time NOW

Event routine begin service(j)
  Q := Q - 1
  S[j] := busy
  Schedule end service(j) event at time NOW + service
    time(j)

Event routine end service(j)
  S[j] := idle
  IF Q > 0 THEN schedule begin service(j) event at time NOW
```

5.5.2 Using Event Parameters in Event Routines

Within the *begin service* event routine, the parameter *j* is used to select which status variable (*S*[1] or *S*[2]) will be assigned the value *busy*; it is also used, through the function `service time (j)`, to determine the parameter used to compute the service time random variable. Recall that if *j* is 1, then the service time random variable is exponential with mean 1.0; but if *j* is 2, then the service time random variable is exponential with mean .5. Thus the routine that computes the service time must know which server is involved. We have not been specific about this; there are many ways to accomplish it. One way is to store the parameters in an array consist ing of two components, *R*[1] and *R*[2], and have the service time routine use *R*[*j*] as the rate value for generating the service time random variable. Finally, the *begin service* event routine passes its parameter, *j*, to the *end service* routine when it schedules the *end service* event. The *end service* event routine must know which server is involved for two reasons: (1) The status of the server ending service must be changed to idle, and (2) if the queue is not empty, a *begin service* event must be scheduled for this server, so the server's index must be passed to the *begin service* event routine.

The final modification to the simulation is in the initialization routine in which both server status variables must be initialized. Figure 5.24 contains a trace that summarizes the operations of this system for 20 events.

5.5.3 Event Parameters in Event Graphs

Event parameters can also be included in the event graph for a model. Figure 5.25 shows two events, A and B, with a single parameter passed from event A (the scheduling event) to event B (the scheduled event). This portion of the event graph can be interpreted as, "When event A occurs, it schedules event B to occur t time units later if condition C holds, and the value p is passed to the event routine for event B." Figure 5.26 shows the event graph for the two-server queueing system just discussed, including event parameters.

5.5.4 Event Parameters and Output Data Analysis

Let's return to the previous section's single-server queue model for which we used data on the number of customers in the queue, measured continuously in time, to estimate the steady state mean number in the queue. We could use the same technique in this system to estimate the same parameter. However, it is frequently the case that we want to use waiting time observations to estimate the steady state mean waiting time per customer. Little's formula (Little, 1961; Heyman and Sobel, 1982), $L = \lambda W$, relates the mean number of customers in the queue, L, to the mean waiting time in the queue per customer, W. In this formula, λ is the arrival rate, or the mean number of arrivals per unit time. In fact, this formula applies to virtually any system in which customers arrive, possibly wait for service, receive service, and leave. Note that in simulation models, the value of λ is known. Thus, the simulator frequently has the option of estimating either L or W and using Little's formula to estimate the other quantity. Here we are using discrete waiting time data to estimate W in order to illustrate the procedure for using event procedures to collect waiting time data. In what follows, we will apply the batch means method discussed in Chapter 4 to estimate this parameter while also illustrating another use for event parameters.

Our first problem is how to compute the waiting time in queue for each customer. To do this, we must record each customer's arrival time. Since the *arrival* event routine is executed at the moment of arrival, the arrival time must be recorded in this routine. Then, when the customer begins service in the *begin service* event, we subtract the arrival time from the current (system clock) time to compute the time the customer spent in the queue. If a customer arrives to find the server idle, then he begins service at the same time as his arrival, and the waiting time in the queue is zero; otherwise, the amount of time in the queue will be positive.

The problem we face is that we must associate each arrival time recorded with the customer to which it belongs. We can think of the arrival time as an *attribute,* or item of data, that belongs to each customer. To solve this problem, we will assign each customer a unique sequence number. Next we will establish an array, $ATIME[\cdot]$, that has one element for each customer, to hold the arrival times. Finally, we will need to pass the customer's index number to the *begin service* event routine to compute the waiting time. The new event routines are shown in Figure 5.27.

FIGURE 5.24
Trace of actions in heterogeneous server queue

```
Simulation of an M/M/2 Heterogeneous Server Queueing System:
                Arrival Rate:     .700
                Service Rates:   1.000       .500

Time        Event                         Action(s)
=======================================================================
                          Run trace for 20 events.
                          System initialized to have 0 customers.
                          Schedule Arrival event at time                .000
  .000    Arrival         ------------ Start event ------------
                          Schedule Arrival event at time 1.787
                          Schedule Begin Serv event at time             .000
  .000    Begin Serv      ------------ Start event ------------
                          Server 1 is busy.
                          Schedule End Serv event at time               .124
  .124    End Serv ------------ Start event ------------
                          Queue length is 0. Server 1 is idle.
 1.787    Arrival         ------------ Start event ------------
                          Schedule Arrival event at time               2.602
                          Schedule Begin Serv event at time            1.787
 1.787    Begin Serv      ------------ Start event ------------
                          Server 1 is busy.
                          Schedule End Serv event at time              4.492
 2.602    Arrival         ------------ Start event ------------
                          Schedule Arrival event at time               3.708
                          Schedule Begin Serv event at time            2.602
 2.602    Begin Serv      ------------ Start event ------------
                          Server 2 is busy.
                          Schedule End Serv event at time              3.886
 3.708    Arrival         ------------ Start event ------------
                          Schedule Arrival event at time               9.377
                          Add customer to queue. Length is 1
 3.886    End Serv        ------------ Start event ------------
                          Take customer from queue. Length is 0
                          Schedule Begin Serv event at time            3.886
 3.886    Begin Serv ------------ Start event ------------
                          Server 2 is busy.
                          Schedule End Serv event at time              4.414
 4.414    End Serv        ------------ Start event ------------
                          Queue length is 0. Server 2 is idle.
 4.492    End Serv        ------------ Start event ------------
                          Queue length is 0. Server 1 is idle.
 9.377    Arrival         ------------ Start event ------------
                          Schedule Arrival event at time              11.819
                          Schedule Begin Serv event at time            9.377
 9.377    Begin Serv      ------------ Start event ------------
                          Server 1 is busy.
                          Schedule End Serv event at time              9.768
 9.768    End Serv        ------------ Start event ------------
                          Queue length is 0. Server 1 is idle.
11.819    Arrival         ------------ Start event ------------
                          Schedule Arrival event at time              13.066
                          Schedule Begin Serv event at time           11.819
```

FIGURE 5.24 (continued)

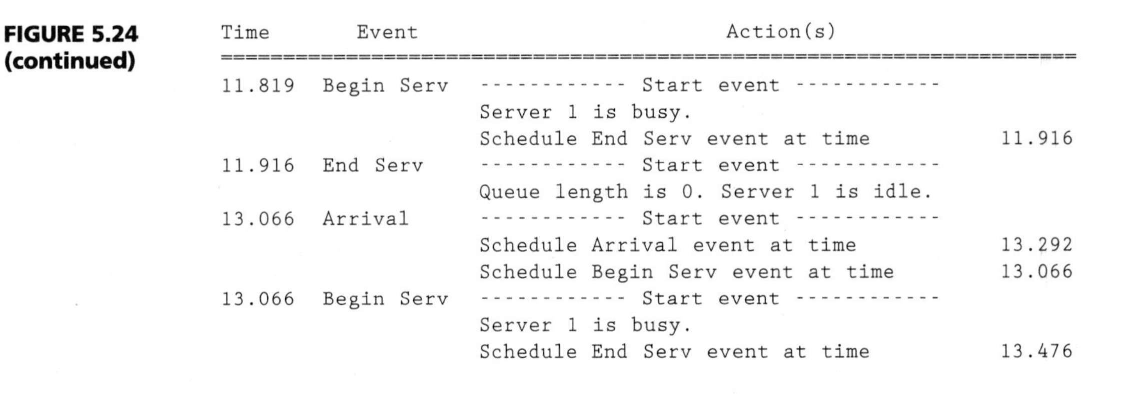

```
Time      Event                            Action(s)
=====================================================================
11.819   Begin Serv   ------------ Start event ------------
                      Server 1 is busy.
                      Schedule End Serv event at time        11.916
11.916   End Serv     ------------ Start event ------------
                      Queue length is 0. Server 1 is idle.
13.066   Arrival      ------------ Start event ------------
                      Schedule Arrival event at time         13.292
                      Schedule Begin Serv event at time      13.066
13.066   Begin Serv   ------------ Start event ------------
                      Server 1 is busy.
                      Schedule End Serv event at time        13.476
```

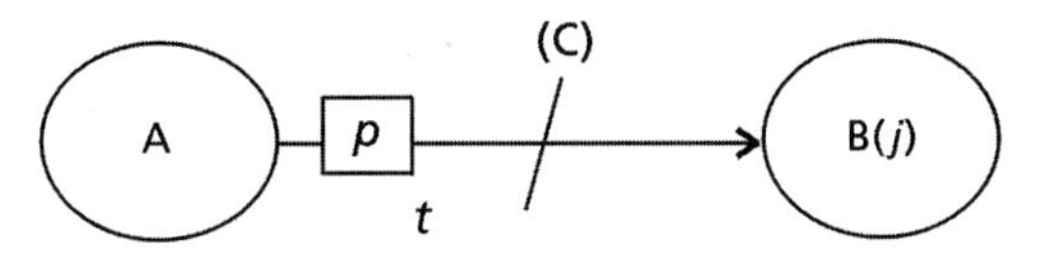

FIGURE 5.25 Event graph with formal and actual parameters

In the *arrival* event, the variable *NA* indexes each customer as he arrives. Each time the *arrival* event routine is executed—that is, each time an arrival occurs—*NA* is incremented by 1. Then *NA* is used to select the element of the array *ATIME*[·] where the arrival time is stored. This is done in the statement "*ATIME*[*NA*] := NOW." The event routines *begin service* and *end service* each have a single parameter, which is the index number of the customer involved. In the *begin service* event routine, this is used to access the arrival time and compute the waiting time in the statement "*W* := NOW – *ATIME*[*N*]." This index is also passed to the *end service* event routine by the *begin service* event routine when the *end service* event is scheduled. The *end service* event routine uses this event parameter in the state-

FIGURE 5.26 Event graph of two-server queue with nonidentical servers

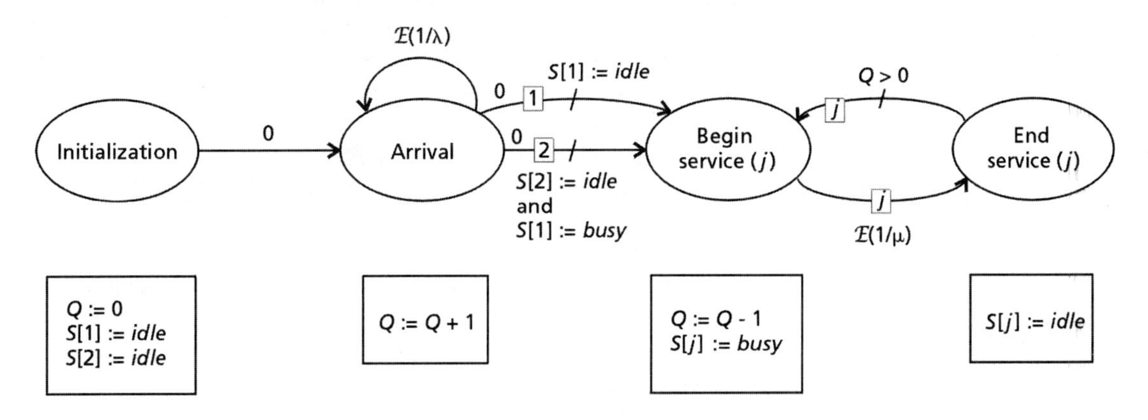

FIGURE 5.27 Event routines for single-server queue with data collection for waiting times

```
Routine Initialize
  Q := 0
  S := idle
  NA := 0
  Schedule arrival at time NOW

Event routine arrival
  Schedule arrival event at time NOW + interarrival time
  NA := NA + 1
  ATIME[NA] := NOW
  Q := Q + 1
  IF S := idle THEN schedule a begin service(NA) event at time
    NOW

Event routine begin service(N)
  Q := Q - 1
  S := busy
  W := NOW - ATIME[N]
  Schedule end service(N) event at time NOW + service time

Event routine end service(N)
  S := idle
  IF Q > 0 THEN schedule begin service(N + 1) event at time
    NOW
```

ment "schedule *begin service*($N + 1$) event at time *NOW*" to specify the index of the next customer to begin service. This works because the queue is operated in a first-come, first-served fashion. If another queue discipline were used, a different method would need to be used to keep track of which customer is involved. Finally, the initialization routine must initialize the customer index variable, *NA*, to zero.

Figure 5.28 contains a trace of the first 20 events for this simulation with exponential interarrival and service times. It is important to understand how the customer's index number is passed from event to event. At time 1.787, customer 2 arrives. At this point, $NA = 2$ and $ATIME[2] = 1.787$. Because the server is idle, a *begin service*(2) event is scheduled at time 1.787. The index number, 2, is kept with the event notice and known by the *begin service* event routine. This event routine then uses this index to compute the waiting time: $W = \text{NOW} - ATIME[2] = 1.787 - 1.787 = 0$. Finally, the index number, 2, is passed to the *end service* event routine when the *end service* event is scheduled to occur at time 4.492 for customer 2. At times 2.602 and 3.708, customers 3 and 4 arrive. Because the server is busy, the queue size, Q, grows to 2. Then, at time 4.492, the *end service* event occurs for customer 2. The event routine knows that customer 2 finished service, so it knows that the next customer in the queue is number $2 + 1 = 3$. A *begin service*(3) event is scheduled to occur at time 4.492. In this event routine, the waiting time for customer 3 is com-

FIGURE 5.28
Trace of 20 events in single-server queue with event parameters

```
          Simulation of an M/M/1 Queueing System: Waiting times.
                    Arrival Rate:     .700
                    Service Rate:    1.000

Time       Event                      Action(s)
======================================================================
                     Run trace for 20 events.
                     System initialized to have 0 customers.
                     Generate new Arrival event at time           .000
  .000   Arrival     ------------ Start event ------------
                     Schedule Arrival event at time              1.787
                     Customer number 1 arrives.
                     Add customer to queue. Length is 1
                     Generate new Begin Serv event at time        .000
  .000   Begin Serv  ------------ Start event ------------
                     Customer number 1. Waiting time:             .000
                     Take customer 1 from queue. Length is 0
                     Server is busy.
                     Generate new End Serv event at time          .124
  .124   End Serv    ------------ Start event ------------
                     Customer number 1 leaves.
                     Queue length is 0. Server is idle.
 1.787   Arrival     ------------ Start event ------------
                     Schedule Arrival event at time              2.602
                     Customer number 2 arrives.
                     Add customer to queue. Length is 1
                     Generate new Begin Serv event at time       1.787
 1.787   Begin Serv  ------------ Start event ------------
                     Customer number 2. Waiting time:             .000
                     Take customer 2 from queue. Length is 0
                     Server is busy.
                     Generate new End Serv event at time         4.492
 2.602   Arrival     ------------ Start event ------------
                     Schedule Arrival event at time              3.708
                     Customer number 3 arrives.
                     Add customer to queue. Length is 1
 3.708   Arrival     ------------ Start event ------------
                     Schedule Arrival event at time              4.625
                     Customer number 4 arrives.
                     Add customer to queue. Length is 2
 4.492   End Serv    ------------ Start event ------------
                     Customer number 2 leaves.
                     Queue length is 2. Server is idle.
                     Generate new Begin Serv event at time       4.492
 4.492   Begin Serv  ------------ Start event ------------
                     Customer number 3. Waiting time:            1.889
                     Take customer 3 from queue. Length is 1
                     Server is busy.
                     Generate new End Serv event at time         8.460
 4.625   Arrival     ------------ Start event ------------
                     Schedule Arrival event at time              5.002
                     Customer number 5 arrives.
                     Add customer to queue. Length is 2
```

FIGURE 5.28 (continued)

```
Time       Event                          Action(s)
===========================================================================
5.002   Arrival      ------------ Start event ------------
                     Schedule Arrival event at time                 7.444
                     Customer number 6 arrives.
                     Add customer to queue. Length is 3
7.444   Arrival      ------------ Start event ------------
                     Schedule Arrival event at time                 8.003
                     Customer number 7 arrives.
                     Add customer to queue. Length is 4
8.003   Arrival      ------------ Start event ------------
                     Schedule Arrival event at time                 9.250
                     Customer number 8 arrives.
                     Add customer to queue. Length is 5
8.460   End Serv     ------------ Start event ------------
                     Customer number 3 leaves.
                     Queue length is 5. Server is idle.
                     Generate new Begin Serv event at time          8.460
8.460   Begin Serv   ------------ Start event ------------
                     Customer number 4. Waiting time: 4.752
                     Take customer 4 from queue. Length is 4
                     Server is busy.
                     Generate new End Serv event at time            8.557
8.557   End Serv     ------------ Start event ------------
                     Customer number 4 leaves.
                     Queue length is 4. Server is idle.
                     Generate new Begin Serv event at time          8.557
8.557   Begin Serv   ------------ Start event ------------
                     Customer number 5. Waiting time:               3.932
                     Take customer 5 from queue. Length is 3
                     Server is busy.
                     Generate new End Serv event at time            8.715
8.715   End Serv     ------------ Start event ------------
                     Customer number 5 leaves.
                     Queue length is 3. Server is idle.
                     Generate new Begin Serv event at time          8.715
8.715   Begin Serv   ------------ Start event ------------
                     Customer number 6. Waiting time:               3.713
                     Take customer 6 from queue. Length is 2
                     Server is busy.
                     Generate new End Serv event at time            9.125
9.125   End Serv     ------------ Start event ------------
                     Customer number 6 leaves.
                     Queue length is 2. Server is idle.
                     Generate new Begin Serv event at time          9.125
===========================================================================
Final r. n. seed:    45875
```

FIGURE 5.29
Results of run of single-server queue to estimate mean waiting time

```
        Simulation of an M/M/1 Queueing System: Waiting times.
                  Arrival Rate:           .700
                  Service Rate:          1.000
            System initialized to have 0 customers.

Results of simulation run:
Batch means method applied to queue lengths.
                          Number of batches:      10
                                 Batch size:     200
                          Initial transient:     100
                           Grand batch mean:       1.781
                     Variance of batch means:       .685
             Standard deviation of batch means:       .828
                  Standard error of batch means:       .262
========================================================
Final r. n. seed:    1486997134
```

puted as $NOW - ATIME[3] = 4.492 - 2.602 = 1.890$. In this way, the waiting time for each customer can be computed.

Figure 5.29 shows the results of running this simulation using arrival rate .7 per minute and service rate 1.0 per minute, and collecting 10 batches of 200 waiting times after an initial transient of 100 customers. A 90 percent confidence interval for the mean waiting time was computed using the batch means method as described in Chapter 4. The resulting confidence interval is $1.781 \pm 1.833(.828/\sqrt{10})$, or from 1.301 to 2.261. In Chapter 4, we saw that the true mean waiting time is 1.633 minutes, so this value is included in this confidence interval. The confidence interval is quite wide, however, and a much larger sample size is needed to produce a precise confidence interval.

5.5.5 Event Canceling

Event graphs can also represent an interaction when one event cancels another. Two basic forms of event canceling are represented in Figure 5.30. In Figure 5.30(a)—which represents immediate, unconditional event canceling at the moment when an instance of event A occurs—any previously scheduled instance of event B is canceled. Figure 5.30(b) shows event canceling with a condition and a delay. In this case, all previously scheduled instances of event B will be canceled t time units after the occurrence of an instance of event A, provided condition (C) holds at the moment when the instance of event A occurs. The use of a time delay with event canceling is rare. Normally, events are canceled immediately. If the canceling arc has an attribute expression for parameter value(s), the condition that only scheduled events with parameter values equal to the values of these arc expressions are canceled is implied. This feature allows, for example, the modeling of the breakdown

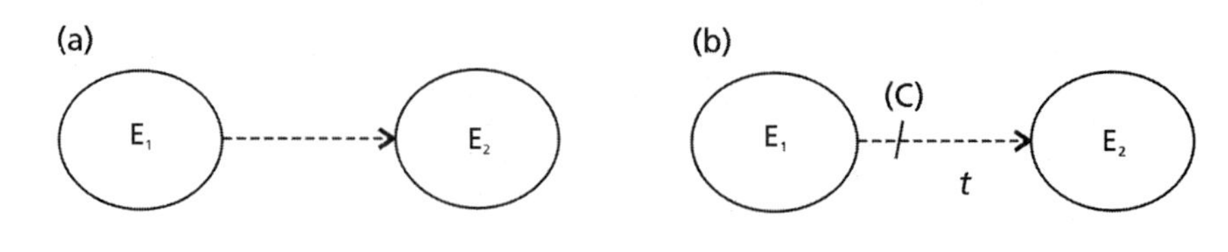

FIGURE 5.30 Representation of event canceling in event graphs: (a) event cancelling; (b) with condition and time

of a particular machine in a job shop having several machines because arc attributes can be used to select the particular machine to break down.

The event graph in Figure 5.31 demonstrates event canceling in a model of a queueing system. In this model, the system has a closing time such that new customers will not be allowed to arrive after the closing time. However, all customers who are in the shop at the closing time will be served. In this model, the Shop Open event schedules the first Customer Arrival event and also a Shop Close event. The Customer Arrival event is scheduled immediately; the Shop Close event is scheduled at the closing time, t. When the Shop Close event occurs, it immediately cancels the current Customer Arrival event, which represents the next arrival. When this event is canceled, the sequence of arrivals is interrupted since each arrival schedules the next arrival. All customers in the shop at closing time will continue to be served because all End Service events are not affected.

The Shop Close event also schedules a Shop Open event at the opening time the next day, allowing the simulation of another working day to begin. If the Shop Close event did not schedule the Shop Open event, then the simulation would finish when the last customer in the system completes service, since there would be no more events in the event list.

The use of events with event parameters is a powerful and flexible modeling tool for representing a wide variety of discrete-event systems. Unfortunately, most systems that operate in the real world are complex and their states cannot reasonably be represented using a few variables and arrays as we did in these examples. The next section presents a modeling framework that, when combined with

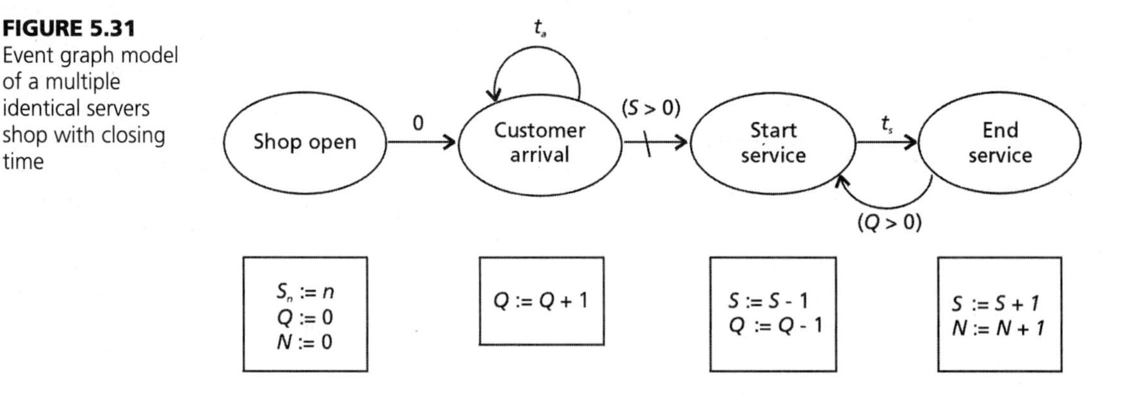

FIGURE 5.31 Event graph model of a multiple identical servers shop with closing time

the event scheduling dynamic representation, provides a powerful tool to model and simulate complex real-world systems.

5.6 STATIC MODEL DESCRIPTION

5.6.1 Entities and Attributes

The entity–attribute–set framework will be used to describe model components. Under this framework, a system is composed of *entities* that have *attributes* and may belong to *sets*. An *entity* is any conceptual component of the system that interacts with other components. It may be a physical component, such as a customer in a checkout line, a cashier serving a checkout line, a patient in a hospital, an inventory of supplies, or a part being manufactured—or it may be a conceptual component such as a message in a communication network or a record to denote that an item is on order. The system operates through some interaction between entities. This interaction may take many forms, but it frequently involves competition for limited resources. For example, customers waiting in a queue all wish to use the server; however, because the server can serve only one customer at a time, all customers except the one being served must wait to begin service. As a second example, an assembly line has two cranes, but several items must use the cranes during assembly. Normally, entities competing for limited resources must wait in a queue, or list, until their needs can be met.

The *attributes* of an entity are any information specific to the entity that are needed for the system to operate. Attributes are normally used to determine how the entity interacts with the system or to record data about the entity. Attributes can be simple, such as the number of items a supermarket customer purchases or his or her method of payment, or a patient's priority in the emergency department of a hospital. They can also be quite complex, as in a part's routing and material requirements in a large manufacturing system. If the purpose of simulating a queueing system is to estimate mean customer waiting time, then each customer entity must have an attribute to record his arrival time so that waiting time can be computed by subtracting the arrival time from the departure time. Because the arrival time is unique for each customer, it must be an attribute of the customer entity. An entity's attributes normally involve data that apply specifically to that entity.

We can distinguish two types of entities: temporary and permanent. *Temporary entities* must be created and are deleted when they are no longer needed. For example, a customer at the supermarket checkout counter begins interacting with the system when she presents herself to the checkout and completes the interaction and departs. The entity is created on arrival and deleted on departure. Most entities that "flow through" the system are temporary. These include parts in a manufacturing system, passengers in a transportation system, and messages in a communication system. Unlike temporary entities, *permanent entities* exist for the entire duration of the simulation. Examples of permanent entities, which usually represent facilities that are a permanent part of the system, are the cashiers in the supermarket checkout system, terminals in a transportation system, machines in a manufacturing system, and rooms in a hospital. There are no fixed rules for determining whether a given entity should be temporary or permanent—this is a decision the modeler must make. However, the structure of the system usually makes it clear which entities should be permanent and which should be temporary.

5.6.2 Sets, Lists, and Queues

A *set* is an ordered collection of entities. Because the term *set* is used in many other contexts, we will use the terms *list* and *queue* to denote an ordered collection of entities. By an *ordered collection,* we mean that each entity in the list except the last has a unique entity following it, and each entity except the first has a unique entity preceding it. Lists are used to hold entities during system operation. For example, in a queueing system, each queue holds customers until a server is free to provide service. If the system has multiple servers, then the servers themselves may be stored in a list. Items in an inventory and messages on a transmission link may be stored temporarily in a queue.

The entire system may be considered a superentity. Then, because all lists and system parameters become attributes of the system, we will call them *system attributes*. System attributes consist of any data necessary to operate the system or to compute data about system operation. For example, in the supermarket checkout system, system attributes would consist of the number of each type of checkout lane, the checkout lane queues and their memberships, the parameters of the service times and interarrival times, and any other parameters necessary to determine arrivals, services, queue jockeying, and all other actions in the system. In a manufacturing system, the number of workers, machine centers, and the production schedule are system attributes.

System attributes (and entity attributes) may be *parameters*—values that remain constant throughout the operation of the system, or they may be *variables*—values that change in response to changes in the system. The *state* of the system is defined to be the collection of all variables whose values are necessary to determine how the system operates over time. Generally, this includes all system variables, including entities and their attributes and list memberships. Any relevant change in the system involves a change in one or more system state variables. Conversely, the system state consists of all variable information relevant to the operation of the system, and any change in one or more state variables reflects a change in the system. The system state for a model of the supermarket checkout system is given in Table 5.2.

TABLE 5.2 System state for the model of a supermarket checkout system

	Lane					
	1	**2**	**3**	**4**	**5**	**6**
Open?	Y	Y	Y	N	N	Y
Scan rate	110	95	80	0	0	160

Waiting Lines

Customer	Lane	Position	Items	Payment Type	In Service?
1823	1	1	63	Cash	Y
1825	2	1	36	Check	Y
1831	2	2	19	Debit	N
1832	1	2	18	Credit	N
1834	6	1	3	Cash	Y
1835	2	3	6	Credit	N

5.7 A MESSAGE-PROCESSING MODEL THAT USES ENTITIES AND LISTS

In this section, we examine a model that utilizes the entity–attribute–set framework but does not use events. This model represents the operations of a computer system that provides message-routing services along with other unspecified services. The objective of the model is to estimate the average delay per message involved with the message-routing operations. The processor checks the message queue every δ milliseconds (ms) and, if the queue is not empty, handles as many messages as possible in the queue. If the queue is empty, the processor immediately goes on to perform other tasks and returns after a delay of δ ms to again check the queue. During its allotted time, the processor can process up to k messages. If the queue has k or fewer messages, the processor processes all messages in the queue. If the queue contains more than k messages, the processor routes the first k messages and leaves the remaining messages for another cycle.

5.7.1 Top-Down Design for Model Building

The following actions occur in the order given in each cycle:

1. The arriving messages, if any, are received and placed in the queue.
2. The processor removes and processes all messages that can be handled in the cycle.

The following algorithm directs the simulation for N cycles:

(1) Initialize the system.
For $n = 1$ to N do:
(2) Receive arriving messages for cycle n.
(3) Route messages for cycle n.
(4) Compute the estimate of mean waiting time.

Of course, to fully specify the model, we must supply the details for each step. Once they are given, this representation will be a correct and complete system model. This approach is usually referred to as "top-down system design." A great deal has been written about this topic, and it is generally covered in introductory computer programming courses. See Wirth (1973) for a description of the process. Although modern programming techniques emphasize object-oriented concepts, top-down design is still a highly recommended method for designing computer programs. Skilled programmers favor this approach because it provides a means to organize and simplify a complex program while maintaining the correct logical structure. This is accomplished by a divide-and-conquer strategy that delays the details of the design until they are really needed. At each step of the process, components or procedures given in the previous step are defined, but details are added only to the extent needed to adequately describe the portion of the program being developed. For this reason, the process is sometimes called "stepwise refinement." The objective of this chapter is not to teach top-down design, but most system models are very complex and top-down design is an especially useful and perhaps necessary methodology for developing correct models. All models in this chapter and later chapters will be developed using this approach. This will make the structure of the models clear and also teach the principles of top-down design by example.

This model will be presented in three steps: First, the operations of the model will be described verbally, as a person who operates the system would describe it. Second, these operations will be formalized as a pseudocode to make it clearer exactly what actions take place and in what order. Finally, the pseudocode will be translated into our generic simulation language (GSL). This final representation defines specific variables, entities, attributes, and lists, and it also defines the specific operations on these elements that constitute the model. We will introduce the elements and statements of GSL as needed.

5.7.2 Top-Level Model Description

The verbal description of the model is as follows. Messages arrive at the processor according to a Poisson process with rate λ. Therefore, the number of messages arriving in a processing cycle of duration δ is a Poisson random variable with mean $\lambda\delta$. For each message received, an entity is created and placed last in an input queue after the arrival time is recorded. We will record the cycle number, rather than the exact time of arrival. Every δ time units, the processor checks the queue. If the number of messages waiting is less than or equal to k, then all messages are removed and processed. Otherwise, the first k messages are removed and processed. The processing of each message consists of recording the waiting time and deleting the message. (Of course, the message in the real system is passed on to the next node in the network. The model deletes the message because it is removed from the part of the system we are modeling.) Initially, the message queue is empty.

This description has provided the logical structure of the model, but to fully specify the model so that it can be run and parameters can be estimated, we must provide parameter values. Also, we must specify the number of cycles in the simulation run. Table 5.3 lists these parameters and some possible values for them.

The following pseudocode summarizes the main program for the message-routing model. The main routine for message-routing simulation is:

Read model parameters and print a model description.

Initialize the system.

For each cycle, do the following:

Receive new messages.

Process messages.

Compute the mean delay and print a report giving the results of the simulation.

Each italicized item refers to a routine that needs to be defined further. For now, the following model definition omits input and output of parameters, computing the mean delay, and printing the final report.

5.7.3 Model Details

Now consider the actions that must be taken to initialize the system. Basically, all we need to do is to set up an empty queue to hold messages waiting to be processed. We can summarize the actions for the procedure to initialize the system in Figure 5.32.

Our GSL must have facilities to initialize an empty list, create an instance of a temporary entity, assign attribute values to an instance of an entity, and put an in-

TABLE 5.3 Parameters in the message-processing model

Notation	Parameter	Value
λ	Mean arrivals per cycle	2.50
k	Maximum messages processed per cycle	3
	Number of transient observations in run	100
	Number of batches in run	10
	Number of messages per batch	200

FIGURE 5.32

```
Initialize
  Initialize an empty list message_queue
```

stance of an entity in a list. The statement "Initialize empty list *name*" creates a new list, initializes it to be empty, and assigns the identifier *name* to it. Thus, the statement "Initialize empty list *message_queue*" creates an empty list that will be referenced with the name *message_queue.* Italics denote identifiers defined for a particular model. The statement "Create entity *name*" creates a new instance of an entity and assigns the identifier *name* to it. Once the entity has been created, its attributes can be accessed with the notation *name.attribute.* Thus, "Create entity *message*" creates a new instance of an entity with the name *message,* and "*message.arrival_cycle* := 0" assigns the value 0 to the *arrival_cycle* attribute of the entity instance referenced by *message.* Finally, the statement "Place *entity-name position* in *list-name*" places the entity referenced by *entity-name* in the *position* in the list referenced by *list-name.* For example, "Place *message* last in *message_queue*" places the entity referenced by *message* last in the list referenced by *message_queue.*

The specific actions in the *Receive new messages* and *Process messages* procedures are shown in Figures 5.33 and 5.34, respectively.

When messages are processed, N will contain the count of the current cycle. So in the *Receive new messages* procedure, the statement *message.arrival_cycle* := N simply records the current cycle count in the *arrival_cycle* attribute of the entity instance *message.* Similarly, in the *Process messages* procedure, the statement W := N – *message.arrival_cycle* just computes the difference between the current cycle count and the cycle count when the entity instance message arrived. If this difference is zero, the message was processed in the same cycle in which it arrived; but if the difference is positive, the message had to wait one or more cycles to be processed.

Many of the languages for discrete-event simulation such as SIMSCRIPT II.5, SLAM II, and SIMAN could be used to implement this model. If appropriate data structures are used and procedures are available to manage creation and deletion of entities, place them in and remove them from lists, and perform other operations,

FIGURE 5.33

```
Receive messages
Generate M ~ 𝒫(λ)
IF M > 0 THEN
  FOR i := to M DO:
    create message
    message.arrival_cycle := N
    place message last in message_queue
```

FIGURE 5.34

```
Process messages
IF length(message_queue) > 0 THEN
  l := min(length(message_queue),k)
  FOR i := 1 TO l DO
    remove first message from
    message_queue
    W:= N - message.arrival_cycle
    accumulate W
    delete message
```

then the model can also be implemented using a general-purpose language such as C or Java.

5.7.4 Model Output

Figure 5.35 contains a sample trace of the model. In this sample, the arrival rate was set to 5.0 per cycle, while the processor's capacity was three per cycle. This results in an unstable system, but these parameters were used just to force some early cycles to have more messages than could be handled in one cycle. This allowed us to make sure the messages were being handled correctly in the queue and subsequent cycles. A long run to estimate the mean waiting time per message was then made and the results are given in Figure 5.36. The parameters of the last run are as follows. Message arrivals per cycle have a Poisson distribution with a mean of 2.5 per cycle. In each cycle, the processor could process up to three messages. The run involved 10 batches of 200 observations per batch and an initial transient period of 100 observations. The mean of the batch means is .639 cycles, and the standard deviation is .317, producing a 90 percent confidence interval of $.639 \pm 1.833(.317/\sqrt{10})$, or from .455 to .823 cycles.

FIGURE 5.35
Sample simulation run for message-routing model

```
                 The parameters for this model are:

           Mean arrivals per cycle:    5.00
                Maximum Batch Size:    3
               Transient Messages:    0
                Number of Batches:    2
               Messages per Batch:    5
        Starting random number seed:    525454956

 Time     Event                         Action(s)
====================================================================
                     Run trace for 1 events.
                     Set up list Message Q 0

                     Cycle: 1
                     RECEIVE NEW MESSAGES
                        0 Messages in queue.
                        5 new arrivals.
                     Create Message        1
                     Insert Message        1 last in Message Q
                     Create Message        2
                     Insert Message        2 last in Message Q
                     Create Message        3
                     Insert Message        3 last in Message Q
                     Create Message        4
                     Insert Message        4 last in Message Q
                     Create Message        5
                     Insert Message        5 last in Message Q
                     Process Messages.
                     5 Messages in queue.
                     Remove Message        1 from Message Q
                     Message 1 waited 0 cycles.
                     Delete Message        1
                     Remove Message        2 from Message Q
                     Message 2 waited 0 cycles.
                     Delete Message        2
                     Remove Message        3 from Message Q
                     Message 3 waited 0 cycles.
                     Delete Message        3

                     Cycle: 2
                     RECEIVE NEW MESSAGES
                        2 Messages in queue.
                        5 new arrivals.
                     Create Message        6
                     Insert Message        6 last in Message Q
                     Create Message        7
                     Insert Message        7 last in Message Q
                     Create Message        8
                     Insert Message        8 last in Message Q
                     Create Message        9
                     Insert Message        9 last in Message Q
                     Create Message        10
                     Insert Message        10 last in Message Q
```

FIGURE 5.35 (continued)

```
Time        Event                           Action(s)
=====================================================================
                                Process Messages.
                                   7 Messages in queue.
                                Remove Message        4 from Message Q
                                Message 4 waited 1 cycles.
                                Delete Message        4
                                Remove Message        5 from Message Q
                                Message 5 waited 1 cycles.
                                Delete Message        5
                                Remove Message        6 from Message Q
                                Message 6 waited 0 cycles.
                                Delete Message        6

                                Cycle: 3
                                RECEIVE NEW MESSAGES
                                   4 Messages in queue.
                                   4 new arrivals.
                                Create Message        11
                                Insert Message        11 last in Message Q
                                Create Message        12
                                Insert Message        12 last in Message Q
                                Create Message        13
                                Insert Message        13 last in Message Q
                                Create Message        14
                                Insert Message        14 last in Message Q

                                Process Messages.
                                   8 Messages in queue.
                                Remove Message        7 from Message Q
                                Message 7 waited 1 cycles.
                                Delete Message        7
                                Remove Message        8 from Message Q
                                Message 8 waited 1 cycles.
                                Delete Message        8
                                Remove Message        9 from Message Q
                                Message 9 waited 1 cycles.
                                Delete Message        9

                                Cycle: 4
                                RECEIVE NEW MESSAGES
                                   5 Messages in queue.
                                   6 new arrivals.
                                Create Message        15
                                Insert Message        15 last in Message Q
                                Create Message        16
                                Insert Message        16 last in Message Q
                                Create Message        17
                                Insert Message        17 last in Message Q
                                Create Message        18
                                Insert Message        18 last in Message Q
                                Create Message        19
                                Insert Message        19 last in Message Q
```

FIGURE 5.35 (continued)

```
Time        Event                           Action(s)
======================================================================
                    Create Message         20
                    Insert Message         20 last in Message Q      0

                    Process Messages.
                       11 Messages in queue.
                    Remove Message         10 from Message Q         0
                    Message 10 waited 2 cycles.
                    Delete Message         10
                    Remove Message         11 from Message Q         0
                    Message 11 waited 1 cycles.
                    Delete Message         11
                    Remove Message         12 from Message Q         0
                    Message 12 waited 1 cycles.
                    Delete Message         12

Results of simulation run:
Cell    Frequency
  0         3
  1         7
  2         1
  3         0
  4         0
  5         0
  6         0
  7         0
  8         0
  9         0
 10         0

Final random number seed:      1592980298
```

5.8 A SINGLE-SERVER QUEUE USING ENTITIES AND EVENTS

In the previous sections, we presented the concepts of events and entities separately. These components will be combined in this section and future chapters to show how they can work together to provide a powerful modeling tool for a broad range of complex systems.

The model that we consider represents the same system as in section 5.5: a queueing system with a single server and a single queue. However, we use entities, attributes, and lists to represent the system state rather than just two variables that represent the number of customers in the queue and the status of the server. The purpose of the model is to estimate the mean waiting time per customer in the queue.

FIGURE 5.36
Results of a long simulation run to estimate mean cycles for message-routing model

```
               The parameters for this model are:
        Mean arrivals per cycle:    2.50
             Maximum Batch Size:    3
             Transient Messages:    100
              Number of Batches:    10
             Messages per Batch:    200
    Starting random number seed:    1948827033

                          Batch mean:     1.270
                          Batch mean:      .420
                          Batch mean:      .700
                          Batch mean:      .570
                          Batch mean:      .455
                          Batch mean:     1.075
                          Batch mean:      .750
                          Batch mean:      .480
                          Batch mean:      .305
                          Batch mean:      .365

Results of simulation run:
Cell    Frequency
 0         1093
 1         636
 2         198
 3         54
 4         21
 5         0
 6         0
 7         0
 8         0
 9         0
10         0
                Overall Mean:                       .63900
       Variance of Batch Means:                     .10031
        Standard Error of BM's:                     .10015
      Final random number seed:         1814465532
```

The system has two types of entities: customer and the server. Customer is a temporary entity that is created on arrival, interacts with the system by joining the queue before being served, and is deleted after service is completed. Each customer has a single attribute—their arrival time—that is assigned a value on arrival and later used to compute the waiting time in queue. The server is a permanent entity with two attributes: status (busy or idle) and parameter(s) of service time distribution. If the service time distribution is exponential, the mean is the only parameter required; if it is normally distributed, the mean and standard deviation are needed. In this example, we will assume that service times are exponentially distributed. Figure 5.37 shows these two entity types with their attributes. The system has a single queue that is assumed to have an infinite capacity.

FIGURE 5.37 Structure of customer and server entities

Customer	
Arrival time	0.912

Server	
Status	idle
Mean service time	1.6

As before, the dynamic model description has three events: arrival of a customer, beginning of service, and end of service. The event routines for these events are shown in Figure 5.38.

These event routines look very similar to those presented in section 5.5, but there are important differences. Because customers are temporary entities, each instance of a customer entity must be created in the `arrival` event routine. The statement "*Customer.arrival time* := *NOW*" assigns the current system time to the *arrival time* attribute of the instance of a customer that was just created. The notation *entity.attribute* denotes the value of a specific attribute of a specific entity. The *customer* entity is then placed last in the *queue*. To see if service can begin on arrival, the *status* attribute of the server entity is checked, and, if its value is *idle,* the *begin service* event is scheduled immediately. There is no need to keep a counter of the number of customers in the queue or in the system because that information is contained in the length of the queue and the server's status attribute.

In the *begin service* event routine, first we remove the first *customer* entity from the queue. The model is set up so that the next customer to be served is always the first customer in the queue. Then we set the server's status attribute to *busy* to show

FIGURE 5.38 Event routines for single-server queue with entities

```
Event routine arrival
Schedule arrival event at time NOW + interarrival time
Create customer
Customer.arrival time := NOW
Insert customer last in queue
IF Server.status = idle
   THEN schedule begin service event at time NOW

Event routine begin service
Remove the first customer from queue
Server.status := busy
Waiting time := NOW - customer.arrival time
Collect waiting time data
Schedule end service (customer) event at time NOW +
   exponential(server.meanservtime)

Event routine end service(customer)
delete(customer)
Server.status := idle
IF length(queue) > 0
   THEN schedule begin service event at time NOW
```

FIGURE 5.39

```
Routine initialize
Initialize empty queue
Server.status := idle
Schedule arrival at time NOW
```

that the server is not available to serve arriving customers. Next we compute the customer's waiting time in queue because this is the moment when the waiting in queue ends. The statement "*Collect waiting time data*" refers to doing whatever data collection is needed for the particular statistical procedure that will be used. (If the batch means method is used to compute a confidence interval for the mean waiting time, then this routine would check for the end of the initial transient and, if the initial transient is passed, accumulate the batch mean for the current batch and finally check for the end of the simulation run.) Finally, we schedule the *end service* event to occur at a time after the service time from now.

The *end service* event routine has a single parameter that identifies the specific customer entity that is completing service. These data must be passed to the *end service* event routine because the routine must delete the customer (and therefore must know exactly which customer entity is completing service), and the customer entity is no longer in a queue. The *end service* event routine begins by deleting the instance of the customer entity that has just completed service. Then the length of the queue is checked to see if another customer is waiting to begin service. If so, the *begin service* event is scheduled immediately.

The initialization routine in Figure 5.39 is similar to those before. This routine assumes the system is to be initialized to the empty-and-idle state, with the queue empty and the server idle. Many simulation languages automatically initialize all queues to be empty at the start of execution, but we explicitly chose to represent this step. If the system were initialized to have three customers in the system, then the initialization routine would look like Figure 5.40.

This routine places two customer entities in the queue first, but leaves the server's status attribute with the value *idle* so that when the first *arrival* event occurs, a customer will go immediately into service and the system will start with one customer in service and two customers in the queue.

Figure 5.41 contains a trace of the first 20 events of a simulation run for this model. The input parameters (arrival rate and service rate) were exactly the same

FIGURE 5.40

```
Routine initialize
Initialize empty queue
For i := 1 to 2 do
   Create customer
   customer.arrival time := NOW
   Insert customer last in queue
Server.status := idle
Schedule arrival at time NOW
```

FIGURE 5.41
Sample trace of M/M/1 queue simulation with entities

```
          Simulation of an M/M/1 Queueing System: Waiting times.
                        Arrival Rate:          .700
                        Service Rate:         1.000

 Time   Event          Action(s)
======================================================================
                       Run trace for 20 events.
                       Set up list Queue 1
                       System initialized to have 0 customers.
                       Schedule new Arrival event at time            .000
  .000  Arrival        ------------ Start event ------------
                       Schedule Arrival event at time               1.787
                       Create Customer 1
                       Insert Customer 1 last in Queue 1
                       Server status is idle.
                       Schedule new Begin Serv event at time         .000
  .000  Begin Serv     ------------ Start event ------------
                       Take Customer 1 (first) from Queue 1
                       Server is now busy.
                       Customer number 1. Waiting time:              .000
                       Schedule new End Serv event at time           .124
  .124  End Serv       ------------ Start event ------------
                       Delete Customer 1
                       Server is now idle.
                       Queue length is 0.
 1.787  Arrival        ------------ Start event ------------
                       Schedule Arrival event at time               2.602
                       Create Customer 2
                       Insert Customer 2 last in Queue 1
                       Server status is idle.
                       Schedule new Begin Serv event at time        1.787
 1.787  Begin Serv     ------------ Start event ------------
                       Take Customer 2 (first) from Queue 1
                       Server is now busy.
                       Customer number 2. Waiting time:              .000
                       Schedule new End Serv event at time          4.492
 2.602  Arrival        ------------ Start event ------------
                       Schedule Arrival event at time               3.708
                       Create Customer 3
                       Insert Customer 3 last in Queue 1
                       Server status is busy.
 3.708  Arrival        ------------ Start event ------------
                       Schedule Arrival event at time               4.625
                       Create Customer 4
                       Insert Customer 4 last in Queue 1
                       Server status is busy.
 4.492  End Serv       ------------ Start event ------------
                       Delete Customer 2
                       Server is now idle.
                       Queue length is 2
                       Schedule new Begin Serv event at time        4.492
 4.492  Begin Serv     ------------ Start event ------------
                       Take Customer 3 (first) from Queue 1
                       Server is now busy.
```

FIGURE 5.41 (continued)

```
 Time   Event        Action(s)
=======================================================================
                     Customer number 3. Waiting time:           1.889
                     Schedule new End Serv event at time        8.460
 4.625  Arrival      ------------ Start event ------------
                     Schedule Arrival event at time             5.002
                     Create Customer 5
                     Insert Customer 5 last in Queue 1
                     Server status is busy.
 5.002  Arrival      ------------ Start event ------------
                     Schedule Arrival event at time             7.444
                     Create Customer 6
                     Insert Customer 6 last in Queue 1
                     Server status is busy.
 7.444  Arrival      ------------ Start event ------------
                     Schedule Arrival event at time             8.003
                     Create Customer 7
                     Insert Customer 7 last in Queue 1
                     Server status is busy.
 8.003  Arrival      ------------ Start event ------------
                     Schedule Arrival event at time             9.250
                     Create Customer 8
                     Insert Customer 8 last in Queue 1
                     Server status is busy.
 8.460  End Serv     ------------ Start event ------------
                     Delete Customer 3
                     Server is now idle.
                     Queue length is 5
                     Schedule new Begin Serv event at time      8.460
 8.460  Begin Serv   ------------ Start event ------------
                     Take Customer 4 (first) from Queue 1
                     Server is now busy.
                     Customer number 4. Waiting time:           4.752
                     Schedule new End Serv event at time        8.557
 8.557  End Serv     ------------ Start event ------------
                     Delete Customer 4
                     Server is now idle.
                     Queue length is 4
                     Schedule new Begin Serv event at time      8.557
 8.557  Begin Serv   ------------ Start event ------------
                     Take Customer 5 (first) from Queue 1
                     Server is now busy.
                     Customer number 5. Waiting time:           3.932
                     Schedule new End Serv event at time        8.715
 8.715  End Serv     ------------ Start event ------------
                     Delete Customer 5
                     Server is now idle.
                     Queue length is 3
                     Schedule new Begin Serv event at time      8.715
 8.715  Begin Serv   ------------ Start event ------------
                     Take Customer 6 (first) from Queue 1
                     Server is now busy.
                     Customer number 6. Waiting time:           3.713
                     Schedule new End Serv event at time        9.125
```

FIGURE 5.41 (continued)

```
 Time   Event        Action(s)
===================================================================
9.125   End Serv     ------------ Start event -----------
                     Delete Customer 6
                     Server is now idle.
                     Queue length is 2
                     Schedule new Begin Serv event at time      9.125
```

as the model presented in section 5.5, and the same initial random number seed was also used. The significance of using the same seed is that exactly the same sequence of random numbers was produced. Comparing Figure 5.39 with Figure 5.28, one can see that the same sequence of events was produced. The only difference between the two versions of this model is the way customers and the server were represented.

5.9 SUMMARY

This chapter has presented two important new ideas: the event view, or event scheduling time advance mechanism, and the entity–attribute–set modeling framework. Together, these concepts allow us to develop models of complex systems and represent their changes over time.

We began by introducing the concept of a discrete-event dynamic system as one whose state changes only at discrete points in time to mark the occurrence of an event. The time-varying behavior of such a system is described by the dynamic model description. Normally, there are a small number of distinct *types* of events but a large number of *occurrences* of each type of event. Each type of event has an associated set of instructions, called an *event routine,* that specifies exactly how the system state changes when an event of the given type occurs. The simulation maintains a system clock that always holds the current system time. Each event occurrence has an associated event notice. The system maintains a list of scheduled events that contains an event notice for each scheduled event occurrence and is kept in the order of increasing time of event occurrence. The timing routine controls the order of execution of events by removing the first event notice from the event list, updating the system clock to the event time, and executing the event routine for the type of event represented by the event notice. This process is repeated until the event list is empty or another signal is given to the timing routine to cease.

Several special simulation languages are available to implement discrete-event simulation models. General-purpose programming languages such as C, Java, and Pascal can also be used if declarations and libraries of routines are available to manage the event list and implement the timing routine. To avoid focusing on one simulation language and to concentrate on the essential concepts, we introduced an abstract generic simulation language (GSL) that allows all of the capabilities of a general-purpose programming language and extends them to include an event list, system clock, timing routine, event routine declaration, and commands to schedule and cancel events.

The first example presented was a single-server queue. The system state was represented by two variables: Q, the number of customers in the queue, and S, the state of the server (busy or idle). Three types of events represented the arrival of a customer, beginning of service for a customer, and completion of service for a customer. Data were collected on the queue length process to estimate the mean number of customers in the queue using the batch means method.

Event graphs visually show the scheduling and canceling relationships between and among events. When data must be passed to an event, event parameters can be used. A second example model of a queueing system having two nonidentical servers demonstrated the use of event parameters to allow the correct server to be engaged. A third example showed how event parameters are useful for collecting waiting time data on each customer.

The static model description specifies what components are in the system and their relationships to one another. The entity–attribute–set modeling framework describes a system as composed of entities that have attributes and may belong to sets (queues or lists). This framework is a widely used representation that can model a broad range of systems in manufacturing, service, transportation, communications, and other areas. A model of a message-processing system illustrated the use of the entity–attribute–set modeling framework exclusive of the event scheduling mechanism.

Finally, the entity–attribute–set framework was combined with the event scheduling time advance mechanism and demonstrated using the single-server queue. Later chapters will show how discrete-event simulations of more complex queueing and service systems can be developed using these two representations in combination.

References

Heyman, D. P., and M. J. Sobel. 1982. *Stochastic models in operations research, Volume I.* New York: McGraw-Hill.

Little, J. D. C. 1961. A proof of the queueing formula: $L = \lambda W$. *Operations Research* 9:383–387.

Sargent, R. G. 1988. Event graph modelling for simulation with an application to flexible manufacturing systems. *Management Science* 34:1231–1251.

Schruben, L. W. 1995. *Graphical simulation modeling and analysis using Sigma for Windows.* Danvers, Massachusetts: Boyd & Fraser.

———. 1983. Simulation modeling with event graphs. *Communications of the ACM* 26:957–963.

Wirth, N. 1973. *Systematic programming: An introduction.* Englewood Cliffs, New Jersey: Prentice Hall.

Problems

5.1 Modify the model of the single-server queue to have only a single state variable, N, the number of customers in the system (including service).

5.2 Modify the single-server queue model so customers remain in the queue while receiving service.

5.3 Develop a model of a two-server queue with identical servers. Do not use arrays to store the system state.

5.4 Develop a model of a two-server queue in which servers are identical and each server has its own queue. Arriving customers choose the shortest queue.

5.5 Develop a model of a heterogeneous server queue that has $s > 2$ servers, where s can be specified as part of the input parameters. Assume that service times for server i are independent and identically distributed exponential random variables with mean $1/\mu_i$.

5.6 Modify the model of the single-server queue so that the queue discipline is last-come, first-served.

5.7 Modify the model of the single-server queue to include a limit, L, on the number of customers who can be in the queue.

5.8 Modify the model of the single-server queue to make the server provide server at a faster rate, μ_f, if there are two or more customers left in the queue when service begins; and a slower rate, μ_s, when the number of customers left is zero or one.

5.9 Modify the single-server queue to serve customers in batches of three. If the number of customers in the queue is less than three, then the server waits until three are present to begin service.

5.10 Modify the single-server queue to serve customers in batches of one, two, or three. If three or fewer customers are in the queue when service begins, the server will serve all who are waiting in one batch. Otherwise, the server takes three and the remainder wait.

5.11 Modify the heterogeneous server queue in problem 5.5 to include a server preference by the customer. When each customer arrives, she specifies a server preference. If a customer waits more than five minutes, she will accept either server, however. When a server becomes free, he searches the queue for the first customer willing to be served by him. If none is found, the server becomes idle until a customer requests him.

5.12 Modify the model of the heterogeneous server queue in problem 5.5 to collect waiting times for customers.

5.13 Modify the initialization routine in problem 5.11 so that the system is initialized with three customers in the queue and one in service.

5.14 Modify the model of the single-server queue so that when a customer finishes service, he has a probability p of leaving and $1 - p$ of rejoining the queue as he did on arrival.

5.15 Given the event graphs in Figure 5.42, simplify them by removing all nodes for events that are scheduled simultaneously with other events.

5.16 In a single-server queueing system, the server is subjected to preventive maintenance on the first end of service after 10 hours of service. The maintenance time is exponentially distributed with a mean of two hours.

a. Specify the variables needed to represent the system state.

b. Specify the events and develop an event graph for this system.

c. Give algorithms for the event routines for all events in the system.

5.17 In a single-server queueing system, the customer will leave (without starting service) if his wait in line is longer than a uniformly distributed random variate between .0 and 3.0 minutes. Such behavior is called *reneging* in the queueing literature. The simulation is to be used to estimate mean waiting time in queue and to estimate the proportion of arriving customers who actually receive service.

a. Specify the variables needed to represent the system state.

b. Specify the events and develop an event graph for this system.

c. Give algorithms for the event routines for all events in the system.

FIGURE 5.42

(a)

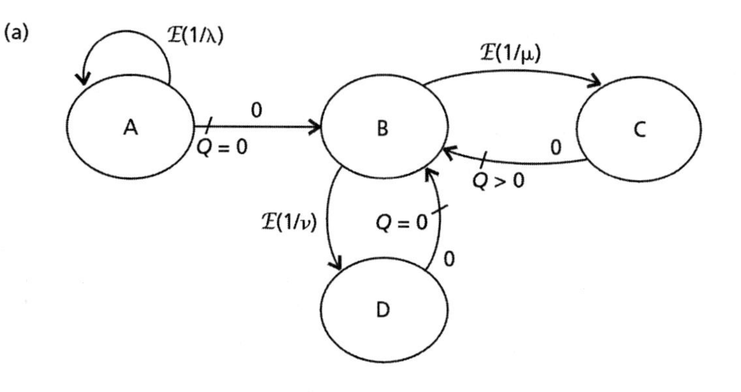

(b)

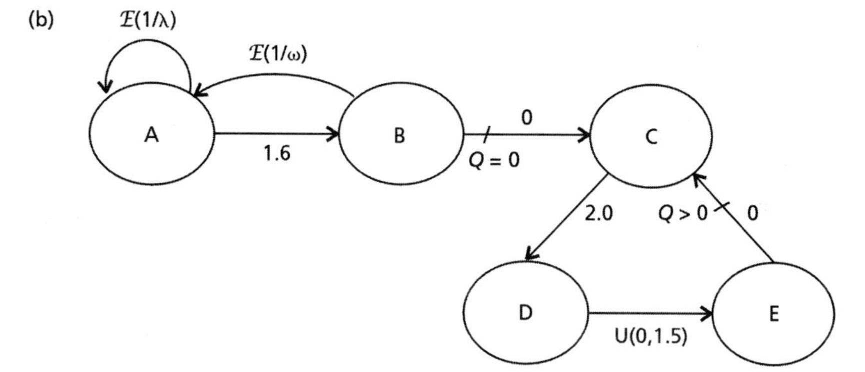

(c)

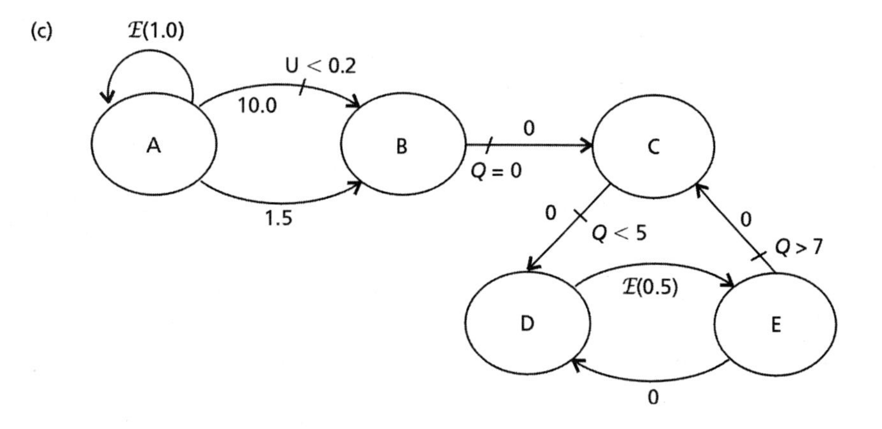

5.18 In the message-processing model in section 5.7, modify the model to let each message have a length in bytes. The length is sampled from a normal distribution with mean 10,000 bytes and standard deviation 2000 bytes. Lengths must be non-negative, however. The router can route up to b bytes per cycle (rather than k messages). Assume that the length of each message is less than b bytes. Modify the model to include these details.

5.19 In the message-processing model in section 5.7, assume that each message has a priority that can be 1 or 2. Priority 1 messages are always routed before priority 2 messages. Modify the model to incorporate this characteristic.

5.20 In the message-processing model in section 5.7, suppose that each message has a length as presented in problem 5.18, but suppose that messages are processed shortest message first. Modify the model to incorporate this characteristic.

5.21 In the message-processing model in section 5.7, assume that the router is adaptive. We would like to utilize the capacity of the router, but we do not want messages to wait too long. Suppose that if the average number of messages processed per cycle for the last three cycles is greater than $.75k$, then the router will increase its capacity per cycle to $1.2k$ by allocating more time per cycle to message routing. If the average number of messages routed per cycle for the last three cycles is less than $.4k$, then the router will decrease its capacity to $.75k$ by allocating more time per cycle to other activities (hence, less time to routing). Incorporate these characteristics in the model.

6

System Modeling Paradigms

This chapter, along with Chapter 5, presents the basic modeling concepts for discrete-event simulation. In Chapter 5, we discussed the event view and the entity–attribute–set representation in some detail. Here we want to expand and extend the discrete-event simulation concepts by first summarizing the basic ideas of Chapter 5 to put them in perspective. Chapter 5 dealt mostly with details. In our summary, we want to look at the "big picture." Then we want to introduce two additional worldviews—the activity view and the process view. We will illustrate these alternatives to the event view using the same single-server queue as in the previous chapter. The purpose of using a single-server queue is to focus on the fundamental modeling concepts without being distracted by the details of a more complex or slightly different model. The reader should be able to develop a model of a single-server queue using any of the three worldviews when they finish section 6.4. In the problems at the end of the chapter, we will ask you to modify the single-server model to have multiple identical servers or two nonidentical servers, and to present the model using one of the three worldviews.

In section 6.2, we briefly review the concepts of the event view and then expand them by introducing the ideas of resources as permanent entities that are used by temporary entities in the course of system operation. Section 6.3 introduces the activity view as an alternative way to arrange the events in the simulation. The ideas of conditional and scheduled events are introduced, and we make the point that the activity view allows the modeler to make events independent and modular. This causes the timing routine to be somewhat less efficient, but the event routines can be simpler and more reliable. The process view carries this a step further by joining a group of events to follow the progress of an entity through the system. All three worldviews will be further illustrated with a more complex example—a manufacturing system. The idea is to now expand the reader's view away from the single-server queue, and to make these concepts have a more general field of application.

6.1 INTRODUCTION

The term *worldview* refers to the way the modeler conceptualizes the operations of a model. We can think of it as the basic framework on which the model is developed. During the development of simulation modeling concepts and software, which began simultaneously with the development of symbolic programming languages, several simulation worldviews have been developed. The three most popular are the *event view,* the *activity view,* and the *process view.* The previous chapter presented the event view in some detail, including the event scheduling timing mechanism and the entity–attribute–set framework for describing the structure of objects in the simulation model. In this chapter, we want to extend these concepts in two directions. First, we will be expanding our worldview to include two additional approaches to system dynamics: the activity view and the process view. These worldviews provide two more ways to organize events in a discrete-event simulation. Second, we will examine a more complex manufacturing model and use it to demonstrate how these techniques can be used to represent models that are much more complex than the single-server queue. Each of the three worldviews is the modeling basis for several simulation languages, and some languages (e.g., SIMSCRIPT II.5) are able to use more than one simulation view. This chapter will therefore provide the background to understand the simulation languages and other software in use today.

6.2 REVIEW OF DISCRETE-EVENT DYNAMIC SYSTEMS CONCEPTS

Basic discrete-event simulation concepts were introduced in Chapter 4. Here we review these concepts and put them in a more general context. A *discrete-event dynamic system* contains discrete components that are usually called *entities.* Examples of entities are customers, vehicles, computer workstations, and bank transactions. Entities can be classified as either *permanent* or *temporary.* Traditionally, permanent entities are those that are part of the system for the entire system lifetime. Machines in a manufacturing line and berths in a port are examples. The terms *resource* and *facility* are also used for permanent entities. Temporary entities are those that enter the system, interact with it or flow through it in some manner, and then leave the system. Customers, cars, messages, and bank transactions are examples of temporary entities. In many systems, temporary entities interact with permanent entities. This interaction can be active, as in the case of a customer (temporary entity) interacting with a permanent entity such as a bank teller, or it can be passive as in the case of a part in a manufacturing line (temporary entity) being processed by a permanent entity such as a machine. In either case, it is usually sufficient from a modeling perspective to think of the permanent entities as resources and the temporary entities as requesting a certain quantity of the resource, using that resource for a period of time, and then relinquishing the resource and proceeding on through the system. We can think of the server in the single-server queue as being a resource. Each customer requests one unit of the resource "server" and waits in a queue until that unit is available. Then the customer uses that unit of the resource during service, relinquishes that unit of the resource after completing service, and leaves the system.

6.2.1 Temporary and Permanent Entities

Temporary entities can be created and destroyed. Because a simulation can involve many thousands of temporary entities, each must be created when it enters the system and destroyed when it has finished interacting with the system. Permanent entities, on the other hand, are not created or destroyed.

This characterization of temporary and permanent entities has its limits, however. In many models, permanent entities can fail or become obsolete. When this happens, they must be discarded and replaced. Thus, these permanent entities behave more like temporary entities. Many contemporary technological systems often use advanced processing components such as robots or computer-controlled systems that possess both some intelligence and mobility. Moreover, humans and computer systems often perform complex services during which a variety of tasks have to be done using other resources (such as computers, communication links, or vehicles) at different locations. Finally, the scope of the model can also influence the character of entities. For example, in a construction model, a crane is usually viewed as a resource used in constructing a building. However, if a group of cranes is used in multiple projects at different locations and the model involves transporting the cranes between projects, as well as their breakdown and replacement, then the crane operates more like a temporary entity. Therefore, the role of each entity in the system must be well understood to be modeled properly. Some simulation software includes provisions to allow resources to perform their tasks dynamically, moving through different system locations where specific tasks are done.

6.2.2 Attributes and Entity States

Entity *attributes* are items of information about specific entities that enable us to distinguish entities of the same type. Thus, we can distinguish passengers in an airport terminal as domestic or international, or ships in a port by their size (small, medium, or large). An entity's attributes usually determine how the entity interacts with the system. For example, each passenger in an airport can have an attribute giving the number of bags he or she has, and the processing time for the passenger at check-in will depend on this attribute. A ship in a port can have an attribute giving its speed, which will in turn determine the length of time to travel from the mouth of the harbor to the dock.

Some entity attributes are variables used to describe the current status of the entity. A given value for such status attributes denotes a specific state for the entity. Two basic types of states are the *active state,* during which the entity is currently interacting with other entities, and the *waiting* (or *passive*) *state,* during which the entity is waiting for the conditions that will allow it to continue through the system. To illustrate, suppose that a part in a manufacturing system is waiting for a machine to be available to do the next step in processing. We would say that the part is in the waiting state. On the other hand, if the part is being painted, we would say it is in the active state. Conditions for activation may include the availability of resources required for an activity as well as the presence of a particular system state. For example, before a ship can leave its berth, a tug must be available (resource needed) and the tide must be high (a particular system state). Usually, temporary entities move through a system by alternating between an active state and a waiting state. Consider customers in a bank lobby. A customer enters the bank and joins a queue to wait for service (waiting state) until a teller is idle. Then she is served by the teller (active state). After service, she may, with a given probability, go to a customer representative and wait until he is free to serve her (waiting state), then discuss a problem with the him (active state). This process continues until the customer's business is finished and she leaves the bank. Normally, managers want to minimize the amount of time customers spend in a waiting state.

Sometimes the goal of the model dictates that only waiting and service be included in the model, while in other circumstances more subtle features of customer behavior must be included. For example, machines in a manufacturing system can be in one of two basic states: idle or busy. However, if the machine is broken or under repair, then these states may have to be taken into account in a more detailed version of the model.

The *system state* is a complete description of all entity states, along with the values of any system variables, at a given moment. For example, in a bank lobby model with four tellers, at a particular point in time, three may be busy and one teller may be taking a break, while three customers are being served by the three busy tellers and five customers are waiting in a common queue. All of this information constitutes the system state. In any model, the system state must include all information needed for the model operations to proceed.

6.2.3 System Dynamics

Up to this point, the discussion has involved only the *static* system description, consisting of the components (entities) in the system, and their properties (attributes). As we noted in Chapter 5, a complete model description must also include the system *dynamics*—that is, the system behavior that results from the *interactions* between entities over time. Recall that there is a close relationship between the static system description and the system dynamics because the static description tells us what data are available and the system dynamics tell us how those data are used as the entities interact over time. A customer waiting to be served in a bank lobby cannot change her state without interacting with a teller, and this interaction can begin only when the teller becomes idle and the customer is at the head of the line. When these conditions are met, then service begins for the customer. The occurrence of this interaction, which causes an instantaneous change in the system state, is called an *event.* In discrete-event simulation, the system state changes only at discrete time points when an event occurs. When the *begin service* event occurs, all entities interacting in this event change their states: The state of the particular customer that is beginning service changes from waiting to being served, the number of customers in the queue decreases by one, and the server state changes from idle to busy.

Although there may be many event occurrences, the number of distinct *types* of events in a given model is generally rather small. Consider a model of a large telephone network. In a short period of time, there may be many events that denote the start of a phone connection and many events that denote the end of a connection. However, there are only two *types* of events: start of connection and end of connection. The number of types of events is not related to the size of the telephone network, but the number of event occurrences is closely related. Rather, the number of event types is related to the level of detail in the simulation model. For example, our model of the telephone system might represent the initiation of a connection with three stages, each having a start and an end. Thus, we might need six additional events if this level of detail is required.

6.2.4 Conditional and Scheduled Events

We can refine the concept of events by classifying them as *conditional* or *scheduled.* A conditional event is one that can occur only when a specified system condition is satisfied. Conditional events typically correspond to the beginning of interactions between system entities, and the conditions usually relate to the states of the entities involved in the interaction. For example, the beginning of a phone conversation requires that the following conditions be met simultaneously: Somebody places a call, there is a free channel between the caller's phone and the called phone, the phone being called is idle, and the person being called is available to answer the phone. When a conditional event occurs, the resources required for the corresponding entity interaction are captured. For example, the event starting a phone conversation engages the phone being called, the channel between the two phones, and the person being called.

Scheduled events, on the other hand, are typically concerned with the end of an interaction between system entities. To determine the moment when a phone connection ends, we add the length of the connect time (which may be a random variable) to the time at which it began. For example, the duration may have a normal distribution with mean 14.2 minutes and standard deviation 4.7 minutes. In the simulation, we model the length of the call by generating a random variate from this distribution with the given parameters. Once the observation has been sampled, its value is known and the end of the connection is scheduled to occur at a time that is this length of time after the start of the connection. Thus, the event denoting the end of the connection does not depend on the availability of certain resources but is scheduled to occur unconditionally at a specific (though random) time in the future. Indeed, when the end of the connection event occurs, resources that were used in the interaction are released and some conditional events that are waiting for the resources can then occur.

We can have scheduled events that are not actually related to entity interactions. Consider, for example, the *arrival event,* which represents the arrival of a temporary entity to the system. In the real system, the causes of arrival events are usually unknown, so arrivals are frequently treated as being drawn from a large population of potential arrivals with a given distribution of time between successive arrivals. The easiest way to accomplish this is to let each arrival event schedule the next arrival event. Another type of scheduled event is a *breakdown event.* The model might specify that the time between breakdowns be a random variable from a specific distribution. In this case, a breakdown event would be scheduled by the previous breakdown event. Actually, in most realistic models, a breakdown would be scheduled to occur after a specific amount of operating time has elapsed since the previous repair. This is a somewhat more complicated modeling problem. These types of scheduled events can be logically represented as cyclic events that do not seize or release system resources. However, they do cause changes in the system state. For example, after each customer arrival to the bank lobby, the length of the customer queue increases by one; and after a machine breakdown, the machine state goes from operational to not functioning. Thus, these types of scheduled events change either the workload on the system or the capacity of resources to handle the workload.

6.2.5 Activities and Processes

The interaction between different entities over a specific length of time constitutes an *activity.* The activity begins with the conditional event that starts the interaction and ends with the scheduled event that concludes the interaction. The length of the activity, or *activity duration,* is usually determined by sampling from a distribution specified by the model. For example, the duration of the service activity in a bank lobby could be represented by a gamma distribution with specific parameter values. Resources needed for the activity are requested as a condition for beginning the activity. At least one teller must be idle, and the customer requiring the service must be the next to be served for the service transaction to begin in the bank lobby, for example. However, as we pointed out earlier, activities do not have to require resources. Arrivals and breakdowns are examples of cyclical activities that spend time but do not require resources for their occurrence. Figure 6.1 shows the relationship of conditional events to scheduled events and activities. Note that although no time elapses during an event, an activity does involve the elapsing of time. We will look at activities in more detail in section 6.3.

A *process* is a sequence of related events and activities. Normally, the process is designed to represent the flow of a temporary entity through the system. Consider, for example, a traveler in an airport terminal. The traveler must proceed through various operations at the airport: After arrival, she first goes to the check-in counter where she waits for service before leaving her baggage. Then she goes through the security check, waits to have her passport checked at the customs counter, and finally waits in the gate area for her plane. During this process, the passenger will move sev-

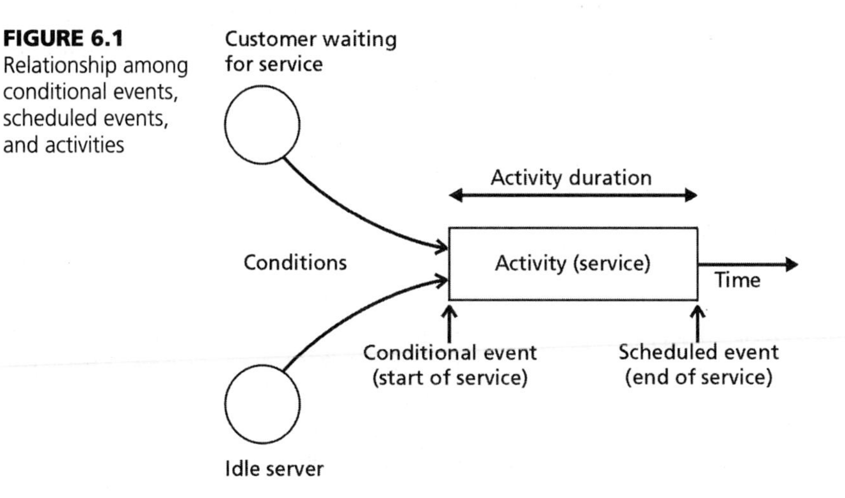

FIGURE 6.1
Relationship among conditional events, scheduled events, and activities

eral times, wait in various queues for service, and finally leave the system. Indeed, she will go through an alternating series of passive states when she is waiting for service followed by active states when she is receiving service or proceeding.

Although simulation of a single passenger would be quite simple, real airports have many passengers who go through basically the same operations, wait for the same resources, and are served at the same facilities. The movement of each passenger in the system depends on interactions with and movements of other passengers. For example, at the baggage check-in, it is only when the previous passenger finishes service that the server becomes idle and the next passenger can be served. At any moment, many activities are in process simultaneously. This parallelism can involve not only different types of activities but also different instances of the same type of activity. For example, several passenger check-ins may occur simultaneously, each at a different check-in counter.

This brief discussion shows that certain capabilities have to be included in the simulation modeling methodology in order to represent the structure and dynamics of complex real-world systems. The discrete-event simulation mechanism must be able to (1) describe entities in the system, (2) represent the dynamics of system behavior, and (3) control the execution of the simulation. We have already discussed the representation of entities in some detail. In the remainder of this chapter, we will examine the last two capabilities.

In every discrete-event simulation, the system dynamics involve progressing from event to event, updating the system state at each event. The event view, which we discussed in Chapter 5, is a direct approach to representing events and the actions that take place in them to change the system state. Two alternatives are the activity view and the process view, which we will discuss in the following sections. Each provides an alternative way to organize and present the events in the model. Regardless of the worldview used, however, every simulation must have a timing routine that finds the next event, updates the system clock to the time of the event, and executes the actions involved in the event. We will see that the timing routine can take on a somewhat different form depending on the worldview that is adopted, but the timing routine still accomplishes the same result: moving the simulation from event to event.

6.3 THE ACTIVITY VIEW

In the activity view, activities are closely related to entities because an activity normally starts when an entity begins using a resource, such as a server or machine, and ends when the entity releases the resource. Thus, entities can be in one of two states: the *active state,* in which resources are being used and the activity is ongoing; and the *passive state,* in which the entity is waiting for the conditions that enable the start of the activity.

Historically, the activity view, which originated in the United Kingdom and is still quite popular there, was developed in two different forms. We will discuss the original form first, even though it is seldom used today, because it will help us understand the nature of the activity view and to see the independent and modular nature of the event routines in this version. The modified form, called the *three-phase approach,* has become popular because it keeps the independent modules but implements the activity view more efficiently. See Pidd (1992) for more on the activity view.

6.3.1 Original Version

The activity view, which was originally implemented in the simulation language CSL (Buxton and Laski, 1962), used activities as its sole dynamic element. Each type of activity is represented as an *activity routine.* Every activity routine consists of a *testhead* and *actions.* The testhead contains a test for conditions that must be satisfied so the activity can start, while the actions describe the changes to the system that occur during the activity.

Consider the single-server queue model first presented in Chapter 5. Using the generic simulation programming language (GSL) that we introduced there, we can describe these activity routines as shown in Figure 6.2.

Recall that the state variables are Q (the number of customers in the queue), S (the status of the server—busy or idle), and N (the total number of customers to complete service). Note that each activity routine starts with an IF statement, which is the testhead. The actions in the activity routine are contained in the THEN clause. Thus, each activity routine will have the following form:

IF *conditions*
THEN *actions.*

This model has two types of activities: *arrival* and *service.* Let's look at the service activity first. It consists of the period during which the customer is actively re-

FIGURE 6.2
Activity routines for single-server queueing model

```
Activity routine arrival:
  IF     next arrival is due at time NOW
  THEN   Q := Q + 1;
         Schedule next arrival at time NOW + interarrival time.

Activity routine begin service:
  IF     Q > 0 AND S = idle
  THEN   Q := Q - 1
         S := busy
         Schedule end service activity at time NOW + service time.

Activity routine end service:
  IF     end of service is due at time NOW
  THEN   N := N + 1;
         S := idle.
```

ceiving service—that is, the service begins when the customer begins service and ends when he completes service. Thus, the two event routines involved with the service activity are the *begin service* routine and the *end service* routine. The *begin service* event is a conditional event. Examine all three activity routines; notice that no statement in any activity routine schedules a *begin service* routine. Instead, the *begin service* event can only begin when the conditions $Q > 0$ (at least one customer waiting in the queue) and $S := idle$ (the server is idle) are met. The first question that comes to mind is, if this event is never scheduled, when are the conditions for its execution checked? We will see shortly that the responsibility for checking conditions is shifted to the timing routine. Basically, every time an event is executed, the timing routine checks all event types to see if conditions for their execution are satisfied.

When the conditions for executing the *begin service* event are met, the actions reduce the number of customers in the queue by 1 ($Q := Q - 1$), set the server's status to busy ($S := busy$), and schedule the *end service* event at a time that is the service time later than the current time (Schedule *end service* activity at time NOW + service time). Thus, the event that ends the service activity is scheduled by the event that starts the activity.

Now let's examine the arrival activity. This activity starts with an arrival and ends with the *next* arrival. So the conditions for execution of the arrival event are really nothing other than that the event is actually an arrival event. Also, the starting event of one arrival activity instance corresponds to the time of the ending event of the previous arrival activity instance. This is a special type of activity that we can think of as being cyclical. In contrast, in the service activity, the end of service for one customer corresponds to the start of service for the next customer *if* there is at least one customer in the queue waiting for service. If not, then the start of service for the next customer will have to wait for the next arrival. Thus, we cannot combine the start of service activity routine with the end of service activity routine because they do not always coincide in time. With the arrival activity, on the other hand, we do not need distinct activity routines to denote the start and end of the arrival activity because these two events *always* coincide in time.

Note that in this model we put the customer in the queue immediately on arrival, regardless of whether he can begin service at this moment. This is useful for the activity view because the test to see if the service activity can begin involves having at least one customer in the queue ($Q > 0$). If we did not put the customer in the queue, then we would need another state variable to denote that a customer has just arrived and must either begin service or go into the queue. In this model, we have chosen to take the customer out of the queue when service begins in the *begin service* activity routine, but we could just leave the customer in the queue during service and decrement the counter, Q, for the number in the queue when the customer ends service. This would affect the collection of data but would not affect the operation of the model.

We used the term *activity routine* to denote the three routines in the activity view for this model. These routines are actually event routines because they describe what happens at a moment of time. The two routines *begin service* and *end service* together constitute the description of the service activity; the routine *arrival* by itself constitutes the description of the arrival activity. All activity routines have a testhead that at least checks the identity of the event to see if the event can occur now. This is the characteristic of the activity view that allows the routines to be independent and modular. Compare these routines to the event routines in Chapter 5 and notice that here we have been able to do away with checking within each event routine for conditions that allow other events to occur; hence, we have created independence among the event routines. For example, the *begin service* event can occur either just after the *arrival* event if the customer arrives to find the queue empty and the server idle or just after the *end service* event if a customer is waiting. Thus, the event routines for these two events are responsible for checking these conditions. The activity view puts these conditions at the start of the *begin service* event routine and transfers responsibility for checking them to the timing routine. Imagine what would happen if the conditions for executing the *begin service* event routine were to change. For example, suppose that the customer were to need two servers for service to be rendered. In the event view, we would have to change two event routines where we check to see if the *begin service* event routine can occur. In the activity view, we need only change the testhead of the *begin service* event routine. In a much more complex model, this localization of testing for conditions for each event to occur can possibly avoid many errors.

Now, let's look at the timing routine for the activity view. The logic of the timing routine is as shown in Figure 6.3. Like the timing routine in the event view, it will first find the next *scheduled* event occurrence. This can be done, as in the event view, by maintaining event notices in a list of scheduled events ordered by time of occurrence; or we could keep a time for the next scheduled event with each entity and search through all entities for the earliest time. Once the time of the next scheduled event is determined, the system time is updated to that value. Basically, in the activity scan we want to execute all events that can be executed at this moment. This includes the scheduled event whose time we used to update the system clock plus all conditional events whose conditions can now be met. The activity scan therefore enters a loop to check all types of activities and execute their event routines if the

FIGURE 6.3 Timing routine for activity view

```
Timing routine simulate:

  WHILE not finished DO

      (Time scan)
      Update the clock to the time of the next scheduled event.

      (Activity scan)
      REPEAT
      FOR each type of activity DO
           IF conditions for any activity is satisfied,
                THEN execute the activity
      UNTIL no activity is executed.

  END DO
```

conditions in the testhead are met. This loop is repeated until no event is executed. Now the activity scan is complete. The entire loop is repeated until some signal is given to stop the simulation.

Now we see the compromise we have made for the simplicity of the event routines. Generally, the timing routine will check many activities to see which can be executed, even though only one or perhaps a few activities will qualify for execution. If any activity is executed, then we must repeat the scan until we go through it without executing an activity. Although the logic of the timing routine is complicated somewhat, it is usually not computationally burdensome because we are just checking conditions.

As in every simulation, the system must be initialized before the timing routine is called. The initialization routine in Figure 6.4 will start the system in the empty and idle state and schedule the first arrival.

We will not go through a detailed example here, but the basic operations of the activity view are illustrated in an example in the next section.

6.3.2 The Three-Phase Approach

The three-phase approach (Tocher, 1963) is a modified version of the activity view that combines the mutually independent activity routines in the original activity view with the efficiency of the timing routine in the event view. It partitions the events into two classes—conditional and scheduled events—and treats them differently in the timing routine.

Look back at the event routines in subsection 6.3.1. The events *arrival* and *end service* have trivial conditions: Check to see if this type of event is to occur at the current time. On the other hand, the *begin service* event has nontrivial conditions associated with it: Check to see if a customer is available and a server is idle so service can begin. For the *arrival* and *end service* events, we do not need to check any conditions. We can just keep track of the identity of the event when we schedule it, and when its time comes, execute the event routine just as we did in the event view in Chapter 5. For the *begin service* event, we will still need to check its conditions after each scheduled event is executed.

Conditional events, which we will call *C-events,* can occur only when certain conditions are satisfied. *Scheduled events,* which we will call *B-events,* must occur only at their scheduled time. Think of B-events as *bound* events—that is, as events that are bound to the event notice for the scheduled event at that time. B-events are unconditional and do not need testheads, but C-events will *always* have testheads. Usually, B-events correspond to the end of activities, such as arrival or end of ser-

FIGURE 6.4 Initialization routine for activity view

```
Initialize system:
  Q := 0;
  S := idle;
  N := 0;
  Schedule first arrival activity at time NOW.
```

FIGURE 6.5
Event routines for three-phase approach

```
B-event routine arrival:
   Q := Q + 1;
   Schedule next arrival event at time NOW + interarrival time.
B-event routine end service:
   N := N + 1;
   S := idle.
C-event routine begin service:
   IF   Q > 0 AND S = idle
   THEN Q := Q - 1
        S := busy
        Schedule end service activity at time NOW + service time.
```

vice. C-events correspond to the start of activities because each activity cannot begin until conditions for its start are satisfied. The event routines for the three-phase version of the single-server queue are shown in Figure 6.5.

Notice that the event routines in this approach are almost exactly the same as in the original activity view version. For the *arrival* and *end service* event routines, only the (trivial) testhead has been removed; in the *begin service* event routine, nothing has changed. The initialization routine is also unchanged for the three-phase approach. The timing routine, however, is modified to avoid having to check conditions for scheduled B-events. The timing routine for the three-phase approach is shown in Figure 6.6.

In this timing routine we have not changed phase A (except to call it "phase A"). But we have changed the rest of the timing routine to separate the execution of B-events from the execution of C-events. In phase A, we selected an event from the event list and updated the system clock to the time on the event notice. In phase B, we execute the event routine for this event. Then, in phase C, we will check *only* C-event types to see if any can now be executed because the conditions for execution have now been satisfied. Thus, when checking conditions for executing events, we need only check for conditional events. This simplifies the logic of the timing routine and eliminates checking that is redundant and nonproductive.

FIGURE 6.6
Event routines for three-phase approach

```
Simulate
WHILE not finished DO
Phase A:  Advance.
          Select next scheduled event from event list;
          Set system clock to time of this event.
Phase B:  Execute B-events.
          Execute the event scheduled for this time.
Phase C:  Execute C-events.
          Scan all C-event types in sequence. IF conditions for any
          C-event are satisfied, execute this event and repeat the
          C-event scan.
END DO
```

Note that in phase C, a C-event might free resources or otherwise cause the conditions for another C-event to be satisfied. So it is important that the scan include all types of C-events and continue checking all C-events, even the types that have already been executed, until no additional C-event can be executed. It is also important to note that the order of event scanning in phase C can influence the selection of the C-event to be executed next. For this reason, C-event types must have a set of priorities. Conditions for executing the C-event types are tested in the order of decreasing priority, which allows higher priority C-events to be the first to seize resources released in the previous phase B (or previous C-events). Attaching priorities to C-event types, which correspond to the start of activities, is equivalent to attaching priorities to the activities. As an example, suppose that a bartender can either serve guests or clean up when she is free. Normally, the priority is to serve guests and to clean up only after all guests have been served.

The three-phase approach is implemented in the simulation software systems ECSL/CAPS (Clementson, 1977) and HOCUS (Poole and Szymankiewicz, 1977). Activity view modeling also lends itself to a graphical modeling method called *activity cycle diagrams,* which will be discussed in the next chapter.

6.3.3 An Example of Simulation Execution Using the Three-Phase Approach

We now look closely at the operation of the activity view by examining how the three-phase approach executes the model of the single-server queue. Assume that the initialization routine has been executed and the system is empty and the server idle: $Q := 0$, $S := idle$, and $N := 0$. The initialization routine also schedules the first arrival event at time 0.0, which is the value of the system time *NOW* before the timing routine takes control. Recall that the initialization routine in any discrete-event simulation must schedule at least one event, otherwise there would be no scheduled events for the timing routine to execute and the timing routine would end without executing any event routines.

To illustrate the operation of the three-phase approach, the first 10 randomly sampled observations of the interarrival times and service times are contained in Table 6.1, which has the same random variates as in Table 5.1. When the timing routine starts after the system is initialized, the value of the system clock is 0.0 and there is one scheduled B-event, which is the first arrival due to occur at time 0.0. The system state and the initial event list as well as the other states the model enters during this execution are presented in Figure 6.7.

TABLE 6.1 Randomly sampled observations for interarrival and service times for the single-server queue model

Interarrival times	.5	1.4	.1	1.7	.8	.1	.1	2.6	3.5	.2
Service times	1.0	.2	.8	.8	.4	.5	.4	2.6	.5	.0

FIGURE 6.7 Three-phase approach showing event list and state variable values for a single-server queue execution

Current Phase	A Time	B Scheduled Events	C Conditional Events	State Variables Q	S	N
Initial	0.0	Arrival at t=0.0	-	0	*idle*	0
A	0.0	**Arrival at t=0.0**	-	0	*idle*	0
B	0.0	Arrival at t=0.5	-	1	*idle*	0
C	0.0	Arrival at t=0.5 End service at t=1.0	Begin service	0	*busy*	0
A	.5	**Arrival at t=0.5** End service at t=1.0	-	0	*busy*	0
B	.5	End service at t=1.0 Arrival at t=1.9	-	1	*busy*	0
A	1.0	**End service at t=1.** Arrival at t=1.9	-	1	*busy*	0
B	1.0	Arrival at t=1.9	-	1	*idle*	1
C	1.0	End service at t=1.2 Arrival at t=1.9	Begin service	0	*busy*	1
A	1.2	**End service at t=1.2**	-	0	*busy*	1
B	1.2	Arrival at t=1.9	-	0	*idle*	2
A	1.9	**Arrival at t=1.9**	-	0	*idle*	1
B	1.9	Arrival at t=2.0	-	1	*idle*	2
C	1.9	Arrival at t=2.0 End service at t=2.6	Begin service	0	*busy*	2
A	2.0	**Arrival at t=2.0** End service at t=2.6	-	0	*busy*	2
B	2.0	End service at t=2.6 Arrival at t=3.7	-	1	*busy*	2
A	2.6	**End service at t=2.6** Arrival at t=3.7	-	1	*busy*	2
B	2.6	Arrival at t=3.7	-	1	*idle*	3
C	2.6	End service at t=3.4 Arrival at t=3.7	Begin service	0	*busy*	3

In phase A, the timing routine in Figure 6.6 finds the arrival event scheduled at time 0.0. The events in the event list will always be listed in order of increasing event time, so the first event on the list will be the next event to execute. In addition, we use boldface in Figure 6.7 to identify the next event found in phase A, which becomes the current event. The value of the system clock is set to the time of the next event, the *arrival* event, which is 0.0: *NOW* := 0.0. Phase A is finished and we move to phase B, where the *arrival* event is executed and the event notice is removed from the event list. The results of executing the *arrival* event routine are to place the first customer in the queue: The value of Q is incremented by 1 to 1, and a new *arrival* event is scheduled at time *NOW* plus the first value of the interarrival time from Table 6.1: 0.0 + .5 = .5. This completes phase B. Now in Phase C, conditions for the C-event *begin service* are satisfied because $Q := 1 > 0$ and $S :=$ *idle*. So we execute the actions in the *begin service* event. As the third row of Table 6.1 shows, the queue now becomes empty because Q is decremented by 1, the server becomes busy ($S :=$ *busy*), and the *end service* event is scheduled at time *NOW* + 1.0 = 1.0. The service time 1.0 was obtained from Table 6.1. We check again to see if any additional C-events can be executed. Because no additional conditions can be satisfied, phase C is completed and we move on to phase A of the next cycle.

The next event in the list of B-events is the *arrival* event scheduled for time .5. The system clock is updated to .5, and phase A is complete. In phase B, this *arrival* event is executed and removed from the event list. After that, the queue has one customer in it, and the next arrival is scheduled for time *NOW* + 1.4 = 1.9. In phase C, conditions for the C-event *begin service* are not satisfied because $S :=$ *busy.* The second cycle of the timing routine is now finished.

In phase A of the third cycle, the next event is the *end service* event scheduled at time 1.0, so the system clock is updated to the time 1.0 and the *end service* event routine is executed in phase B and then the event notice is removed from the event list. The actions in the *end service* event routine produce the results shown in row 8 of Figure 6.7. The total number of customers served, N, is increased by 1 to 1, and the server becomes idle. In phase C, the server is idle and $Q := 1$, showing that there is at least one customer in the queue, so the C-event *begin service* can be executed. After executing the *begin service* event routine, the number of customers in the queue is decremented by one to zero, the status of the server is set to busy, and a new *end service* event is scheduled at time *NOW* + .2 = 1.2. Another cycle of the timing routine is now finished.

The timing routine continues cycling through all three phases until the conditions for ending the simulation are satisfied. Several additional phases in the execution of this model are shown in Figure 6.7. As we discussed in Chapter 5, the conditions for ending the simulation could involve reaching a certain time, serving a specified number of customers, the system entering a certain state, or the event list becoming empty.

6.4 THE PROCESS VIEW

The process view implements the concept of a *process* to describe a discrete-event simulation model. Recall that a process consists of a sequence of events that are related in some way, usually by involving a given entity. For example, consider the events that relate to a customer in a queueing system as he arrives, progresses through the system, and finally exits. First, he experiences the *arrival* event where he joins the queue and checks to see if a server is free so service can begin immediately. Next he sees the *begin service* event, where the server's status is changed to busy and the *end service* event is scheduled. Finally, he experiences the *end service* event, where the server's status is changed to idle and the *begin service* event for the next waiting customer (if any) is scheduled. Note that these events occur in a defined order for each customer: *arrival, begin service, end service.*

The process view seeks to represent this sequence of events, with appropriate time delays between them, in terms of a single *process routine* for the customer. To accomplish this, first imagine that each customer instance has entity attributes and also owns an event notice. The event notice is associated with the customer so that, for example, when an *end service* event occurs we will know that the event "belongs" to the customer. Thus, the entity with its attributes and event notice can be thought of as a process. Every process has a sequence of events and an order of execution, which can depend on the system state, defined for that process. To keep track of which events have occurred and which event is next to occur, each process instance must also have a *local event pointer.* When the process starts, this pointer points to the first event, and as the process moves from event to event this pointer moves to the next event in the sequence.

Now let's define some new things we can do with the process. We will say that the process is "active" if the event notice is in the event list, meaning the next event is scheduled to occur at a definite time in the future. The process is active, for example, during service because the *begin service* event schedules the *end service* event. Note that this corresponds to the definition we gave in section 6.2 for an active entity. We will call the process "passive" if the event notice is not in the event list. If the process is passive, then it is waiting for a system state that will allow the next event to occur. For example, a process is passive while the customer waits in the queue for a server to provide service. This also matches the definition of a waiting, or passive, entity in section 6.2. Thus, a process is active before a scheduled event and passive before a conditional event. We can *activate* the process at a specific time in the future. Activation of a process simply means to schedule the next event for the process, that is, put the event notice in the event list with a specific time on it. If the process is active, then we can *passivate* the process, meaning that we will remove the event notice from the event list. This is generally accompanied by putting the entity in a list. Normally a process is passive while waiting for system resources. Passivation also causes the process routine to suspend execution at the point where the passivation occurs and control to be returned to the timing routine to move to the next scheduled event. A process can also *suspend* itself when it is not passive. When a process does so, it simply ceases executing the statements in its process routine and returns control to the timing routine. When a process is passive or suspended, the local event pointer marks the next statement to be executed when the process continues.

There is another thing we need to do with a process—get it started. We will call this *initiation.* To initiate a process, we create the entity and event notice involved in the process, initialize the local event pointer to indicate the initial event in the process, and activate the process at a specific time.

The timing routine operates in the process view the same basic way it operates in the event view or the activity view. Events are always stored in order of increasing time of occurrence. When an event notice becomes first on the list, its event routine is next to be executed. In the process view, each event notice belongs to a process, so there is an associated entity and local event pointer. It is the local event pointer that tells the process routine which statement to execute the next time the process is activated. Thus, when an event notice is first on the event list, the timing routine executes the process routine for the associated process, using the local event pointer to determine the next statement to execute.

6.4.1 The Process View of the Single-Server Queue

Now we can examine how these concepts are applied to implement the single-server queueing simulation using the process view. The process routine for this system is shown in Figure 6.8. This process routine has the labels *Arrival:, Begin Service:,* and *End Service:* to denote the sequence of three events associated with the customer process. The local event pointer will always point to one of these events. Initially, it points to the *arrival* event, where the process first initiates another (instance of a) customer process to represent the next customer to arrive. This process is created and activated at a time that is the interarrival time later. Next the customer is placed in the queue, and the server's status is checked. If the server is busy, this

FIGURE 6.8 Process routine for single-server queue

```
Process Customer
Arrival:
     INITIATE new customer at time NOW + interarrival time delay.
     Insert this customer last in queue.
     IF server.status = busy THEN PASSIVATE.
Begin Service:
     server.status := busy.
     Remove this customer from the queue.
     ACTIVATE this customer at time NOW + service time.
     SUSPEND.
End Service:
     server.status := idle.
     IF queue not empty
          THEN ACTIVATE first customer in queue at time NOW.
```

process passivates, meaning that the event notice is removed from the event list, the process routine is suspended at this point, and the local event pointer is placed to point to the next event, which is the *begin service* event. At this point, control is returned to the timing routine and the next event on the event list is executed, that is, the next event for the process whose event notice is first on the event list is executed. If the server's status is anything other than busy, the process routine continues with the *begin service* event.

In the *begin service* event, the server's status is set to busy and the customer is removed from the queue. The next statement activates the customer process at a time in the future that is the service time later than the current time. The activate statement schedules the next event to occur at the indicated time, sets the local event pointer to the next event in the sequence (*end service*), and then suspends the process routine. Once again, control returns to the timing routine, which selects the next event.

When the event notice for the *end service* event becomes first on the event list, the server's status is set to idle and the status of the queue is checked. If there is at least one customer in the queue, the first customer is selected and activated at time *NOW.* Because the only time a customer can be in the queue is between arrival and the start of service, only customers who have not begun service will be in the queue. Every customer in the queue will have his or her local event pointer pointing to the *begin service* event, so when they are activated the actions that occur next will be those in the *begin service* event.

Note that the sequence of actions in this process view version of the single-server queue is exactly the same as that in the event view or the activity view. This is simply a different way to present the actions that occur as the customer moves through the system. Some actions are implicit and not provided in the process routine. For example, we do not explicitly create or delete a customer. Initiation of a process creates the entity involved, and when the last event in the process routine is completed, the entity, event notice, and all other data items related to this instance of the process are deleted. Thus, we can think of the entity as being created when the process starts and deleted when it ends.

We can represent the process routine for the single-server queue without designating the three events. The events are still there; we have simply not labeled them. The result is shown in Figure 6.9. The events occur at the start of the process routine and after

FIGURE 6.9
Process routine for single-server queueing model

```
Process Customer
    INITIATE new customer at time NOW + interarrival time delay.
    Insert this customer last in queue.
    IF server.status = busy THEN PASSIVATE.
    server.status := busy.
    Remove this customer from the queue.
    ACTIVATE this customer at time NOW + service time.
    SUSPEND.
    server.status := idle.
    IF queue not empty
        THEN ACTIVATE first customer in queue at time NOW.
```

any passivate or activate statement. This form of process routine is much easier to read and gives a much better idea of the flow of the customer through the system.

6.4.2 Process Interaction

The dynamics of discrete-event simulations can also be represented as multiple processes that interact with one another. In the example just presented, we represented the model as a single process that uses a resource, the server. Thus, the server was represented as a passive resource. The process interaction approach treats the server as an active process with its own process routine that interacts with the customer process to move the system through time. The example model of a single-server queue we have been examining is used to demonstrate the process interaction approach. Figure 6.10 shows the result.

Although this is the same single-server queueing model, the current description differs from the previous ones in two important respects. First, we have created a *process routine* for the server—that is, a sequence of events that relate to the same entity. Now imagine the sequence of actions the server experiences in the course of operating the system. She waits until a customer presents herself for service, serves the customer for the service time, and checks to see if another customer is waiting to be served. If so, she begins service for that customer. After the service time, she completes service for that customer and again checks to see if another customer is waiting to begin service. This process of starting service, ending service, starting

FIGURE 6.10
Process interaction view of single-server queueing model

```
Process Customer
Arrival:
    INITIATE new customer at time NOW + interarrival time delay.
    Insert this customer last in queue.
    IF server.status = busy THEN PASSIVATE.
Begin Service:
    ACTIVATE server at time NOW.
    SUSPEND.
End Service:
    ACTIVATE server at time NOW.

Process Server
End Service:
    server.status := idle.
    IF queue not empty
        THEN ACTIVATE first customer in queue at time NOW. SUSPEND.
        ELSE PASSIVATE.
Begin Service:
    server.status := busy.
    Remove first customer from the queue.
    ACTIVATE customer at time NOW + service time.
    SUSPEND.
    Loop to End Service.
```

service, and so on continues indefinitely. The server is active except when the queue is empty and she is waiting for a customer to arrive to begin service. In contrast to the customer process, the server process is an infinite sequence consisting of a *begin service* event followed by an *end service* event.

This process is implemented in the server process routine above. It may seem a little strange to start the server process with an *end service* event, because the first event the server sees is really a *begin service* event. However, note that this sequence works correctly if the system is started empty. In that case, when the *end service* event executes, the server's status is set to idle and the server process passivates, waiting for the first arrival. When the server process is activated again, the *begin service* event is the next to execute. When this happens, the server sets her status to busy, removes the customer from the queue, and activates this customer at the time denoting the end of service. At this point, control is passed to the timing routine. Note, however, that when the process routine starts again, the *end service* event is the next to execute. Thus, the server alternates between *begin service* and *end service* events.

The second difference between this version and the previous one involves the way the two processes interact with one another. Examine the new process routine for the customer process. When the process starts, we do exactly the same actions as in the previous versions—initiate the next customer process, place the customer in the queue, and check to see if the server is free to start service. If the server is busy, we passivate the customer process. If the server is not busy, we go on to the next event, which is the *begin service* event for the customer. Here we just activate the server process. If the server process is passive, its next event will be the *begin service* event, so we let the server process carry out all system changes that occur when service begins. When the customer is next activated, it will be at the end of service. In the *end service* event, the customer once again passes control to the server process, where the server sets her status to idle and checks to see if there is another customer to begin service. Because the actions in the current customer's process are completed, that process including the entity, event notice, and all other attributes is deleted.

This version of the single-server queue model is somewhat more complex than the first version using the process view because the customer process simply "hands off" control to the server process at the beginning and end of service. Why would one want to model a system in this way? This approach divides the actions in the model between the two entities that are carrying them out: That is, the modeler can develop the model by examining what each entity does in the system and writing a process routine for that entity. Even though this may not lead to the model with the smallest number of processes, it is often most natural and closest to the way we see the real system operating. In a real queueing system, we do not normally think of the server as a passive resource. Instead, the server is usually an active entity that does something to provide service. The process interaction modeling approach captures this characteristic. When modeling much more complex systems, it is important to represent complex system interactions correctly. By focusing on each process individually, the complexity can often be managed much better. In the example model, the customer process just activates the server process at the start and end of

FIGURE 6.11 Process interaction version of single-server queueing model without labels

```
Process Customer
    INITIATE new customer at time NOW + interarrival time delay.
    Insert this customer last in queue.
    IF server.status = busy THEN PASSIVATE.
    ACTIVATE server at time NOW.
    SUSPEND.
    ACTIVATE server at time NOW.

Process Server
    REPEAT:
        server.status := idle.
        IF queue not empty
            THEN ACTIVATE first customer in queue at time NOW.
            ELSE PASSIVATE.
        server.status := busy.
        Remove first customer from the queue.
        ACTIVATE customer at time NOW + service time.
        SUSPEND.
```

service. In more complex models, this would not be the case—actions would be taken to secure resources, collect data, monitor system activity, or other things. Thus, the process interaction approach appears somewhat artificial in this case because the system is quite simple.

In this description, the customer process and the server process have identical event labels for the *begin service* and *end service* events. This helps identify corresponding points in time; that is, the customer's *begin service* event and the server's *begin service* event occur at the same time, as do the *end service* events. We could have used different labels for these events because each label is local to its process routine, or we could have used no labels, except that we need to tell the server's process to repeat. Another version of this model without labels is shown in Figure 6.11. Here we see that the points that mark an event are after a SUSPEND or a PASSIVATE action.

6.4.3 Processes, Events, and Activities

The process, event, and activity views are not mutually exclusive and can be used together. Suppose, for example, we need to collect data on server utilization every eight hours. The easiest way to do this is to define a data-collection event and have it schedule itself at eight-hour intervals. Figure 6.12 shows one way to implement the single-server queue using the process view for the customer along with an event to collect server utilization data.

FIGURE 6.12 Combined process and event view of single-server queueing model

```
Process Customer
     INITIATE new customer at time NOW + interarrival time delay.
     Insert this customer last in queue.
     IF server.status = busy THEN PASSIVATE.
     server.status := busy.
     Remove this customer from the queue.
     ACTIVATE this customer at time NOW + service time.
     SUSPEND   server.status := idle.
     IF queue not empty
          THEN ACTIVATE first customer in queue at time NOW.

Event Collect Data
     Schedule Collect Data event at time NOW + 8 hr.
     Update data variables.
```

6.5 A MANUFACTURING EXAMPLE

So far, the simulation worldviews have been presented using a simple single-server queue model. In this section we illustrate the worldviews using a more complex model of a manufacturing system. The purpose is twofold: (1) to show how the three worldviews can present different but equivalent models of the same system, and (2) to progress toward modeling more realistic systems and show some of the issues involved in developing models of more complex systems. The system consists of two subsystems, each processing separate input parts and having its own individual parallel lines consisting of a series of stations (see Figure 6.13). We will label the subsystems A and B. Each subsystem has its own input buffer for arriving parts. We assume that both input buffers have infinite capacity. Subsystem A has two identical parallel lines, each having two phases in the manufacturing process. Thus, a part is processed by first going through phase 1 and then going through phase 2. An in-process buffer holds parts that have completed phase 1 and are waiting to start phase 2.

Subsystem B has three identical parallel lines, each with two phases in the manufacturing process. This subsystem has no in-process buffer to hold parts between the two phases. If a part completes phase 1 but the machine operating on phase 2 on the same line is busy, then the part is *blocked* from proceeding and can only proceed when the second machine on the same line is finished.

The two subsystems do not operate independently. They are, in fact, coupled by a group of three tools that are shared between the manufacturing lines. These tools are used in phase 2 in subsystem A and in phase 1 in subsystem B. Thus, before starting phase 2 of the manufacturing process in subsystem A, both the machine and one of the tools must be available. Similarly, one tool must be available to perform the operations of phase 1 in each line in subsystem B. The system has three identical tools for use by all processes.

FIGURE 6.13 Manufacturing system with two coupled subsystems

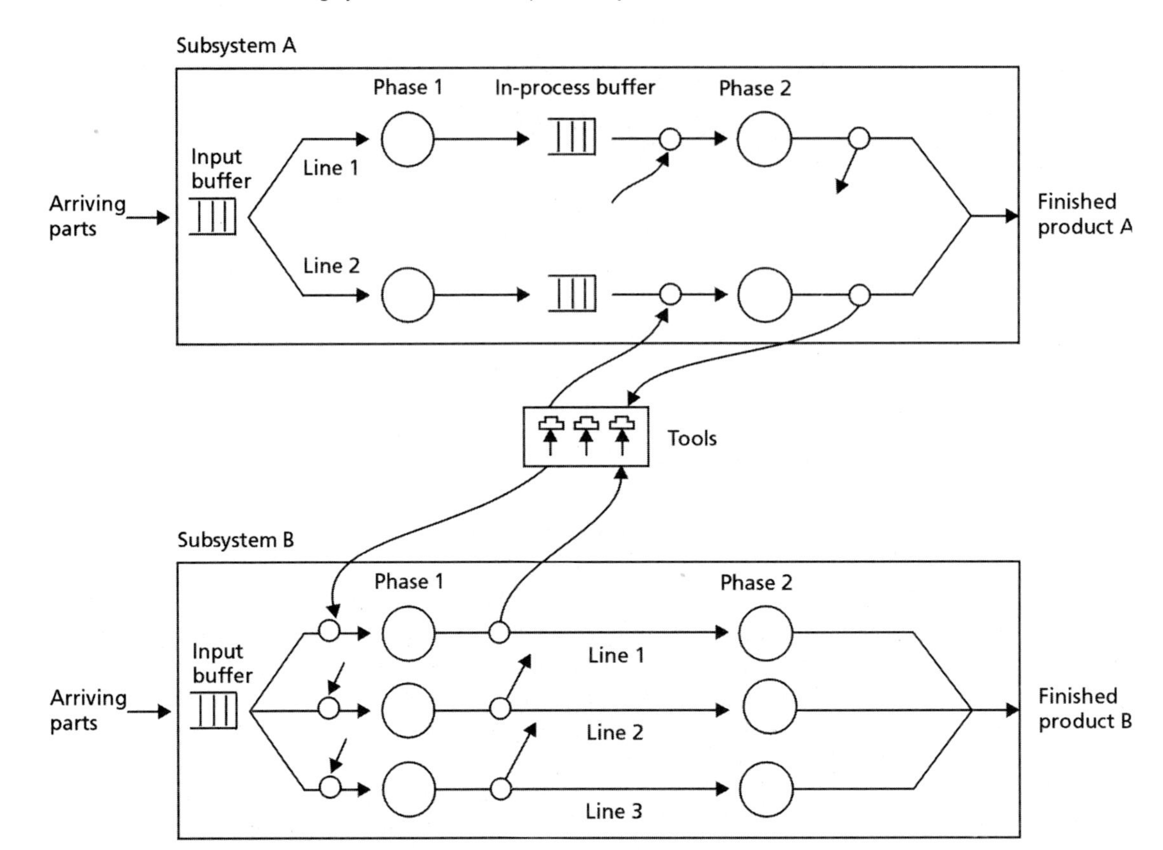

This model has four machines that act as "servers" for parts in subsystem A. There are two lines (1 and 2), and each line has two phases. We will use the notation $S_{nA}[m]$ to denote the status of the machine (idle or busy) used for phase n in line m of subsystem A. Thus, $S_{2A}[1]$ denotes the status of the machine used for phase 2 in line 1 of subsystem A. Six machines process parts in subsystem B. We will label the nth machine in line m of subsystem B as $S_{nB}[m]$. Thus, $S_{1B}[3]$ refers to the first machine in line 3 in subsystem B. This notation seems a little cumbersome, but we will need to index the machines by the line number in order to search across the lines to find a machine that can use a tool when one is released.

When a machine relinquishes a tool after completing phase 2 in subsystem A, the tool will first attempt to be used to process the next part on this same machine, then to be used for processing on the other machines in phase 2 of subsystem A, and then finally to be used for a machine in phase 1 of subsystem B. After a tool is relinquished following phase 1 in subsystem B, the search for another machine to use the tool is done differently. The tool will first attempt to be used on any machine in phase 1 of subsystem B, with the search starting with machine 1 and proceeding incrementally. Preference is not given to the machine that just released the tool. If the necessary conditions for processing are not satisfied for any of the machines in subsystem B, then the tool searches the machines in phase 2 of subsystem A for one that can use it to begin processing. Here the policy is to not give preference to the same machine that relinquished the tool because it is possible that the part in the machine in the first phase will be blocked if the processing time at the machine in the second phase of the same line is longer. Frequent blocking deteriorates system performance, so a different approach to allocating the relinquished tool is expected to improve performance.

We have said nothing about the distributions of interarrival times or processing times at each machine. Since it is trivially easy in a simulation model to change these, it is not necessary to be specific about the distributions at this point. We will assume that the machines in a given phase have identical processing time distributions in lines 1 and 2 of subsystem A, and similarly for subsystem B. The interarrival time for parts in subsystem A is denoted by t_A. The processing times for parts in phases 1 and 2 in subsystem A are denoted by t_{1A} and t_{2A}, respectively. Interarrival times and processing times for subsystem B are denoted similarly.

6.5.1 The Event View Model

In the event view, we will just use numerical variables to represent all state variables, that is, the number of parts in a queue or the status of a machine. Let Q_A be the variable containing the number of parts in the input queue to subsystem A, and Q_B be the corresponding variable for subsystem B. The variables containing the numbers of parts in the input buffers for lines 1 and 2 of subsystem A are $Q_{2A}[1]$ and $Q_{2A}[2]$, respectively. For each machine in the system, let the value 1 denote that the machine is idle and 0 denote that it is busy. Thus, if $S_{1A}[1] = 1$, then that machine is idle; and if $S_{1A}[1] = 0$, then that machine is busy. It is easiest to think of $S_{1A}[1]$ as the number of idle machines in phase 1, line 1 of subsystem A. If a machine is blocked because a part cannot proceed, we will use the value –1 to denote the status of that machine. This only applies to machines $S_{1B}[1]$, $S_{1B}[2]$, and $S_{1B}[3]$, the three machines in phase 1 of subsystem B, lines 1, 2, and 3, since these are the only machines that can be blocked because they do not have an input buffer following them. Thus, these three machines have three states: $S_{1B}[i] = 1$ means the machine is idle and ready to begin processing the next part, $S_{1B}[i] = 0$ means the machine is actively working on a part, and $S_{1B}[i] = -1$ means the machine has completed processing one part but cannot start processing a new part because it cannot move the completed part to the next phase of processing. N_{TOOL} is a variable containing the number of tools available for use. Thus, if $N_{TOOL} = 2$, two tools are idle and can be put to use in the manufacturing system.

In this model there are 10 events, 5 for each subsystem. The events correspond to (1) arrival of a part, (2) start of processing in phase 1, (3) end of processing in phase 1, (4) start of processing in phase 2, and (5) end of processing in phase 2. This

FIGURE 6.14

```
Event Arrival_A
    Schedule Arrival_A at time NOW + t_A
    Q_A := Q_A + 1
    i := 1
    WHILE S_1A[i] = 0 and i < 2 DO i = i + 1
    IF i ≤ 2 THEN schedule Start_1A(i) at time NOW.

Event Start__1A(i)
    Q_A := Q_A - 1
    S_1A[i] := 0
    Schedule End__1A(i) at time NOW + t_1A

Event End_1A(i)
    S_1A[i] := 1
    Q_2A[i] := Q_2A[i] + 1
    IF N_TOOL > 0 and S_2A[i] > 0
        THEN schedule Start_2A(i) at time NOW.
    IF Q_A > 0 THEN schedule Start_1A(i) at time NOW.

Event Start_2A(i)
    Q_2A[i] := Q_2A[i]: - 1
    S_2A[i] := 0
    N_TOOL := N_TOOL - 1
    Schedule End_2A(i) at time NOW + t_2A

Event End_2A(i)
    N_A := N_A + 1
    S_2A[i] := 1
    N_TOOL := N_TOOL + 1
    IF Q_2A[i] > 0
        THEN schedule Start_2A(i) at time NOW.
        ELSE j := 1
            WHILE (Q_2A[j] = 0 or S_2A[j] =0) and (j ≤ 2) DO j =
              j + 1
            IF j ≤ 2
                THEN schedule Start_2A(j) at time NOW
                ELSE IF Q_B > 0
                    THEN k := 1
                        WHILE S_1B[k] ≤ 0 and k ≤ 3 DO k =
                          k + 1
                    IF k ≤ 3 THEN Schedule Start_1B(k) at time
                      NOW.

Event Arrival_B
    Schedule Arrival_B at time NOW + t_B
    Q_B := Q_B + 1
    i := 1
    WHILE S_1B[i] ≤ 0 and i ≤ 3 DO i = i + 1
    IF i ≤ 0 AND N_TOOL > 0
        THEN schedule Start_1B(i) at time NOW.
```

FIGURE 6.14 (continued)

```
Event Start_1B(i)
    Q_B := Q_B - 1
    S_1B[i] := 0
    N_TOOL := N_TOOL - 1
    Schedule End_1B(i) at time NOW + t_1B
Event End_1B(i)
    S_1B[i] := -1 /*Mark machine i blocked*/
    N_TOOL := N_TOOL + 1 /*Return tool*/
    IF S_2B[i] > 0 /*Line is not blocked*/
        THEN schedule Start_2B(i) at time NOW.
    k := 1
    WHILE (S_1B[k] ≤ 0 or Q_B = 0) and (k ≤ 3) DO k = k + 1
    IF k ≤ 3
        THEN Schedule Start_1B(k) at time NOW
            ELSE j := 1
                WHILE (S_2A[j] ≤ 0 or Q_2A[j] = 0) and j ≤ 2
                    DO j := j + 1
                IF j ≤ 2 THEN Schedule Start_2A(j) at time NOW.
Event Start_2B(i)
    S_2B[i] := 0
    Schedule End_2B(i) at time NOW + t_2B
    S_1B[i] := 1
    IF (Q_B > 0 and N_TOOL > 0) THEN schedule Start_1B(i) at time NOW.
Event End_2B(i)
    N_B := N_B + 1
    S_2B[i] := 1
    IF S_1B[i] < 0
        THEN schedule Start_2B(i) at time NOW.
```

number can be reduced; see the problems at the end of this chapter. Each event for a start of processing activity or end of processing activity has a single parameter, denoting the line involved. Figure 6.14 shows the event routines for the 10 events.

Note that in these event routines, we generally make some system state changes and then check to see if additional event(s) can be scheduled immediately. However, when we do the checking, we just check for the events that we know are possibilities to be scheduled. For example, in event *End_1B(i)*, we set the machine's status to idle ($S_{1B}[i] = 1$), return the tool to the tool crib ($N_{TOOL} = N_{TOOL} + 1$), and then check to see if this part can begin processing at the next phase in this line. We also check to see if another part can begin processing in phase 1 in this line, because we know the tool is available and the machine is now idle. If not, we check to see if any other parts are waiting to use the tool just released; if so, we begin processing for the first part that is ready to begin. Thus, in the event view, it is our responsibility to make all checks and schedule all new events that begin.

6.5.2 The Activity View

The three-phase version of the activity view's worldview makes all the same checks and schedules all the same events, but, as you will recall, the organization of the event routines is different. Each B-event is scheduled at a specific point in time, and the event routines for these events are simply written to make the system state changes that occur in the event. Each C-event has an event testhead that checks for the conditions for the event to occur. It is here that we check to see what additional events can be scheduled immediately. However, because the timing routine moves on to execute every C-event routine, without information about which B-event routine was just executed, the testhead in each C-event routine must be more comprehensive and check all system states to see if the event is ready to execute.

The following is the activity view version of the manufacturing model. All system state representations are the same as in the event view version to facilitate comparison between the two versions. The event names are also the same as in the event view version, but they have been labeled as B- or C-events. Note that, for example, the event *End_1A(i)* is a simple event with a single parameter denoting the line involved as before. There is no checking at the end of this event to see whether the part can start processing in phase 2A in the same line. Now, examine the conditional event *Begin_2A*. The testhead in this event checks to see if it can begin. The conditions needed for processing in phase 2A to start are for at least one tool to be available, and for at least one line; and both at least one machine in phase 2A to be idle and at least one part to be in the in-process inventory.

We have not specified a priority scheme for executing C-events. Note that in subsystem B, in each line the machine in phase 2 can block the machine in phase 1. Thus, to facilitate unblocking, we need to first check the beginning of service for phase 2B. If that can begin, it will free the machine in phase 1 of the same line to process another part (if one is available in the input buffer and a tool is available). Thus, the conditional event *Begin_1B* should be checked after the conditional event *Begin_2B*. The priority order, therefore, will be:

Priority	**Event**
1	Begin_2B
2	Begin_1B
3	Begin_2A
4	Begin_1A

and as shown in Figure 6.15.

6.5.3 The Process View

For the process view version of this model, we will switch to an entity–attribute–set modeling framework. There are two processes in this model: Part_A and Part_B. This choice is natural for this model because these are the temporary entities that

FIGURE 6.15

```
B-event Arrival_A
     Q_A := Q_A + 1
     Schedule Arrival_A at time NOW + t_A

B-event End_1A(i)
     Q_2A[i] := Q_2A[i] + 1
     S_1A[i] := 1

B-event End_2A(i)
     N_A := N_A + 1
     S_2A[i] := 1
     N_TOOL := N_TOOL + 1

C-event Begin_1A
     IF Q_A > 0 and (S_1A[1] > 0 or S_1A[2] > 0) THEN
          IF S_1A[1] > 0
               THEN i := 1
               ELSE i := 2
          Q_A := Q_A - 1
          S_1A[i] := 0
          Schedule End_1A(i) at time NOW + t_1A

C-event Begin_2A
     IF ((Q_2A[1] > 0 and S_2A[1] > 0) or (Q_2A[2] > 0 and S_2A[2] > 0))
               and N_TOOL > 0 THEN
          IF (Q_2A[1] > 0 and S_2A[1] > 0)
               THEN i := 1
               ELSE i := 2
          Q_2A[i] := Q_2A[i] - 1
          N_TOOL := N_TOOL - 1
          S_2A[i] := 0
          Schedule End_2A(i) at time NOW + t_2A

B-event Arrival_B
     Q_B := Q_B + 1
     Schedule Arrival_B at time NOW + t_B

B-event End_1B(i)
     (*Machine in phase 1, line i is blocked*)
     S_1B[i] := -1
     N_TOOL := N_TOOL + 1

B-event End_2B(i)
     N_B := N_B + 1
     S_2B[i] := 1

C-event Begin_1B
     IF Q_B > 0 and N_TOOL > 0 and (S_1B[1] > 0 or S_1B[2] > 0 or S_1B[3] > 0)
       THEN
          IF S_1B[1] > 0
          THEN i := 1
          ELSE IF S_1B[2] > 0
               THEN i := 2
               ELSE i := 3
```

FIGURE 6.15 (continued)

```
        Q_B := Q_B - 1
        N_TOOL := N_TOOL - 1
        S_1B[i] := 0
        Schedule End_1B(i) at time NOW + t_1B

C-event Begin_2B
        IF (S_2B[1] > 0 and S_1B[1] < 0)
                or (S_2B[2] > 0 and S_1B[2] < 0)
                or (S_2B[3] > 0 and S_1B[3] < 0) THEN
            IF (S_2B[1] > 0 and S_1B[1] < 0)
                THEN i := 1
                ELSE IF (S_2B[2] > 0 and S_1B[2] < 0)
                    THEN i := 2
                    ELSE i := 3
            (*Deblock a machine in phase 1, line i*)
            S_1B[i] := 1
            S_2B[i] := 0
            Schedule End_2B(i) at time NOW + t_2B
```

are created, flow through the system, and are deleted at the end. Since the actions in each subsystem differ, we need a different process for each subsystem. Each process has an attribute, called *current_line,* that contains the number of the line that is currently processing the part.

In the event view and activity view versions of this model, we used the variables Q_A, Q_B, and so on to denote the queue lengths. In this version, we must use the entity–attribute–set modeling paradigm, so it makes sense to alter our notation to avoid confusion with the earlier two versions. The system has four queues: The input buffers for subsystems A and B are Queue_A and Queue_B, respectively; and the input buffers for lines 1 and 2 in subsystem A are Queue_A_in[*i*], *i* := 1 or 2. For subsystem B, we will introduce three additional queues that we will call Blocked_Queue[*i*], *i:* = 1, 2, or 3. These queues will be used to hold a process in subsystem B that is blocked because the machine downstream on the same line is busy processing a part. Each queue will always hold either zero or one part. We will use the status of these queues to indicate whether a part is blocked and ready for processing in the next phase.

There are 10 machines in the system: 4 in subsystem A and 6 in subsystem B. We will call these Machine_1A[1], Machine_2A[1], and so on, where Machine_*m*A[*i*] refers to the machine performing phase *m* for process A in line *i.* Similar definitions apply to Machine_1B[1], Machine_2B[1], and so forth. Each machine has a status attribute that can have the values *idle* and *busy* to indicate that the machine is idle or busy.

There are five events for process Part_A: *arrival*, *begin_1A, end_1A, begin_2A*, and *end_2A.* Process Part_B has a similar group of five events. These events, which are encountered in the order given, are labeled in the process routines for the two processes. This will facilitate comparison with the two previous versions of this model. The process view model is shown in Figure 6.16.

FIGURE 6.16

```
Process Part_A
Arrival:
     /*Start the next part after an interarrival time delay.*/
     Initiate new Part_A at time NOW + t_A
     Insert this Part_A last in Queue_A
     /*Find an idle machine to process this part*/
     IF Machine_1A[1].status = idle
          THEN current_line := 1
          ELSE IF Machine_1A[2].status = idle
          THEN current_line := 2
          ELSE current_line := 0
     /*If no machine is idle, stop the process here until one is
       idle.*/
     IF current_line = 0 THEN PASSIVATE

Begin_1A:
     Remove this Part_A from Queue_A
     Machine_1A[current_line].status := busy
     ACTIVATE this Part_A at time NOW + t_1A
     SUSPEND

End_1A:
     Machine_1A[current_line].status := idle
     IF Queue_A is not empty
          THEN ACTIVATE the first Part_A in Queue_A at time NOW
     Place this Part_A last in Queue_A_in[current_line]
     /*If the part cannot continue to process in phase 2, stop the
       process here*/
     IF N_TOOL = 0 or Machine_2A.status = busy THEN PASSIVATE

Begin_2A:
     Machine_2A[current_line].status := busy
     Remove this Part_A from Queue_A_in[current_line]
     /*Take one tool.*/
     N_TOOL := N_TOOL - 1
     ACTIVATE this Part_A at time NOW + t_2A
     SUSPEND

End_2A:
     Machine_2A[current_line].status:= idle
     */Return one tool.*/
     N_TOOL := N_TOOL + 1
     /*See if the tool can be used to activate a suspended process
       that is waiting for a tool.*/
     IF Queue_A_in[current_line] not empty
          THEN ACTIVATE the first Part_A in Queue_A_in[current_line]
          ELSE j := current_line(mod 2) + 1
          IF Queue_A_in[j] not empty and Machine_2A[j].status = idle
               THEN ACTIVATE the first Part_A in Queue_A_in[j] at
                  time NOW
               ELSE IF Queue_B not empty
```

FIGURE 6.16 (continued)

```
                THEN k := 1
                    WHILE Machine_1B[k] = busy and k ≤ 3
                        DO k := k + 1
                    IF k ≤ 3 THEN ACTIVATE the first
                        Part_B in Queue_B at time NOW

Process Part_B

    Arrival:
    INITIATE new Part_B at time NOW + t_B
    Insert this Part_B last in Queue_B
    /*Search for an idle machine to begin processing in phase 1.*/
    k := 1
    WHILE Machine_1B[k].status = busy and k ≤ 3 DO k = k + 1
    /*If none is found, stop the process now.
        Otherwise, record the line with the idle machine.*/
    IF (k > 3) or (N_TOOL = 0)
        THEN PASSIVATE
        ELSE current_line := k

Begin_1B:
    Remove this Part_B from Queue_B
    Machine_1B[current_line].status := busy
    /*Take one tool.*\
    N_TOOL := N_TOOL - 1
    ACTIVATE this Part_B at time NOW + t_1B
    SUSPEND

End_1B:
    /*Return one tool.*/
    N_TOOL := N_TOOL + 1
    IF Machine_2B[current_line].status = busy
        THEN /*Put the part into the blocked part queue for this
           line.*/
            Insert this Part_B into Blocked_Queue[current_line]
            PASSIVATE

Begin_2B:
    /*Set phase 1 machine idle and phase 2 machine busy.*/
    Machine_1B[current_line].status := idle
    Machine_2B[current_line].status := busy
    ACTIVATE this Part_2 at time NOW + t_2B
    IF Queue_B is not empty
        THEN ACTIVATE the first Part_B in Queue_B at time NOW

End_2B:
    Machine_2B[current_line].status := idle
    IF Blocked_Queue[current_line] not empty
        THEN Remove the first Part_B in Blocked_Queue[current_line]
            ACTIVATE the first Part_B in
                Blocked_Queue[current_line] at time NOW
```

We should note three things about this version of the model. First, when a part arrives, we first check to see if a machine is idle to begin processing the part. This is done by searching the available machines for phase 1 to see if one with the status attribute has the value *idle.* Second, note that when we check for the resources for a part to begin processing at a given phase, we check to see if the process *cannot* start processing; if so, we turn the process off (passivate). The default for a process is to continue operating unless we stop it, so we have to check to see if *insufficient* resources are available and, if so, stop the process. Finally, note how the blocking in subsystem B is handled. After phase 1, we check to see if the second machine on the current line is busy. If so, the part is blocked. In this case, we put the part in the queue for blocked parts and leave the server's status as *busy.* Thus, no new parts will be able to start processing because the server is still busy. The next time this part is activated, it will continue processing in phase 2. This is checked at the end of the process, in the *end_2B* event. We check the blocked queue for the current line and, if it is not empty, a part is waiting to begin processing in phase 2. In that case, we remove this part from the blocked part queue and activate it immediately. Note that we do not set the status of the machine in phase 1 to idle until the part starts processing in phase 2 (see Figure 6.16).

6.6 SUMMARY

This chapter, along with Chapter 5, has presented the basic modeling concepts for discrete-event simulation. In Chapter 5, we discussed the event view and the entity–attribute–set representation in some detail. In this chapter, we expanded and extended these concepts by introducing two additional worldviews: the *activity view* and the *process view.* An *activity* consists of a starting event followed by an ending event that is scheduled by the starting event. In most activities, the starting event is a conditional event; that is, the system state must satisfy some condition before the event can be executed. Normally, an activity involves the interaction of entities in the system over a period of time. Service in a queueing simulation is an example of an activity. The activity view allows the modeler to create events that are modular and independent. This forces the timing routine to be somewhat less efficient, but the event routines can be simpler and more reliable. The activity view was further refined in the three-phase approach in which events are divided into scheduled events (or B-events) and conditional events (or C-events). Conditional events have a testhead that tests to see if the system state allows the event to occur. The timing routine proceeds through three phases: In phase A, the clock is updated to the time of the next scheduled event. In phase B, the next scheduled event is selected from the event list and executed. Finally, in phase C all conditional events are executed until no more conditional events qualify to be executed. Activity scanning must also involve *priorities* for conditional events so that resources will be captured by the highest priority activities first. After phase C checks all conditions and fails to find any C-events that can be executed, the entire cycle repeats.

The process view represents system dynamics by joining a sequence of events to follow the progress of a temporary entity through the system. For example, all events and activities involving a customer as she flows through a queueing system constitute a process. A *process* consists of an entity record with all attributes for the entity, along with a process routine that describes the sequence of related events and activities for the entity. A process must also have a *local event pointer* to mark the next event in the sequence to occur. A process can be initiated, passivated, and activated. *Initiation* creates the entity record and schedules the first event for the process. *Passivation* removes any scheduled event from the event list and stops execution of the process routine. *Activation* schedules the next event for the process to occur at a specific time in the future. Two versions of the process view were presented. In the first, processes operate by competing for the same resources. In the second version, called the *process interaction approach*, processes interact by activating one another.

The event, activity, and process views are not mutually exclusive but can be combined. An example was presented showing the use of the event view along with the process view. All three worldviews were further illustrated with a more complex example—a manufacturing system. This manufacturing system consisted of two subsystems with two and three lines in each subsystem. Moreover, the manufacturing lines compete for a limited set of tools to do their work. This model demonstrated techniques for sharing resources among processes and for representing blocking in queueing systems.

References

Buxton, J. N., and J. G. Laski. 1962. Control and simulation language. *Computer Journal* 5:194–199.

Clementson, A. T. 1977. *Extended control and simulation language—Computer aided programming system.* University of Birmingham, England: Lucas Institute for Engineering Production.

Pidd, M. 1992. *Computer simulation in management science.* 3d ed. Chichester, England: Wiley.

Poole, T. G., and J. Z. Szymankiewicz. 1977. *Using simulation to solve problems.* London: McGraw-Hill.

Tocher, K. T. 1963. *The art of simulation.* London: English Universities Press.

Problems

6.1 Each activity can be represented as a short process with two events: the event starting the activity and the event ending the activity. Rewrite the single-server queueing model with an arrival event and a service activity that is represented by a process with two events.

6.2 Suppose that the single-server queueing model includes a resource, paint, that is needed to provide service. The amount of paint needed for each customer is a random variable with a uniform distribution between 10 and 200 gallons. The paint is replenished when the remaining amount drops below 400 gallons, and the replenishment fills the inventory to 2000 gallons after a random delay that is normally distributed with mean 24 hours and standard deviation 4 hours. Arrivals are Poisson at rate four per day, and service times are exponential with mean 4 hours. Write this model using the event view.

6.3 Model the system in the previous problem using the activity view.

6.4 Model the system in problem 6.2 using the process view.

6.5 Using the activity view, modify the single-server queueing model so the server breaks down after a random length of time having an exponential distribution with mean two hours. After a breakdown, the server is not able to provide service for a random length of time that is uniformly distributed between 30 minutes and three hours. Customers who are being served when the server breaks down leave the system immediately. Customers in the queue when the server breaks down continue to wait in the queue until the server is up and running again. Arrivals are Poisson with rate four per hour, and services are exponentially distributed with mean 10 minutes. Use the activity view to develop this model.

6.6 In the single-server queueing model, arrivals are Poisson with rate four per hour, and service times are exponentially distributed with mean 10 minutes. Suppose that the server breaks down after a random amount of cumulative time spent in service. This cumulative time has an exponential distribution with mean two hours. Modify the activity view version of this model to accommodate this change. Assume that once the server has broken down, it remains broken for a random length of time having an exponential distribution with mean one hour. Customers who are being served when the server breaks down leave the system immediately, and customers waiting when the server is not working wait in the queue until the server is up and running again.

6.7 Use the event view to express the model in the previous problem.

6.8 Use the process interaction approach to express the model in problem 6.5.

6.9 Consider again the system in problem 6.5. Suppose also that the system receives preventive maintenance periodically. Preventive maintenance is scheduled 100 hours after the last preventive maintenance or after the last breakdown, whichever occurred most recently. Breakdowns occur after an exponentially distributed number of hours after the last preventive maintenance or the last repair, whichever occurred most recently. The mean time to breakdown is 200 hours. Repair times are exactly 2 hours. Assume arrivals are Poisson, and service times are exponentially distributed. The arrival rate is four per day; the service rate is six per day. Use the event view to model this system.

6.10 Use the activity view to model the system in the previous problem.

6.11 Modify the single-server queue so the customer is not placed in the queue on arrival unless the server is free to provide service immediately.

6.12 Modify the single-server queue model to have k identical servers. Assume arrivals are Poisson with rate four per hour, and service times are exponentially distributed with mean 20 minutes. Write the model to work for any value of k. Use the activity view to express the model.

6.13 Use the process view to express the model in the previous problem.

6.14 Modify the single-server queue model to have three servers, each with a different service time distribution. Use the event activity view to express the model.

6.15 Modify the single-server queue model to have two different types of customers—A and B—with different service time distributions. Customers are served first-come, first-served. Use the event view.

6.16 Use the activity view to solve the previous problem.

6.17 Modify the single-server queue model to have two different types of customers, A and B, and two servers, 1 and 2. The system has a single queue. Type A customers prefer server 1 if both servers are idle, and type B customers prefer server 2 if both servers are available. If only one server is available on arrival, then the customer will be served by that server; otherwise, if all servers are busy, the customer will join the queue and be served in first-come, first-served order. The service time distribution depends on both the customer type and the server.

6.18 Currently, the manufacturing model in section 6.5 releases the tool immediately when the part finishes processing in phase 1 in subsystem B. Modify this model to require the tool to stay with the part at station 1 in each line of subsystem B until the part starts processing in phase 2. In the modified model, if the part is blocked because the station downstream is busy, then the tool continues to be unavailable.

6.19 Modify the manufacturing model in section 6.5 so that the buffer between phases 1 and 2 in subsystem A has finite capacity k. Be sure the model works correctly if $k = 0$. Use the activity view.

6.20 Use the event view to solve the previous problem.

6.21 Write the process view version of the manufacturing model in section 6.5 to use the state variables defined in the event view and activity view versions in Chapter 5 and section 6.3.

7

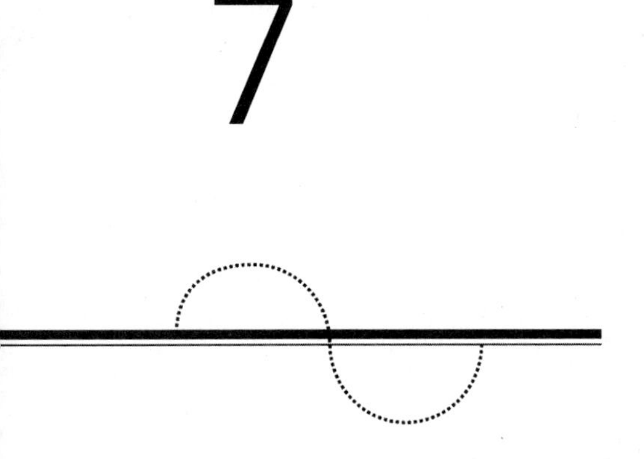

Visual Interactive Simulation and Arena

In this chapter we present the concepts of visual interactive simulation and illustrate them using Arena, a software tool that implements this approach. *Visual interactive simulation* (VIS), which is the leading simulation paradigm, uses a graphical user interface and an interactive approach in all stages of a simulation project, from model building to animation to analysis of the simulation experiment. This approach is popular because it facilitates intuitive model building and makes easier model verification possible. To make our introduction to Arena interesting and easy to follow, we will demonstrate the software features by building, running, and analyzing two models: a simple single-server queueing system and a somewhat more complex bank lobby system.

7.1 VISUAL INTERACTIVE SIMULATION

Until the early 1980s, simulation models were developed by writing simulation programs using a simulation language and a text editor. General-purpose simulation languages as well as specialized languages for specific application areas such as manufacturing were available. Some general-purpose simulation languages were GPSS, SIMULA, and SIMSCRIPT. SIMAN, SIMFACTORY, and XCELL are application-oriented simulation languages that have been used extensively. Simulation programs were also written using general-purpose programming languages such as FORTRAN, C, and Pascal and supplemented by routines for discrete-event simulation.

Using this approach, simulation programs first had to be typed and then compiled and executed. The compilation step detected syntax errors before the executable code was produced; if errors were found, they had to be corrected and the program recompiled before it could be executed. The output produced by the simulation usually consisted of one or more large text or binary files containing observations of system output variables as the simulation executed over time. If the programmer programmed it, the output could also include the results of statistical analyses applied to these variables.

One step in model building is *verification* of the model. Verification involves determining that the model implemented is the model intended. The simulation output had to be examined carefully to detect logic errors in the model, which often were revealed to be present, indicating that the model did not correctly represent the intended system's structure, entities, interactions, or operational rules. The simulation program code then had to be rechecked in detail, corrected, recompiled, output analyzed, program code checked, and so on until the verification was completed and the analyst was assured that the model was operating correctly.

This procedure, however, had three main serious drawbacks:

1. Model development time could be overly long.
2. The process of detecting and locating program errors, especially logic errors caused by an incorrect representation of system operation, was slow, tedious, and often difficult.
3. Neither simulation models, in the form of program code, nor simulation output, in the form of output data, were transparent—that is, the models could only be understood by the person who wrote the code.

The third drawback meant that the users of simulation—managers, engineers, and others—were not able to either understand the simulation model or determine the correctness of the simulation output. Documentation of the model was laborious and often poor. This resulted in inadequate communication between modelers and users. Users saw the simulation model as a "black box," and they were not ready to place much confidence in the results produced or to implement decisions recommended by analysts using the models.

7.1.1 Simulation Hardware and Graphical User Interfaces

In this section we give an overview of the VIS approach along with a short summary of visual interactive simulation software. In section 7.2 we introduce Arena and demonstrate the basic capabilities by modeling a simple single-server queueing system. In section 7.3 Arena is used to model a bank lobby system with more realistic features such as time-varying arrival rates during the day, simultaneous arrivals of two or more customers, a variable number of banking transactions for customers, and customer balking. Finally, in section 7.4 we evaluate visual interactive simulation tools as representative tools for interactive simulation.

The rapid development of personal computers and computer graphics in the late 1980s and early 1990s powerfully influenced the whole area of computer-oriented problem solving and decision making, and simulation modeling was no exception. Starting with the pioneering work of Hurrion (1976), simulation began to incorporate new hardware and user interfaces that completely changed the world of simulation computing in the time span of just a few years. Graphical user interfaces (GUIs) and interactivity have become part of the new generation of simulation tools that has radically changed both the technology of working with computers as well as the nature of problem solving and decision making using computers.

As a result of the rapid growth of computer power and the movement toward the use of graphics, GUIs have become the primary type of interface for interaction with the user (Mandelkern, 1993). Operating systems such as Microsoft Windows for IBM PC-compatible computers as well as the Mac OS are graphically oriented. Unix and Unix-like operating systems such as Linux are not natively graphically oriented, but these systems all include the X window system for workstations, which is a GUI. Virtually all currently available operating systems are also able to run multiple programs and perform multiple tasks simultaneously through multitasking and multithreading. Word processors, spreadsheets, databases, and other types of application software have also adopted graphical interfaces. Parameters are changed, menus are activated, files are copied, and other actions are taken by moving the mouse or other pointing device and clicking to select, the so-called point-and-click approach.

The benefit of the point-and-click approach is that the user needs less instruction and skill to use the software. This allows people who understand the problem they are trying to solve and who know how the software will solve their problem to apply it to their problems without having to spend a lot of time learning to use it. As more and more computer users have become accustomed to GUIs, they expect and even demand its implementation into the other more specialized software tools. It is understandable, therefore, that simulation software developers have also incorporated modern graphical user interfaces in their products.

VIS is a software tool that combines a GUI with the ability to work interactively in all phases of the simulation (Bell and O'Keefe, 1987; Hurrion, 1989). The term *visual interactive simulation* was originally coined because, in the early years of its evolution, graphics and interaction were primarily used during the simulation execution to animate the model and provide interactive control (Hurrion,

1976). In the United States, VIS was originally used to refer to just the animation of simulation experiments (Bell and O'Keefe, 1987). After the initial period of VIS development, graphics and interaction were also incorporated into the model-building phase.

7.1.2 Software for VIS

SEE-WHY (Fiddy, Bright, and Hurrion, 1981), which was developed in the United Kingdom in 1979, was the first commercially available VIS package. It had a simple visual display with entities represented by letters in colors, the ability to represent static backgrounds, and simple windowing for interaction. A large class of VIS tools has been developed for specific application areas. For example, manufacturing-oriented VIS systems include SIMFACTORY, XCELL, and OPTIK Process Line Simulator. These programs use icons to represent specific tools in the manufacturing system. Each icon is related to a set of parameters whose values can be changed during modeling. The icons are then formed into a network to create the graphical model.

In some cases, visual parts were added to existing simulation tools. Usually, animation was added, but sometimes user interactivity was included. Examples are SIMSCRIPT II.5, which was augmented with SIMGRAPHICS; GPSS/H, which was extended with Proof Animation; and the HOCUS package, which was also extended with animation. Actually, Proof Animation is a stand-alone software package that can be used to animate the simulation by reading a text file that describes the geometry and changes in the system. Because the input to Proof Animation is a text file, it can be used with any simulation software that is capable of producing the appropriate text output to drive the animator. The SLAM II simulation package was integrated into SLAMSYSTEM. SIMAN was first extended using the CINEMA animation software and later integrated into Arena.

Several completely new simulation tools were later developed around the VIS paradigm. ProModel for production systems simulation, and two closely related software tools, ServiceModel for service systems simulation and MedModel for medical systems simulation, are graphically based. In these packages, the model is developed by selecting icons that represent system entities, defining their characteristics via tables, and connecting them in a network to represent the system structure and operation. The graphical model is also used directly for animation.

Design/CPN and SIGMA for Windows are other software tools that use a graphical modeling approach. Design/CPN is based on colored Petri nets and is available for Unix workstations and Macintosh computers. SIGMA for Windows uses event graphs and is available for IBM-compatible personal computers. The latter software also implements animation. A more thorough review of available simulation software, including VIS software, can be found online at <http://www.lionhrtpub.com/surveys/Simulation/Simulation.html>.

Regardless of the specifics, virtually all VIS tools rely on a graphical version of the simulation model. Design/CPN and SIGMA for Windows implement standard graphical model representations. Other packages such as ProModel create their own

FIGURE 7.1 Arena model containing several flowchart elements (modules)

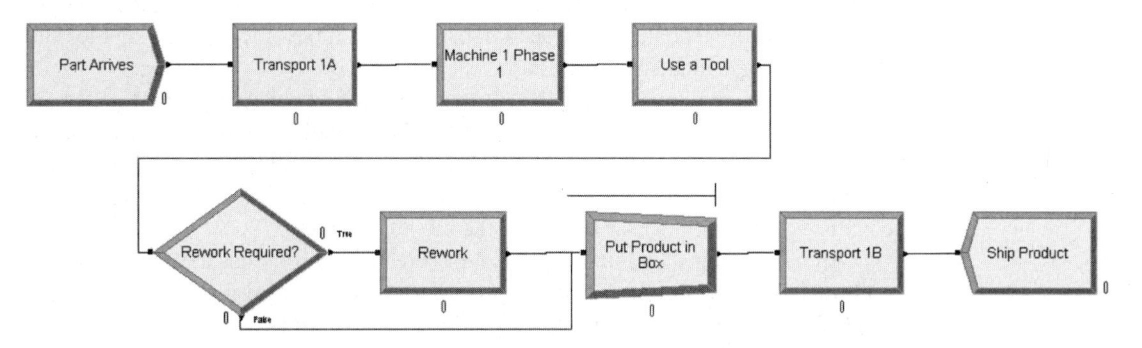

graphical representation. The graphical representation implemented by Arena is a familiar flowchart diagram. The approach in all cases is to allow the user to build the graphical simulation model interactively at the computer, then provide specific data and other information to the model by using dialog boxes, which are usually displayed by double-clicking on parts of the graphical representation. The benefit of this approach is that it is intuitive and allows quick model development. Since the human visual system is much more developed than the language or verbal communication system, VIS uses this system to allow the user to quickly recognize and interpret models. VIS is to simulation analysts what electrical network diagrams are to electrical engineers. Because graphical models can be quickly understood, they also provide an efficient way to document and communicate a model.

7.2 A FIRST LOOK AT ARENA

Arena is a VIS tool that supports the modeling of dynamic business systems using discrete-event simulation. With Arena, the user can interactively develop the model, create an animation of the system, run the simulation, collect output data from the simulation, and create and view statistical reports for the run results. Arena also includes **Input Analyzer,** which is a module for input data analysis.

The graphical model used by this software is based on a simple flowchart that presents the system as a logical network of related activities. Figure 7.1 shows an Arena model that contains a variety of flowchart elements, or *modules.** Arena

*Some of the text in the Arena graphics in this chapter and Chapters 8 and 9 have been enhanced for readability. If you create the same modules, they might appear slightly different.

supports hierarchical model building that enables each model element to represent a submodel; submodels can contain deeper submodels. Hierarchical modeling is an important feature that allows the decomposition of complex models into smaller and more easily understandable model units.

When the simulation is run, the Arena model is first converted into the SIMAN programming language and then compiled and executed using the SIMAN simulation engine. This process happens automatically and is not interrupted unless syntax errors are detected during the model-checking and compilation phase, a runtime error is detected, or the user interrupts the model execution.

Arena contains a Quick Preview set of screens that provides a brief instruction on how to build a simple model. It also contains two useful model libraries, the SMARTs Library and the Examples Library. The SMARTs Library is a collection of numerous simple models that demonstrate different model-building techniques. The Examples Library contains several models of varying complexity that demonstrate model building and animation for several types of systems.

Arena is a Microsoft Office–compatible product—that is, its toolbars and menus are similar to those used by Microsoft Office. This also means that images from the Arena model can be easily transferred via clipboard to Office documents prepared by Word, Excel, or PowerPoint, and that Arena uses ActiveX automation (formerly known as OLE automation) and Visual Basic for Applications (VBA) Windows technologies to enhance the integration of desktop applications. A built-in interface to the process-mapping tool Visio allows process drawings prepared with Visio to be automatically converted into Arena models.

Written documentation for Arena is limited to a brief *Guide to Arena Business Edition* (1999). Extensive documentation is in the interactive help menu, which contains detailed descriptions of all Arena features. Kelton et al. (2002) also provide an extensive tutorial on Arena that complements this and the next two chapters.

7.2.1 Build the Model

Drawing the Arena Model Flowchart

During this tutorial, it will be useful for the reader to follow along on a computer as we develop this model. Start Arena from the Windows Start menu by selecting **Programs, Arena,** and again **Arena.** The Arena modeling environment will open with a new model, as shown in Figure 7.2. We have chosen **Large Buttons** in the toolbars to make them more visible. This option is selected in the **Customize** dialog under the **Toolbars . . .** option in the **View** menu. This modeling environment consists of three regions: the project bar on the left, the model window on the top right, and the spreadsheet window on the bottom right. The project bar contains three panels: the **Basic Process** panel, which has modules used in the basic modeling; the **Reports** panel, which contains different types of simulation reports; and the **Navigate** panel, which allows the user to display different views of the model.

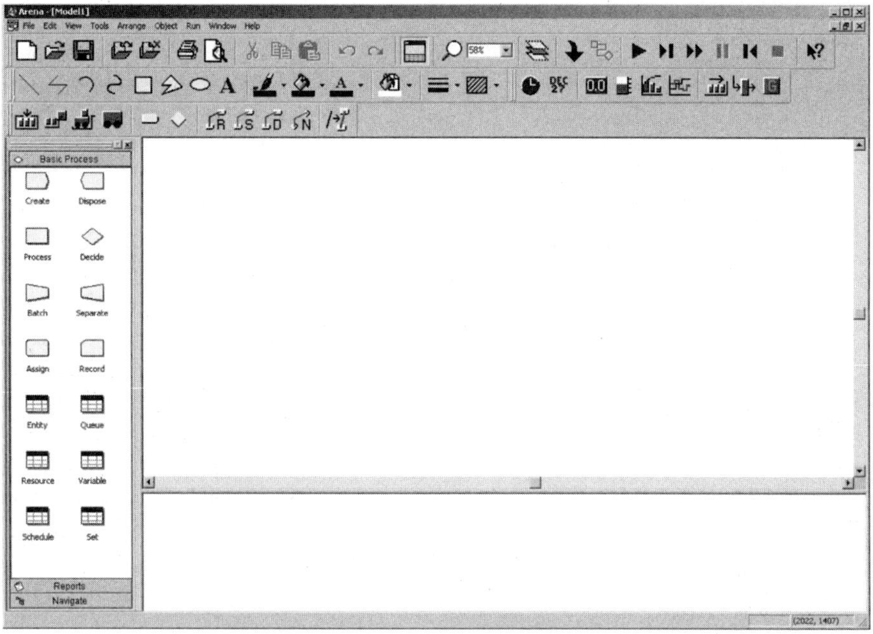

FIGURE 7.2
Arena s modeling environment

The model window contains all model graphics, including the process flowchart and animation, while the spreadsheet window displays model data for the flowchart or data modules.

Before we start building a model, it is useful to check whether the *automatic connection* feature for flowchart modules is selected. This feature causes each module to automatically connect to the previously selected module as the new module is placed, thus avoiding the need to make the connections manually. To enable the automatic connection feature, make sure the **Auto-Connect** option in the **Object** menu is checked.

We will first demonstrate model building with Arena by developing a model of the simple queueing system with a single server that we examined in Chapters 5 and 6. Assume that customers' interarrival times have an exponential distribution with a mean of 2.5 minutes, and service times have a normal distribution with a mean of 1.85 minutes and standard deviation of .35 minutes. We want to simulate 8 hours of system operation and compute the following system performance measures: mean queue length, mean customer cycle time, and server utilization. Customer cycle time is the length of time from arrival to departure.

The entities flowing through the model are customers, and they are served by a single resource, a server. The flowchart is a visual representation of the customer's process. The finished model, shown in Figure 7.3, consists of three modules: a **Create** module representing customer arrivals, a **Process** module representing service, and a **Dispose** module representing customers leaving the system. Arena's graphical

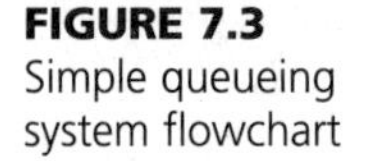

FIGURE 7.3
Simple queueing system flowchart

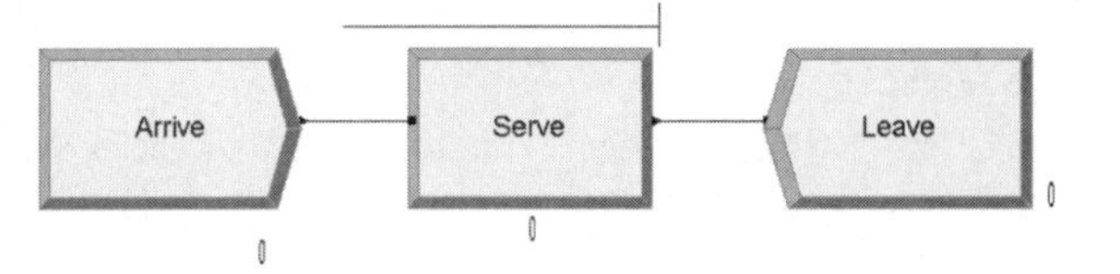

model represents the process view, as defined in Chapter 6. Each process involves a sequence of activities and events experienced by a temporary entity as it proceeds to interact with the system. For this model, Arena's graphical model first represents the customer's creation; the customer then attempts the service activity in the process module; finally, after the service activity, the customer is deleted. In any Arena model, the entity must first be created, and the last activity for every entity is to be destroyed.

Model building starts with drawing the model flowchart. Our approach will be to first draw the model diagram and then edit each of the modules to enter the specific parameters. First we need to place a **Create** module in the model window. Drag the **Create** module from the **Basic Process** panel and drop it onto the model window. The **Create** module will appear in the model window with the default name **Create 1,** as shown in Figure 7.4.

We continue drawing the model flowchart by connecting a **Process** module to the **Create 1** module. So that a connection will automatically be drawn between these two modules, be sure that the **Create** module is selected before putting the **Process** module in the model window. Now drag the **Process** module from the **Basic Process** panel to a position on the right of the Create 1 module; Arena will automatically connect the two modules. Finally, drag the **Dispose** module from the **Basic Process** panel to a position on the right of the **Process** module. After these operations the model window will appear as in Figure 7.5.

Sometimes you will need to edit a model by deleting or adding modules. You can delete a module or just a connection by clicking once on the element to select it and then pressing the Delete key on the keyboard. You also can manually connect modules. You will need to do this if you edit the model by deleting or creating modules when the automatic connection feature is turned off. To manually draw a connection, select the **Connect** option in the **Object** menu or click on the **Connect** toolbar button. The cursor will change to a crosshair. Each module (except the **Create** and **Delete** modules) has an entry point, which is a small square on the left side of the module; and an exit point, which is a small arrow on the right side. You can manually draw the connection between two modules by clicking on the exit point of the first module and then clicking on the entry point of the second module.

Arena's snap and grid features can be used to arrange the modules in the model neatly. These features work by defining a grid of equally spaced points in the model window. When the **Snap** option is operating, each module will be placed on the nearest grid point, allowing modules to be aligned. To use the snap and grid feature, check the **Snap** option on the **View** menu. To realign modules that have already been placed, select the modules you want to realign by holding the Control key and

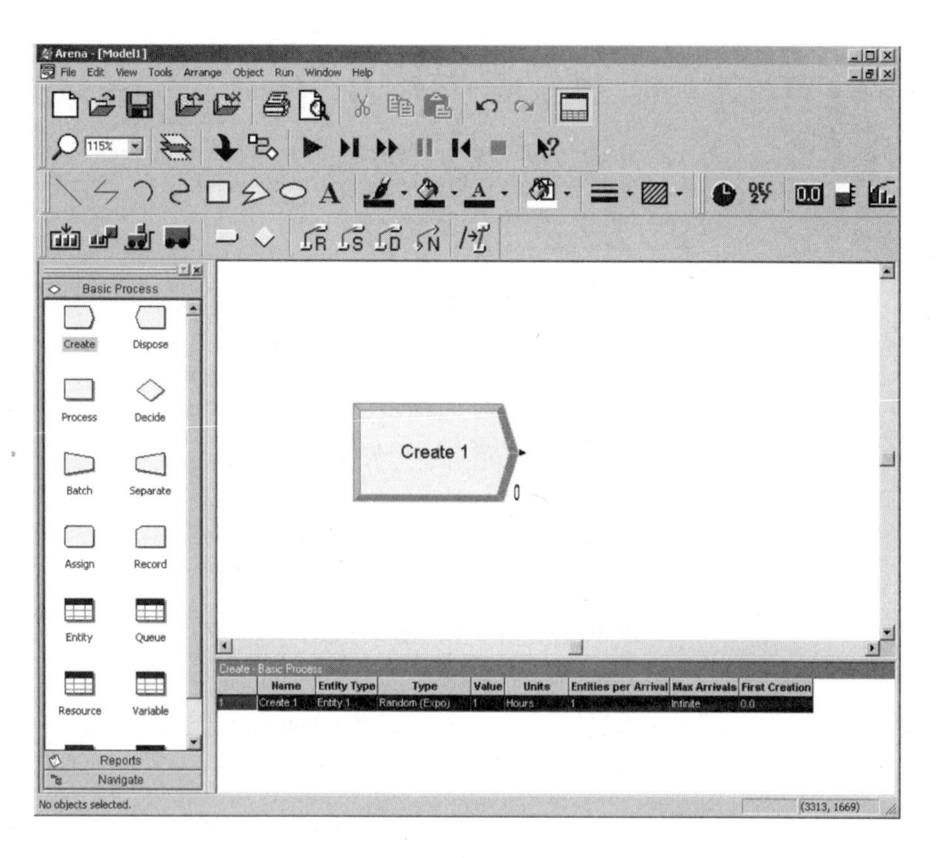

FIGURE 7.4 Placing the Create module in the model window

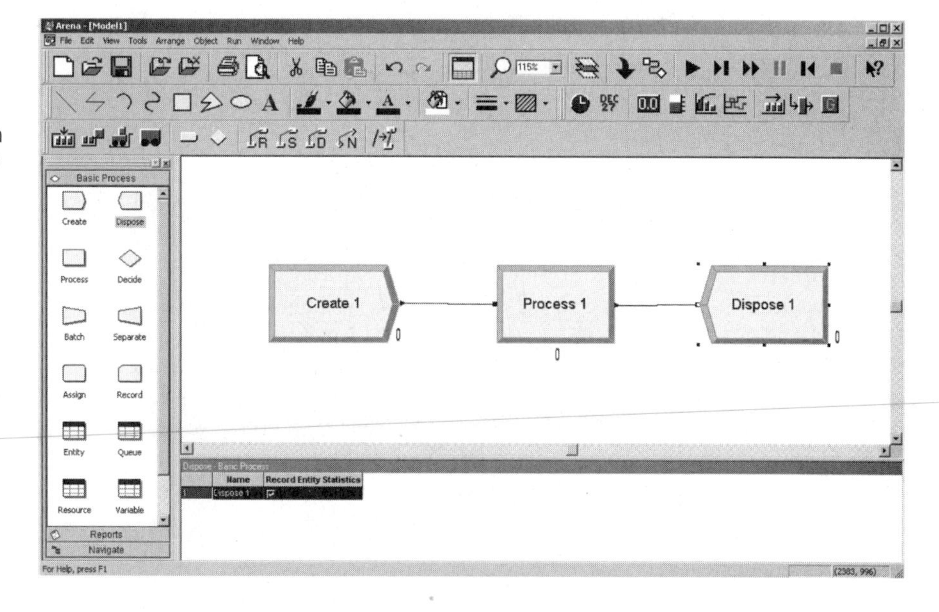

FIGURE 7.5 Model flowchart after placing the Process and Dispose modules in the model window

clicking on each module or by clicking and dragging a rectangle that includes all modules. Then select the **Snap to Grid** option on the **Arrange** menu to align these modules with grid points.

You may also zoom in on the model window by pressing the plus (+) key on the keyboard or zoom out by pressing the hyphen key (-). These keys will allow you to work on specific parts of a large model.

Enter the Modules Data

After the model diagram is created, our next task is to customize the modules by entering specific parameters for them. First, we will name the **Create** module and provide parameters that describe the interarrival time distribution. Double-click on the **Create** module to open its property dialog. Type "`Arrive`" (without the quotes) in the **Name** field. Whatever we enter in this field will appear as the name of the module in the model window. Descriptive names are useful for model documentation and understanding. Next, give the entity we are creating a name by entering "`Customer`" in the **Entity Type** field. The **Time Between Arrivals** parameters define the arrival stream. If you click the down arrow in the **Type** option, you will see four options: **Random, Schedule, Constant,** and **Expression.** These can be used to define arrival patterns for Poisson arrivals, arrivals according to a schedule, equally spaced arrivals, and interarrival time distribution as given by a mathematical expression, respectively. We want Poisson arrivals, so select **Random** for the Type option. Poisson arrivals have an exponential interarrival time distribution. Enter "`2.5`" for the **Value** option and select minutes for **Units** to specify a mean interarrival time of 2.5 minutes. Leave the default values for the last three parameters and click **OK** to close the dialog. The **Create** module property dialog with the data entered is shown in Figure 7.6.

Next, double-click on the **Process** module to open its property dialogue. Enter "`Serve`" in the **Name** option and leave the default **Standard** for the process type. Define the **Logic** parameters. For the **Action** option, select **Seize Delay Release.** You will see additional options under the **Logic** area. The **Seize Delay Release** ac-

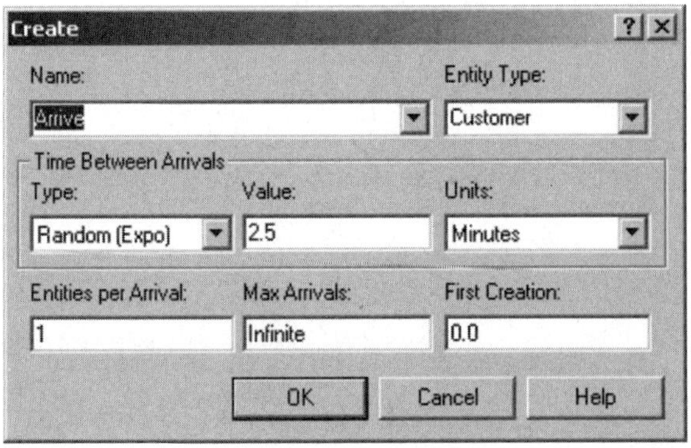

FIGURE 7.6 Create module property dialog with the data for the simple queueing system with a single server

Create
Name: Arrive
Entity Type: Customer
Time Between Arrivals
Type: Random (Expo)
Value: 2.5
Units: Minutes
Entities per Arrival: 1
Max Arrivals: Infinite
First Creation: 0.0
OK Cancel Help

FIGURE 7.7
Resource dialog with the data for the simple queueing system with a single server

tion means that the entity will attempt to seize one or more resources and will wait in a queue until all resources are available. When they are available and have been seized, the entity will hold them for the defined time delay and then release them. In our model, we need only one resource. Define it in the **Resources** field by clicking the **Add** button on the right side of the field. The dialog is shown in Figure 7.7.

When the **Resources** dialog appears, leave the default **Resource** as the **Type** of resource, enter "Server" in the **Resource Name** field, and leave default value of 1 for the resource **Quantity.** We have just defined a new resource, "Server," and specified that each customer will use one unit of this resource during this activity. Click **OK** to close the **Resources** dialog. The remaining edit fields refer to the delay associated with this process module. For the **Delay Type,** choose **Normal** to specify a normally distributed delay. Select minutes for **Units.** The **Allocation** field refers to options for computing costs. We will discuss this later, so for now leave the default **Value-Added** for **Allocation.** Type "1.85" for the **Value (Mean)** field; for **Std Dev** (standard deviation), enter ".35." Click **OK** to close the dialog. The whole **Process** module property dialog with the data we entered is shown in Figure 7.8. As soon as you close the **Process** dialog you will notice a horizontal line over the **Process** module. This line represents the queue for entities waiting for the resource. This queue will be used in the model animation.

Finally, double-click on the **Dispose** module and enter the data in its property dialog. Type "Leave" in the **Name** option; leave the **Record Entity Statistics** option checked. Click **OK** to close the dialog. The **Dispose** dialog is shown in Figure 7.9. Now the model has been fully defined. This is a good place to save the model by selecting **Save** in the **File** menu or by clicking on the **Save** button in the **Standard** toolbar.

Standard toolbar

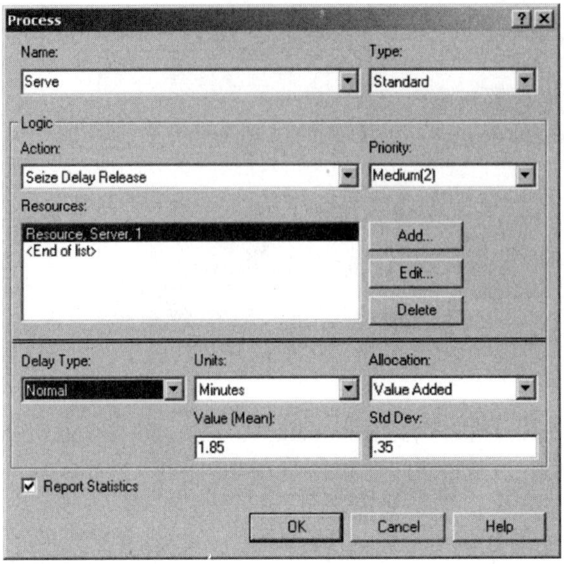

FIGURE 7.8
Process module property dialog with the data for the simple queueing system with a single server

Change Some Model Data

Arena modeling elements are entities, resources, sets, queues, and variables. Information about these elements can be defined by clicking on the appropriate module in the **Basic Process** panel or by editing data in the related sheet in the spreadsheet view.

Let's first update information about the server resource used in the model. Arena automatically added a resource with this name in the **Resource** spreadsheet when we defined the resource "Server" in the **Process** module. Select the **Resource** sheet in the spreadsheet view by clicking on the **Resource** module in the **Basic Process** panel. This sheet lets us not only change the name of the resource but also define the number of units of the resource and other parameters that we will discuss later. To change the number of units of the resource, click on the **Capacity** cell and enter the new value. Since we will use the default number of units, 1, we will not change this

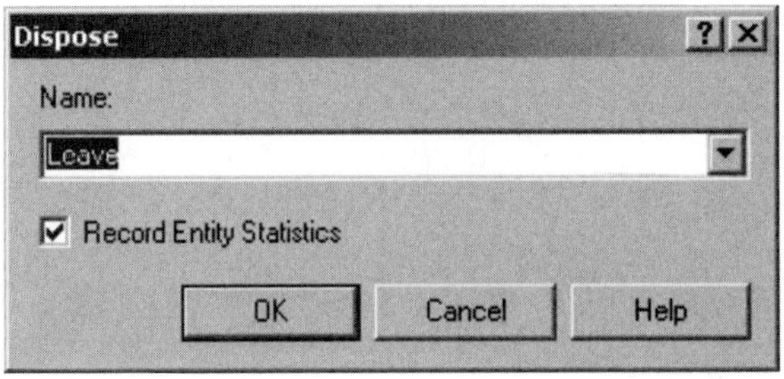

FIGURE 7.9
Dispose module property dialog with the data for the simple queueing system with a single server

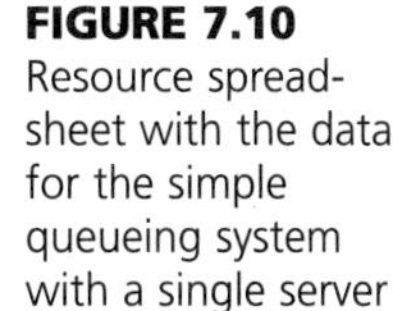

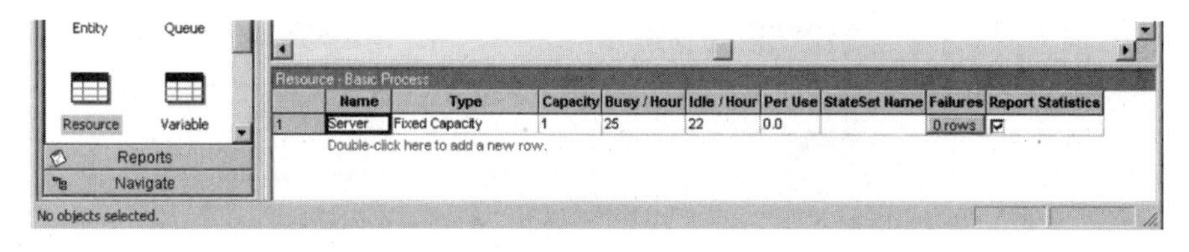

FIGURE 7.10 Resource spreadsheet with the data for the simple queueing system with a single server

value. (Clicking on the cell in the **Busy/Hour** column allows us to enter the cost rate of 25 when the server is busy. Clicking on the cell in the **Idle/Hour** column allows us to enter the cost rate of 22 when the server is idle.) Information in the **Resource** spreadsheet is shown in Figure 7.10.

We would also like to update information about the customer entity used in the model. Click on the **Entity** module in the **Basic Process** panel to update the **Entity** spreadsheet information in the spreadsheet view. Let's click on the cell in the **Initial Picture** column and select `Picture.Woman` as the icon that will represent customers in the animation.

7.2.2 Preparing a Simulation Run

After the model is defined we have to provide the information required to run the simulation—that is, we have to tell the model how long to run the simulation as well as other details for the model run. Select **Run → Setup** from the main menu to open the **Run Setup** dialog. This dialog has several tabs. Under the **Project Parameters** tab, you can give the project a name and enter the analyst's name. In the **Project Name** field, type "`Simple Queueing System`," enter "`8`" in the **Replication Length** field, and select minutes for the **Time Units** option. Leave all other parameters with their default values. The complete simulation module property dialog is shown in Figure 7.11, and the completed Arena model is shown in Figure 7.12.

Sometimes the model window is hard to read because the fonts are too small. We would like to provide some useful advice regarding readability of text and numbers that appear in the model window. By default, text and numbers associated with modules are written in a small font (Arial) that is difficult to read. We suggest you make the following changes. For the text associated with modules such as "True" or "False" in a Decide module, double-click on the text, choose size 9 in the **Text Font** window, and click **OK.** For numbers associated with modules (e.g., `Service.WIP` under the Process block named "Service"), double-click on the number, select **Font . . .** in a **Variable** window, choose a Univers font or other font that has medium size in the **Variable Font** window, click **OK,** and click **OK** again on the **Variable** window.

Save the model again by selecting the **Save** option in the **File** menu or by clicking the **Save** button. Arena stores all of the model definition including the flowchart and the model data in the model file.

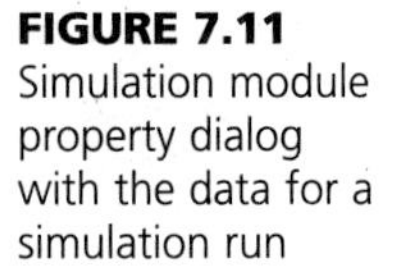

FIGURE 7.11
Simulation module property dialog with the data for a simulation run

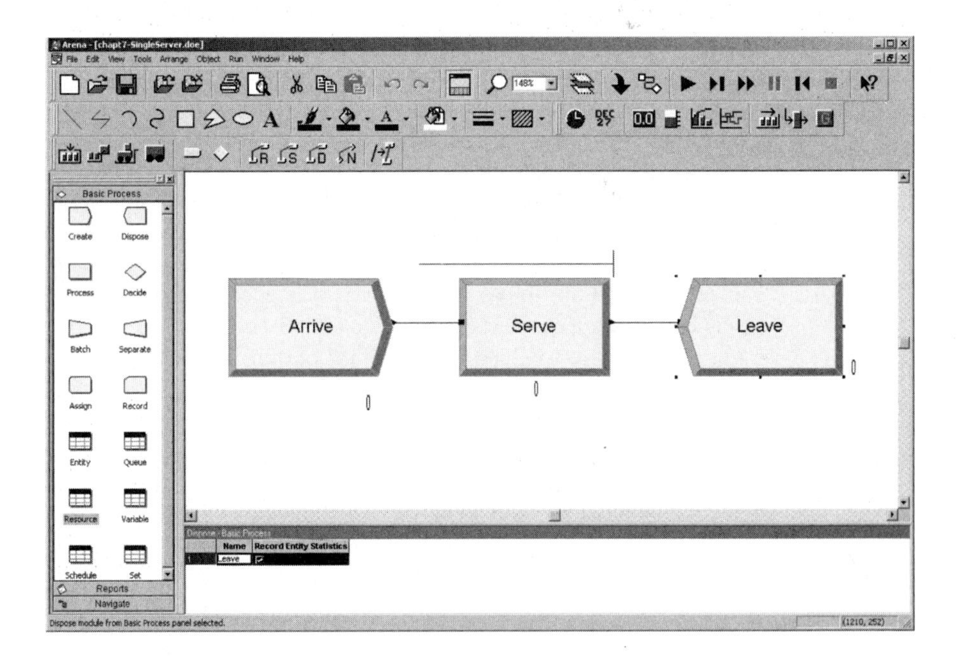

FIGURE 7.12
Completed Arena model of the simple queueing system with a single server

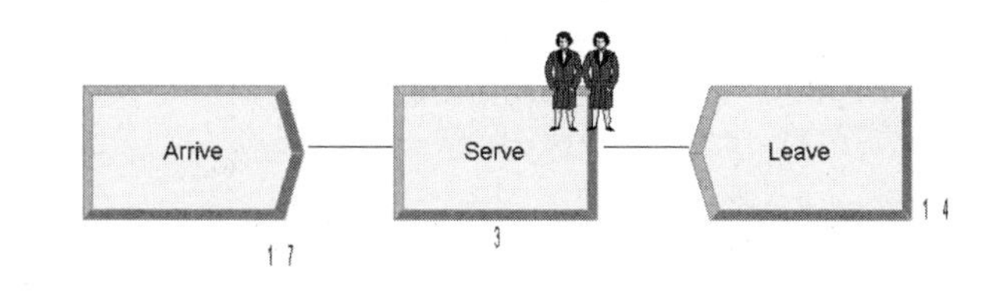

FIGURE 7.13 Animation of the simple queueing system with a single server

7.2.3 Interactive Model Runs

The model is now ready to be executed. Start the simulation by selecting the **Go** option in the **Run** menu or by clicking the **Go** button in the **Standard** toolbar. The **Standard** toolbar has a group of buttons (which we will call the "Run buttons") that look like the controls on a videocassette recorder. The **Go** button looks like a VCR's play button. When the model executes, Arena first checks the model for errors and then initiates the simulation run to see if the model is free of errors. Arena simulations are animated by default. During the simulation, you will see an animation with small entity graphics moving along the flowchart connectors and waiting in the queue positions. Variables shown near modules will also change as the simulation progresses.

Figure 7.13 shows the animation as the model executes. The values near each module represent the number of entities created, the number of entities currently in process (either being serviced or waiting for service), and the number of entities that have been deleted. The animation also shows entities in various stages of processing in the model. You can slow the animation by pressing the "<" key repeatedly, and you can speed it up by pressing the ">" key. The simulation can be paused by clicking the **Pause** button in the Run buttons; clicking the **End** button will stop it. If you want to step through simulation one event at a time, you can pause the simulation and then click the **Step** button). Sometimes this is useful to debug or verify a model.

During the animation, entities first appear when they are created. They then move through the flowchart and disappear when they are being processed in process modules. They also disappear permanently when they enter a dispose module. This effect, combined with a careful analysis of changes in variable values, is useful in verifying the model logic. When the model has been verified and you are ready to make full-length runs, you can run the simulation faster by turning off the anima-

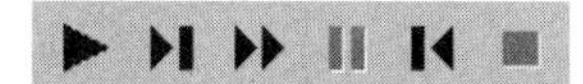

Run buttons on the Standard toolbar

Go button

Step button

Fast forward button

Pause button

Reset button

End button

tion. To turn off the animation and run the simulation at full speed, either select the **Fast Forward** option in the **Run** menu or click the **Fast Forward** button in the **Standard** toolbar.

7.2.4 Reports

Of course, the purpose of any simulation run is to report on the system's performance. Arena automatically collects data on such things as waiting times, queue lengths, and operating costs. When the run is finished, Arena will ask whether you want to view reports. Click **Yes** and the default report titled "Category Overview Report" will be displayed in a report window as shown in Figure 7.14. A tree showing all of the types of information available in the report is displayed on the left side of the report. The project title, which in this case is "Simple Queueing," is listed at the top of the tree.

FIGURE 7.14 Category Overview Report for the simple queueing system with a single server

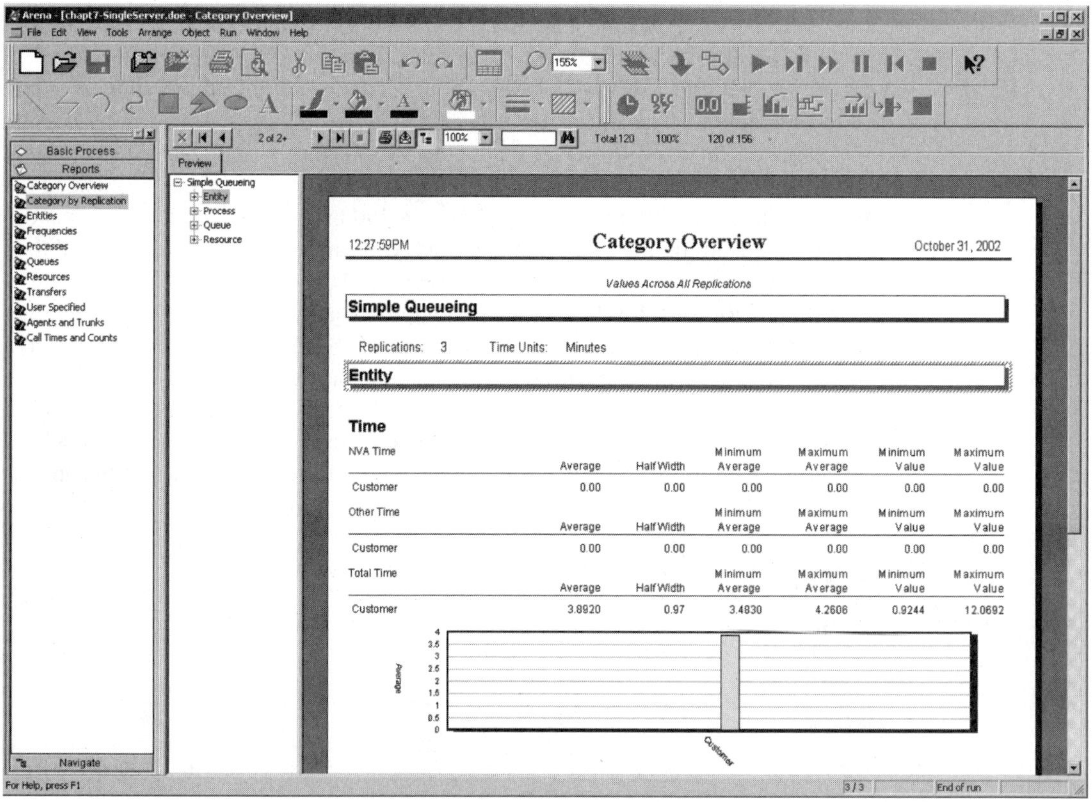

12:27:59PM **Category Overview** October 31, 2002

Values Across All Replications

Simple Queueing

Replications: 3 Time Units: Minutes

Entity

Time

NVA Time	Average	Half Width	Minimum Average	Maximum Average	Minimum Value	Maximum Value
Customer	0.00	0.00	0.00	0.00	0.00	0.00

Other Time	Average	Half Width	Minimum Average	Maximum Average	Minimum Value	Maximum Value
Customer	0.00	0.00	0.00	0.00	0.00	0.00

Total Time	Average	Half Width	Minimum Average	Maximum Average	Minimum Value	Maximum Value
Customer	3.8920	0.97	3.4830	4.2606	0.9244	12.0692

The Category Overview Report gives an overview of simulation results for entities, processes, queues, and resources. The following are useful statistics relating to *entities*:

- Customer *cycle time.* This is the time each customer spends in the system, that is, the time between the customer's creation and deletion. In this case, the average of these observations is 3.71 minutes, the minimum time is 1.07 minutes, and the maximum time is 12.09 minutes.
- Customer's *value added (VA) cost.* This is a value calculated for each customer as the product of the customer's VA time and the cost of busy resources per unit of time. The average for this simulation run is 46.01, with a minimum of 25.91 and maximum of 67.02.
- Customer's *waiting time* for the available resource. This is the length of time the customer waits in a queue for the resource. For this run, the average is 1.87 minutes, with minimum time of 0 minutes and maximum time of 9.87 minutes.
- *Work in progress (WIP).* This is the number of customers in the system at any point in time. The time average for this run is 1.46, with a minimum of 0 and a maximum of 6.0.

In all of these reports, "**(Insufficient)**" appears under the **Half Width** column. Arena checks to see if the amount of data generated is sufficient to compute a reliable confidence interval for the parameter. If the amount of data is not sufficient to compute a reliable interval, then this notation will appear; otherwise, the half-width of the interval will be printed.

The output results related to *processes* are the same as those for entities, since all that entities are doing in this simple single-server queueing system is waiting for the service and being served in the **Process** module.

If you click on the "**Detail on Queues**" under the **Reports** panel on the Project bar, you will get a detail report for all queues in the model. In this model, there is only one queue. The report shows that the average waiting line length is .73 customers, with a minimum of 0 and a maximum of 5 customers in the line. Over the course of the simulation run, the number of customers in the waiting line varied between 0 and 5, and the time average of this function was .73, indicating that it was close to 0 far more often than it was close to 5. You can also get a detailed report on the resources in the model by clicking on "**Detail on Resources.**" In this report, you will observe that the average utilization of the server is .72, with a minimum of 0 and a maximum of 1, as can be expected because there is only one resource in the system and it can be either idle or busy. Thus, the server was busy 72 percent of the time in this simulation run.

Other Types of Reports Category by replication, detail reports, and summary reports for entities, processes, queues, and resources can be found by opening the **Reports** panel on the Project bar and clicking on the appropriate line. The **Category by Replication** report displays information about entities, processes, queues, and resources for each replication, as well as information on user-assigned variables or items. The **Detail Reports** give detailed information per replication for all entities, processes, queues, and resources, and the **Summary Reports** give summary information per replication for all entities, processes, queues and resources. Since we made only one replication in this simulation run, these reports contain no additional information.

7.2.5 Modeling Costs in Arena

Arena has facilities for modeling two types of costs: *value added* and *non-value added*. The concept behind this feature is that in any business process there are two types of time spent by entities in the system. Time that is spent delivering a service for which the customer is paying is considered value added time. Thus, a patient who is being examined by a physician in a clinic is spending value added time because she will pay the clinic for the physician's time. Time that is spent waiting for service or in some activity that could be omitted without reducing the effectiveness of the service is considered non-value added time. For example, time spent by a patient waiting for laboratory tests in a medical clinic is non-value added time because it does not increase income to the clinic. Generally, organizations want to minimize non-value added time and maximize value added time, depending on their objectives. The time spent in each process can be categorized as value added or non-value added.

Arena allows the user to associate a cost-per-unit time with value added time and a different cost-per-unit time with non-value added time. We will return to this topic again in Chapter 8.

7.2.6 Modeling Elements

Now that we have developed a model using Arena, we want to look more carefully at the modeling facilities provided by this software. Arena's basic modeling elements are entities, resources, sets, queues, and variables. These elements are called *data modules* in Arena, and all can be accessed in the **Basic Process** panel. In the following, we describe each element:

Entities are temporary system objects that flow through the flowchart, such as customers, documents, or parts in a manufacturing process. There can be various types of entities, and each can have different attributes and a different graphic representation in the animation.

Resources represent system assets used for service or to process system entities. Examples are clerks, operators, and machines. In the Arena worldview, entities flow through the system, requesting resources at various nodes, and waiting in queues until the resource requirements can be met. The term *capacity* for resources refers to the amount of the resource that is available for use by entities. A resource can have either a fixed capacity or a capacity that changes in time according to a given schedule. Several types of costs can be associated with entities and resources, such as entity holding cost or cost of busy or idle resources per hour.

A **set** of resources is a collection of multiple resources of identical type, such as receptionists or machines that perform identical operations. Resource sets are used when an entity can use any of the resources from the set rather than a specific resource. Usually, resource sets are needed when an entity can use a resource and needs to return to that same resource later.

Queues are waiting lines for entities that are automatically created by Arena in the places in the model where resources may be seized. Several queue disciplines can be specified—first-in, first-out; highest-attribute-first; and so on.

Variables can be defined that are related to resources, queues, entity attributes, simulation statistics, and so forth. Their value can be initialized, changed, and referenced during simulation. Variables are useful for storing information while the simulation is running.

7.2.7 Modeling Blocks

Arena basic modeling blocks are the **Create, Process, Decide, Assign, Batch, Separate, Dispose, Record,** and **Simulate** modules. These blocks, called *flowchart modules* in Arena and detailed below, are the components of the Arena graphical model and can be found in the **Basic Process** panel.

The **Create** module is the generator of entities that enter the system—for example, a customer's arrival in the bank lobby. Every Arena model must have at least one **Create** module, and every entity must originate from a **Create** module.

The **Process** module models an activity—that is, a delay in the entity's progress through the system, possibly involving seizing and releasing of one or more resources or sets required for the activity. An example is the service activity in the queueing system we just examined. Process time can be value added or non-value added. The delay can be zero, in which case no time elapses during the activity.

The **Decide** module represents decision making in the system—that is, a choice of paths an entity can take through the system. For example, a **Decide** module can be used to model different actions for parts depending on whether or not a defect is present. The decision can be two-way or multiway. The decision is represented as a two-way or multiway branch out of the decision module, and the choice can be based on a condition determined by the value of one or more variables or it can be randomly sampled with a given probability.

The **Assign** module is used to assign a value to a variable or entity attribute.

The **Batch** module allows a specified number of entities or entities with a specific value of an attribute to be collected into a single group or batch. For example, it can be used to accumulate a certain number of parts before they are packed together for shipping.

The **Separate** module is used either to duplicate an entity in order to create multiple identical entities or to separate previously batched entities. This can be useful in systems where entities are processed in batches and then later processed individually, and it can also be useful to create parallel processes in a model.

The **Dispose** module permanently removes entities from the model. In most models, every entity eventually enters a **Dispose** module.

The **Record** module collects statistics in the simulation model—for example, recording the time spent by customers in a queue.

7.3 A BANK LOBBY MODEL IN ARENA

We will now use Arena to model a bank lobby with more realistic features. During this modeling task we will learn how to use some other useful modeling constructs of Arena.

7.3.1 Bank Lobby System

A bank lobby has four tellers—Alice, Mary, Jeff, and Doris—with similar working characteristics. The customer arrival pattern varies over time. The average number of arrivals per hour is 10, 20, 40, 36, 27, 32, 18, and 4 for each of the 8 one-hour periods from the opening of the bank lobby until the closing time. During each period, the arrival process is Poisson. In addition, customers can arrive in groups of more than one. For each arrival instance, there is a 75 percent probability that it is a single customer, a 20 percent probability that the group size consists of two customers, and a 5 percent probability that three customers are in the group.

The number of banking transitions for each customer is sampled from the distribution in Table 7.1, which was obtained from historical data.

A single queue serves all four tellers. When a customer enters the lobby, she will join the queue if the total number of customers in the lobby—that is, the number of customers being served plus the number of customers waiting in the queue—

TABLE 7.1 Probability distribution of number of transactions per customer

Number of transactions	1	2	3	4	5	6
Probability (%)	20	30	22	15	8	5

FIGURE 7.15
Bank lobby flowchart

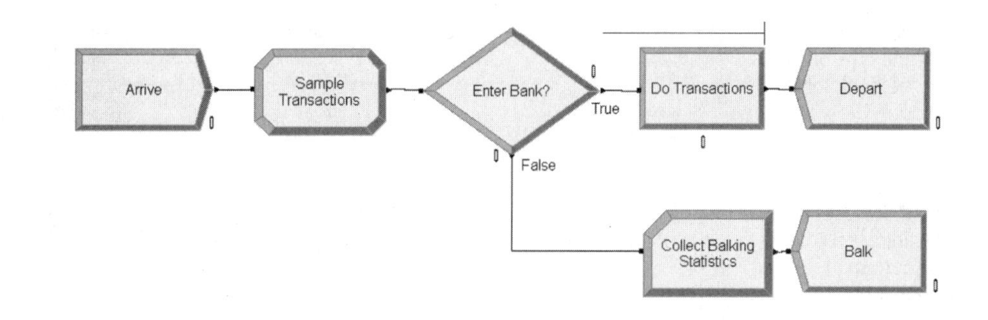

is less than 10. Otherwise, she will balk. The service time for the customer depends on the number of transactions to be processed for the customer. Processing time for each transaction has an Erlang distribution with mean 1.08 minutes and the number of stages (parameter k) equal to 2.

Eight hours of the bank lobby operation will be simulated: from 9 A.M. to 5 P.M.

7.3.2 Model Building

We begin by creating an Arena flowchart, as shown in Figure 7.15. First place a **Create** module named "Arrive" to serve as a generator of customers arriving at the bank lobby. The next module, which is an **Assign** module named "Sample Transactions," is used to assign a value to the user-created attribute "Number of Transactions" for the customer who just arrived. Now the customer checks to see if he will enter the bank using a **Decide** module named "Enter Bank?" This module simulates the decision to enter the bank or balk by testing whether the number of customers in service or waiting for service is less than 10. If this condition is true, then the customer will take the right branch from the decision module; otherwise, she will take the bottom branch. If the customer decides to enter the bank, she will wait and then be served in the **Process** module named "Do Transactions." After the service is finished, the customer enters the **Dispose** module named "Depart" and is removed from the simulation. Customers who did not want to wait enter the **Record** module named "Collect Balking Statistics," which saves statistics about the number of customers who balked. After that, these customers enter the **Dispose** module named "Balk" and are removed from the simulation. Creating this flowchart should be straightforward after your experience with our first Arena model.

Continue developing the bank lobby model by defining the model resources used in the bank lobby. There are four identical resources called Alice, Mary, Jeff, and Doris. To create them, first click on the **Resource** module in the **Basic Process** panel and then double-click on the text "**Double-click here to add a new row**" under the **Resources** spreadsheet that automatically opens. Fill this spreadsheet with the four rows of data shown in Figure 7.16. If you insert an extra row by mistake, you can delete it by right-clicking on it and selecting **Delete Row.**

FIGURE 7.16 Defining resources in the bank lobby model

	Name	Type	Capacity	Busy / Hour	Idle / Hour	Per Use	StateSet Name	Failures	Report Statistics
1	Alice	Fixed Capacity	1	0.0	0.0	0.0		0 rows	☑
2	Mary	Fixed Capacity	1	0.0	0.0	0.0		0 rows	☑
3	Jeff	Fixed Capacity	1	0.0	0.0	0.0		0 rows	☑
4	Doris	Fixed Capacity	1	0.0	0.0	0.0		0 rows	☑

Double-click here to add a new row.

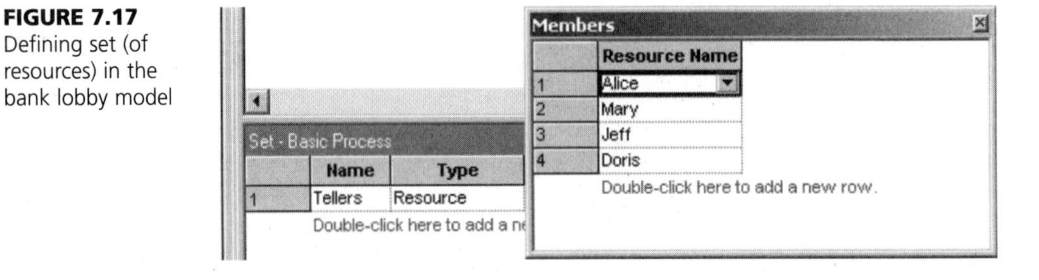

FIGURE 7.17 Defining set (of resources) in the bank lobby model

The bank lobby uses four identical resources with a single queue in front of them, so we need to create a set (of resources). To do this, click on the **Set** module in the **Basic Process** panel and then double-click on the text "**Double-click here to add a new row**" under the **Set** spreadsheet. In the **Name** field, type "`Tellers`" and click on the row under the **Members** field to open a window where you can enter the resource names for the members. In the **Members** window, add four new rows and select the names of our four servers for the **Resource** names, as shown in Figure 7.17.

Now that the teller resources have been defined and a set created, we need to model customer arrivals to the bank. Because customer arrival rates vary according to a specified schedule, the schedule will need to be specified. Click on the **Schedule** module in the **Basic Process** panel and then double-click on the text "**Double-click here to add a new row**" to add a new row in the **Schedule** spreadsheet. Name the schedule "Arrivals schedule" and select **Arrivals** as the type of schedule as shown in Figure 7.18. To insert the schedule, right-click in the row under the **Durations** column and select the **Edit via Spreadsheet** option. When the spreadsheet window appears, fill eight rows as shown in Figure 7.19. The schedule can be viewed graphically. To see the graphical view, close the **Duration** spreadsheet window and double-click on the cell under the **Duration** column in the **Schedule** spreadsheet. The window shown in Figure 7.20 will appear (it is just part of the window that contains the graphical representation of the schedule).

More than one customer can arrive at any arrival event, with the number of customers given by the probability distribution in Table 7.1. To enter these probabili-

FIGURE 7.18 Initiating the arrivals schedule in the bank lobby model

Schedule - Basic Process

	Name	Type	Time Units	Scale Factor	Durations
1	Arrivals schedule	Arrival	Hours	1.0	8 rows

Double-click here to add a new row.

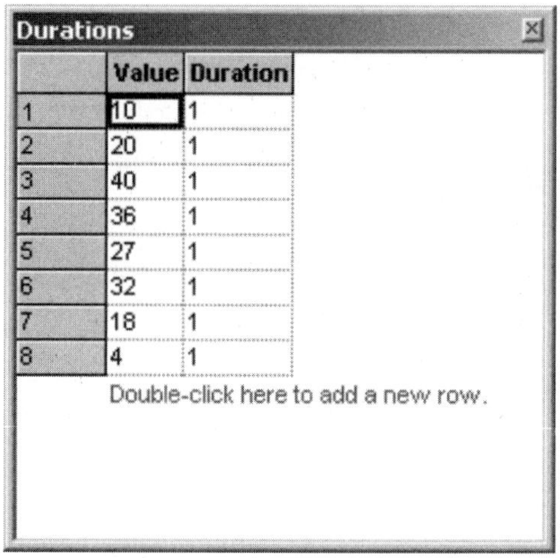

FIGURE 7.19 Numerical definition of the arrivals schedule in the bank lobby model (edited via spreadsheet)

Durations

	Value	Duration
1	10	1
2	20	1
3	40	1
4	36	1
5	27	1
6	32	1
7	18	1
8	4	1

Double-click here to add a new row.

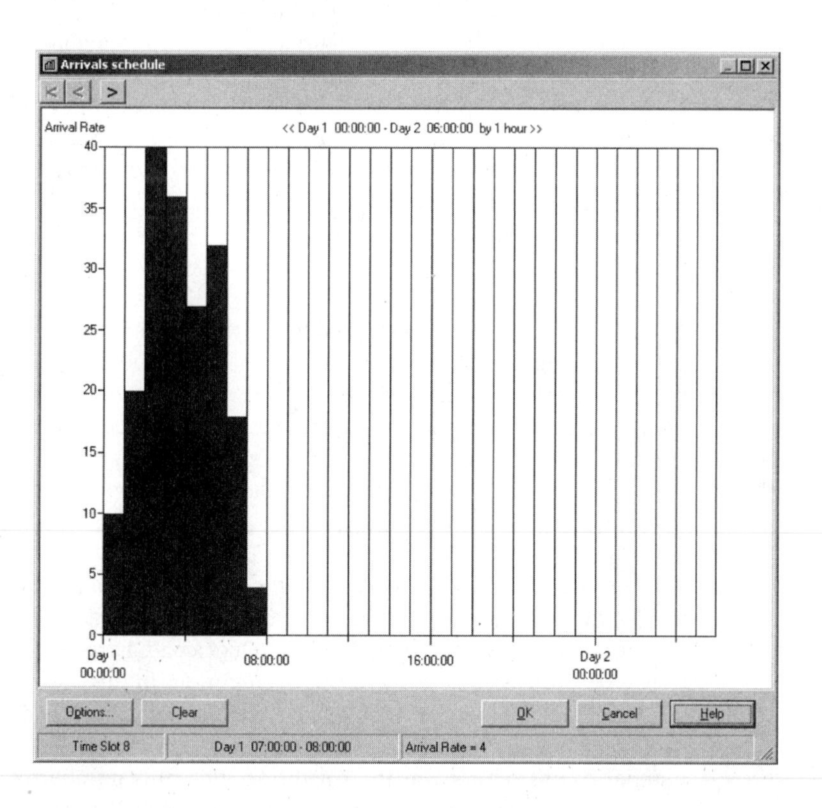

FIGURE 7.20 Graphical view of the arrivals schedule in the bank lobby model

ties, double-click the **Create** module named "Arrive" and enter the data as shown in Figure 7.21. The time between arrivals is determined by an exponential distribution with the mean time between arrivals given by the "Arrivals schedule" we just de-

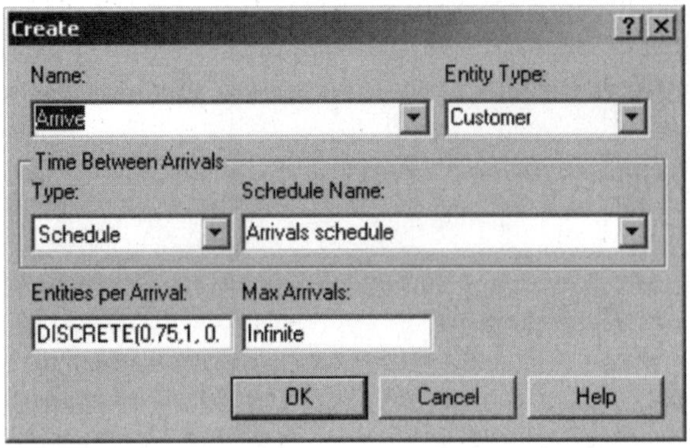

FIGURE 7.21
Defining the probability of simultaneous arrivals in the bank lobby model

Create
Name: Arrive
Entity Type: Customer
Time Between Arrivals
Type: Schedule
Schedule Name: Arrivals schedule
Entities per Arrival: DISCRETE(0.75,1, 0.
Max Arrivals: Infinite
OK Cancel Help

fined. The **Entities per Arrival** field specifies the number of entities created each time an arrival occurs. This field contains the discrete distribution used for a user-defined discrete probability distribution. This function requires parameters that are sequences of *cumulative* probabilities, followed by the value of the random variable. Thus, the sequence .75, 1, .95, 2, 1.0, 3 would specify that the value 1 would be sampled with probability .75, the value 2 would be sampled with probability .95 – .75 = .20, and the value 3 would be sampled with probability 1.0 – .95 = .05. Note that the final probability in this function call must be 1.0. The entry in the **Entities per Arrival** field will look like

```
DISCRETE(.75, 1, .95, 2, 1.0, 3)
```

This completes the **Create** module "Arrive." Our next task is to define the number of banking transactions per customer. Start by clicking on the **Assign** module named "Sample Transactions" and fill the data as shown in Figure 7.22. In the **Type** field, select **Attribute** and define the attribute name as "No. of transactions." This creates a new attribute for each customer that we can refer to as "No. of transactions." For the value, use a discrete distribution again with the following data:

```
DISCRETE(.2, 1, .5, 2, .73, 3, .87, 4, .95, 5, 1.0, 6)
```

Now double-click on the **Decide** module "Enter Bank?" to enter the parameters for the customer's decision to enter the bank or not. Fill in the data as shown in Figure 7.23. The **Decide** module can select a path based on either sampling with a specified probability or by testing a condition. In this case, we want to test the condition that the number of customers already in the bank is greater than 10. This decision is made using the variable `Do Transactions.WIP`, which is the total number of customers waiting for the service plus number of customers being served in the **Process** module "Do Transactions." WIP means *work in progress.* If `Do Transactions.WIP` is less than 10, then customers will wait to be served; otherwise they will balk. Arena has many built-in variables that are updated automatically as the simulation runs. A complete list of these variables is provided in Arena help. Select **Help** in the main menu, then **Arena Help Topics.** From the **Con-**

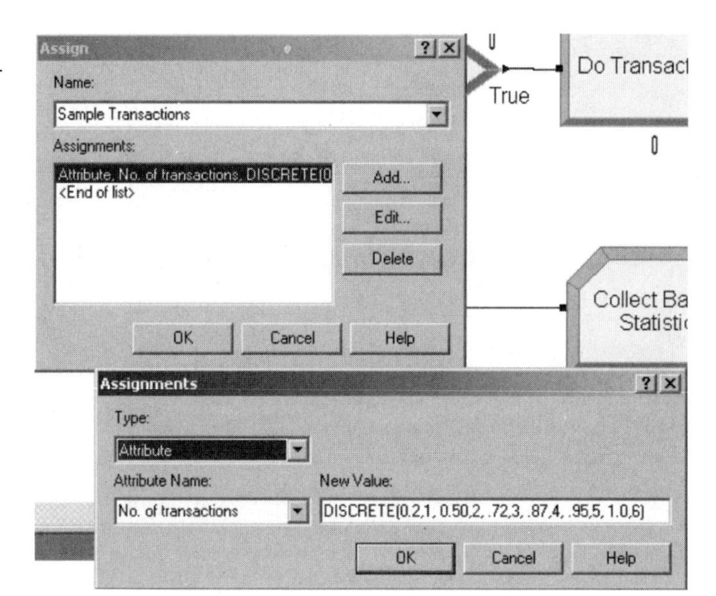

FIGURE 7.22 Defining the probability that the customer will request specific number of banking transactions

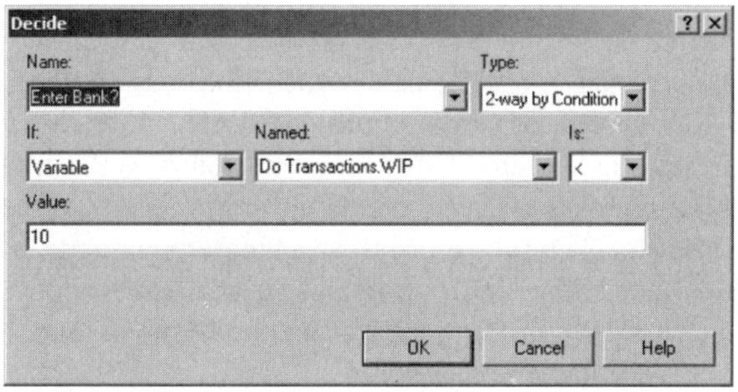

Decide
Name: Enter Bank?
Type: 2-way by Condition
If: Variable
Named: Do Transactions.WIP
Is: <
Value: 10
OK Cancel Help

FIGURE 7.23 Defining the balking decision in the bank lobby model

tents tab, double-click on "**Using Variables, Functions and Distributions.**" Then double-click on "**Variables**" to see a page with links to all built-in variables.

We have two branches leading from the **Decide** module "Enter Bank." First define the branch going to the "Do Transactions" **Process** module. Double-click on the **Process** module "Do Transactions." In the **Action** field of the module, select **Seize Delay Release** to indicate the complete activity consisting of waiting for the resource(s), seizing the resource(s), performing the time delay required for the activity to be processed, and releasing the resource(s). Click on the **Add . . .** button in the **Resource** window and select **Set** as the type of resource, "Tellers" as the name of the set, and **Cyclical** as the selection rule for resources in the set. We have already defined the set Tellers. The **Cyclical** selection rule selects the next resource from the available resources in a set by cycling through the resources—1, 2, 3, 4, 1, 2, 3, 4, and so on—until a free member is found, starting with the resource after the last se-

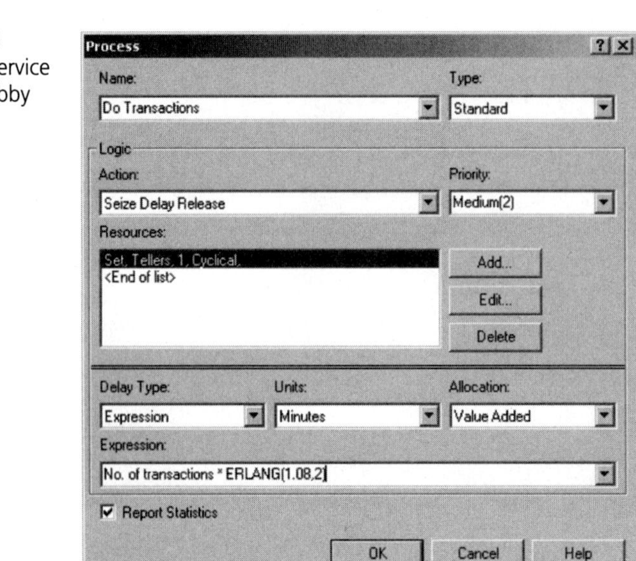

FIGURE 7.24
Defining the service in the bank lobby model

lected resource each time. This rule tends to utilize each resource approximately equally as the system runs. The delay type is **Expression,** the units are minutes, and the expression is "**No. of transactions * ERLANG(1.08,2).**" This expression computes the time delay required for serving a customer as the product of the number of banking transactions that have to be done for him and the time needed to process a single transaction. Time for processing a single banking transaction is distributed according to the Erlang distribution and written as ERLANG(1.08,2). The dialog should look as shown in Figure 7.24. For the **Dispose** module "Depart," the only item to enter is the name of the module.

The second branch leading from the **Decide** module "Enter Bank?" describes the customer's process when balking. In this case, the customer actually does nothing but leave. However, we also want to collect extra statistics not normally collected by Arena. This branch starts with a **Record** module "Collect Balking Statistics" that is used to define statistics about the number of customers who balked. Double-click this module to open the **Record** dialog and select **Expression** for the type. Enter "1" in the **Value** field. The **Record** window should appear as shown in Figure 7.25. For the **Dispose** module "Balk," only the name of the module needs to be entered.

The model is ready to run at this point. However, it is useful to add some other features to the model to improve the amount of information we receive as the simulation runs.

A graphical clock can be added to display the value of the simulation time as the simulation runs. This clock is on the **Animate** toolbar. If the **Animate** toolbar is not showing, then you can display it by selecting **Toolbars . . .** in the **View** menu

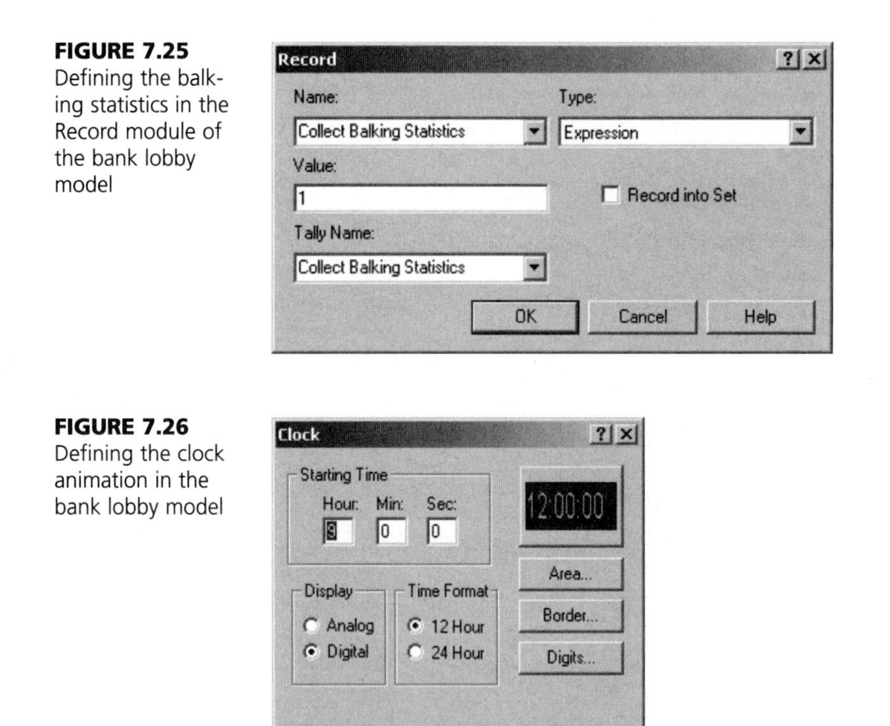

FIGURE 7.25 Defining the balking statistics in the Record module of the bank lobby model

FIGURE 7.26 Defining the clock animation in the bank lobby model

and clicking **Animate** in the **Toolbars** tab. On the **Animate** toolbar, click the **Clock** button to bring up the **Clock** dialog. Enter the data to this dialog as shown in Figure 7.26 and click **OK.** The cursor will now change to a crosshair, and you can draw the clock in the model window by clicking and dragging to set the clock location and size. Click again to finish placing the clock and return to normal editing mode. Figure 7.27 shows the result. This drawing was captured during a simulation run, so the clock has a nonzero value.

7.3.3 Simulation Run and Results

Before we start the simulation execution, we have to define the simulation run parameters. Select **Settings** from the **Run** menu to bring up the **Run Settings** dialog.

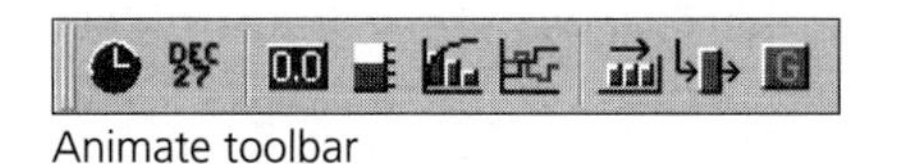

Animate toolbar

Clock button

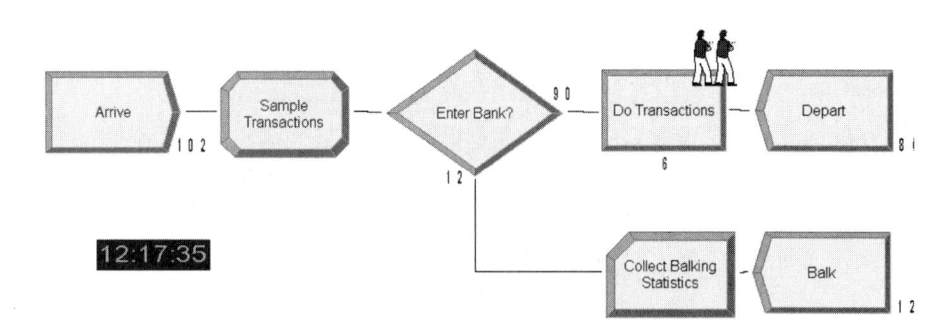

FIGURE 7.27 Clock animation picture inserted in the bank lobby model (screen snapshot during simulation run)

Name the module "Bank Lobby Analysis," define the run length as 8, and time units as hours, number of replications as 5, and simulation time units as minutes. Click **OK.** We are now ready to start the simulation run.

Start the simulation run by clicking the **Go** button on the **Standard** toolbar. Then run the simulation one step at a time by clicking on the **Step** button in the **Standard** toolbar. Each time this button is clicked, the simulation advances one event. An event executes a single module except when the module is a **Process** module. For the **Process** module, there are two events: the starting event and the ending event. Because the **Process** module actually represents an activity, the starting event will place an event notice on the event list and the ending event will not be executed until the time of the ending event is realized. As we step through the simulation, we notice that often something is happening on the screen (e.g., customers are moving) but the value of the simulation time is unchanged. This is not a mistake! This only means that Arena's clock is moving in discrete steps rather than continuously. The time displayed on the simulation clock does not change until all events at the current simulation time have executed and the simulation clock is updated to the time of the first event on the event list. Now click the **Go** button on the **Standard** toolbar to run the animation. If the animation is running too fast to observe everything you want to see, slow it down using the "<" key; if it is running too fast, speed it up using the ">" key. As the simulation runs, observe it to assure yourself that the system is operating as expected. In particular, customers should take both paths out of the **Decide** module "Enter Bank?".

We will perform five independent replications of the simulation, and in each replication the simulation time starts at 9 A.M. and finishes at 4 P.M. As the simulation runs, the main Arena window will show the progress in the bar at the bottom of the main window. If you have a fast computer, this simulation might proceed too fast to monitor. However, larger simulations will proceed at a slower pace. Click the **Fast Forward** button in the **Standard** toolbar to run the simulation at full speed and turn off the animation. At the end of the simulation run, Arena will ask if you want to view results. Click **Yes,** and the default **Category Overview Report** will be displayed in a report window. You can choose any of the 10 reports offered; let's select the **Category by Replication** report and click on **Replication 1** and then on the **Process** folder. The resulting report is shown in Figure 7.28. You can see that be-

FIGURE 7.28 Simulation reports by replication for the bank lobby model (Replication 1, Process, WIP)

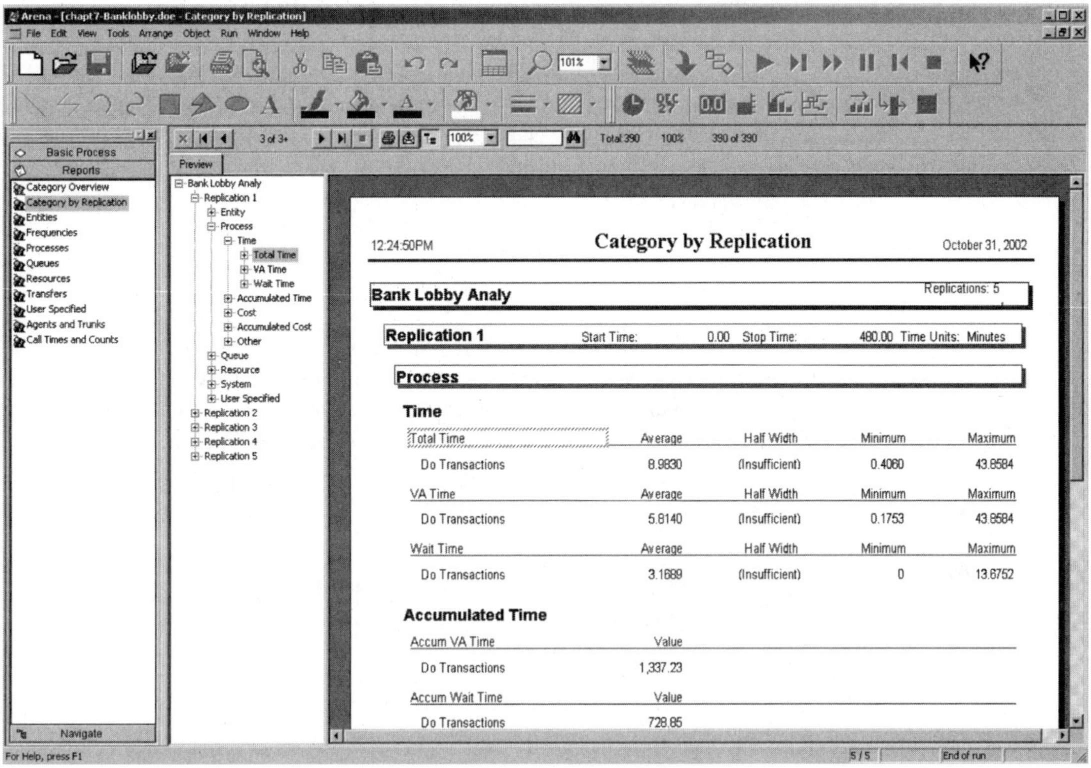

sides the usual four elements (entities, processes, queues, and resources), the replication report contains the user-specified report of type **Expression** related to the **Balking Statistics** module we defined earlier.

It is interesting to see how much the simulation output data vary between replications. To see an example, look at Table 7.2. We chose four simulation output variables and for each one we took the total of the averages for all replications as well as the averages for each of the five replications. The results show how much the simulation output varies between replications and demonstrates the danger in executing just one replication unless it is a sufficiently long. In this case, each replication involves only one day of operation of the bank.

The maximum value of the **Balking Statistics** in the user-specified report gives the number of customers who balked in each replication. We took the maximum value because the number of customers who balked during each replication grows from the beginning until the end of the replication, and we are only interested in the total number of customers who balked in each replication. The **Balking Statistics** gives the following values for five replications: 23, 10, 15, 32, and 19. We will leave

TABLE 7.2

	Total Average	Average 1	2	3	4	5
Entity cycle time	7.63	7.87	7.06	7.44	7.04	8.74
Process WIP	3.49	3.63	3.18	3.01	3.58	4.06
Queue: number waiting	1.09	1.09	1.00	.77	1.11	1.49
Resource: utilization of Alice	.60	.59	.54	.53	.67	.67

it to you as a problem at the end of this chapter to compute similar statistics for the numbers of customers who entered the **Dispose** module "Depart" and therefore actually entered the bank each day. Using these two variables you can calculate the proportion of customers who balk in each replication of simulation.

7.4 VISUAL INTERACTIVE SIMULATION: A RECAP

From the preceding sections, we can see that the basic features of VIS can be summarized as the ability to:

1. build and modify simulation models on the screen,
2. execute graphical simulation models,
3. animate models as they execute,
4. present simulation output graphically, and
5. interactively interrupt the model when it is executing to change model parameters, structure, or operation logic and then continue execution with these modifications. (Arena does not support such changes but other VIS packages do.)

This approach to simulation has strongly influenced both model builders and model users. Using VIS, model builders can intuitively develop models graphically by visually representing model entities and their interactions instead of writing simulation code in a programming language. Communication between modelers and users is greatly facilitated, because users can more easily understand a model in graphical form. Using a VIS tool, it is easier to detect and localize some of the logic errors since model behavior can be viewed using animation rather than a long, complex text output file. For users, the model is a more understandable system representation. Animation also allows a much faster assessment of the general level of change in system behavior caused by modifying the system's parameters or structure.

At least as important is that the user can now become a part of the model-building and experimentation process by interacting directly with and observing the model during its execution. Interactive simulation allows a much faster and more effective search for changes that improve system operation. As a consequence, users often regard VIS models as their own and are able to work with the models by themselves and with the model builder's help to make changes. Users can therefore have an active role in experimenting with a simulation model. In the words of Hurrion (1989), "The visual interactive simulation approach tends not to feed the manager with results, but allows him to search for alternative solutions himself." This is an important consequence of using a VIS approach and gives a new impetus to more intensive use of discrete-event simulation in decision making. Simulation software that implements VIS is, in fact, a *decision support* tool rather than just a model-building tool. Confidence in the simulation is increased, and simulation study results are implemented more often.

VIS has been applied in many different areas, including flexible manufacturing systems, tree cutting and logging, the chemical industry, health care, and rail locomotive service centers. Better decision making and acceptance of simulation from managers involved was reported in a large majority of these and other reported applications of VIS.

There is another side of the coin, though. One difficulty with interactively executing simulations is the problem of precisely planning simulation experiments and analyzing their results. One should be careful in interpreting the output of a simulation run that was interrupted or when model features were changed. Because the simulation is stochastic, all output is random. Often the output has an enormous amount of variation such that the "signal" in a system change can be masked by the "noise" in the output data. Recall that we showed that one cannot normally determine the length of the initial transient period by just observing a single simulation output sequence. Thus, new statistical methodology is needed for VIS that is different from the traditional methodology in which a simulation model is used as a tool for statistical experimentation (Bell and O'Keefe, 1987).

We suggest the following VIS procedure. First, graphics and interaction can be used in an intuitive way to ensure that models are logically correct and to filter out alternatives that obviously perform badly (unstable systems, large queues or backlogs, low throughput, flow discontinuities, low equipment utilization, low utilization-to-cost ratio, deadlocks, etc.). Then when alternatives are found with behaviors that are visually acceptable, they should be simulated using long runs. The resulting system performance measures should be computed and analyzed, using appropriate statistical methodology, and compared to ensure confidence in the simulation results.

7.5 SUMMARY

Visual interactive simulation (VIS) evolved as a consequence of the dramatic improvements in computer hardware and software in the last decade, particularly the appearance of powerful graphical hardware, graphical user interfaces (GUIs), and interactivity. These improvements radically changed both the technology of working with computers and the nature of computer-supported problem solving and decision making. Graphics and interaction are now incorporated in all phases of modeling and simulation, from model building to execution, animation, and the graphical analysis of simulation experiments. Numerous VIS software tools were developed for that purpose such as Arena, ServiceModel, and SIMSCRIPT II.5.

We demonstrated the abilities of VIS tools using Arena, a modeling and simulation tool that models discrete-event systems, simulates their behavior through time, animates the simulations, and automatically produces statistical reports about system performance. Modeling with Arena is based on a simple flowchart approach where the model is presented as a logical network of related activities. Arena supports hierarchical model building where each model element can be represented a submodel, and submodels can contain deeper submodels. Arena is also compatible with Microsoft Office.

Besides presenting Arena's main features, we described in detail the process of model development and simulation, and we demonstrated them with two models: a simple single-server queueing system and a bank lobby system. Although the single-server queueing system is highly simplified and developed with just a few basic modeling constructs, the bank lobby incorporates several more realistic features such as customers arriving with dynamics that vary through the day, two or more customers arriving simultaneously, customers having different numbers of banking transactions, and customer balking if the queue is too long. Additional modeling and reporting features of Arena were also demonstrated in the bank lobby model.

The basic features of VIS can be summarized as the ability to build and modify simulation models on-screen, execute graphical simulation models, animate models as they execute, present simulation output graphically, and interact with the model during execution. VIS enables model builders to develop models graphically by visually representing the model instead of writing simulation code in a programming language. It facilitates communication between modelers and users, makes it easier to detect and localize logic errors, and produces a model that users can understand. This allows users to have an active role in model building as well as interact directly with the model and increases their confidence in the model as a tool for evaluating decisions. Simulation software that implements VIS is a decision support tool rather than just a model-building tool. However, VIS users should be careful when interpreting the output of a simulation run that was interrupted and where characteristics of the model were changed.

References

Bell, P. C., and R. M. O'Keefe. 1987. Visual interactive simulation—History, recent developments, and major issues. *Simulation* 49:109–116.

Fiddy, E., J. G. Bright, and R. D. Hurrion. 1981. SEE-WHY: Interactive simulation on the screen. *Proceedings of the Institute of Mechanical Engineers* C293/81:167–172.

Hurrion, R. D. 1976. The design, use, and requirements of an interactive visual computer simulation language to explore production planning problems. Doctoral thesis, University of London.

———. 1989. Graphics and interaction. In *Computer modelling for discrete simulation,* ed. M. Pidd. Chichester, England: Wiley.

Kelton, D., R. Sadowski, and D. Sadowski. 2002. *Simulation with Arena*. New York: McGraw-Hill.

Mandelkern, D. 1993. Introduction to special issue on graphical user interfaces. *Communications of the ACM* 36:37–39.

ProModel. 1995. *ServiceModel user's guide.* Orem, Utah: ProModel Corp.

Schruben, L. 1994. *Graphical simulation modeling and analysis: Using SIGMA for Windows.* Danvers, Massachusetts: Boyd & Fraser.

Systems Modeling. 1999. *Guide to Arena business edition.* Sewickey, Pennsylvania: Systems Modeling Corp.

Problems

7.1 Discuss the drawbacks of traditional simulation procedures.

7.2 What are the main features of the VIS approach?

7.3 What are the most important benefits of VIS for modelers and users?

7.4 What is a potential problem with VIS?

7.5 What is the recommended approach to overcome problems with VIS?

7.6 Develop an Arena model of the two-server system with separate queues in front of each server. Customers who enter the system will join the shorter queue; if both queues are of the same length, then the customers will join the queue in front of the first server.

Suggestion: Use a **Decide** module to model queue selection.

7.7 Add the departure statistics in the bank lobby model presented in this chapter. After performing the simulation experiments, calculate the percentage of customers who balk in each of five simulation runs as the ratio of the number of customers who balk to the sum of the number of customers served. Plot these statistics to show their variation graphically. Also calculate the average percentage balking over all five simulation experiments.

7.8 Determine whether the bank lobby simulation would work with three tellers instead of four. If we run five simulation experiments, what would be the overall average of the entity cycle time, process WIP, number of customers waiting in the queue, utilization of Alice, and balking percentage? Compare these figures with those obtained in a simulation of a bank lobby with four tellers.

7.9 Find out what happens when bank lobby simulation experiments include two hours of warm-up. If we run five simulation experiments, what would be the overall average of the entity cycle time, process WIP, number of customers waiting in the queue, utilization of Alice, and balking percentage? Compare these figures with those obtained in a simulation of a bank lobby without warm-up time.

7.10 Change the bank lobby model presented in this chapter in such a way that servers have a one-hour break one after another: Alice has the break on the fourth hour of bank lobby operation, Mary on the fifth hour, and so on. Run five simulation experiments and find the overall average of process WIP, the number of customers waiting in the queue, and the utilization of Alice when compared with the bank lobby simulation with four tellers.

Suggestion: Use the resource schedule to model server breaks.

7.11 Change the balking decision in the bank lobby model. The new balking decision is based on the following logic: When service WIP is less than 10, all customers will wait for service. If service WIP is greater than or equal to 10 but less than 15, 50 percent of customers will wait for service while the other 50 percent will balk. When service WIP becomes greater than or equal to 15, all customers balk.

Suggestion: Use several **Decide** modules, each resolving a part of the customer decision logic—one outcome of a **Decide** module leads to a decision, while the other leads to another **Decide** module. (There is a simpler way to model this kind of decision, but it requires the use of logical operators in expressions that have not yet been demonstrated. A search for "logical operators" using Arena's help feature will give you ideas about how to simplify the model.)

7.12 Develop a model of a manufacturing system with rework. The manufacturing system consists of a set of three machines that process incoming parts arriving each 10 ± 3 minutes. (The notation 10 ± 3 means uniformly distributed between 10 – 3 and 10 + 3.) Parts can be processed on any machine, and processing time has a normal distribution with mean time of 27 minutes and standard deviation of 11 minutes. Three percent of processed parts have to be reworked. Rework can be accomplished on any machine, and processing time has a normal distribution with mean time of 9 minutes and standard deviation of 4 minutes. Rework has priority over regular processing. One simulation run with run length of 60 hours should be performed.

Suggestion: Place a **Decide** module after a **Process** module for regular processing to direct the routing of parts either to the end of processing or to the new **Process** block that simulates rework. The priority of parts can be declared in **Process** modules.

7.13 Develop a model of a two-phase manufacturing system with in-process storage between the two manufacturing phases. Each manufacturing phase uses one machine. Parts arrive each 29 ± 5 minutes, and manufacturing times in both phases have normal distribution. Manufacturing on the first machine has a mean time of 22 minutes and a standard deviation of 6 minutes, while manufacturing on the second machine has a mean time of 17 minutes and a standard deviation of 5 minutes. Execute five simulation experiments and find the in-process storage capacity between the two manufacturing phases that guarantees that during manufacturing there will always be enough storage space for parts that finish the first phase.

Suggestion: The required in-process storage space is the maximum value of maximum number of parts waiting in the in-process storage during each of the five simulation runs.

7.14 A supermarket has eight regular counters and one express counter. Customers arrive randomly to counters with a mean time of 1 minute, and 15 percent of all customers use the express counter. Service time on the regular counter has a normal distribution with a mean time of 7 minutes and a standard deviation of 5 minutes with a minimum time of .2 minutes, while service time on the express counter is 3 ± 1.5 minutes. Run 10 simulation experiments each eight hours long and find the average and maximum queueing lengths for the regular counters and for the express counter.

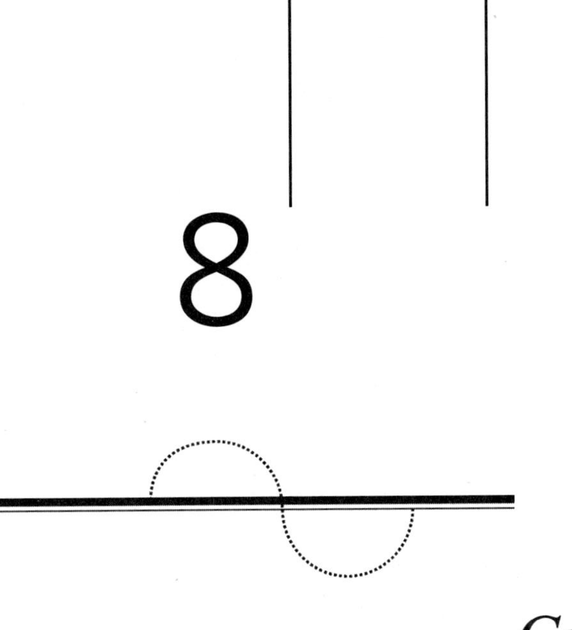

8

Graphical Simulation Modeling

This chapter explores the graphical representation of a model as a powerful tool to conceptualize system logic and operations. Graphical simulation models are two-dimensional network representations that describe the system using nodes connected by arcs. They allow the modeler to quickly develop an understanding of the model structure and operations and to communicate that understanding to other modelers and clients. Three graphical modeling techniques—event graphs, activity cycle diagrams, and Petri nets, representing each of the simulation worldviews presented in Chapter 6—are presented briefly. Event graphs were introduced in Chapter 5. We discuss graphical modeling with Arena and introduce two new flowchart modules: the **Separate** module, which produces multiple identical entities from an input entity; and the **Batch** module, which collects entities into a batch using various rules.

Graphical modeling methods support hierarchical modeling, which is a technique to organize complex models into functional components or submodels. The benefits of hierarchical modeling include the ability to compartmentalize the model, simplifying changes by localizing them to only a few submodels, and thus reducing the possibility of introducing errors and increasing model reliability. We also describe and demonstrate Arena's hierarchical modeling abilities and present its animation facilities. Animation is an important visual feature that enhances the power of graphical models by visually representing the changes in entities, system resources, and system performance measures during the simulation run.

8.1 GRAPHICAL MODELING

In Chapters 5 and 6 you learned how to model a discrete-event dynamic system. Modeling involves using one of the simulation worldviews to capture the basic elements of the system as well as their dynamic interactions in the form of event, activity, or process routines. For simple systems this can be easily done, especially if the modeler is experienced. However, when confronted with a more complex system, even experienced modelers can have serious problems with a text-based approach. The simulation routines will too often contain errors that are difficult to detect and correct.

A second problem with the traditional approach to model development is that of understanding the procedural, or algorithmic, representation of the system. This understanding is important because simulation model development for complex systems is usually a joint effort by a team of modelers that includes specialists involved with the actual system operation. Real systems such as flexible manufacturing systems, airport terminals, and computer systems are so complex that modelers cannot understand all operations just by reading and observing. Instead, they must rely on the direct knowledge of specialists who design and operate the system. On the other hand, it is important that specialists in system operation are able to understand the model; otherwise, they cannot see possible modeling errors early in the model development process. The cost of an error increases rapidly with the length of time before it is detected. This same communication problem can also appear among members of the modeling team.

Graphical modeling provides a solution to these problems. This methodology is a communication tool that allows the model to be understood even by persons who are not modeling specialists, and it also can be used to show that the model is a correct representation of the intended system—that is, for model verification. Furthermore, this approach can be used to directly execute the simulation model or automatically generate a simulation program, thus avoiding the need for manual program coding and the inherent errors in such coding.

Graphical modeling techniques (Pooley, 1991a; Ceric and Paul, 1992) describe the system structure and logic of operation using diagrams in a way that is similar to schematics describing the structure and function of electronic components. All graphical model diagrams consist of a network of connected symbols. Each type of symbol has a specific meaning that relates either to a basic concept in discrete-event simulation (such as an event, resource, activity, condition, or temporary entity) or to a specific simulation object or action such as a collection of servers, a generator of temporary entities, or the act of capturing resources, depending on the type of graph.

Graphical techniques can be classified into two categories depending on the meaning of the symbols they use: *generic graphs* and *process graphs.* The most familiar generic graphs are event graphs, activity cycle diagrams, and Petri nets. Process graphs include queueing network graphs, activity diagrams, GPSS block diagrams, SLAM networks, and Arena flowcharts.

Graphical models are intended to be intuitive and show the fundamental nature of each discrete-event simulation model as the imitation of the objects, structure, and operation logic of a real system. This increases model transparency and provides a significant advantage that makes the model easier to understand and maintain. Graphical modeling techniques utilize our visual information-processing system to quickly assimilate and understand system operations. Human visual information-processing capabilities have been developing much longer than verbal skills because they were needed for food gathering, hunting, and defense. Thus, our visual information-processing system is a highly developed and powerful tool that can help us summarize the structure and operation logic of the system.

Most graphical modeling techniques have a small number of symbols and simple rules for building the model. We would like to be able to build models without having to refer to a lot of tables or apply complex rules. Complexity is best handled by modularizing the model components and combining them in a hierarchical structure. This approach has the added benefit that system modifications can often be limited to one or a few modules, thus making model changes easy and reliable. We will see that several graphical modeling techniques support hierarchical modeling.

Graphical modeling also allows us to simulate model operation manually—that is, perform state changes by manually executing events, activities, or processes. This is a useful tool for visualizing and understanding how the simulation operates through time.

Many of the graphical modeling techniques have been implemented as interactive computer programs that can be used to both develop and run the simulation. Arena falls in this category. As noted in Chapter 7, these programs are *visual interactive simulation* (VIS) software. As a result, graphical modeling techniques have become the basis for attractive and productive computerized tools for model development, execution, and analysis.

In section 8.2 we will examine three popular graphical modeling techniques: *event graphs,* which are based on the event worldview; *activity cycle diagrams,* which are based on the activity worldview; and *Petri nets,* which are based on the process worldview. Each technique will be demonstrated by modeling a queueing system with multiple servers. In section 8.3 we will discuss and demonstrate additional Arena graphical modeling abilities: duplication and collection of entities. Later, in section 8.4, we will examine how Arena supports hierarchical modeling. Finally, we will present some of Arena's animation features in section 8.5 and discuss how these are useful to verify models.

8.2 SOME GRAPHICAL MODELING TECHNIQUES

8.2.1 Event Graphs

As we saw in Chapter 5, event graphs represent a graphical representation of the event scheduling worldview by visualizing the scheduling and canceling relationships among events. The modeler can look at an event graph and quickly understand which events are scheduled by other events, under what conditions, and with what parameters being passed. We also saw that event graphs allow one to quickly determine which events are "unnecessary" in the sense that they are always scheduled simultaneously with other events, and how these "unnecessary" events can be removed from the event graph to produce a simpler representation with fewer events. If you have not read Chapter 5 recently, it would be a good idea to review that chapter now to remind yourself how event graphs function so you can easily compare them with the other two graphical modeling methods presented here: activity cycle diagrams and Petri nets.

Figure 8.1 is an event graph model of a queueing system with multiple identical servers. The model uses two state variables, S and Q, to designate the number of idle servers and the number of customers waiting to be served, respectively. The Initialization event initializes the state variables to represent a model with n idle servers ($S := n$) and an empty queue ($Q := 0$). The Initialization event also schedules a Customer Arrival event immediately and unconditionally. As soon as the Customer Arrival event occurs, the queue length, Q, is increased by 1, and the Customer Arrival event recursively schedules itself with a delay equal to the interarrival time t_a. The Customer Arrival event also schedules the Start Service event if at least one server is idle ($S < 0$).

The other events operate similarly, as described in Chapter 5. The important thing to note about event graphs is that they show the events in the model and the scheduling relationships as well as the data transfer between the events in the form of parameters passed to the event routines. When the state changes that occur in each event are included, the entire model can be shown in a single event graph. You

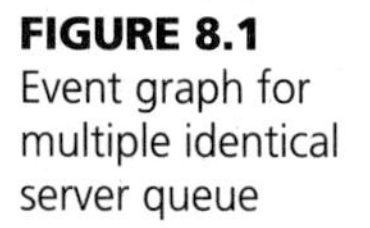
FIGURE 8.1
Event graph for multiple identical server queue

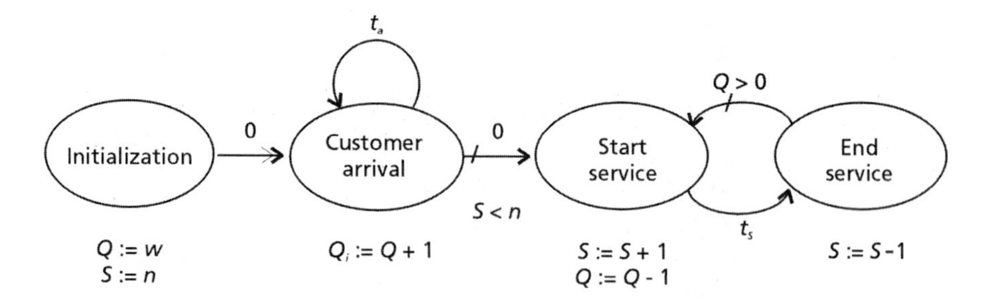

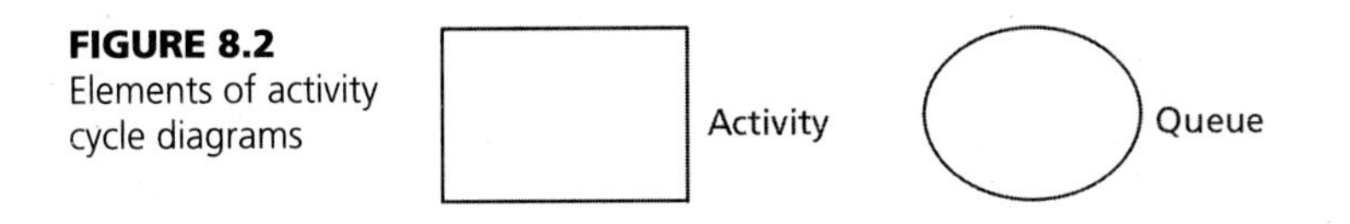

FIGURE 8.2
Elements of activity cycle diagrams

will see that event graphs display the model at a greater level of detail and lower level of abstraction than process graphs; this forces the modeler to pay attention to detailed operations of the system. In process graphs and process modeling techniques, which have largely replaced event graphs, the detail is frequently hidden or defaulted to standard behavior. If the modeler is not careful, incorrect behavior can be incorporated by default. This is less likely if event graphs are used.

8.2.2 Activity Cycle Diagrams

Activity cycle diagrams (Tocher, 1964) provide a graphical modeling method for discrete-event simulation using activities as the key dynamic elements. Activity cycle diagrams were developed to be an informal graphical modeling method using only two symbols: activities and queues, as shown in Figure 8.2. Activities denote active entity states in which different types of entities interact, while queues denote passive entity states in which entities are waiting to enter active states. These definitions correspond to the active and passive states for processes as discussed in section 6.4. An active entity waits a definite (though possibly random) length of time; a passive entity waits an indefinite length of time until conditions are met for it to proceed with its cycle.

The logical behavior of each type of entity is described in its entity life cycle. An *entity life cycle* is a closed cycle that alternates between waiting in a queue and engaging in an activity with other types of entities. Each type of entity cycles through its own life cycle. Entities cooperate by temporarily joining their life cycles in activities. An activity can have several input queues and several output queues, and it can start only when all of its input queues have at least one entity in them. When the activity finishes, after the specified activity time, all entities engaged in it are released and put in the output queues belonging to each of the participating entity life cycles.

The activity cycle diagram for the single-server queueing system is shown in Figure 8.3. This diagram combines life cycles of three entities: customers, servers, and arrivals. Customers come from an infinitely sized Outside queue that represents the system environment, and then are engaged in the Arrival activity, which is characterized by specific interarrival times for customers. After arrival they wait in the Wait queue before being serviced in the Service activity. After the Service activity they join the Outside queue again. Servers cycle between the Idle queue and the Service activity. Finally, a single logical entity, the auxiliary arrival entity, is introduced to produce the rhythm of customers arrivals. This entity cycles through the

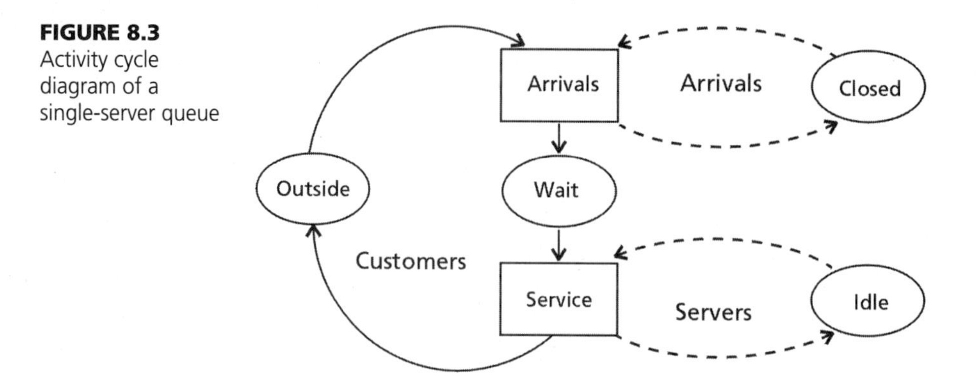

FIGURE 8.3 Activity cycle diagram of a single-server queue

Closed queue and the Arrival activity. We can think of this as a switch that closes each moment an arrival occurs.

These three life cycles are joined in the Arrival and Service activities. Thus the activity Service is an interaction point for the customer and server entities. This activity can begin only when there is at least one customer waiting for service and at least one idle server—that is, at least one customer in the Wait queue and at least one server in the Idle queue. The activity Arrival is an interaction point for the customer and the logical arrival entities. It can start when there is at least one customer in the Outside queue (which is always true because, by definition, it has an infinite membership) and when the logical arrival entity is in the Closed queue. When the simulation begins, a single arrival entity is in the Closed queue so that the Arrival activity can start immediately. The Closed queue will then be empty until the interarrival time passes. After that, the customer who arrived will be delivered to the Wait queue, while the single arrival entity will be delivered to the Closed queue. Then the new Arrival activity starts again, repeating its cycle. In this way the mechanism allows the Arrival activity to deliver new arrival customers according to their interarrival time process.

Activity cycle diagrams were originally developed as an extremely simple method to support intuitive modeling of dynamic phenomena. Several additional symbols have been added to extend their modeling power. Activity *priorities* resolve conflicts if several activities must start at the same moment. The activity with the highest priority will be executed first, and activities with lower priorities will be executed only if previously started activities with higher priorities did not engage the resources required for these lower priority activities. Entities in extended activity cycle diagrams can have attributes. Two types of decision rules are also introduced: *Conditional decision rules* involve conditions that can use entities' attributes or state variables, and *probability decision rules* use a random number to evaluate the decision. Finally, *state variables* are global system variables that can be used anywhere in the model and modified with an assignment statement.

FIGURE 8.4
Elements of elementary Petri nets

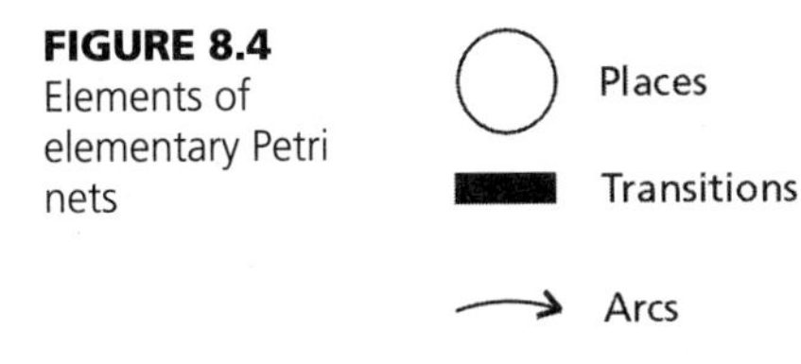

8.2.3 Petri Nets

Petri nets (Reisig, 1985, 1992; David and Alla, 1994) were originally developed to model and analyze concurrent processes in computers. Later they were extended and applied to the modeling of a wider variety of phenomena. Several extensions to the original Petri net formalism were developed to satisfy the needs of discrete-event simulation. The Petri net representation has much in common with the process view of simulation.

An elementary Petri net is a directed graph that consists of *places, transitions,* and *directed arcs.* In the Petri net model, places and transitions alternate. Each directed arc connects either a place to a transition or a transition to a place. Each place can contain one or more tokens. Places are represented by circles, transitions by bars, and tokens by black dots (see Figure 8.4). A configuration of tokens in a Petri net is called a *marking*. It is important to recognize that tokens in Petri nets are *not* equivalent to entities. Rather, they are indicators of the system state; that is, a marking denotes a specific system state. The concept of time is not part of the elementary Petri net representation. Elementary Petri nets simply represent the logic of system operation and the possible sequence of system state changes.

An input place for a transition is a place connected by a directed arc to the transition. A transition is said to be *enabled* when each of its input places has at least one token in it. Petri net models are executed using a simple mechanism called *transition firing.* A transition is ready to fire when it is enabled—that is, when each input place to the transition has at least one token in it. Tokens that are involved in enabling a transition are called *enabled tokens.* When a transition is enabled it will not necessarily fire immediately because other transitions may also be enabled at the same time and may fire before this one. Their firing could result in this transition no longer being enabled because the other transition firing removed all tokens from one of the input places. Elementary Petri nets provide no rule that determines the order of firing of transitions.

If an input place contains multiple tokens, only one token in the input place will be the enabled token because the input place needs only one token to enable the transition. When a transition fires:

1. All enabled tokens are removed from their input places, and
2. one token is deposited into each output place for the transition.

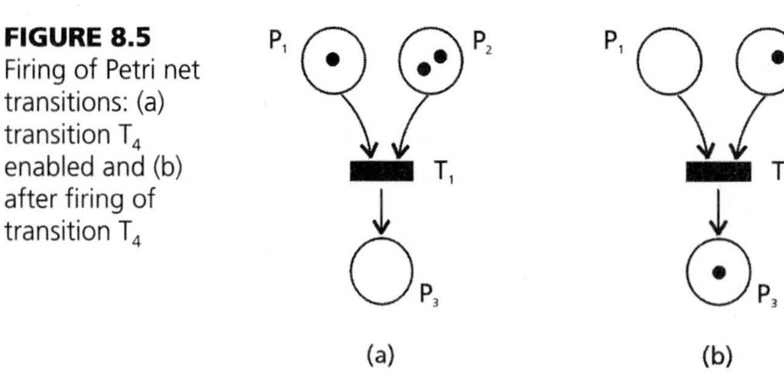

FIGURE 8.5 Firing of Petri net transitions: (a) transition T_4 enabled and (b) after firing of transition T_4

The important concept here is that tokens deposited to transition output places are in no way related to tokens removed from input places. Moreover, firings neither have to nor usually do conserve the total number of tokens in a Petri net.

An example of a transition firing is presented in Figure 8.5. Transition T_1 in Figure 8.5a is enabled because both of its input places, P_1 and P_2, have tokens. Figure 8.5b shows the model after a transition T_1 fires. Now one token has been removed from each input place, and one token was then deposited to the output place P_3. After firing, transition T_1 is not enabled because one of its input places, P_1, is now empty.

Note that transition firing is a local mechanism for changing a model state. Each transition depends only on the state of the input places that are directly connected to the transition by input arcs. When a transition fires, only the places directly connected to the transition—only the local state around the transition—is changed. Each input place loses one token, and each output place gains one token. The graphical elegance of Petri nets and the power of the applicable mathematical analysis results from the localization of state changes. Petri nets do not use global system variables. Mathematical analysis can be used to determine such things as which states can be reached from the initial system state, whether the number of tokens in every place of the net is bounded or not, and whether the net can have a deadlock condition. The interested reader can consult the book by Reisig (1985) and the review paper by Murata (1989) for mathematical analyses of Petri nets.

Building a realistic simulation model requires modeling power beyond that offered by elementary Petri nets. Several extensions have been proposed for this purpose (Törn, 1981, 1985). When these extensions are added, modeling power is increased but the elegance and ability to analyze the model mathematically is decreased. A fundamental extension for discrete event simulation is the concept of *timed transitions.* The timed transition symbol, shown in Figure 8.6a, is a rectangle with the time delay marked inside the rectangle. A timed transition fires with a time delay after the transition is enabled. In fact, firing occurs in two phases for a timed transition. First, the firing is initialized by having enabling tokens in the transition's input places. This is shown in Figure 8.6b. After the delay, the firing is finalized and the state change caused by the firing is executed. Enabled tokens are removed from the transition input places, and new tokens are deposited in the output places as shown in Figure 8.6c.

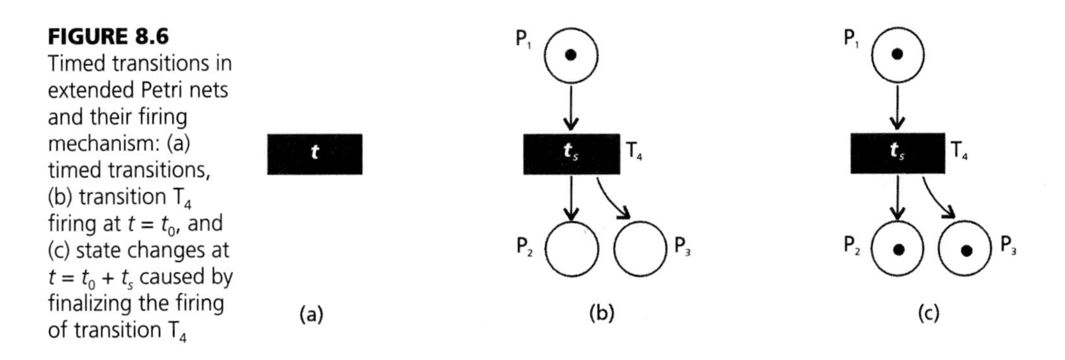

FIGURE 8.6 Timed transitions in extended Petri nets and their firing mechanism: (a) timed transitions, (b) transition T_4 firing at $t = t_0$, and (c) state changes at $t = t_0 + t_s$ caused by finalizing the firing of transition T_4

The elements of Petri nets with timed transitions have the following correspondence with the concepts presented in Chapters 5 and 6 (extended from Evans, 1988):

Place—condition, state, activity, or waiting of one or more type of entities

Token—indicator of state of one or more type(s) of entities

Directed arc—route for changing state of one or more type(s) of entities

Immediate transition—event

Immediate transition firing—event occurrence

Timed transition—activity

Initialization of timed transition firing—occurrence of the start of the activity event

Finalization of timed transition firing—occurrence of the end of the activity event

Petri nets with timed transitions can be used to develop simulation models. Figure 8.7 shows a Petri net model of the queueing system with multiple identical servers. This model consists of a customer process and a server cycle connected through the service activity. Customer arrivals are modeled by the timed Arrival transition that has a place PA with the single token as its input–output place. The arrival mechanism is functionally equivalent to the arrival mechanism in activity cycle diagrams. After the firing of the Arrival transition is initialized, the interarrival time must pass before firing is finalized. The enabled token is then destroyed and a new token is put in the Wait place and in the input–output place PA of the Arrival transition. This enables firing of the Arrival event again.

The Start of Service transition can fire when there is at least one token in both the Wait and Idle Server places. This represents the condition that there must be both a waiting customer and an idle server for service to begin. After the End of Service transition firing is finalized, one token is deposited into the Customer Leaving place and another in the Idle Server place, indicating that the previously engaged server is again idle and can begin service if the Wait place contains at least one token.

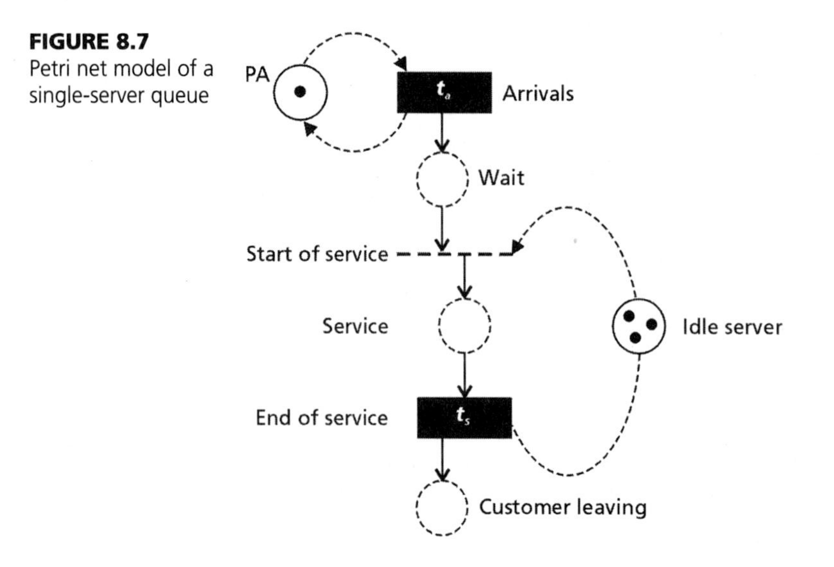

FIGURE 8.7
Petri net model of a single-server queue

Note that tokens are actually used as flags or semaphores to indicate which events can occur. The mechanism of the server cycling through the Idle Server and Service places visually indicates the state of the server and allows conditions for the start of service to be easily identified in the model diagram. Using event graphs, conditions for scheduling events were embedded in the logic of event routines, because event graphs do not represent system resources explicitly.

Other extensions of Petri nets have also been proposed for discrete-event simulation. One of the most important extension is the *colored Petri net approach* (Jensen, 1991) that allows more compact models to be built and token identity to be modeled. In this extension token color sets represent different types of tokens, such as ships or customers, while token colors represent different classes of tokens from the same token color set, such as slow and fast machines or small, medium, and large orders. Transitions can have a condition attached to them, and these conditions constrain the values of token colors that can enable the transition. Arc labels can also constrain the token color values that can enable a transition.

8.3 GRAPHICAL MODELING WITH ARENA

As we saw in the previous chapter, Arena's graphical modeling technique represents the logic of system operation using one or more flowcharts. Flowchart modules contain system logic and actions such as creating temporary entities, decision making, processing entities using system resources, and assigning a value to an entity's at-

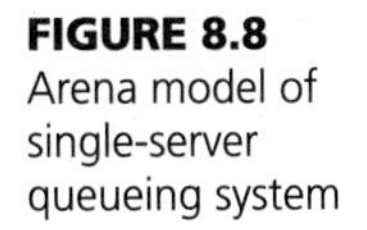
FIGURE 8.8
Arena model of single-server queueing system

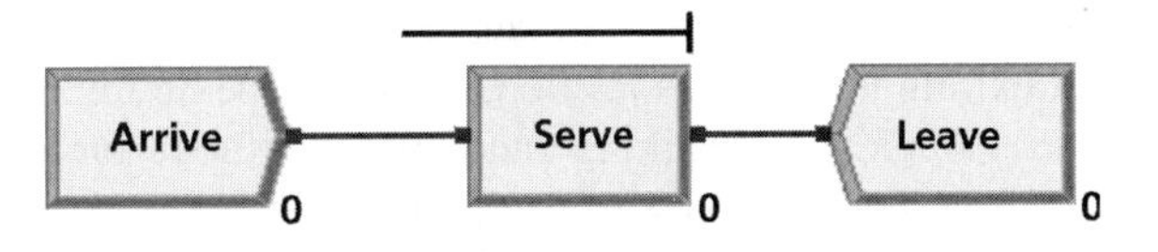

tributes or a system variable. Arena flowcharts can be categorized as a *process graph* type of graphical model because they represent the processes in the system.

Consider the graphical model in Figure 8.8, which is the simple system introduced in Chapter 7. This simple graphical model represents the life cycle of a customer in the queueing system. The customer's process consists of the following:

- create a new customer entity,
- request a server,
- take server and start service (when a server is available),
- delay the length of the service time,
- release the server, and
- dispose of the customer entity.

All of these actions involve the customer entity that was created in the first step. Most of this process happens in the **Process** module, where the server resource is requested, the entity waits until it is available, the resource is seized, the service time delay is inserted, and the server resource is released (and other processes waiting for the resource are now activated). Part of the action of the **Process** module is to create an indefinite delay while the customer waits for a server, and to schedule an end of service event when the service time has completed. Most of this action is not shown explicitly, but it is important for the modeler to understand that it is all happening.

The bank lobby model in Chapter 7, which is also shown in Figure 8.9, is a little more complex process. However, all modules in the graphical model involve the same temporary entity, a bank customer. Because there is a **Decide** module in this model, the customer has more than one possible path through the process, and these paths are shown explicitly in the graphical model. If the number of customers

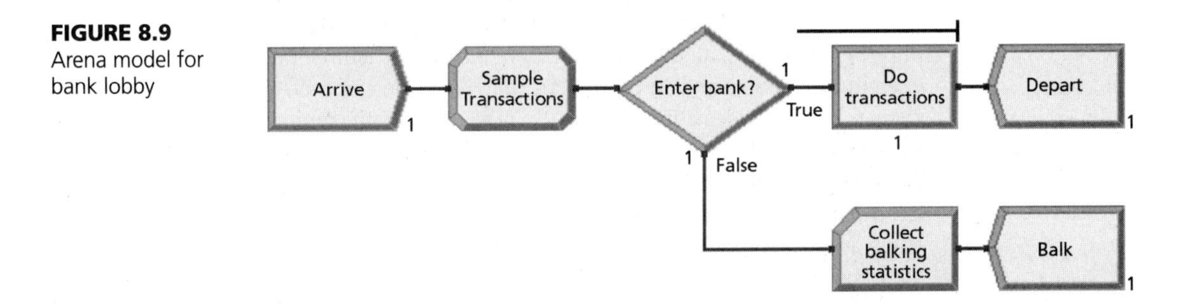

FIGURE 8.9
Arena model for bank lobby

waiting is less than 10, then the top path, which involves doing the banking transactions and exiting the bank, is taken. Otherwise, the bottom path, which involves just collecting statistics on balking and leaving without doing the transactions, is taken. Thus, a process can have different paths depending on the state of the system as the entity moves through it. The important idea here is that all actions in all modules involve the same entity. Thus, the model is a description of the process for the entity.

Since each module concerns the process for a specific entity, any references to attributes are references to the attributes of the specific entity instance. Thus, when assigning an attribute value within a module, we do not need to specify which entity is involved; it is the entity whose process is currently executing. On the other hand, it is not possible for one entity to directly change the attributes of another. To do this, the first entity must activate the second, and that entity then can change its own attributes.

8.3.1 The Batch and Separate Modules

The **Batch** and **Separate** modules give us the ability to collect entities into batches and process the batches together (batch processing) as well as divide a batch of entities into their individual components for separate processing (parallel processing). This is especially important in manufacturing applications where groups of parts are processed together, but it is equally important in other models where we want to cause actions to proceed in parallel.

The **Batch** module collects a specified group of entities into a batch. A batch can be formed with either a specified number of entities entering into the module or by matching entering entities based on the value of a specified attribute. Batches can be grouped permanently or temporarily. Temporary batches will be split later using the **Separate** module. When entities enter a **Batch** module, they are placed into a queue until the batch is formed. After the batch is formed, each entity in the batch is updated with the wait and cost data accumulated while in the module and a new entity is created to represent the whole batch. The representative entity's attributes are calculated on the basis of the individual entities' attributes; for example, waiting time is calculated as the sum of waiting times of the entities within the batch.

As we mentioned before, the **Separate** module is used to split a batched entity into its components. It can also be used to duplicate entities, thus creating multiple identical entities. This gives us a way to model processes that occur in parallel. When duplicating entities, all attribute values, including the animation picture, are copied to each duplicated entity. In particular, they have the same **Entity.SerialNumber** attribute. This provides a means for us to rebatch the same entities at a later time using the **Batch** module. The original entity and all of its copies leave the module, so that (number of duplicates + 1) leave the module. When splitting existing batches, the temporary representative entity is destroyed and the original entities that formed the batch are recovered. These entities leave the module in the same order in which they were joined to the batch.

8.3.2 A Copy Center Model

The **Batch** and **Separate** modules will be demonstrated using a model of a copy center that has one fast copier and one slow copier. In the model, the copy time per page for the fast copier is normally distributed with mean 1.6 seconds and standard deviation .3 seconds. The copy time per page for the slow copier is normally distributed with mean 2.8 seconds and standard deviation .6 seconds. The arrival process is Poisson, so the interarrival time distribution for customers is exponential, with mean 3.0 minutes. The number of copies requested by each customer is uniformly distributed between 10 and 50 copies.

The policy for selecting a copier is as follows: If the number of copies requested is less than or equal to 30, the slow copier will be used. If the number of copies exceeds 30, the fast copier will be used, with one exception: If no jobs are in progress on the slow copier and the number of jobs waiting for the fast copier is at least two, then the customer will be served by the slow copier. After the customer gives the originals for copying, she proceeds to the service counter to pay for the copying. The time to complete the payment transaction is normally distributed with mean 2.1 minutes and standard deviation .6 minutes. As soon as both the payment and the copying job are finished, the customer takes the copies and departs the copying center. The copy center works 10 hours per day.

Management has requested the model to be developed because they are concerned that customers have to wait too long for copies. Recently, several customers complained about long waits. Their standard is that customers' waiting time should average no more than 3 minutes. If mean waiting time is too long, several options are available: The policy for allocating jobs to the fast copier could be modified or the company could purchase an additional copier which could be either a slow copier or a fast copier.

8.3.3 The Arena Model

First, let's think about the model a bit. Arriving customers will first sample the number of copies they wish to make, then they will decide whether to use the slow copier or the fast copier. Once the copier is selected, two activities will be done in parallel: Make the copies and pay for the copies. This calls for a separation of the customer process into two coordinated subprocesses. The customer cannot leave the copy center until both are complete—that is, until both the copying and the payment activities have finished. The way to make this happen is to separate the original customer process into two identical (i.e., duplicate) processes using a **Separate** module: one for the copying activity and one for the payment activity. After these activities, we will use a **Batch** module to put the entities back together and then dispose of the batched entity. The **Separate** and **Batch** approach is dictated by the need to have these activities in parallel and to make sure that both are finished before the customer entity leaves the system.

We want to develop this model as we would in a real environment. Normally, we do not have a finished model to examine before building it. Before putting modules in the model window, we need to establish a strategy for building the model. It is generally a good idea to start by defining the entities and resources that we plan to use in the model. At this point, we know that we have the following resources: a slow copier, a fast copier, and a clerk who receives payment for the copies. We also know that we need a temporary entity—customer—to represent the people who arrive with copying jobs. These elements can be set up by clicking on the **Entity** and **Resource** modules in the **Basic Process** panel and using the spreadsheet window.

Next, a good strategy is to create the structure of the model by placing modules on the model window and connecting them, but without editing their dialogs yet except to give each module a name. We wait to enter all module information because sometimes a module that is placed earlier in the process will need to use data from a later module that has not been created yet. If all modules are placed and named, then the module identifiers are known. This is the case in the **Decide** module where the arriving customer decides whether to use the slow copier or the fast copier. This decision depends on the number of jobs waiting at the fast copier, which is determined by a later module in the process.

The basic structure of the copy center customer process is:

1. Create a new customer.

2. Sample the number of copies for the customer.

3. Decide whether the customer will use the slow copier or the fast copier.

4. If the slow copier is selected:

   4A. Start parallel processes to make copies on the slow copier and make payment.

5. If the fast copier is selected:

   5A. Start parallel processes to make copies on the fast copier and make payment.

6. Collect the copies (after payment and copying).

7. Exit the system.

Note that the payment process is the same, regardless of whether the slow copier or the fast copier is selected, and that both the copying process and the payment process must be completed before the customer exits. The Arena flowchart is shown in Figure 8.10.

This flowchart includes two **Separate** modules (**Duplicate Slow Jobs** and **Duplicate Fast Jobs**) as well as one **Batch** module (**Collect Copies**). We introduced these **Separate** modules to create the two parallel processes for doing the copying and making payment. The payment process is the same for all customers, so we need only one **Process** module for payment, and we can connect it to one of the outputs of the **Separate** modules.

FIGURE 8.10 Copy center model

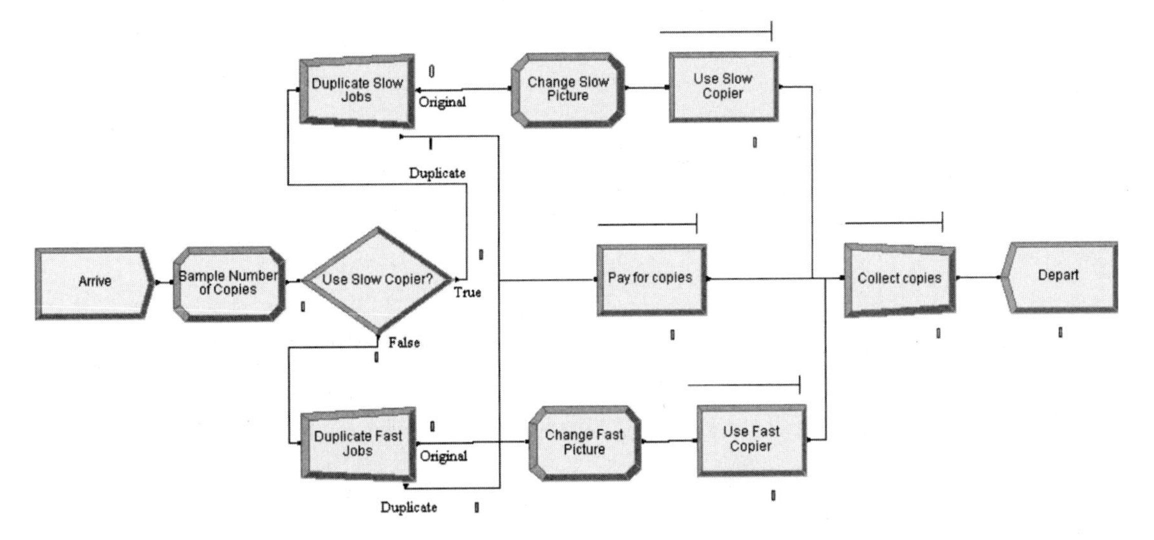

Once all of the modules have been placed, we can enter the data into each module via the modules' dialogs. Enter the data with just the information at the start of this section. Table 8.1 (pages 294–295) contains the data for the dialogs, which you can use to check your entries.

We used entity type **Customer** to represent customers, and set the initial picture to **Picture.Woman.** These data can be entered via the spreadsheet view in the **Entity** data module, as we demonstrated in the previous chapter. Alternatively, if we right-click in any field of the **Customer** entity in the **Entity** spreadsheet view and select **Edit via Dialog,** we will get the **Entity** window that can be used to enter and change data for this module. With the same technique we can edit the windows for the other data modules—**Resource, Set,** and so on. Specifically, we will use this approach to define three resources—**Slow Copier, Fast Copier,** and **Clerk**—each with fixed capacity of 1. **Clerk** is the person who receives payment for the copies.

Notice that two additional modules not mentioned in the outline were added in Figure 8.10: the **Assign** modules **Change Slow Picture** and **Change Fast Picture.** To enhance the animation, we used these modules to change the graphic used to animate the copy processes to that of a piece of paper. When you run the simulation, notice that the graphic moving through the payment process is a person, but the graphic moving through the copy process is a piece of paper. In addition, the graphic in the fast copy process is a red piece of paper while the graphic in the slow copy process is blue paper.

Here is a brief description of the Arena model: After the entity **Customer** arrives in the **Create** module Arrive, an **Assign** module **Sample Number of Copies** is used to sample the number of copies and store the value in the **Entity** attribute **No.**

TABLE 8.1 Parameters in copy center model

Module	Parameter	Subparameter	Value
Create	Name		Arrive
	Entity Type		Customer
	Time Between Arrivals	Type	Random(Expo)
		Value	3
		Units	Minutes
	Entities per Arrival		1
	Max Arrivals		Infinite
	First Creation		0
Assign	Name		Sample Number of Copies
	Assignments—Line 1	Type	Attribute
		Attribute Name	No. Copies
		New Value	"UNIFORM(10, 50)"
Decide	Name		Use Slow Copier
	Type		Two-way by Condition
	If		Expression
	Value		(No. copies.LE. 30) .OR. ((No. copies .GT. 30). AND. ((Use Slow copier. WIP .EQ. 0) .AND. (Use Fast copier.WIP .GE.2)))
Separate	Name		Duplicate Slow Jobs
	Type		Duplicate Original
	Percent Cost to Duplicates		50
	# of Duplicates		1
Assign	Name		Change Slow Picture
	Assignments—Line 1	Type	Entity Picture
		Entity Picture	Picture.Blue Page
Process	Name		Use Slow Copier
	Type		Standard
	Logic	Action	Seize Delay Release
		Priority	Medium(2)
		Resources—Line 1	Resource, Slow Copier, 1
	Delay Type		Normal
	Units		Seconds
	Allocation		Value Added
	Value(Mean)		2.8*No. copies
	Std Dev		.6*SQRT(No. copies)

TABLE 8.1 (continued)

Module	Parameter	Subparameter	Value
Separate	Name		Duplicate Fast Jobs
	Type		Duplicate Original
	Percent Cost to Duplicates		50
	# of Duplicates		1
Assign	Name		Change Fast Picture
	Assignments—Line 1	Type	Entity Picture
		Entity Picture	Picture.Red Page
Process	Name		Use Fast Copier
	Type		Standard
	Logic	Action	Seize Delay Release
		Priority	Medium(2)
		Resources—Line 1	Resource, Fast Copier, 1
	Delay Type		Normal
	Units		Seconds
	Allocation		Value Added
	Value(Mean)		1.3*No. copies
	Std Dev		.2*SQRT(No. copies)
Process	Name		Pay for Copies
	Type		Standard
	Logic	Action	Seize Delay Release
		Priority	Medium(2)
		Resources—Line 1	Resource, Clerk, 1
	Delay Type		Normal
	Units		Minutes
	Allocation		Value Added
	Value(Mean)		2.1
	Std Dev		0.6
Batch	Name		Collect Copies
	Type		Permanent
	Batch Size		2
	Save Criterion		Last
	Rule		By Attribute
	Attribute Name		Entity.SerialNumber
Dispose	Name		Depart

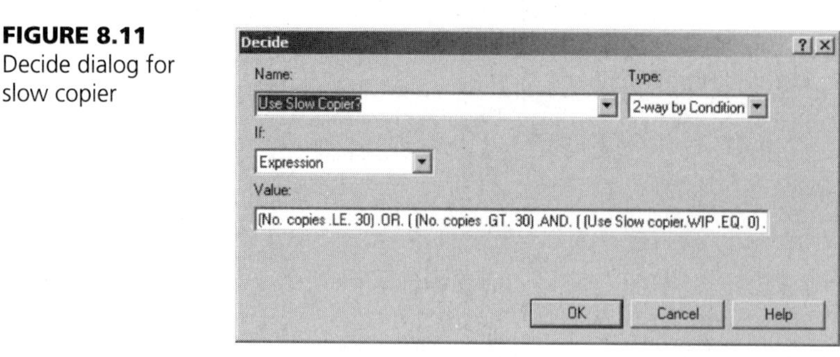

FIGURE 8.11
Decide dialog for slow copier

copies. Next, the type of copier is selected in the **Decide** module **Use Slow Copier?** with the expression

```
(No. copies .LE. 30) .OR. ((No. copies .GT. 30) .AND. (Use
Slow Copier.WIP .EQ. 0).AND.(Use Fast Copier. WIP .GE.2))
```

as is shown in Figure 8.11. (Note that only a part of this expression can be seen in the window.) The operators AND, OR, LE, and GT are used in this expression. AND and OR have the obvious meanings; LE is the "less than or equal to" comparison operator, and GT is the "greater than" comparison operator. More information about these operators can be found in the Arena Help window under the index topic "Logical operators." The built-in variable **Slow Copier.WIP** holds the work in process for the resource **Slow Copier.** At this point, the process divides into two branches—one for customers using the slow copier and one for customers using the fast copier. We will examine the branch for customers using the slow copier; the other branch is identical in structure.

The next module is the **Separate** module **Duplicate Slow Jobs** where the **Customer** entity is duplicated, as seen in Figure 8.12. The purpose here is not to create

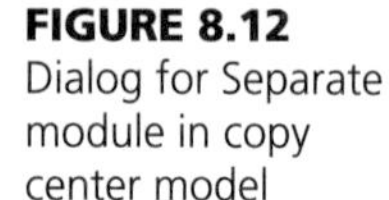

FIGURE 8.12
Dialog for Separate module in copy center model

a second **Customer** entity but to allow two processes—making the copies and paying for the copying—to operate in parallel. We can think of the duplicate entity not as a duplicate entity but as a second process for the same entity.

The original entity then proceeds to the **Assign** module **Change Slow Picture,** where a new graphic is assigned to the entity as discussed before. Next we have the **Process** module **Use Slow Copier,** where the entity requests a **Slow Copier** resource for copying. The total copying time per customer is sum of *n* independent normal random variables, each being a copying time per one copy. Here *n* is number of copies per customer, represented in the model by the attribute **No. copies.** In the case of a slow copier, where the mean time per copy is 2.8 seconds and the standard deviation is .6 seconds per copy, total copying time per customer is represented by a normal distribution with parameters

```
NORMAL( No. copies * 2.8, SQRT(No. copies) * .6)
```

Note that using a normal distribution to sample a processing time can lead to a problem because the normal distribution always has a nonzero probability of generating a negative value; this clearly is impossible and will lead to an error when the simulation is executed. In this case, the problem is not likely to occur because the standard deviation is much smaller than the mean. Since the number of copies is at least 10, the mean of the distribution is at least 28 seconds and the standard deviation is approximately 1.90, so the mean is almost 15 standard deviations more than zero. The probability of generating a value this far in the tail of a normal distribution is extremely small. Hence, we have chosen in our implementation to ignore this problem. If you cannot ignore the possibility of sampling a negative value from the normal distribution, it can be handled by automatically truncating any negative value to a value near zero—for example, to .001 seconds. Here is an expression that will do this computation:

MAX (.001, NORMAL(No. copies * 2.8, SQRT(No. copies) * .6))

Note that this must be entered as an expression in the **Process** module dialog. The finished **Process** dialog is shown in Figure 8.13.

Following the second path from the **Separate** module, we have the **Process** module **Pay for Copies** representing the payment activity. This is a standard process that uses one unit of a single resource, **Clerk.** Following the two process modules is a **Batch** module **Collect Copies,** where the original and duplicate customer entities are again put together using the **Entity.SerialNumber** attribute. When one of these processes finishes before the other, the entity will wait in the **Batch** module until the other arrives; they are then batched and sent to the **Dispose** module **Depart,** where they will be destroyed.

There are two reasons for directing entities from both the slow copier and the fast copier into the same **Process** module: **Pay for Copies.** They use the same resource, **Clerk,** and they wait in the same queue when waiting for the resource. Thus, to collect the correct data on customers waiting to pay, we need to have them to use the same facility and wait in the same queue. The entities are collected using a **Batch**

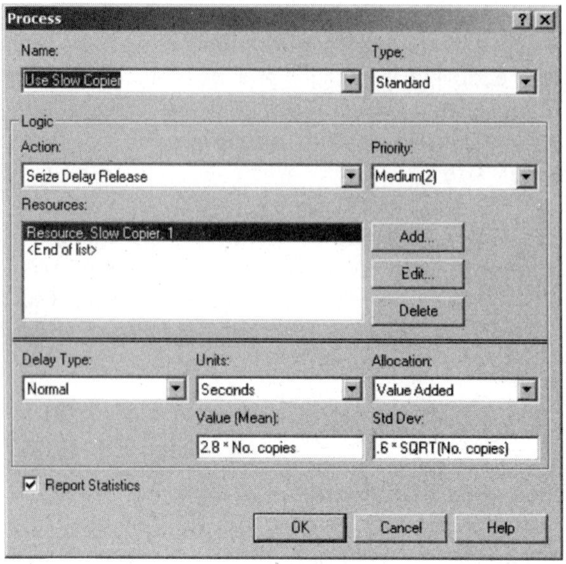

FIGURE 8.13
Process dialog for slow copier

module. Figure 8.14 shows the dialog for this **Batch** module. Entities are grouped in a batch of size 2, and the entities in each batch must have the same value of the attribute **Entity.SerialNumber.** When the entities were duplicated, they were assigned the same value of **Entity.SerialNumber,** so in this module, the original pairs will be reunited.

The behavior of the entities in the model, particularly their duplication and batching, can be analyzed by running the model. Figure 8.15 shows a snapshot of a model execution in which both entity pictures can be seen. By watching the animation closely, you can gain confidence that the model operates correctly and therefore contributes to model verification.

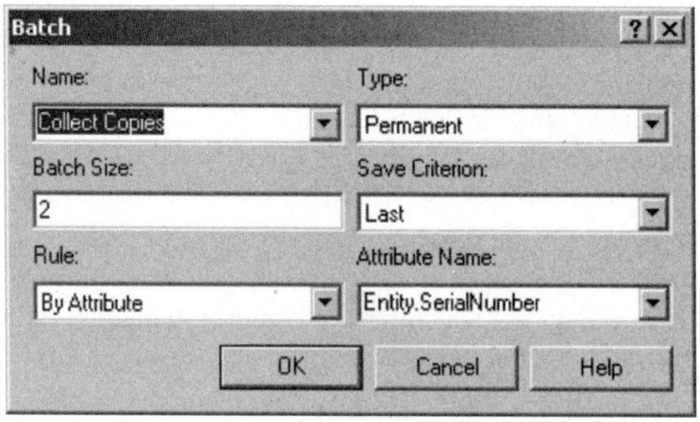

FIGURE 8.14
Dialog for Batch module in copy center model

FIGURE 8.15 Animation of copy center in Arena

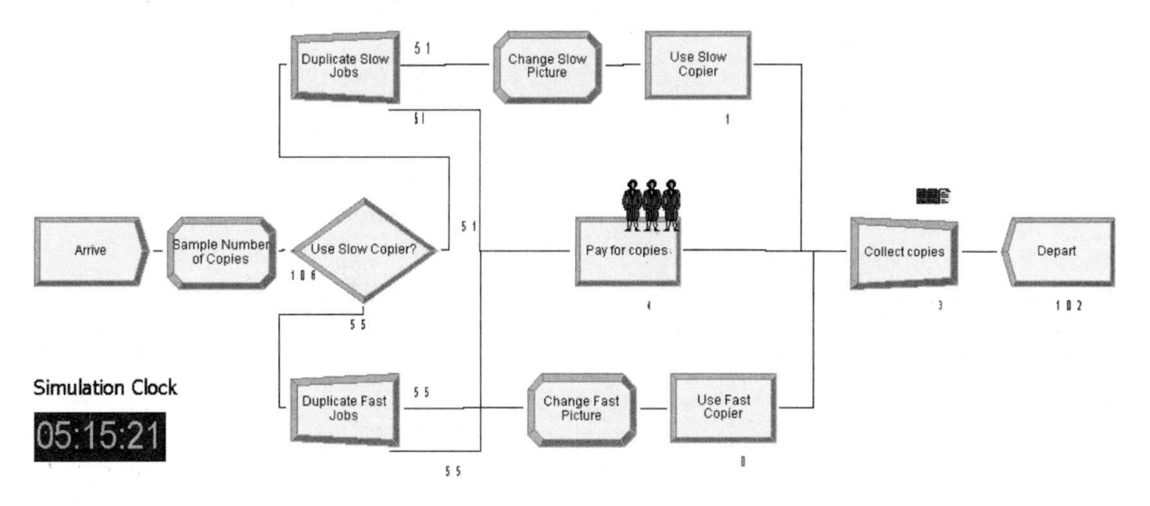

8.3.4 Running the Simulation

Before making the simulation run, we need to think about the performance measures we want to estimate using the simulation. The purpose of this model is to determine if the mean waiting time per customer exceeds 3 minutes. Thus, the obvious performance measure is mean waiting time per customer.

The next step is to set up the experimental design and run the simulation. Select the **Setup** option under the **Run** menu. Figure 8.16 shows the **Run Setup** dialog with the warm-up period, number of replications, length of each replication, and the time units in the output reports. The copy center runs for 10 hours per day and then shuts down after serving all waiting customers. We have a choice about how to design the experiment: We could just run one long replication, ignoring the 10-hour operating period each day. If the parameters are such that the system is stationary and reaches stationarity early each day, we can compute the stationary performance measures as in Chapter 4. On the other hand, if the system does not achieve steady state behavior during the day, we will need to run some replications and use their data to estimate the performance measures. Normally, we are not certain whether steady state behavior will be achieved during the day, so we will assume it is not and run some replications with each replication consisting of a day's operation.

Once you are assured that the model runs correctly, you can run all 10 replications as quickly as possible by executing the simulation in *fast forward* mode. Do this by clicking the **Fast Forward** button in the standard toolbar. If you have the sta-

Fast forward button

FIGURE 8.16
Run setup dialog

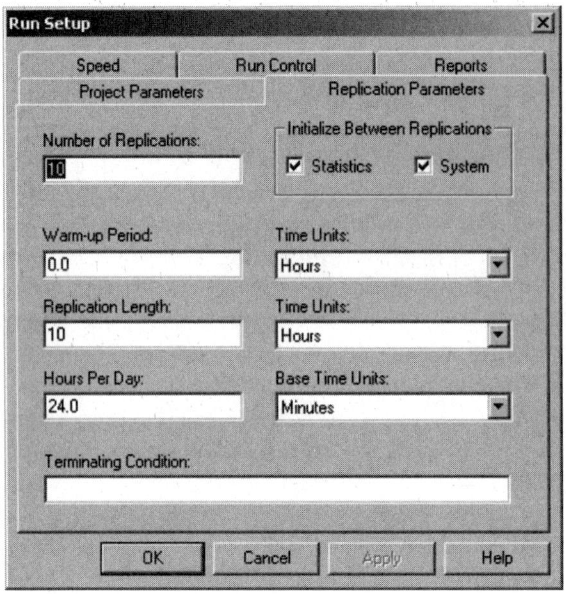

tus bar turned on at the bottom of the Arena Window (using **View → Status Bar**), you can observe the replications as they run. After the simulation has run to completion, Arena will present a dialog stating that the simulation is finished and asking if you want to view the results. (This behavior can be changed by setting appropriate options; see the Arena Help system for details.) Click **Yes** to view the output report.

We discussed the output report format in Chapter 7, so we will not go into detail about the report here. Figure 8.17 shows the output report tree and selection of the **Entity Time** report. The actual report is shown in Figure 8.18.

As you can see from this report, the average waiting time for customers (over all customers and all replications) is 2.32 minutes, which is well below our standard of 3 minutes. Moreover, this report provides an estimate of the statistical error. The half-width of a 95 percent confidence interval for the mean waiting time is .39, so the true mean waiting time is between 1.93 minutes and 2.71 minutes, with 95 percent confidence.

FIGURE 8.17
Tree view of copy center simulation output

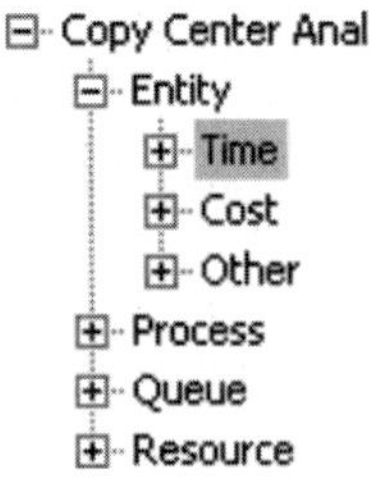

- Copy Center Anal
  - Entity
    - Time
    - Cost
    - Other
  - Process
  - Queue
  - Resource

FIGURE 8.18 Entity Time report for copy center model

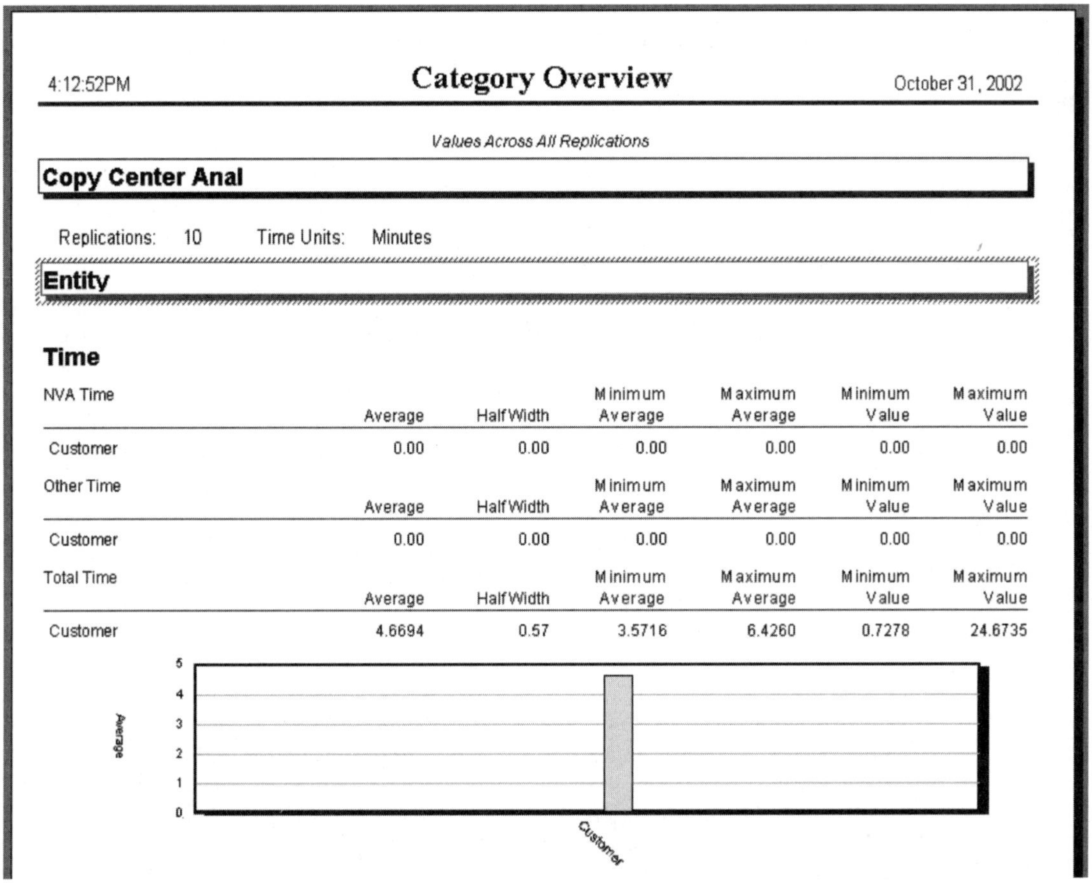

4:12:52PM **Category Overview** October 31, 2002

Values Across All Replications

Copy Center Anal

Replications: 10 Time Units: Minutes

Entity

Time

NVA Time	Average	Half Width	Minimum Average	Maximum Average	Minimum Value	Maximum Value
Customer	0.00	0.00	0.00	0.00	0.00	0.00

Other Time	Average	Half Width	Minimum Average	Maximum Average	Minimum Value	Maximum Value
Customer	0.00	0.00	0.00	0.00	0.00	0.00

Total Time	Average	Half Width	Minimum Average	Maximum Average	Minimum Value	Maximum Value
Customer	4.6694	0.57	3.5716	6.4260	0.7278	24.6735

We can also examine the waiting time in more detail. Open the **Queue** branch of the **Report** tree, then open the **Time** branch and click on **Waiting Time.** Figure 8.19 shows the **Waiting Time** report for the three processes.

It is clear from this report that the fast copier poses no problem: The average waiting time is only .1 minute, or six seconds. The maximum waiting time over all replications was less than 2 minutes. The slow copier had an average waiting time of less than .16 minutes, or 10 seconds, so it also was quite uncongested. The half-widths for these estimates were small (.02 and .03 minutes, respectively), so we can be confident of the estimates. Waiting time at the payment process was longer. This delay averaged approximately 1 minute and varied up to almost 10 minutes. Although this was the most extreme waiting time observed, customers clearly would not be happy waiting 10 minutes to pay for their copying. We also see that the waiting time in the **Collect Copies.Queue** was rather long. This is the queue in the

FIGURE 8.19 Waiting Time detail report for all replications

Queue

Time

Waiting Time	Average	Half Width	Minimum Average	Maximum Average	Minimum Value	Maximum Value
Collect Copies.Queue	1.8434	0.29	1.2930	2.7289	0.00	23.7250
Pay for copies.Queue	2.5711	0.57	1.5569	4.3159	0.00	20.8541
Use Fast Copier.Queue	0.06688069	0.01	0.04816052	0.08853799	0.00	1.3482
Use Slow Copier.Queue	0.0920	0.02	0.04936435	0.1289	0.00	2.2606

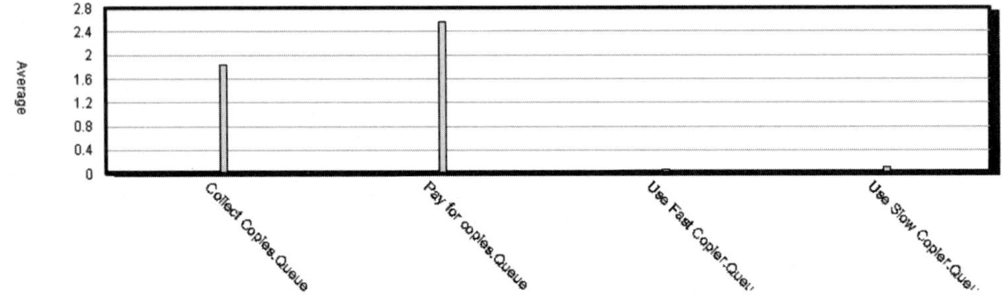

Process module **Collect Copies** where customers wait for their copying job to be completed. Because the waiting times in the copy machine queues were short, we can assume that the length of this queue is due mainly to the process times to make the copies and could only be reduced by having faster copiers. See the problems at the end of this chapter for ways to improve this waiting time.

As you have no doubt noticed, Arena offers many reports. We have only looked at summary reports for certain system characteristics, but Arena provides summary reports for all entities, processes, queues, and resources that use all replication data to compute estimates of delay times and utilizations for each system component. These reports also compute cost estimates based on value added and non-value added allocations. The concept of value added process time and non-value added process time fits some types of models such as manufacturing models and health care systems but does not make sense for others. In the copy center model in particular, customers pay by the copy, so total income to the center depends only on the total number of copies. If the machines run faster or slower, total income is not affected.

8.4 HIERARCHICAL MODELING

8.4.1 Basic Concepts

Graphical models of complex systems tend to be large. This complicates the model development process and makes the models difficult to read. It also causes problems when models must be modified, and it is well known that model modification is a rule, not an exception, during both the model's development period and its lifetime. Finally, complex models must be organized somehow to manage the construction process. Some order must be followed as the modules are created and the model is built, so the model builder must adopt a strategy for building the model.

These problems can be managed using a hierarchical approach (Simon, 1981) to model building. *Hierarchical modeling* provides a means for managing system complexity by dividing the system into smaller units that can be modeled and manipulated independently and have well-defined relationships to one another. Each unit, which we will call a submodel, can be further decomposed into smaller units. Decomposition is done in a hierarchical manner in which units can be related only between adjacent levels: Each unit can be a part of the decomposition of its "parent" unit and may be decomposed itself into its "child" units (Luna, 1993). Of course, the unit is also related to other units on its own level.

Hierarchical models have several advantages over monolithic models (Gordon et al., 1990). Recall from the discussion in Chapter 4 that top-down design is a hierarchical methodology. The advantages of using a hierarchical methodology in graphical modeling echo those for using top-down designs for program development and are worth repeating here. Hierarchical methods allow the model to be created using a stepwise refinement process that starts with a simplified top-level system consisting of one or more submodels; the methods then add detail by decomposing each submodel into additional submodels. Many model modifications can be localized to a submodel level without having to involve the entire model. This improves model reliability because new errors are less likely to be introduced if the changes can be localized. By using parameters, a single submodel can represent multiple identical or similar model parts. The submodels can also be reused in other hierarchical models with similar structures.

The basic symbols used in a graphical modeling technique are called *atomic symbols,* while *compound symbols* refer to symbols that represent connected groups of atomic symbols or other compound symbols (Pooley, 1991b). Models that use compound symbols are called *aggregated* models; those that contain only atomic symbols are called *flat* models. Up to this point, all Arena models we have presented have been flat models. Because compound symbols can be grouped to form a higher level of compound symbols, there may be several levels of model aggregation. There are many different ways to organize flat models into aggregated ones. One common way is to group model elements into subsets that perform specified identifiable functions in the model.

In hierarchical modeling, each compound symbol has inputs and outputs that connect it with other model elements represented by symbols. A change to the internal structure of a submodel represented by a compound symbol does not affect other parts of the model if the inputs and outputs of that submodel are not changed. This important property of hierarchical models leads to robust models and allows the model to be developed by a team of people.

8.4.2 Hierarchical Modeling with Arena

Arena implements hierarchical modeling through the use of *submodels*. Each model can have one or more submodels, and each submodel can have submodels. There is no limit to the amount of submodel nesting. In Arena, a **Process** module can be designated as a submodel, so a submodel can go anywhere a **Process** module can go. Because a **Process** module has exactly one input and one output, a submodel must receive entities from the higher-level model through exactly one input, and it delivers entities to the higher-level model through exactly one output.

Submodels can contain any object that can appear in the main model—that is, flowchart logic, static graphics, and animation. Each submodel is represented in its own view (window). Submodels can be connected to either modules or other submodels.

8.4.3 A Fuel Depot Model

We will introduce another model to show additional modeling techniques with Arena and to demonstrate hierarchical modeling. This model consists of a fuel depot, which has two types of fuel—regular and premium—and a stream of trucks arriving to take fuel. Truck arrivals form a Poisson stream with mean time between arrivals of 20 minutes. The trucks' fuel requests are 65 percent regular fuel (**Fuel 1**) and 35 percent premium fuel (**Fuel 2**). The amount of fuel requested is uniformly distributed between 10,000 and 14,000 gallons for both fuels. Each pump can deliver fuel at 30 gallons per second or 1800 gallons per minute. Thus, the time, in minutes, to pump the fuel is the amount of fuel requested divided by 1800.

A complicating factor of this model is the fact that fuel is a consumable resource, so it must be replenished periodically. Each fuel has a *trigger level* so that a replenishment is triggered when the amount of fuel remaining falls below that level (inventory managers call this a *reorder point*). The trigger level for **Fuel 1** is 12,000 gallons, and the trigger level for **Fuel 2** is 50,000 gallons. Once a replenishment is ordered, there is a delay before the fuel is available. For **Fuel 1,** the delay has a triangular distribution with minimum, most likely, and maximum values of 1, 2, and 3 hours, respectively. For **Fuel 2,** the delay has a triangular distribution with parameters 1, 1.5, and 3 hours. We assume that none of the arriving fuel is available until the end of this delay.

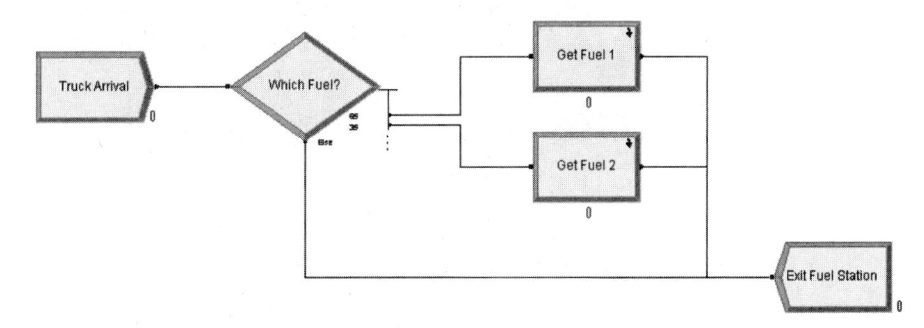

FIGURE 8.20
Top-level hierarchical model for fuel depot

This is a continuously operating model. Management is concerned that trucks are being delayed too long at the depot. Like the copy center, the depot's income depends on the amount of fuel sold, and the model assumes that all arriving trucks will be served. The queues that hold the trucks are assumed to have infinite capacity.

Before proceeding to build the graphical model, we need to define three resources: **Pump Station** with a capacity of 3 and **Fuel 1** and **Fuel 2** with (initial) capacities of 100,000 and 80,000, respectively, using the **Resource** module.

Figure 8.20 is the model window for the highest level of this model.

This graphical model shows the model logic at its most aggregated level: A truck arrives, selects its fuel, gets the fuel, and exits the system. Truck arrivals are modeled using a **Create** module, and the selection of the fuel is implemented in the model using a **Decide** module. Figures 8.21 and 8.22 show the dialogs for these two modules.

Once a fuel has been selected, the truck entity moves into a submodel labeled **Get Fuel 1.** This can be inserted by selecting a **Process** module from the **Basic Process** panel of the project bar. Once this **Process** module has been placed on the model window and connected to the other modules, you can double-click it and assign it a name. In the input on the right of the **Process** dialog, click the down arrow

FIGURE 8.21
Create module for trucks in fuel depot model

FIGURE 8.22
Decide module for top level in fuel depot model

Decide
Name: Which Fuel?
Type: N-way by Chance
Percentages:
65
35
<End of list>
Add...
Edit...
Delete
OK Cancel Help

and select **Submodel.** This option will turn the module into a container for a submodel. Click **OK** to record the changes and return to the model window.

You are now ready to create a submodel that contains the model logic for trucks that request **Fuel 1.** To create the submodel, right-click on the **Submodel** module and select **Edit Submodel.** You will now see a blank model window with an entry connector (a triangle) on the left and an exit connector (a square) on the right. You can create any portion of a model in this window, but it must have only one input coming from the entry connector on the left and it must connect all outputs to the output connector on the right.

Let's think about what happens to the truck once it has selected **Fuel 1.** The next action is to select the amount of fuel. Now two processes occur in parallel: We attempt to begin pumping by selecting a pump and the appropriate amount of fuel while also checking to see if the fuel needs to be replenished and, if so, placing an order for the fuel. Just as we did for the **Copy Center** module, we use a **Separate** module to duplicate the truck entity. Of course, the duplicate entity is not a second truck; it just works to create a separate, parallel process for checking the fuel level and possibly placing an order. Figure 8.23 shows the **Process** submodel for **Get Fuel 1.**

The original entity proceeds to a submodel, **Pump Fuel 1,** where the fuel is pumped. The duplicate entity goes first to a **Decide** module, **Replenish Fuel 1?**, where the replenishment rule is applied. The decision rule is that the fuel will be replenished if the amount of fuel is less than the trigger level and no flag has been set to indicate that a resupply is already in progress. Figure 8.24 shows the dialog box

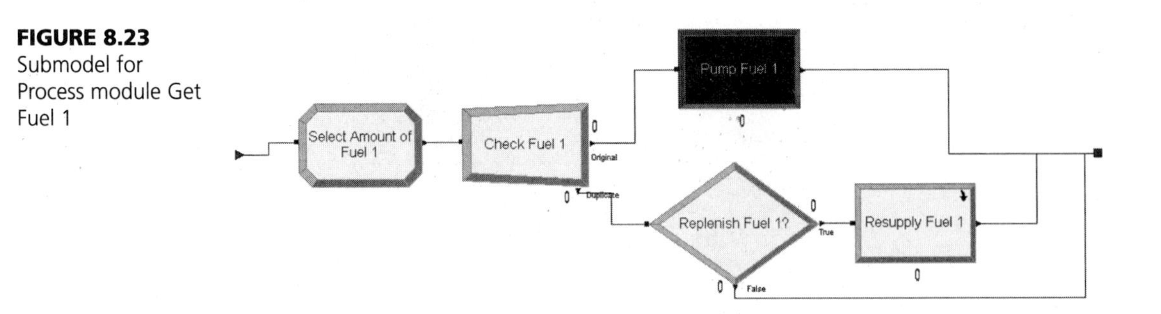

FIGURE 8.23
Submodel for Process module Get Fuel 1

FIGURE 8.24 Fuel depot model replenishment Decide module dialog

for this module. In this **Decide** module, the entity is discarded if the replenishment rule is not satisfied; otherwise, it enters a submodel **Resupply Fuel 1** where the resupply takes place.

A similar submodel is created for **Get Fuel 2,** with the only change being that all references to a fuel are to fuel 2. One way to do this is to copy the submodel **Get Fuel 1** in Figure 8.20 and paste it at the place for submodel **Get Fuel 2** and then edit the submodel to change references to **Fuel 1** to **Fuel 2.** Other elements might also need to be changed, but the logic structure is identical.

The next step in the refinement process is to create submodels **Pump Fuel 1** and **Resupply Fuel 1.** Let's look at submodel **Resupply Fuel 1** first. Right-click on module **Resupply Fuel 1** to create the submodel. Figure 8.25 shows the submodel. First, an **Assignment** module **Flag Fuel 1** assigns the value 1 to **FuelFlag1,** indicating that the flag is set for **Fuel 1** to denote that a resupply operation is underway. Then a **Process** module implements a delay while the resupply progresses. Figure 8.26 (next page) shows the dialog for this module. We see that no resources are used; a delay is simply incurred. Finally, at the end of the delay, the fuel is added to the reservoir using an **Assign** module. Two assignments are made here. The flag is turned off by assigning the value 0 to **Fuelflag1,** and the fuel is added to the capacity of resource **Fuel 1** by assigning the value **FullLevel1** to variable **mr(Fuel1). FullLevel1** is the parameter we created to denote the capacity of the tank for **Fuel 1** (or the level to which we fill the tank when replenishments are made). The Arena built-in variable **mr(Fuel1)** denotes the capacity of resource **Fuel 1.** The Arena Help system can show you more built-in variables. In this model, the only two built-in variables we will use are **mr( )** and **nr( ),** the number of units of the resource that are currently in use.

The submodel **Pump Fuel 1** is shown in Figure 8.27 (next page). This submodel consists of four modules:

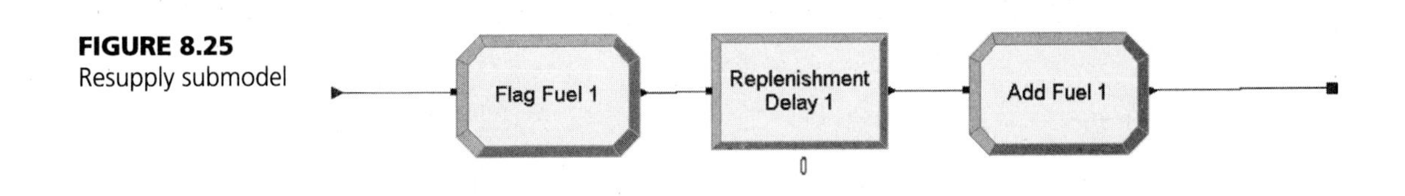

FIGURE 8.25 Resupply submodel

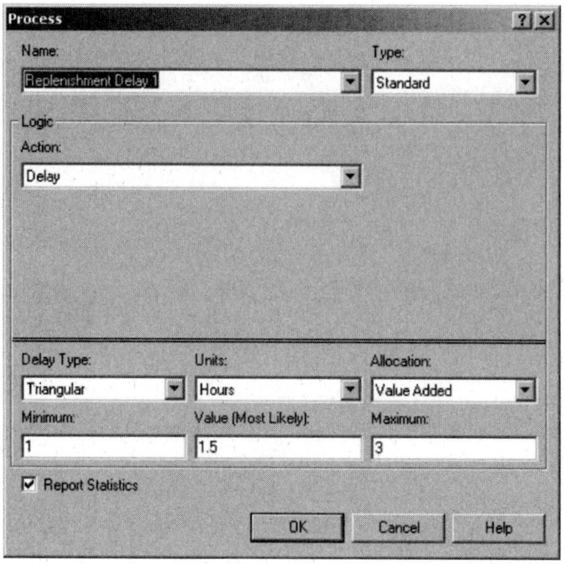

FIGURE 8.26
Fuel 1 resupply Process module dialog

1. a **Process** module where the pump and fuel are seized,
2. a **Process** module where the pump is released,
3. an **Assignment** module where the capacity of the fuel is reduced, and
4. a **Process** module where the fuel is released.

This seems rather convoluted. You might ask why we cannot just release the fuel and pump, and then reduce the fuel by the amount the truck took. The problem is that if another truck is waiting for the pump and fuel, it will take the fuel before we have a chance to reduce the amount. However, if we release the fuel after reducing the capacity of the fuel, the waiting truck will not have a chance to take it.

This completes our discussion of hierarchical modeling. In the next section, we will show how to create a more realistic model of animation using another type of submodel. Both process submodels and object submodels can be used to modularize models.

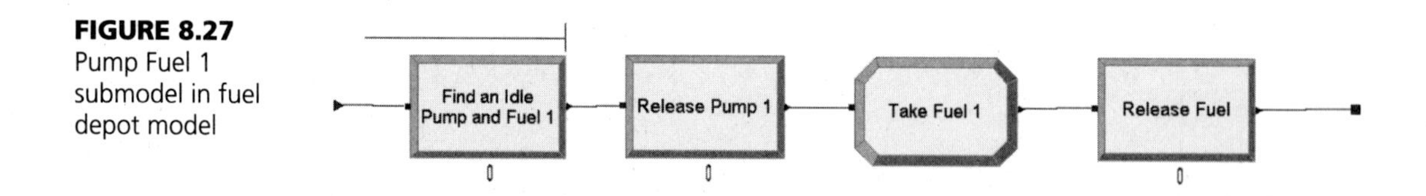

FIGURE 8.27
Pump Fuel 1 submodel in fuel depot model

8.5 ANIMATION

An *animation* is a graphical depiction of the system that displays the movements of entities and other changes to the system and elements of the system as the simulation executes. Animations are useful because they draw on our intuitive understanding of system dynamics to demonstrate that the model does or does not faithfully represent the intended system.

If the simulation does faithfully represent the system, then an animation is a useful tool to convince management and other decision makers that the model can reliably be used to evaluate system decisions. Thus, it helps build trust in the model. Without trust, decision makers will not be willing to use the model, and all of the time and expense to develop the model is wasted. High-level decision makers cannot be expected to understand the programming details of the model or deal with complex tables or other outputs that show model behavior. An animation allows them to observe program operations using their current understanding of system operations.

If the model does not behave correctly, then an animation can identify aberrant behavior and locate the portion of the model exhibiting the problem. Thus, an animation is a useful tool for model verification and validation. For example, if an emergency department (ED) animation shows that patients who are critical are triaged at the same level as less serious cases, then both the modeler and the emergency department personnel will see that the model is not following the protocols of the actual system; it will also be clear that the problem involves the assignment of severity codes or their use in the logic that routes patients to triage or directly to treatment.

Finally, for models that are completely developed and in use, an animation is helpful to identify where problems can occur in system operations and what conditions and events lead up to problem behavior. For example, where do bottlenecks appear? What causes periods of excessive congestion? In a set of runs for an emergency department at a hospital, we might observe excessively long average waiting times for patients. An animation might show that when two critically injured patients arrive, as can happen in an automobile crash, all of the ED staff are occupied with these patients and other patients must wait excessively long times. This also suggests a solution: Keep backup personnel available for when this happens.

Animations can have various levels of detail. Normally, the level of detail corresponds to the level of detail in the model. But it is possible to provide an animation with less detail to follow model operations at a macrolevel. It is difficult to animate a model at a level of detail greater than the model itself, however. Animations consume a lot of computer resources because they must provide code to move entities and make other system changes visually. They must also add events to the event list to provide intermediate state changes to move entities smoothly, for example. These additional events add more computational burden to the entire model. Thus, animation should be used to demonstrate the model and to investigate specific behavior, but it should be removed or turned off for most runs.

8.5.1 Animation with Arena

Most graphical simulation packages provide some form of animation, and it is usually available by default—that is, it does not have to be turned on. Arena has two types of animation: *connector animation* and *facility-based animation*. We have already seen connector animation in action. Here entities flow from module to module along the connectors in the Arena model as the simulation runs, stopping in **Process** modules and elsewhere when they need to wait for resources. Connector animation runs automatically when a model is run in the normal mode, and it is turned off if the model is run in fast-forward mode. Connector animation does not provide a faithful representation of actual system timing and physical operations, however, because entities are unconstrained in their movement from module to module and the movement does not represent the actual time to go from one location to the next. Thus, connector animation is useful to check model logic, but it is not useful to represent physical system operations.

Facility-based animation uses a new graphical object called a *station* to represent the locations where entities experience events. Routes and other objects (such as conveyors) constrain the movement of entities from station to station, and the actual travel times are represented in the movements. Thus, facility-based animation is designed to be a more accurate representation of the physical movement of entities in a physical system. Although facility-based animation works best for physical systems, it can also be used to represent the logical or conceptual movement of entities that are not physical. For example it could be used to represent the movement of messages in a telecommunications system or electronic documents in a business process model.

8.5.2 The Fuel Depot Model and Facility-Based Animation

In this section, we will examine another model of the fuel depot system we just developed and include a facility-based animation. To demonstrate such animation more fully, we will change the model slightly by (1) adding several stations to the depot, (2) including travel times from station to station, and (3) representing the three pumps as three individual resources. Otherwise, the model is identical to the one presented in section 8.4.

We will also introduce two new Arena modules: **Station** and **Route,** which are found on the **Advanced Transfer** panel. If the **Advanced Transfer** panel is not displayed, right-click on the panels and select **Attach** to display a dialog with the available panels. Select **AdvancedTransfer.tpo** to attach the **Advanced Transfer** panel.

In the **Advanced Transfer** panel, you will see that each module is a rectangle with one of three colors: red, green, or yellow. The red modules define stations and move entities from station to station along defined paths. The green modules model the behavior of conveyors and are useful primarily in production and transportation models. The yellow modules are used to model transporters that operate indepen-

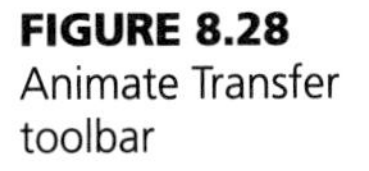

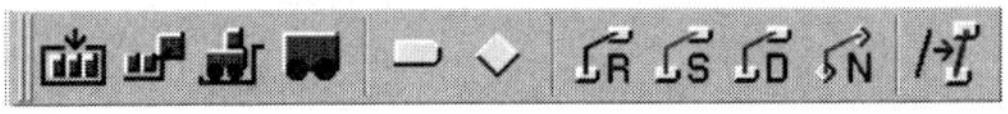

FIGURE 8.28 Animate Transfer toolbar

Route tool button

Station tool button

Submodel tool button

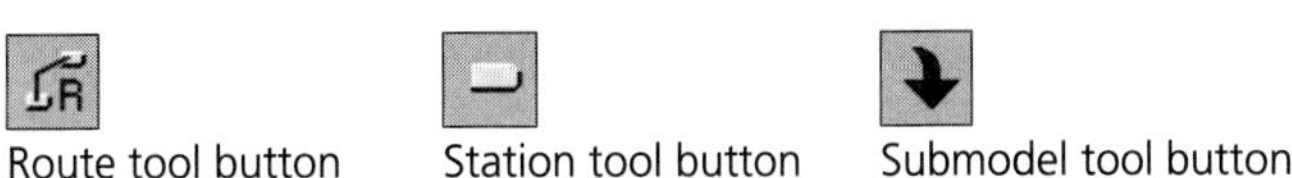

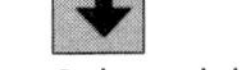

dently of one another to transport one or more entities from station to station. In this section, we will confine ourselves to the red modules for constrained movement between stations.

Before we actually build the model, it is helpful to establish a strategy for constructing the model. Since this model includes an animation, our strategy will be a little different than the strategy in section 8.4.3. The model will consist of both an Arena flowchart and a separate animation graphic. The animation graphic will represent the entrance to the depot, check-in station, each pump, checkout station, and exit. Each element is a station, and we will connect the stations with animation routes. We will first create a space for the facility-based animation where we will place the stations and routes. Then we will create the logic of the simulation model and link it to the animation by identifying the stations in the model logic.

Check to be sure the **Animate Transfer** toolbar is present. Figure 8.28 shows the **Animate Transfer** toolbar, which contains all of the tools needed to do facility-based animation. We are going to use only the **Station** and **Route** tools from this toolbar in the present model.

Normally, a background image is used to create interest. It would show the overall system layout along with graphics to represent the static elements of the system such as buildings or facilities. Our background image will simply be a rectangle that encloses the system; text will identify each station. We created this using the drawing and text tools in Arena. This is shown in Figure 8.29.

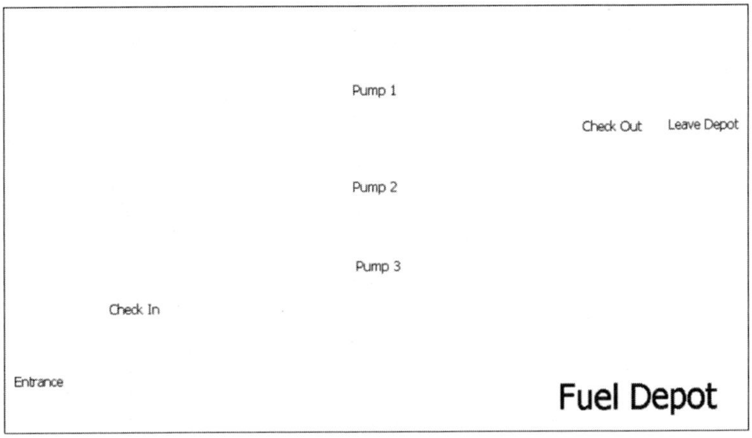

FIGURE 8.29 Background for fuel depot animation

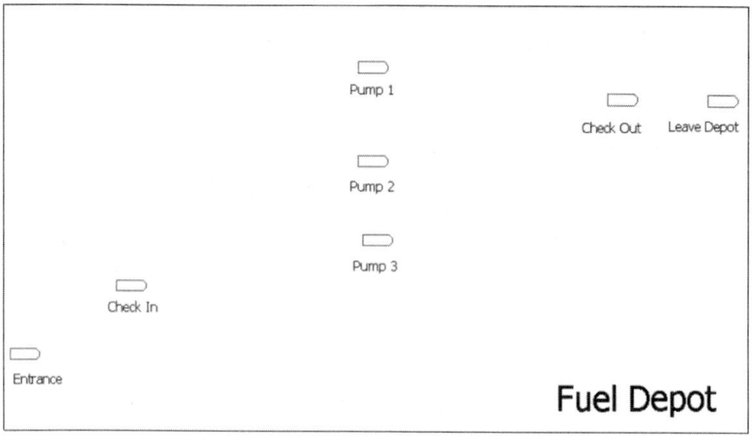

FIGURE 8.30
Stations for fuel depot animation

Begin by placing station elements from the **Animate Transfer** toolbar for each station as shown in Figure 8.30. Note that the term *station* denotes two closely related objects: (1) graphical elements on a facility-based animation and (2) logical elements in the model. We will call the former an "animation station" or "station location" and the latter a "**Station** module" when there is a possibility of confusion. When you place each animation station, a dialog like the one in Figure 8.31 will appear. Enter an identifier for each station and remember them. You will use them to identify **Station** modules when you build the simulation model.

After the stations have been placed, use the **Route** tool to create routes between the stations that represent the movement of trucks through the depot. Each route you place will connect one station to the next. Thus, you will have one route from

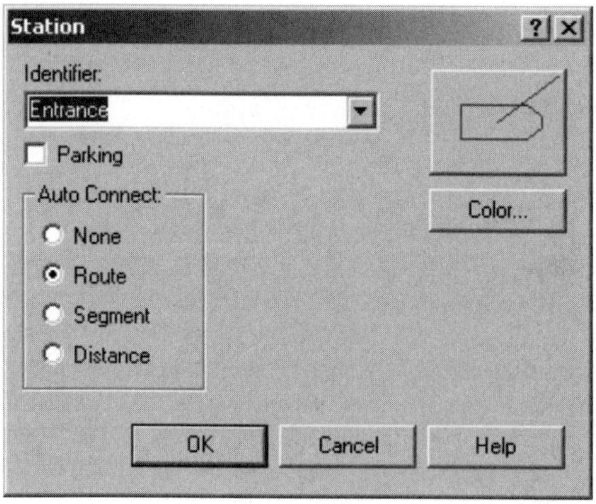

FIGURE 8.31
Dialog for animation station at Entrance location

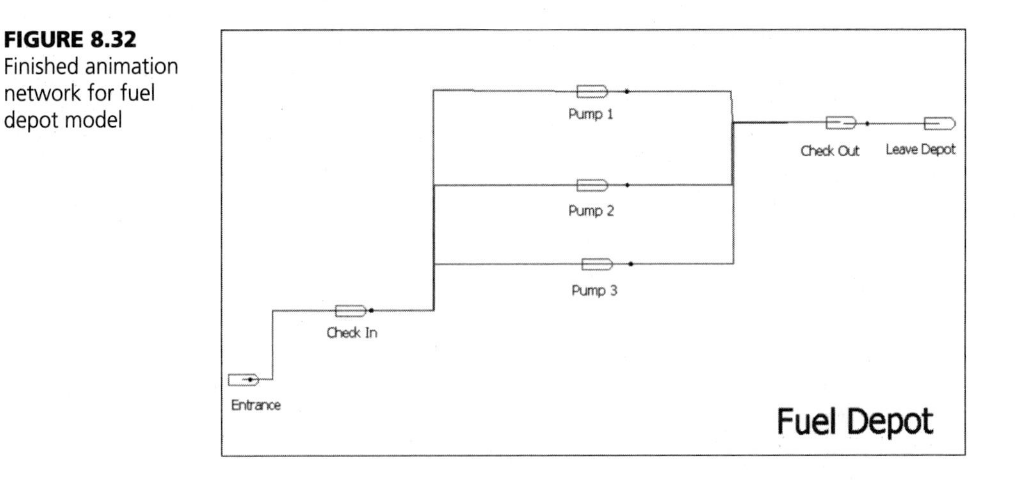

FIGURE 8.32 Finished animation network for fuel depot model

Entrance to **Check In,** one from **Check In** to **Pump 1,** and so forth. The finished animation network looks like Figure 8.32.

Now we are ready to create the model logic. Our strategy is to create a submodel for each station in the animation. Each submodel except the one for the **Entrance** station will start with a **Station** module. Each submodel except for the **Leave Depot** station will end with a **Route** module. The **Route** module moves the entity to a **Station** module while also moving the entity's graphic on the animation to a station location along a route and possibly with a delay time for the travel. The design is to provide the logic for each station location in a submodel, including the logic to move the entity to the next station or dispose of the entity.

First, find an area in the top-level model window for the submodels. We put a rectangle in this area to border the area where the submodels will be placed (see Figure 8.33).

To create each submodel, either select **Submodel** from the **Object** menu and choose **Add Submodel** or simply click on the Submodel tool in the **Standard** toolbar. This will produce a new submodel icon. To give the submodel a label and set properties, right-click on the submodel icon and select **Properties** from the **Context** menu. This will open the **Submodel Properties** dialog, as shown in Figure 8.34. Enter the label for the submodel; in this case, also enter "0" for the number of entry

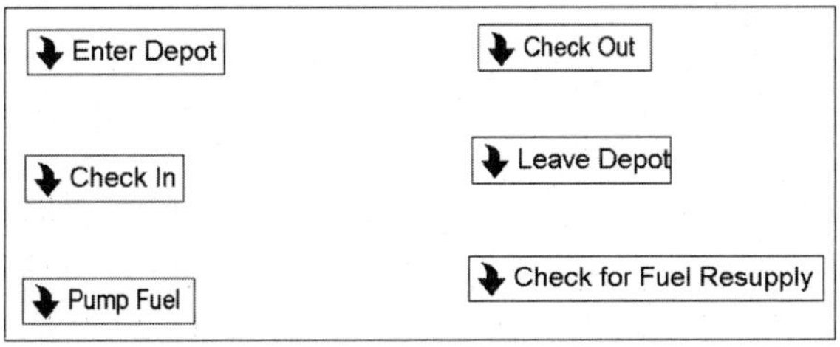

FIGURE 8.33 Fuel depot submodels

FIGURE 8.34
The Submodel Properties dialog for Enter Depot submodel

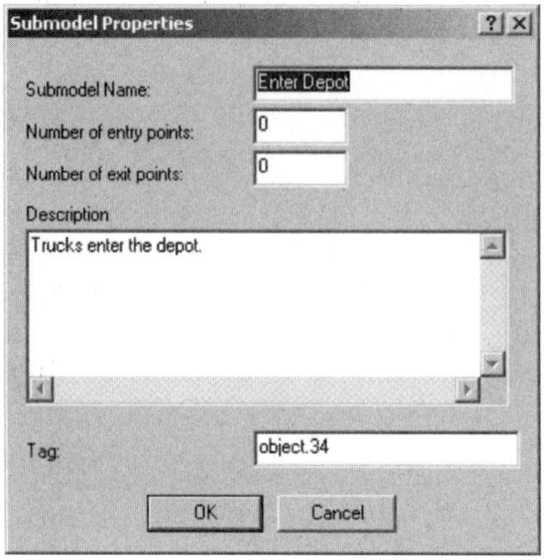

and exit points. Entry points allow entities to enter the submodel through connectors, and exit points allow entities to leave the submodel and go to other modules or submodels through connectors. In this case, we do not need entry and exit points because we will be transferring entities to other submodels using **Route** and **Station** modules. You can also provide a description for the submodel. This is a good place to put comments. Click **OK** to close the dialog.

Now you can enter the actual modules in the submodel by either double-clicking on the submodel icon or by right-clicking on it and selecting **Edit Submodel.** You will be presented with a blank model window where you can place modules as you would in any model. Figure 8.35 shows the logic for the **Enter Depot** submodel. This submodel creates each truck as it arrives and defines the station entrance in the **Station** module **Enter Fuel Depot.** It selects the type of fuel (**Fuel 1** or **Fuel 2**) and the amount, and sets the entity picture to **Picture.Truck.** The submodel ends with a **Route** module that moves the entity along a defined route to the next station.

The dialog for the **Enter Fuel Depot** station is shown in Figure 8.36. In addition to the module name (every module has one), this module has a **Station Type,** which can be either **Station** to indicate an individual station or **Set** to indicate a station in a set. We selected **Station** for the station type and **Entrance** for the **Station Name.** The **Station Name** *must* match the name given to the station location in the graphical animation. Recall that **Entrance** is the name of the first station where trucks arrive at the depot.

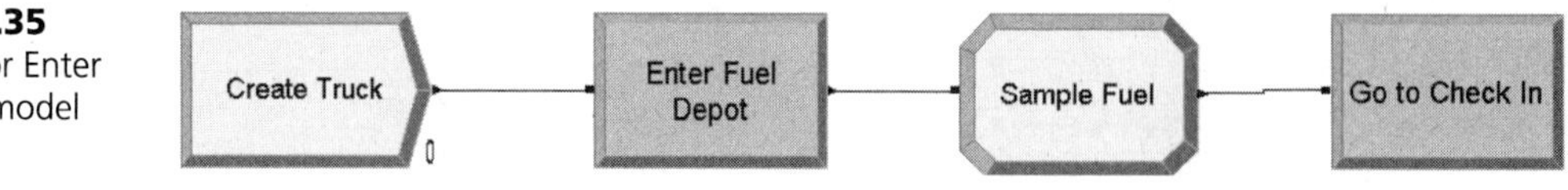

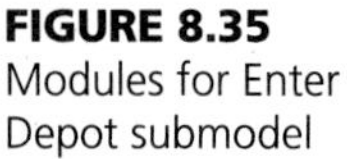
FIGURE 8.35
Modules for Enter Depot submodel

FIGURE 8.36
Dialog for Entrance station module

Figure 8.37 shows the dialog for the **Route** module in the **Enter Depot** submodel. The **Route** module has places to enter the transit time used by the entity to move to the next station. This time can be a constant or be sampled from a distribution. The dialog also provides two methods to denote the next station under **Destination Type:** We can specify **Station** for the destination type and enter a specific station or we can specify **Sequence** and the entity will be routed to the next station in its sequence. We will discuss station sequences a little later. In this case, we just want to route the entering truck to the check-in station, so we specify **Station** as the destination type and **Check In** as the station name.

Note once again how the station locations are connected to the **Station** and **Route** modules. Each **Station** module defines the name of a station and locates that station in the model logic. Each **Route** module defines when an entity moves to a station from the current station and how much time this movement takes.

FIGURE 8.37
Dialog for Route module in Enter Depot submodel

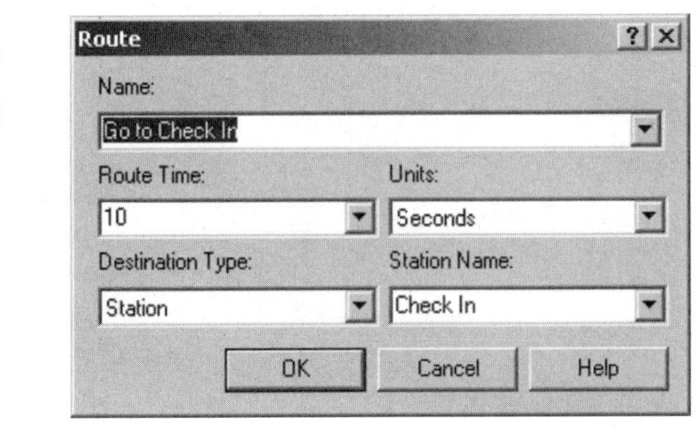

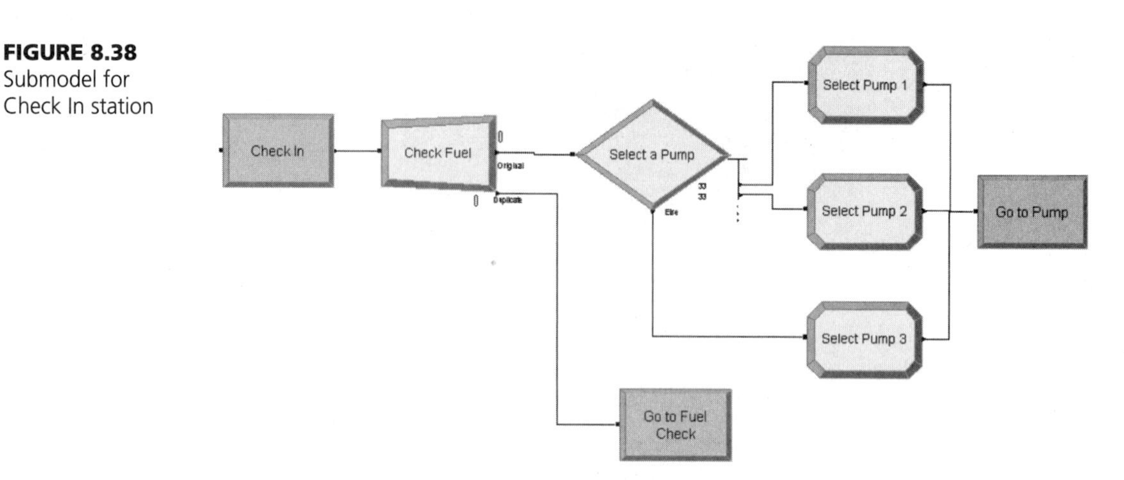

FIGURE 8.38
Submodel for Check In station

Now that the logic for the Entrance station has been created, let's look at the logic for the **Check In** station. The submodel for this station is shown in Figure 8.38. The first module in this submodel is the **Station** module for station **Check In.** Because we used a **Route** module in the **Enter Depot** submodel to route entities to the **Check In** station module, this guarantees that all entities will arrive at the **Check In** station module. Next, a **Separate** module duplicates the entity so we can perform two processes in parallel: (1) check the fuel level and, if necessary, order a resupply, and (2) proceed to select a pump and attempt to pump the fuel. The process of checking the fuel and placing an order for a resupply is the same as before except that the model logic is placed in a submodel and the duplicate entity is routed to the first station in that submodel. Since the duplicate entity is not a physical entity but instead is simply created to allow two processes to proceed in parallel, we do not include it in the animation.

We have changed the model a little so that entering trucks randomly select a pump, giving each pump an equal probability of selection. This is done in the **Decide** module **Select a Pump.** Figure 8.39 shows the dialog for this module. There are three pumps, so each of the first two has a probability of .33 of selection, and the default is to select the last pump, which has a probability of .34.

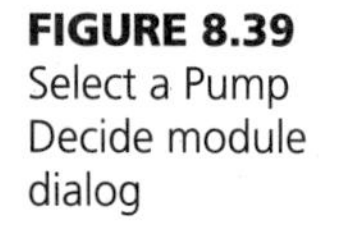

FIGURE 8.39
Select a Pump Decide module dialog

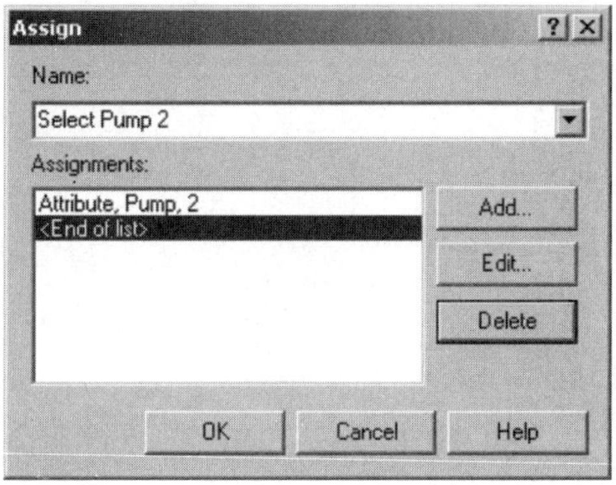

FIGURE 8.40
Assign module dialog for Pump 2 in fuel depot model

Trucks move to one of three **Assign** modules after selecting a pump. We will examine the **Assign** module **Select Pump 2.** The other **Assign** modules are similar. The dialog for this module is shown in Figure 8.40. In this module, we assign the value 2 to the **Pump** attribute to indicate that this truck will use pump 2. The next assignment sets the station sequence for this truck and requires a little explanation.

8.5.3 Station Sequences

We can create a sequence of stations, name the sequence, and assign it to an entity's **Entity.Sequence** attribute. Then in a **Route** module we can specify that the route should follow the entity's sequence. Arena has a built-in marker for each entity to track which station in the sequence was last visited and forward the entity to the next station in the sequence when it enters a **Route** module that specifies **Sequence** as the destination type. So we must first establish some sequences. Figure 8.32 shows that all entities follow one of three sequences of stations:

Entrance, Check In, Pump 1, Check Out, Leave Depot
Entrance, Check In, Pump 2, Check Out, Leave Depot
Entrance, Check In, Pump 3, Check Out, Leave Depot

These sequences differ only in the pump used. The truck has already visited the Entrance station and the Check In station, so we need to establish three sequences:

Pump 1, Check Out, Leave Depot
Pump 2, Check Out, Leave Depot
Pump 3, Check Out, Leave Depot

We will name these sequences **Pump 1 Trucks, Pump 2 Trucks,** and **Pump 3 Trucks.** Do this by clicking on **Sequence** in the **Advanced Transfer** panel to show the sequence spreadsheet view. Figure 8.41 shows a portion of the Arena window

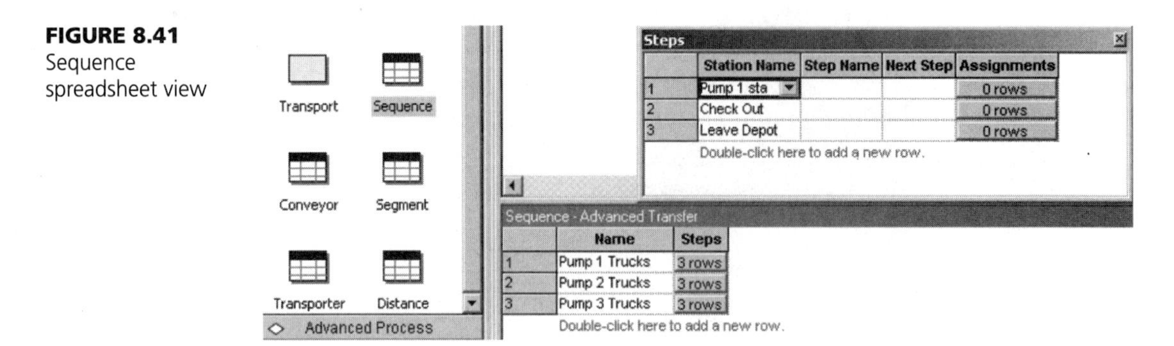

FIGURE 8.41 Sequence spreadsheet view

with the sequence icon and the sequence spreadsheet view. After assigning a name to each sequence, you can assign the station locations by clicking the button under the **Steps** column and entering the stations in the pop-up dialog. Figure 8.41 shows the stations in the sequence **Pump 1 Trucks.**

After the sequences are set up, we can assign them to entities in an **Assign** module. Figures 8.42 and 8.43 show the dialogs that assign the sequence **Pump 2 Trucks** to the **Entity.Sequence** attribute.

FIGURE 8.42 Assignment of sequence to trucks

FIGURE 8.43 Pump 2 Assign module dialog

FIGURE 8.44 Route module dialog in Check In submodel

Now we are ready to use the sequences we just defined to route the truck to the stations defined in the sequence. Figure 8.44 shows the dialog for the **Route** module **Go to Pump.** Here we simply need to tell Arena that the destination type is sequential. Arena then uses the information in the assigned sequence to determine where the entity will go next. This approach is especially useful if entities follow certain prescribed paths through the system, depending on an attribute of the entity. In this case, the pump number determines the path through the system.

The **Pump Fuel** submodel is shown in Figure 8.45. In this submodel, three stations represent the three pumps. The logic of this submodel is quite similar to that of the earlier version except that the pumps are organized into a set and the truck requests one unit of a specific member of the resource set **Pumps.** Note that trucks waiting for a pump or fuel (or both) will wait at the station they currently occupy, but the model logic provides that they are in a single queue and are serviced on a first-come, first-served basis as pumps and fuels become available. A more refined version of this model would want to modify this to serve trucks in a more efficient order.

FIGURE 8.45 Pump Fuel submodel

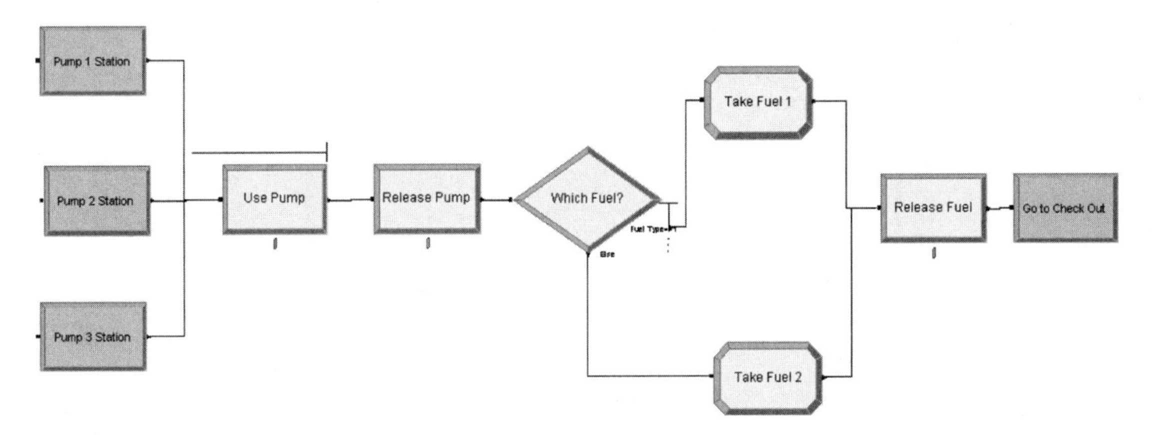

8.5.4 Named Views and Navigation

The previous section shows that a complex model can have many parts, and navigating these parts can become difficult and confusing. Often we want to go directly to a submodel or even a portion of a submodel. Arena provides two handy tools to let us do this: the **Navigate** panel and **Named Views.** The **Navigate** panel lets us jump directly to a view of the model. The concept of a named view is that we can name a particular view of a portion of the model and return instantly to that view by referencing its name.

First, let's look at the **Navigate** panel. Click on its tab in the **Panels** window to open it. Figure 8.46 shows the **Navigate** panel for the model in Section 8.5. Arena automatically creates a view for each submodel as well as the top-level model. You can click on any submodel or the top-level model to instantly show that view. In Figure 8.46, the **Enter Depot** submodel would be showing in the model window.

To create a named view, first display the portion of the model you want to show. Then select **Named Views** from the **View** menu to bring up the **Named Views** dialog in Figure 8.46. Using this dialog, you can manage named views by creating, editing, and deleting them. We will create a named view. First, position the animation we created in section 8.5 in the middle of the model window and enlarge it to approximately fill the window. Then select **Named Views** from the **Named Views** menu to display the **Named Views** dialog. Click **Named Views** to add a named view and enter a name such as "`Depot Animation.`" Click **OK** to add this view. It will be placed on the navigate menu; you can access it by simply clicking on it. Figure 8.47 shows the **Navigate** panel after adding the **Depot Animation** named view.

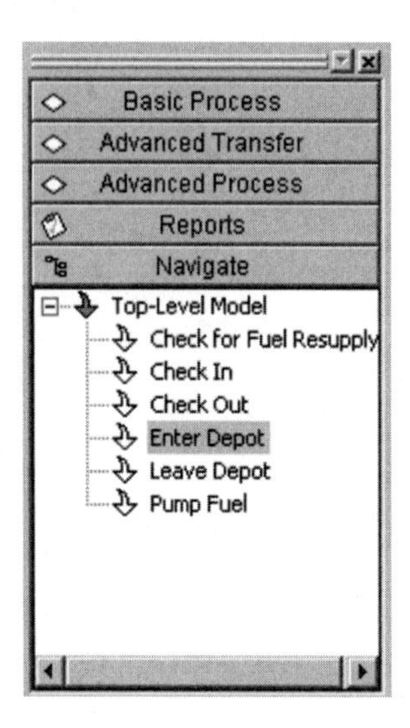

FIGURE 8.46
Navigate panel in Fuel Depot model

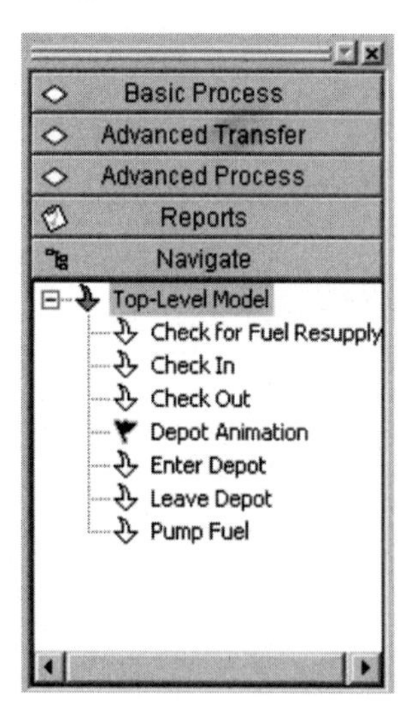

FIGURE 8.47
Navigate panel with Named View added

8.6 SUMMARY

This chapter has presented techniques for graphically representing the structure and dynamics of discrete-event systems and demonstrated those techniques using Arena. Graphical modeling allows the simulation model to be developed by persons who are not simulation specialists, and it facilitates communication about model operations between and among simulation specialists and persons involved with system operations. Other benefits of graphical modeling include the following:

- Conceptually close objects can be represented as physically adjacent, making their relationship apparent.
- Interactions between objects are shown in two dimensions rather than the one-dimensional sequential ordering as in procedural representations. This uses the enormous processing power of the human visual system to make model comprehension quicker and easier.
- The number of symbols in generic graphical techniques is small, and rules for graphical modeling are typically simple.
- Graphical modeling techniques support modular and hierarchical modeling. This allows complex models to be developed in stable and easily modifiable forms.
- Graphical modeling techniques allow the system to be simulated manually. Manual simulation is a valuable tool for learning how the simulation operates through time.

Graphical modeling techniques can be divided into two categories: generic graphs and process graphs. Examples of process graphs include GPSS block diagrams, SLAM networks, and Arena flowcharts. In this chapter, we presented three examples of generic graphs: event graphs, activity cycle diagrams, and Petri nets. All three techniques use a small number of symbols but have powerful modeling capabilities.

Event graphs, a graphical modeling technique that implements the event view of discrete-event simulation, were introduced in Chapter 5. Activity cycle diagrams provide a simple but powerful graphical means to implement the activity view of discrete-event simulation. These diagrams employ only two symbols: activities and queues. In an activity cycle diagram, entities cycle between two states, active and passive, in a closed entity life cycle. Petri nets constitute a graphical technique that corresponds closely to the process world view. In their simplest form, Petri nets consist of two types of nodes: places and transitions, and directed arcs. Tokens are put into places to indicate the state of the system. State changes occur through a process called *firing,* in which tokens are removed from the input places for a given transition and placed into the output places for the transition. Timed transitions are an extension of Petri nets that were added to capture the discrete-event nature of system dynamics.

Graphical modeling with Arena was demonstrated by introducing two flowchart modules that were not used in previous chapter: the **Separate** module, which creates multiple identical entities; and the **Batch** module, which collects a specified number of entities or entities with specific values of attributes. These modules are useful not only in manufacturing and business process models where entities are batched and separated during processing but also where we want to cause multiple processes to proceed simultaneously. We demonstrated these and other Arena modules with a model of a copy center. In this model, the copying process was performed simultaneously with the customer's payment process.

Hierarchical modeling provides a set of techniques for organizing complex models. Because most discrete-event simulations are complex, hierarchical modeling or a similar methodology is necessary to organize the model components into units in order to keep error to reasonable levels and promote model reliability. Hierarchical modeling starts with a simplified top-level model consisting of one or more submodels. Then each submodel is decomposed into atomic elements and compound elements. Compound elements represent other submodels. This process continues until all submodels consist only of atomic elements. The benefits of hierarchical modeling include the ability to limit modifications of the model to only one or a few submodels, thus reducing the possibility of introducing errors and increasing the reliability of the model.

The hierarchical modeling abilities of Arena were presented using a fuel depot model. In this model, the process of pumping fuel into a tanker truck was represented by a submodel. This submodel was further divided into submodels to pump the fuel and to resupply the fuel.

Graphical simulation tools such as Arena provide animation by default. Animation is extremely useful as a verification and validation tool, and it is an attractive means for demonstrating the model to users because they tend to accept simulation much faster when they see the animation that resembles the real system they are working with. An animation can also be a useful tool for examining detailed system operations and identifying problem behavior.

Arena has two animation methods: connector animation and facility-based animation. Connector animation moves entities along the connectors that link modules in the Arena model. Facility-based animation depicts the physical movement of entities in a visual model of the system. We used a modified version of the fuel depot model to demonstrate facility-based animation. This model also introduced two new Arena modules—the **Station** and **Route** modules. Facility-based animation uses these modules along with animation stations and routes to define the locations to which entities move and the times required to move them.

When complex models are developed and animated, the modeler often needs to examine specific parts of the model or animation. Arena provides a convenient way to move quickly between these views using **Named Views** and the **Navigate** panel. These work together by letting the modeler give specific views names and then display the names in the **Navigate** panel so the user can display each of them with a single click. We demonstrated **Named Views** and the **Navigate** panel using the fuel depot model.

References

Ceric, V. 1995. Petri nets for discrete event simulation: Would the standard extension be beneficial? *Proceedings of the EUROSIM 95 Congress,* Wien: 1169–1174.

Ceric, V., and R. J. Paul. 1992. Diagrammatic representation of the conceptual model for discrete event systems. *Mathematics and Computing in Simulation* 34:317–324.

David, R., and H. Alla. 1994. Petri nets for modeling of dynamic systems—A survey. *Automatica* 30:175–202.

Evans, J. B. 1988. *Structures of discrete event simulation: An introduction to the engagement strategy.* Chichester, England: Ellis Horwood.

Gordon, R. F., E. A. MacNair, K. J. Gordon, and J. F. Kurose. 1990. Hierarchical modeling in a graphical simulation system. *Proceedings of the 1990 Winter Simulation Conference,* New Orleans: 499–503.

Jensen, K. 1991. Colored Petri nets: A high level language for system design and analysis. In *Advances in Petri nets 1990,* ed. G. Rozenberg, 342–416. Berlin: Springer-Verlag.

Luna, J. J. 1993. Hierarchical relations in simulation models. *Proceedings of the 1993 Winter Simulation Conference,* Los Angeles, California: 132–137.

Murata, T. 1989. Petri nets: Properties, analysis and applications. *Proceedings of the IEEE* 77:541–580.

Pooley, R. J. 1991a. Towards a standard for hierarchical process oriented discrete event simulation diagrams. Part I: A comparison of existing approaches. *Transactions of the Society for Computer Simulation* 8:1–20.

———. 1991b. Towards a standard for hierarchical process oriented discrete event simulation diagrams. Part III: Aggregation and hierarchical modelling. *Transactions of the Society for Computer Simulation* 8:33–41.

Reisig, W. 1985. *Petri nets: An introduction.* Berlin: Springer-Verlag.

———. 1992. *A primer in Petri net design.* Berlin: Springer-Verlag.

Simon, H. A. 1981. *The sciences of the artificial.* 2d ed. Cambridge: MIT Press.

Tocher, K. D. 1964. Some techniques of model building. *Proceedings of the IBM Scientific Computer Symposium on Simulation Models and Gaming,* New York: 119–155.

Törn, A. A. 1981. Simulation graphs: A general tool for modeling simulation designs. *Simulation* 37:187–194.

———. 1985. Simulation nets, a simulation modeling and validation tool. *Simulation* 45:71–75.

Problems

For the following problems, you may add any details such as distributions for service times needed to build the models.

8.1 Develop an informal event graph model for a queueing system with two servers having different service times and a common queue of customers.

8.2 Develop a formal event graph model for the system described in the previous problem.

8.3 Develop a formal event graph model for a queueing system with two servers having different service times and a separate queue for each server. Arriving customers select the server with the shorter queue.

8.4 Develop a formal event graph model for the system described in the previous problem using a one-dimensional array for the system variable(s).

8.5 Develop a formal event graph model of a production system with one machine having three processing phases. In each phase, one part can be processed at a time. The first and third phase require a special tool, and there is only one such tool in the system. The machine has in-process buffers before the second and third processing phases. Arriving parts are stored in an input buffer.

8.6 Develop a formal event graph model of a production system with one machine and periodic stochastic interrupts. Use the event canceling mechanism for this purpose.

8.7 Extend the model described in the previous problem to include calculation of the total interrupt time for the machine.

8.8 Develop an event graph for the operation of a parking lot with a limited number of parking places. Cars that arrive at the lot when it is full leave the system immediately.

8.9 Modify the model from the previous problem to include waiting for parking. Cars that arrive at the lot are willing to wait if the queue contains fewer than five vehicles.

8.10 Modify the model from the previous problem by introducing two types of parking places in the parking lot: one for small cars and another for big cars. Eighty percent of arriving vehicles are small; the rest are big.

8.11 Develop an event graph of a one-way street with a signal for a pedestrian crossing. The signal has a red and green light cycle with a 30-second duration for red and green lights.

8.12 Modify the model from the previous problem by including two-directional traffic on the street.

8.13 Develop an event graph of a port with three types of ships bringing freight to the port. The port has two tugs for ships berthing and deberthing, and it has three docks for unloading freight.

8.14 Modify the model from the previous problem to include the tide cycle. When the tide is low neither berthing nor deberthing is allowed to start.

8.15 Develop an event graph of a bar serving two types of drinks, each from a different type of glass. There are three servers, and glasses must be washed after each use. The drinking time has a different distribution for each type of drink. Use any distribution you feel is appropriate.

8.16 Develop an Arena model of the service counter at a post office. The lobby has a single queue that is served by three clerks. Two clerks handle routine duties; the third clerk handles special mailings such as certified mail. Develop both a logical model and facility-based animation.

8.17 Develop an Arena model for the operation of a parking lot with a limited number of parking places. Cars that arrive at the lot when it is full leave the system immediately. Include a facility-based animation for this model.

8.18 Modify the model from the previous problem to include waiting for parking. Cars that arrive at the lot are willing to wait if the queue contains fewer than five vehicles.

8.19 Modify the model from the previous problem by introducing two types of parking places in the parking lot: one for small cars and another for big cars. Eighty percent of arriving vehicles are small; the rest are big.

8.20 Develop an Arena model of a one-way street with a signal for a pedestrian crossing. The signal has a red and green light cycle with a 30-second duration for red and green lights. Include a facility-based animation for this model.

8.21 Modify the model from the previous problem by including two-directional traffic on the street.

8.22 Develop an Arena model of a port with three types of ships bringing freight to the port. The port has two tugs for ships berthing and deberthing, and three docks for unloading freight. Treat tugs as resources in this model.

8.23 Modify the model from the previous problem to include the tide cycle. When the tide is low neither berthing nor deberthing is allowed to start.

8.24 Develop an Arena model of the graduate admissions process at a university. In this process, applicants request application material according to a Poisson process and receive the material after a random delay. The application consists of the following items: an application form, transcripts from the student's undergraduate institution, three letters of recommendation, and test scores from the Graduate Record Exam. Each item requires a random delay before receipt at the graduate school. The application is complete when all items are received. Applications are batched and reviewed by a committee before letters of acceptance and denial are sent.

8.25 Create a facility-based animation of the copy center model in section 8.3.

8.26 Create a model and facility-based animation of a walk-in medical clinic. The clinic has a receptionist, two doctors, and two nurses. A patient registers with the receptionist and then is seen by the nurse, who takes him to an examination room. He is then seen by the doctor. After being seen by the doctor, the patient leaves (releases) the examining room and again checks with the receptionist before leaving. For this model, create at least four named views.

9

Problem Solving Using Simulation

This chapter focuses on service systems that involve waiting-line models. This is perhaps the single most important area of application for simulation. We want to give the reader an overall understanding of how waiting-line models are represented and what measures of performance are usually used. The discussion and examples proceed from simple queueing models to more complex models involving conveyors and batch processing. As in the other chapters, the emphasis is on model representation; however, examples will include an analysis of output data and decision-making implications.

9.1 INTRODUCTION

In Chapters 4 and 5, we learned how to set up a model of a simple waiting-line system consisting of servers, customers, and queues of customers having infinite capacity. In Chapter 8, we developed more complex models using Arena. These models can be used to represent some real-world systems; however, most service systems are much more complex and have characteristics that we have not yet discussed. In this chapter, we will study the characteristics of many realistic service systems and how to model them.

The world is full of service systems. By *service system,* we mean a system that provides some sort of service to a finite or infinite stream of "customers." These are queueing systems in which customers compete for a limited amount of a scarce resource: server time. We have already seen examples of a service system in the checkout counter at the grocery store in Chapter 1 and the manufacturing system in Chapter 5. In the grocery store, shoppers are the customers, cashiers are the servers, and service consists of totaling the bill and collecting the money for the items purchased. In the manufacturing system example, servers can be people or machines, and service involves manufacturing and assembly operations on parts at each manufacturing station. Consider the following familiar examples of service systems:

- a registration system at a college or university,
- a cafeteria,
- a fast-food restaurant,
- a loading dock at a truck terminal,
- an auto tune-up or oil change facility,
- a court system,
- a medical clinic,
- a hospital emergency room,
- an operating room suite at a hospital,
- the intensive care unit in a hospital,
- a system for transporting oil from oil fields to refineries,
- a telephone network,
- a time-shared computer system,
- a rapid transit system,
- an airline reservation system,
- an airport runway,
- an airport ticket counter,
- the claims processing department of an insurance company,
- a ride at a theme park such as Six Flags or the Magic Kingdom,
- a golf course,
- a movie theater, and
- the procurement process at a business.

All of these systems consist of some sort of customer who requests service from some type of server and may wait in a queue until the server is available. In most of these systems, the customer is a person; however, in a telephone network, the customer is a call, fax transmission, or other network traffic such as a data packet. On an airport runway, the customer is an airplane waiting to take off or land. In many of these systems, the server is also a person; however, one can frequently consider the server to be some other component of the system. For example, in a telephone network, the servers are the switches that make the connection from the source to the destination for a call. At a ride in a theme park, the server is the ride, which may serve many customers simultaneously. In an intensive care unit, the server can be thought of as the bed or room that must accommodate the patient. Finally, the type of service and its characteristics (duration, number of customers served simultaneously, etc.) depend on the system. In a time-shared computer system or telephone network, the service time is measured in milliseconds; in an intensive care unit at a hospital, the service time is measured in days or weeks.

There are many other service systems that we may not immediately recognize as such. Manufacturing systems, for example, are service systems. "Customers" consist of the raw materials that are input into the production process. Service consists of processing the raw materials into finished products. Of course, even the smallest manufacturing systems are quite complex compared to the systems that we have discussed so far in this text. In this chapter, we will have an opportunity to look at manufacturing systems in more detail. In transportation systems such as mass-transit bus and subway systems, the "customers" are the people who travel on the system, and service is provided by moving customers from their departure point to their destination. Similarly, package and mail delivery systems (including electronic mail) are also service systems, as are most inventory systems. All of these systems have many common characteristics: a stream of customers, a service mechanism, and one or more queues where customers may wait for service.

9.2 SERVICE SYSTEMS

9.2.1 Characteristics of Service Systems

To effectively develop models of service systems, we must recognize their common characteristics. In this section, we will discuss these characteristics in more depth to give the reader an understanding of the variety of models used to represent service systems. A *customer* is an entity that requests service; if service cannot be provided immediately, then the customer may wait in a queue. In some systems and under certain circumstances, a customer denied service may leave the system either for a finite period of time or permanently. In many systems, customers are not identical. Instead, their attributes distinguish them from one another and determine how they interact with the system. For example, a customer may have an attribute, *priority,* that is a numerical value, and system rules provide that service is provided first to customers with the smallest value of this attribute.

A *server* is an entity (person or thing) that provides service. Implicit in this definition is the concept that service is provided to an individual customer or group of customers and, while this service is being provided, other customers requesting service may be denied and thus have to wait or leave the system. A service system may have more than one server, and servers may have attributes that determine which customers they can serve and the characteristics of the service such as the length of service time. For example, an intensive care unit at a hospital may have six beds for surgery patients and two beds for medical patients (nonsurgery). If a surgery patient arrives, he will be served by the surgery bed if it is available; if it is not available, he will be served at a lower level of service by one of the medical beds. In addition, the length of stay (service time) for surgery is typically shorter than that for medical patients. In many models, especially those involving multiple identical servers, it is convenient to model servers as a resource that is used by customers.

A *queue* is a collection of customers waiting for service, or otherwise waiting to proceed through the system. Usually a queue is an ordered collection, but it does not have to be. Customers can be served in random order. Often a queue is a physical entity such as a line of persons waiting to enter a movie theater, or it may be a conceptual entity such as patients waiting for an organ transplant or messages waiting in a communications system to be transmitted. A queueing system may have multiple queues, as in the manufacturing model of Chapter 5. For example, a fast-food restaurant may have three queues for customers eating in the restaurant and another queue for drive-through customers.

Other characteristics of queueing systems include the arrival process, customer behavior on arrival, customer behavior after joining the queue, and the service process. We will examine each element individually.

The *arrival process* is determined by the sequence of interarrival times, or the times between arrival events. In the models considered up to this point, interarrival times have been statistically independent, identically distributed (iid) random variables. Such a process is called a *renewal process.* If the distribution of interarrival times is also exponential, then the process is a Poisson process. Otherwise, interarrival times may have any other appropriate distribution such as a gamma, normal, or uniform distribution. They can also have a discrete distribution if arrivals may only come at discrete points in time. A renewal arrival process is the simplest type of arrival process. Interarrival times may also be *autocorrelated.* It is beyond the scope of this chapter to discuss autocorrelated processes, but in this case each interarrival time is dependent on the previous interarrival times. Autocorrelated interarrival times often provide a good model for arrivals that may occur in groups such as a catastrophe process. Arrivals may also be governed by much more complex processes. For example, consider a dentist's office, where most patients are given appointments but some patients may walk in with acute problems. Most arrivals occur at discrete scheduled times (which may be rescheduled according to the logic of the model), but superimposed on these are arrivals that occur according to a Poisson process.

Upon arriving, customers may request service and, if denied, join a queue and wait for service. This is the normal mode of operation. Alternatively, customers may *balk*—that is, refuse to join the queue. They may balk if refused service, if the queue is too long, or for other reasons. If refused service, customers may also try again later after a fixed or random length of time. A *retrial* is simply another arrival by the same customer at a later time. All of these behaviors have been observed in real service systems.

After joining the queue, customers may wait until service can be provided. In some systems, however, customer behavior may involve *reneging*—that is, leaving the queue before service begins. Reneging is normally a characteristic of impatient customers. After a customer reneges, he may leave the system permanently or schedule a retrial after a fixed or random length of time. If more than one queue is present, customers can also *jockey* to another queue. In the supermarket example in section 4.1, customers jockeyed to a shorter queue as soon as another queue had at least two fewer customers. If a customer engages in jockeying, he usually jockeys to a shorter queue, because customers' objectives usually involve completing service and leaving the system as soon as possible.

Service can be provided to customers individually or in groups. If customers are served in groups, the group size can be random or fixed. From the server's perspective, the service process consists of a sequence of service times: one for each customer or group of customers. Service times may be constant (e.g., the time to switch a packet in a communications network) or random (e.g., the time to unload a truck at a terminal). In the simple queueing systems we have seen so far, the service time process is a *renewal process*—that is, the sequence of service times consists of iid random times. However, just as the arrival process can be autocorrelated, so can the service process. This can be the case when the server is subject to deterioration or when services involve a switchover between customers such that services for similar customers can be provided faster than services for dissimilar customers. In addition, service times can (and usually do) depend on customer attributes. For example, the time to unload a truck depends on the truck's weight or the number of parcels in the truck, or the time to treat a patient in the emergency room depends on the severity of the patient's illness or injuries. Interestingly, more seriously injured patients often require shorter treatment times in the emergency department. Can you explain why?

The order in which customers are served is specified by the queue discipline. Queue discipline can be FCFS (first come, first served), LCFS (last come, first served), SPT (shortest processing time), LPT (longest processing time), priority, random, and possibly others such as random within priority groups. These disciplines define the order in which servers select customers from the queues for service. Normally, customers are maintained in the queue in the order in which they will be served; however, the model can also specify that the server select the next customer from a position other than the front of the queue, searching the queue, for example, for a customer who satisfies a set of criteria based on the customer attributes.

If priority order is specified for the model, then we must also provide instructions about the server's actions if a higher priority customer arrives while a lower priority customer is being served. Service can be *nonpreemptive* or *preemptive.* In nonpreemptive service, a customer currently being served will finish before service is started for a higher priority customer who arrived during the service time. For example, in an airport, military flights are given priority over civilian flights. However, if a civilian flight is in the process of landing, it is allowed to finish landing before an arriving military flight begins its landing approach. Preemptive service provides that, when a higher priority customer arrives during the service of a lower priority customer, the server cease service for the lower priority customer and begin service for the higher priority customer. For example, suppose an emergency room physician is dressing a minor wound for a patient when a critically injured patient arrives. She would stop attending to the patient with the minor injury and immediately attend to the incoming patient, who has higher priority because his injuries are critical. If service for a customer is preempted, we must also specify the disposition of that customer's service when it resumes. The customer can be placed back in the queue to resume service when all higher priority customers have been served. This mode of operation is called *preemptive resume.* Alternatively, the customer can be released from the system without completing service, a mode of operation called *preemptive nonresume.* Under preemptive resume operation, the customer can resume service where it was interrupted or she can resume service from the beginning. In the emergency room example, if only one physician is working, then the patient with a minor injury can be asked to wait until the physician can return to finish dressing the wound (preemptive resume) or a nurse can finish dressing the wound (preemptive nonresume).

9.2.2 Resources

Recall that a *resource* is a permanent entity that represents a constraint on the ability to provide service. In this regard, a server is a resource. If a service system had an infinite number of servers, then each customer could be served immediately on arrival. For example, the collection of seats in a theater is also a resource. In queueing systems, customers arrive and request a certain amount of specific resources. If the amount requested is not available immediately, the customer waits in a queue until it is available. Resources can exist in discrete or continuous units. Servers and theater seats are in discrete units. In a model of a manufacturing plant, oil is a resource in continuous units. Frequently, a resource that is available in discrete units is called a *facility.*

A resource can be considered consumable or nonconsumable. A *consumable* resource is consumed or depleted in the process of delivering service. For example, in a model of an electric-transformer manufacturing plant, paint and oil are consumable resources because they are used in the manufacturing process. A *nonconsumable* resource is used and returned by entitities in the process of receiving service. In the same model, a crane and electric power are nonconsumable resources. Electric power is considered nonconsumable because it is being delivered at a fixed rate per unit time to the plant during the course of operation, and it can be allocated to a selection of operations but cannot be stored for later use. The main difference between a crane and electric power is that the crane is discrete and electric power is in continuous units.

Consider a multiple server queue with three servers. We can think of the servers as a (nonconsumable) resource. Upon arrival, the customer requests one unit of the server resource. If a unit is available, meaning at least one of the servers is free, then a begin service event is scheduled immediately; otherwise, the customer is placed in the queue. Similarly, in the begin service event, the customer "takes" a unit of the server resource. When the end of service event executes, the customer relinquishes this unit of the server resource. Thus, we think of this system's operation as a process of customers arriving, requesting certain resources, possibly waiting until the resources are available, using the resources for a period of time, and finally relinquishing the resources.

9.2.3 Multiple Servers and Multiple Queues

In this section we want to extend the discussion of models in earlier chapters to include queueing systems that are somewhat more complex and more realistic than the simple systems presented there. Many service systems in the real world are composed of components that consist of multiple servers, or multiple queues, or both. We will examine models of these rather simple systems as building blocks for more complex systems. It is important to know how to build realistic models of these simple systems so that larger models that have them as components will function correctly.

Consider a queueing system with three servers and a single queue. Such a system might exist, for example, in a bank lobby with three tellers and a single waiting line for customers. For now, assume that all servers are identical and have the same service time distribution. To be specific, we will assume that the service times are uniformly distributed between .5 and 2.5 minutes, and interarrival times are iid exponential random variables with mean 1.0 minute.

As we saw in Chapter 7, this type of model is easy to build using Arena. It simply consists of three modules as shown in Figure 9.1: a **Create** module to create customer entities, a **Process** module to provide the service portion, and a **Dispose** module to dispose of the customer. To provide three servers, we simply define the resource **Server** that is seized in the **Service** module and set its capacity to 3.

It is simple to add such features as bulk arrivals, scheduled arrivals, multiple or random numbers of servers, priority service, different queue disciplines, and multiple resources by just changing the parameters in the **Create** and **Process** module

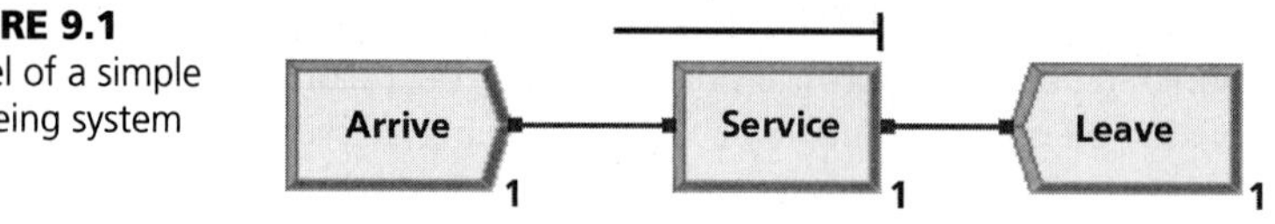

FIGURE 9.1
Model of a simple queueing system

FIGURE 9.2
A tandem queue

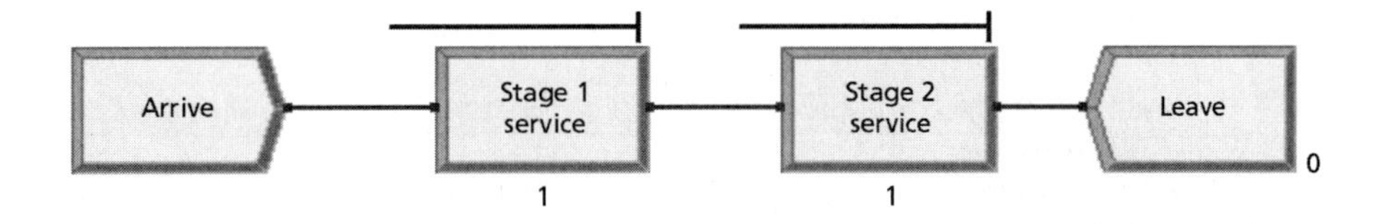

dialogs. For priority service, we would have to add an **Assign** module to assign a priority for each customer. We can also create a tandem queue by adding another **Process** module as shown in Figure 9.2. Here service is delivered in two stages. In the first stage, customers request a unit of `Server Type 1`; in the second stage, they request one unit of `Server Type 2`. These resources could be the same, in which case the same servers would serve both stages.

In this model, we assume that the capacities of the two queues at stages 1 and 2 are infinite, so the stages are effectively uncoupled if they do not use the same resource. Another way the stages can be coupled is through blocking. *Blocking* is the condition in which the stage 2 queue has a finite capacity and customers who complete stage 1 cannot proceed because the queue is full at stage 2. Using Arena, we must substantially modify the model to incorporate this behavior. Our approach is to treat places in the stage 2 queue as units of a resource. Then we will not release the stage 1 server until we have secured a unit of the stage 2 queue. The completed model is shown in Figure 9.3.

How would we model a multiserver queue if the servers were not identical? In this case, some means must be used to identify each server. Suppose we have a system with two servers, which we will label `Server 1` and `Server 2`. `Server 1` has service times that are exponentially distributed with mean 1.0 minute, while `Server 2` has service times that are uniformly distributed between 1.5 minutes and 4.5 minutes.

Arena has a handy facility—*resource sets*—that can be used to collect resources into a set. Our strategy will be to create a resource set, `Server Set`, that has two

FIGURE 9.3 Complete tandem queueing system

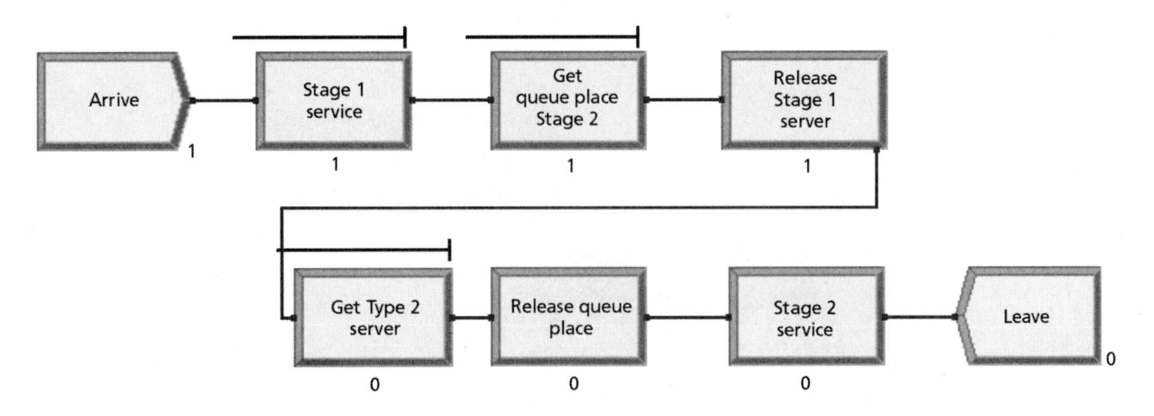

FIGURE 9.4 Resource spreadsheet

	Name	Type	Capacity	Busy / Hour	Idle / Hour	Per Use	StateSet Name	Failures	Report Statistics
1	Server 1	Fixed Capacity	1	0.0	0.0	0.0		0 rows	☑
2	Server 2	Fixed Capacity	1	0.0	0.0	0.0		0 rows	☑

Double-click here to add a new row.

servers. When customers start their service process, they will request a member of `Server Set` and their service time will be determined by which member of `Server Set` was assigned. Start with a new model and click on **Resource** in the **Basic Process** panel. To create a model in Arena with nonidentical servers, do the following: In the **Resource** spreadsheet, create two resources—`Server 1` and `Server 2`, each with 1 unit capacity as shown in Figure 9.4. Now click on **Set** in the **Basic Process** panel and create a set consisting of these two resources. Enter the name of the set and then click on the button under the **Members** column. Put the two resources in the set by double-clicking under the rows and selecting the resources from the drop-down list. These resources are available in the list because we have already defined them. Figure 9.5 shows the completed spreadsheet with the window showing the two resources. Close the **Members** window by clicking on the **Close** button in the upper right corner.

The next step is to build the model, which consists of the modules in Figure 9.6. The model has three **Process** modules. In the **Get a Server** module, the entity requests one unit of a resource in the `Server Set` and, if granted, delays for 0 time units—that is, the customer proceeds to the next module. The dialog for the **Process** module **Get a Server** is shown in Figure 9.7. Note that we selected **Largest Remaining Capacity** as the selection rule in the **Resources** dialog. This dialog is shown in Figure 9.8. Because the capacity of each server is 1, if a server is idle, its remaining capacity is 1. Thus, this rule will select the next idle server. Note also in Figure 9.8 that we entered `ServerNum` in the **Save Attribute** field. The server assigned will be saved in an attribute `ServerNum` for each customer.

Now that the server has been selected, we must cause the service time to be sampled from the correct distribution for the server that was assigned. This can be

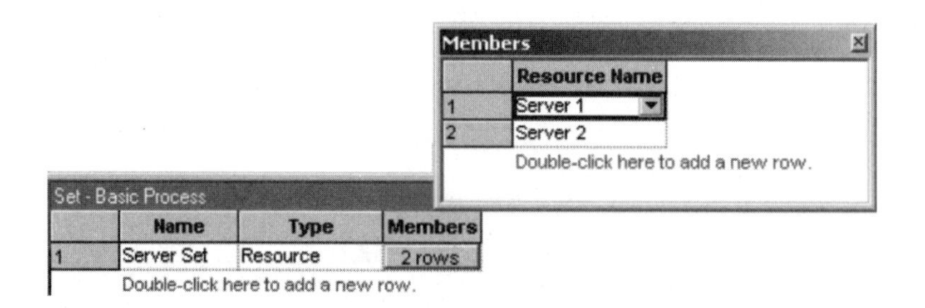

FIGURE 9.5 Completed Resource spreadsheet

FIGURE 9.6 Building the model

Arrive
1
Get a Server
1
Server 1 serving?
True
1
False
1
Use Server 1
1
Use Server 2
1
Leave
1

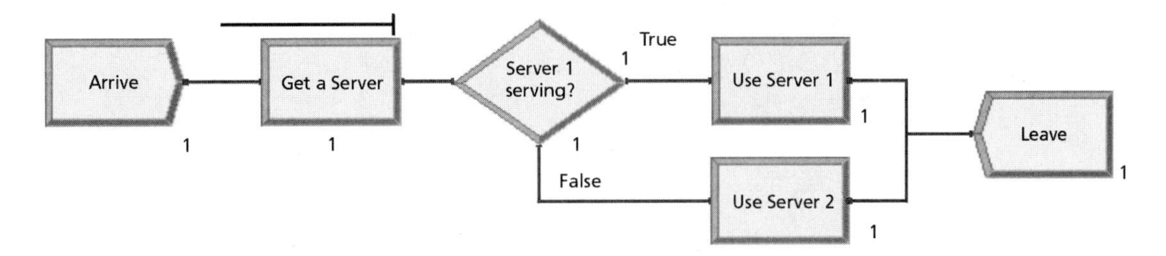

FIGURE 9.7 Process dialog for Get a Server Process module

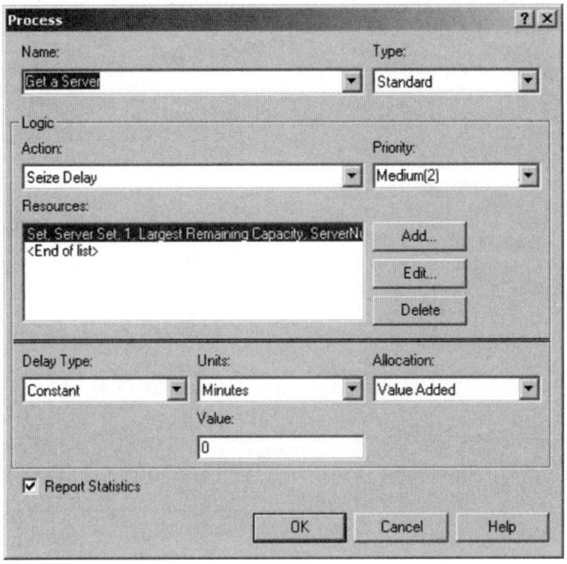

Process
Name:
Get a Server
Type:
Standard
Logic
Action:
Seize Delay
Priority:
Medium(2)
Resources:
Set, Server Set, 1, Largest Remaining Capacity, ServerNu
<End of list>
Add...
Edit...
Delete
Delay Type:
Constant
Units:
Minutes
Allocation:
Value Added
Value:
0
Report Statistics
OK
Cancel
Help

FIGURE 9.8 Resources dialog for Server Set

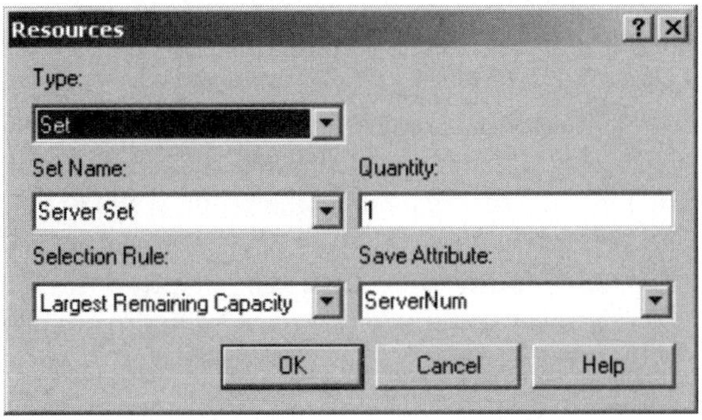

Resources
Type:
Set
Set Name:
Server Set
Quantity:
1
Selection Rule:
Largest Remaining Capacity
Save Attribute:
ServerNum
OK
Cancel
Help

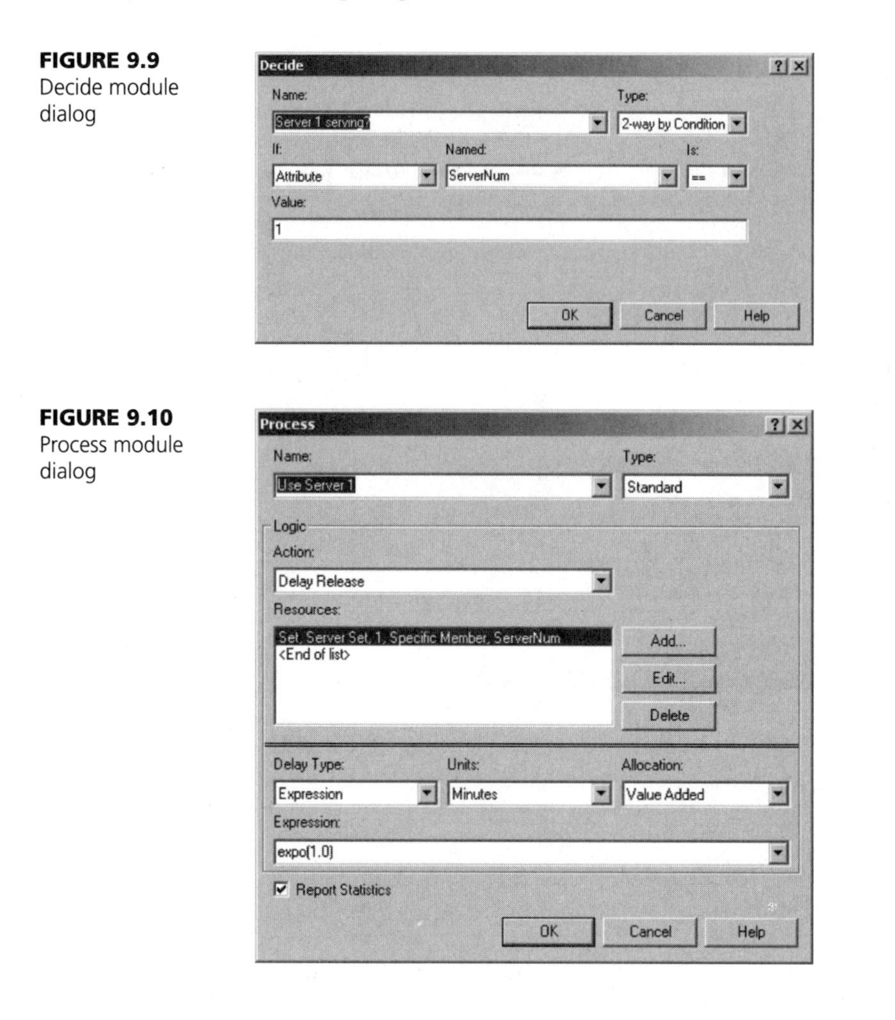

FIGURE 9.9
Decide module dialog

FIGURE 9.10
Process module dialog

done in several ways in Arena. We will use a straightforward approach. We add a **Decide** module that determines whether `Server 1` or `Server 2` was assigned. The `ServerNum` attribute contains this information. The dialog for the **Decide** module is shown in Figure 9.9. If `Server 1` was selected, then we move to **Process** module **Use Server 1.** In this module, we delay the appropriate length of time for `Server 1` and then release it. The dialogs for these modules are shown in Figures 9.10 and 9.11. Note that in this **Process** module, the action is **Delay Release.** The delay time is expo(1.0), as is correct for `Server 1`; in the **Resources** dialog, we release a specific member of the `Server Set`, denoted by `ServerNum`. Thus, the save attribute is used to specify which server will be marked idle and available. Of course, the entries in the dialogs for **Process** module **Use Server 2** are similar.

FIGURE 9.11
Resources module dialog

9.2.4 Networks of Queues

Many service systems have the structure of a network of queues. A *network* is a system consisting of nodes connected by arcs. For example, the manufacturing system considered in Chapter 5 is actually a network of queues. Figure 9.12 shows a sample network. The nodes are represented by circles and the arcs by lines connecting the nodes. Networks are used to represent many different models in operations research, especially models of transportation systems where the structure of the network has a physical resemblance to the real system. For example, in a transportation network, the nodes might represent cities, and the arcs might represent the available routes between cities. In a queueing network, the nodes represent simple—or at least simpler—queueing systems such as those presented in the previous section, and the arcs represent the traffic patterns for customers entering and leaving these systems. When a customer finishes service, he either leaves the network or attempts service from another server (node). Thus, the output of one server becomes the input to another server in the system.

Figure 9.13 shows a small queueing network that represents a model of a polling place. In the 2000 presidential election in the United States, much controversy resulted from the voting procedures not only in Florida but also in other

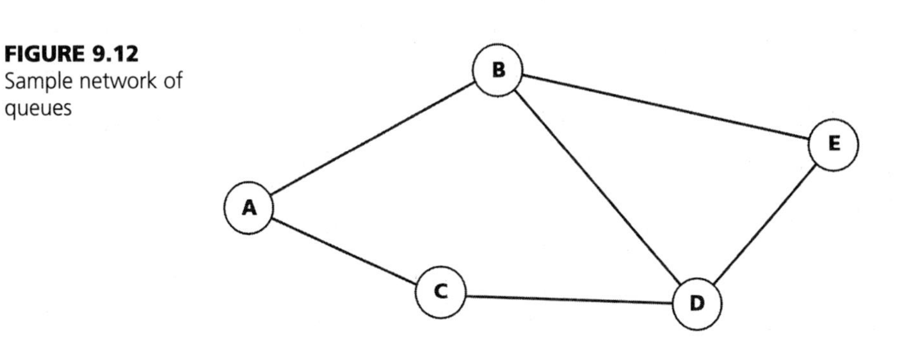

FIGURE 9.12
Sample network of queues

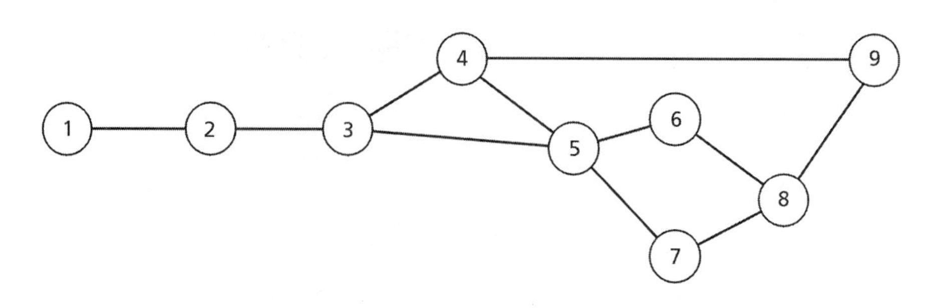

FIGURE 9.13 Network model of an election poll

states. This produced many proposals to reform the system for voting (CNN, 2001). One proposal involved a *provisional* voting system in which a voter could vote even if his or her registration was not verified. Another recommendation concerned better systems for recording votes and confirming to the voter that he or she had voted for his or her intended candidate. The report did not address a related issue, namely, the effect on waiting time that these changes would cause. If these changes were to swell the waiting lines, many voters will decide not to vote rather than wait for hours in line. One way to address this issue is, of course, to develop a model of the voting process at a polling place. Figure 9.13 is just such a model, and the purpose of the model is to determine, given the characteristics of the processes, how many poll workers and voting booths are needed to keep the waiting time within reasonable limits.

The network in Figure 9.13 is, in fact, eight coupled queueing systems. Nodes 1 and 9 represent customers entering and exiting the poll, respectively. Nodes 2 through 8 are described in Table 9.1. After arriving, the voter joins a queue waiting to fill out a form giving her name and address. In node 3, the voter hands this form to a poll worker who checks the registration list to see that the voter is registered and eligible to vote. If the voter's name is not on the list, the voter proceeds to node 4 where she consults with another poll worker. If the voter is not eligible to vote or is registered at another precinct, then she exits the polling place (node 9). If she is

TABLE 9.1 Summary of nodes in election poll network model

Node	Description
1	Arrival at polling place
2	Fill out voting form
3	Turn in form to poll worker and check registration list
4	Name not on list; consult with poll worker
5	Receive ballot
6	Vote nonhandicapped
7	Vote handicapped
8	Turn in ballot
9	Exit polling place

able to vote provisionally, she proceeds to node 5 where she receives a ballot along with the voters who were on the registration list. If she is not handicapped, she proceeds to node 6 and takes a voting booth for nonhandicapped voters. If the voter is handicapped, she proceeds to node 7 and uses a voting booth designed for voters who are physically handicapped. After voting, the voter takes her ballot to node 8 where she feeds it into an optical scanner and turns in her stub in another collection box before exiting the poll (node 9). There are several paths through the network that the voter could take. The path depends on whether the voter is registered in the precinct and whether she is physically handicapped or not.

Each node in Figure 9.13 represents a simple queueing system with a queue having possibly finite waiting room, one or more servers, and other specific characteristics. For example, node 2 is a single-server queue with infinite waiting room because voters can line up outside the building. Node 3 actually represents a tandem queueing system because the first poll worker takes the form and files it while the second checks the registration list. The capacity of the first queue in this node is 3; the capacity of the second is 1. Node 6 is a system having many servers but a single queue with capacity 5. Node 7 (handicapped voting) is a similar system in structure with a different number of servers and different queue capacity. Note that the service times in node 7 would be different than those in node 6. This shows the variety of simple queueing systems we find in these nodes.

Building a model of this queueing network is a simple process of creating each node as a simple queueing system (as we did earlier in this section) and connecting these nodes. The connections can be made with connectors in Arena or, if there are multiple paths out of a node, a **Decide** module will have to be inserted to select the next module. Creating a model for this election poll is left as an assignment at the end of the chapter.

9.2.5 Performance Measures for Service Systems

The purpose of modeling and simulating service systems is to measure the performance of candidate designs of the system in order to select the system design that best meets our objectives. In this section we discuss the various parameters that can measure system performance as well as how Arena estimates these parameters.

In previous models we estimated parameters that are specific to queueing systems, including the mean number of customers in the queue and the mean waiting time per customer. These are important performance measures for queueing systems because they tell us how long the customers were delayed, on average, when seeking service from the system. These parameters are related by Little's Law: $L = \lambda W$, where L is the mean queue length, λ is the arrival rate, and W is the mean waiting time.

Another important measure of system performance is the *server utilization,* which is defined as the average proportion of busy servers when the system is in steady state. If the system has a single server, then the server utilization is the proportion of time the server is busy or the probability the server is busy at an arbitrary moment in time. Suppose the simulation is run for 1000 hours in steady state, and during the run the server is busy 693 hours. Then an estimate of the server utilization is 693/1000 = .693. The batch means method, as presented in section 4.4 to estimate the mean number in queue, can be used to compute a confidence interval for the server utilization.

If a system has more than one server, the server utilization is defined to be the mean number of busy servers divided by the total number of servers, or the mean proportion of busy servers. Suppose that, for the system in section 9.2.3 with three servers, we estimate the mean number of busy servers to be 2.32. Then the estimate of server utilization is 2.32/3 = .773. We can use the same method that was described in section 5.3.5 for estimating the mean number in the queue to estimate the mean number of busy servers.

As we have seen, servers are resources in queueing systems. The same concepts can apply to define resource utilization as a measure of system performance. Analogous to server utilization, *resource utilization* is the mean proportion of the resource that is in use when the system is operating in steady state. If the resource consists of a single discrete unit, then its utilization is just the proportion of time the resource is being used. If it consists of multiple discrete units, then its utilization is analogous to server utilization in a multiserver queue: the mean number of units being used divided by the total number of units available. For resources measured in continuous units, utilization is still the mean number of units being used divided by the total number of units available, but at any point in time the number of units in use can be any value between zero and the total number of units available. If the number of units available varies over time, as in the fuel resources in the fuel depot model in Chapter 8, the same concepts apply: We can estimate the mean proportion of the resource in use. However, this is probably not a useful performance measure in the case of consumable resources. A more useful performance measure is probably the mean amount of the resource available.

The mean number of customers in the queue and the mean waiting time are performance measures that reflect the customers' perspective. Customers benefit most when these performance measures are as small as possible. Server utilization and other resource utilization are performance measures that reflect the system management perspective. The owner or manager of the system wants all resources to be utilized as much as possible because these resources represent a cost and high utilization means high return on investment. In virtually all systems, mean waiting time and resource utilization are inversely related: When one increases, the other decreases.

When the system is more complex, as in the case of a queueing network, there are many other possible performance measures. Customers arrive and flow through the system, taking various paths, waiting in various queues, and using various resources. Thus, we can seek to know the mean length of any queue or the mean waiting time in any queue or collection of queues. We can also wish to know the mean number of customers in any part of the system or the mean length of time a customer spends in any part of the system. For example, in the election poll model discussed in the previous section, we could be interested in the mean number of customers who are voting or the mean number of customers waiting for handicapped booths. In many simulations of queueing systems the objective is to locate bottlenecks—that is, places in the network where the workload is too much for the server(s), and consequently the number of customers waiting is more than a desirable amount. In the election poll model, the bottleneck could be the number of voting booths or it could be in nodes 2 and 3 where the voter's name is checked against registration lists. With knowledge of where the bottleneck is located, money and effort can be targeted to the specific solution that will improve throughput and reduce waiting time.

Because resource utilization is an important measure to system administrators, we could be interested in measuring the utilization of any number or combination of resources in the system. For example, in the election poll model, we could wish to know the utilization of the poll workers in nodes 2 and 3 who handle the voting forms and check registration lists or the utilization of the voting booths—that is, the average proportion of booths that are being used. This is another way to identify bottlenecks: Resources that experience higher utilization are impeding the flow of entities through the system.

9.2.6 Cost Parameters

In many systems, management's objective is to select the system configuration that minimizes cost. When this is the case, useful performance measures are the mean cost per unit time and the mean cost per entity. For example, if a system is operated for 10 hours, resulting in a total cost of $43,400 and servicing 100 customers, the estimate of the mean cost per hour is $43,400/10 = $4340, and the mean cost per customer is $43,400/100 = $434.

Arena has a rather robust data-collection and parameter-estimation system that can collect both time and cost data. Let's start with discrete observations on each entity. Imagine watching an entity as it flows through the system. The entity will experience a series of time delays. Some delays are waiting times while the entity sits in a queue waiting for resources. Others are activity times while the entity is engaged in some process. Still others are concerned with waiting for a transporter or conveyor, or with being transported from one location to another. Arena's built-in data-collection facilities will not only accumulate these times but also allocate them to one of five cost categories: *wait, value added, non-value added, transfer,* and *other.*

All time spent in a queue is automatically allocated to the wait category. For time spent in an activity, the categories can be selected from a list in the **Process** module dialog or in the **Process** module spreadsheet, as shown in Figure 9.14. The names are fairly self-explanatory, and the selection is entirely up to the modeler. Value added and Non-value added are categories that derive from manufacturing and some service operations. In the process of operating the system, some operations are core to the manufacturing or service delivery operation. Examples are an assembly operation in manufacturing an automobile engine and time spent treating a patient in an emergency department. These operations result in an added value to the product or a bill being sent to the patient. On the other hand, some activities can

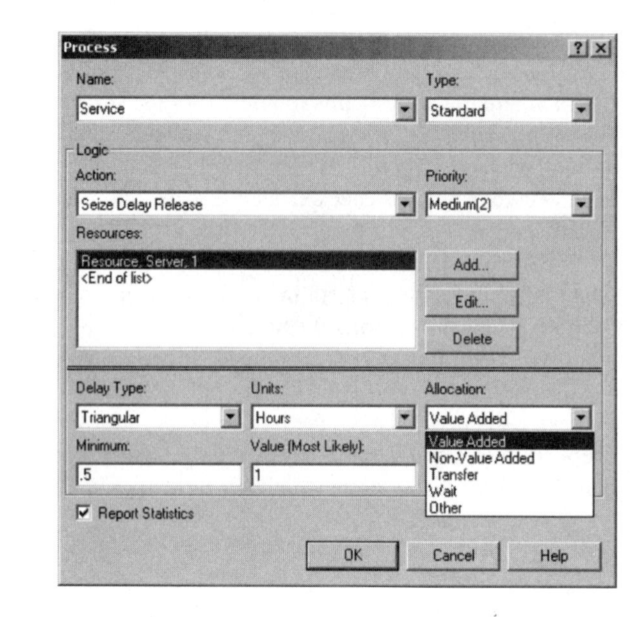

FIGURE 9.14
Process module dialog

be considered to be collateral to the system operations. Such an activity might be the time needed to start a machine or the time during an emergency department visit when laboratory results are retrieved. If these times could be reduced or eliminated, then efficiency would increase at no decrease in revenue. Thus, these times do not add value to the product. Any time spent transferring an entity from one location to another—for example, the time to move a patient from the emergency department to radiology—is generally allocated to transfer time. This would not include time waiting for a transporter. The *other* category is available for cases where the four categories just described will not fit. Of course, with the exception of *wait,* you can use these categories as you wish. Since all waiting times in queue are allocated to the wait category, you do not have complete control over observations applied to this category.

When an entity experiences a delay, two costs are incurred: a holding cost for having the entity in the system and a cost specific to the category of the delay. The category-specific cost includes two components: a fixed cost and a resource-usage cost that is proportional to the length of time allocated to the category. The holding cost rate and fixed costs are specified in the **Entity** module spreadsheet. Figure 9.15 shows an example. The resource-usage cost is specified in the **Resource** module spreadsheet, as shown in Figure 9.16. While the resource is busy, the rate at which cost is incurred is given in the **Busy/Hour** column. While it is idle, the rate is given in the **Idle/Hour** column; each time it is allocated or seized, the cost in the **Per Use** column is incurred. Thus, if a customer spent 30 minutes in the service activity using resource `Server`, the service activity is allocated to the value added category and the costs are as in Figures 9.15 and 9.16. The cost would be \$10.5 + \$2.2(.5) +

FIGURE 9.15 Entity spreadsheet with cost parameters

Entity - Basic Process

	Entity Type	Initial Picture	Holding Cost / Hour	Initial VA Cost	Initial NVA Cost	Initial Waiting Cost	Initial Tran Cost	Initial Other Cost	Rep
1	Customer	Picture.Repor	132	10.5	3.1	1.6	4.2	0.0	☑

new Double-click here to add a new row.

FIGURE 9.16 Resource spreadsheet with cost parameters

Resource - Basic Process

	Name	Type	Capacity	Busy / Hour	Idle / Hour	Per Use	StateSet Name	Failures	Report Statistics
1	Server	Fixed Capacity	3	360	150	1.00		0 rows	☑

Double-click here to add a new row.

\$6.6(.5) = \$14.90. In general, we can express total cost during a simulation run for an entity as

$$\text{Total cost} = \sum_{i=1}^{n} \left[f_{e,k_i} + \left(h_e + \sum_{j=1}^{m_i} r_j \right) T_i \right]$$

where

n = number of delay periods

k_i = category assignment for period i

f_{e,k_i} = initial (fixed) cost for entity type e in category i

h_e = holding cost per hour for entity type e

m_i = number of resources held by entity in period i

r_j = variable busy cost per hour for resource j

T_i = length of period i.

Now imagine watching a particular resource as the system operates. The resource will have units allocated to various processes; those units will then be released and sit idle (possibly for zero time) until the next process seizes them. Thus, each unit will go through a busy period followed by an idle period. The cost model in Arena provides for three types of costs for resources: a busy cost, an idle cost, and a per use cost. Each unit of a resource incurs cost at a particular rate per unit time when it is busy and another rate per unit time when it is idle, and it also incurs a cost each time it is put into service. This last cost could be considered the cost to prepare the resource for use, for example. So if the server resource in Figure 9.16 is busy for 7 hours and idle for 3 hours during a 10-hour simulation run, and it is used

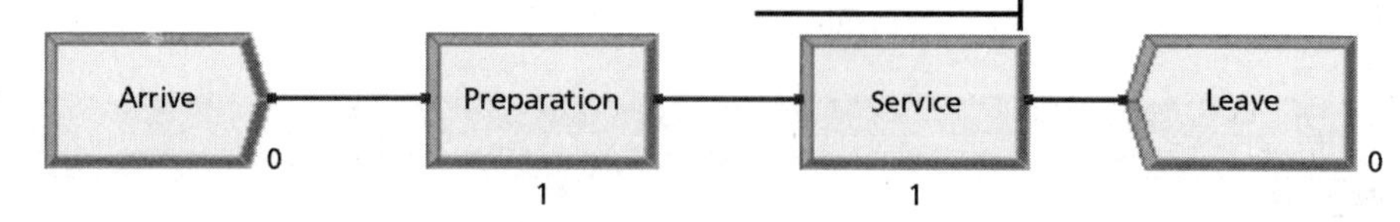

FIGURE 9.17 Demo model to show how costs are allocated

50 times during the run, the total cost would be \$360(7.0) + \$150(3.0) + \$1.(50) = \$3020. If there were multiple servers in the system, the same computation would be performed for each server and the totals added to get the total cost for all servers in the system. Arena accumulates these costs as the system operates rather than waiting until the end of the simulation to compute the resource cost.

As an example to illustrate the Arena cost model, the simple model shown in Figure 9.17 was developed. In this model, arrivals are Poisson with mean interarrival time 1.5 minutes. The preparation process does not use any resources, has a delay that is uniformly distributed between .5 and 1.5 minutes, and a cost allocated to the non-value added category. The service process uses one unit of the resource Server. Its delay is uniformly distributed between .7 and 1.3 minutes, and its cost is allocated to the value added category. The simulation was run for 1000 minutes, and the results are summarized in Tables 9.2 and 9.3. All figures in Tables 9.2 and 9.3 were taken from the output report for the simulation but reformatted to save space.

To understand how these cost estimates were computed, let's look at the value added cost per entity, 19.6219. This cost was only incurred in the process Service, and the simulation output showed that 647 entities completed service. Each entity

TABLE 9.2 Entity costs for cost demo model

Cost Category	Number Out	Total Cost (\$)	Time per Entity	Cost per Entity (\$)
Non-value added	650	1065.93	.7451	4.7392
Other	0	0	0	0
Transportation	0	0	0	4.2000
Value added	647	5902.88	.9905	19.6219
Wait	648	1477.26	1.0378	3.8833
Total		8446.08	2.7734	32.4444

TABLE 9.3 Resource costs for cost demo model

Cost Category	Total Cost (\$)
Busy*	3854.04
Idle	896.28
Usage*	648.00
Total	5389.32

*These costs are included in the totals in Table 9.2.

incurred an initial cost of \$10.50 in value added cost. The average time spent in service was .9905 minutes. For each minute spent in service, a cost of \$132/60 = \$2.20 was incurred in holding costs and \$360/60 = \$6.00 was incurred in resource busy costs. Also, for each service period, a \$1.00 usage cost was incurred. Thus, the total value added cost per entity is

$$10.5 + .9905(2.20 + 6.00) + 1 = 19.6219$$

The total value added cost does not include the initial value added cost, \$10.50. The total value added cost for the replication, as shown in the Table 9.2, can be computed from

$$647[.9905(2.20 + 6.00) + 1] = 5902.00$$

The small difference between 5902.00 and the value 5902.88 in Table 9.2 is due to either round-off error or the portion of service time used by entity number 648. It is not clear why Arena does not include the initial value added cost per entity in the total, but you can easily add it if you wish.

As we discussed in Chapters 4 and 5, two types of data can be produced: discrete observations from each temporary entity as it exits the system and is deleted, and continuous observations as the simulation progresses over time. Discrete observations are generally times and costs associated with each entity. Continuous observations concern system measures such as queue lengths, inventory levels, or resources available that can be measured at any point in time.

Resource busy costs and usage costs are included in the totals for both entities and resources. The total cost for the system, therefore, is the sum of the total entity costs plus the resource idle cost:

$$8446.08 + 896.28 = 9342.36$$

9.3 MANUFACTURING SYSTEM MODELS

Manufacturing systems are an important application area for discrete-event simulation. Because the structure and operations of most manufacturing systems are so complex, this family of systems has attracted the attention of simulation software designers since the beginning of discrete-event simulation. Today, simulation is the only tool available to study the operational performance of most stochastic manufacturing systems. As more flexibility, automation, and computerization are incorporated into manufacturing processes, they will become much more complex.

In this section, we will examine the structure of some manufacturing systems and their operations and discuss the problems of design and reengineering that require more powerful simulation methodology. We will also examine how to model manufacturing systems for discrete-event simulation, what performance measures are appropriate for these systems, and how simulation can be used in decision support and implementation of system changes.

9.3.1 Characteristics of Manufacturing Systems

Manufacturing systems involve the processing, transportation, and storage of materials in order to produce items for a market in a timely, efficient, and profitable manner. A wide variety of products such as steel, consumer electronics, computers, cars, oil, and beverages are produced, each using its own special and distinct production process. However, there are basic elements, structures, and organizational characteristics that are common to all production processes.

The most important elements of manufacturing systems are: parts or materials, workstations, load and unload stations, transport equipment, and buffer storage. *Parts* or *materials* are temporary entities that enter the manufacturing system and are processed and transported through it until the final product is finished and sent out of the manufacturing facility. Usually, parts are discrete, such as disk drives for computers, but we will also use this terminology to refer to parts that can be continuous such as molten iron or oil being refined.

Workstations are machines that are used to process parts. The machines can be as simple as a machine to cut a piece of sheet metal or they could be as complex as a machine to assemble and test a circuit board. Machines are usually but not always served by human operators for loading, setup, and unloading operations. *Load* and *unload* stations appear at those points in the system where parts are moved into or out of the system or loaded onto or unloaded from a machine. It is important to include these stations in the model when they require the use of system resources to accomplish their tasks. For example, if a crane or transport vehicle is needed to load or unload the parts, these resources normally need to be included in the model. Loading and unloading operations also often require human operators to place parts on or remove them from pallets.

Transport equipment refers to resources needed to transport parts between load and unload stations and workstations. This equipment also provides temporary storage for parts and thus is an important part of the inventory subsystem. There are two basic types of transport equipment: transporters and conveyors. *Transporters* such as forklift trucks and railcars are normally controlled by human operators and can move some quantity, often one item, of the part or product independently of other vehicles. These transporters can also be constrained to travel specific paths in the physical system. *Conveyors,* on the other hand, carry parts or finished goods in a steady stream. When the conveyor is moving, all parts on the conveyor are moving at the same speed. Physically, conveyors can be overhead, roller, and belt conveyors, but they all move a sequence of parts along a predetermined path from one operation to the next. Before a part can gain entry to a conveyor, it must have access to sufficient free space on the conveyor. Conveyors can be classified into two categories: nonaccumulating and accumulating. *Nonaccumulating conveyors* maintain a fixed spacing between each pair of adjacent parts on the conveyor. Sometimes, these conveyors must stop to allow parts to be placed on the conveyor or removed from it. *Accumulating conveyors,* such as roller conveyors, will let parts accumulate at the end and remain stopped while waiting to be removed. Thus, we can think of accumulating conveyors as having a conveyor with variable spacing between parts and a queue at the end that accumulates parts that have completed the trip on the conveyor.

Buffer storage refers to a place where in-process parts are stored between operations at workstations. Many manufacturing systems use buffers to protect against parts and materials shortages that would cause the output of the system to be erratic due to unequal processing rates at workstations, setup times, and uncertainties involving machine breakdowns. Thus, buffer storage allows the system to operate more smoothly. Workstations often have a buffer for input and another buffer for output, although the output buffer for one workstation is frequently also the input buffer for the next workstation in the process.

9.3.2 A Manufacturing System Model

We have already seen an example of a manufacturing system in Chapters 5 and 6. Figure 9.18 shows another manufacturing system that consists of a production unit and an assembly unit. Parts arriving at this system are unloaded in the unload area and stored in an input buffer common to the two production machines. After processing at machine 1 or machine 2, parts are placed on a common output conveyor and transported to an inspection station where they are inspected. Parts that do not

FIGURE 9.18 Manufacturing system with production and assembly units

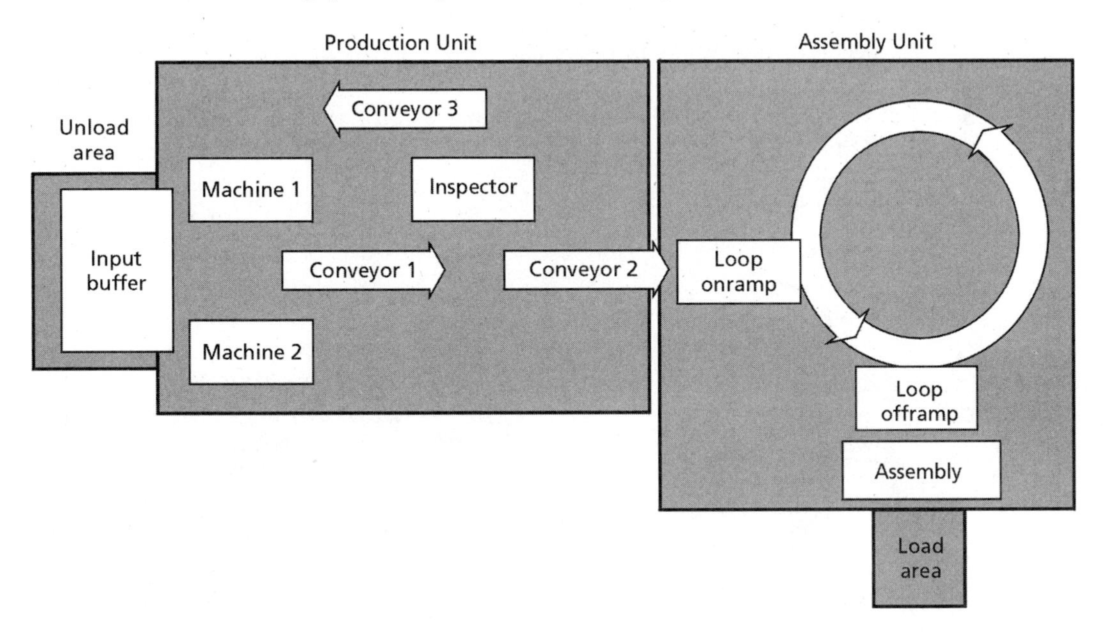

pass inspection are sent back on a conveyor for rework on machine 1, while parts that pass inspection are placed on a conveyor that feeds a nonaccumulating loop conveyor. All other conveyors in the system are accumulating conveyors. The loop conveyor moves at a constant speed. When there is a free space on the conveyor at the pickup station and an available part in the input buffer, the part is placed in the space on the conveyor. When a part traveling on the loop conveyor reaches the drop-off station, it enters the input buffer of the assembly station if the buffer is not full. If the input buffer of the assembly station is full, the part remains on the loop conveyor until it can be placed in the assembly station input buffer. After the assembly operation, the finished product is removed from the system.

We will develop a model of this system using Arena. However, first we will examine how Arena models conveyors and how to model the blocking that occurs when an output conveyor is full so that a workstation cannot offload the part that has just finished processing.

9.3.3 Conveyor Modeling

First consider accumulating conveyors. There are many types of accumulating conveyors, but we will limit this discussion to a roller conveyor that accumulates parts at the end of the conveyor. Figure 9.19 shows an accumulating conveyor. In this conveyor, gravity or perhaps some other mechanism moves parts along a path using rollers. The parts move independently along the conveyor path but their order is maintained: a part cannot pass another on the conveyor. A detailed physical representation of the dynamics of this conveyor could be developed to describe the acceleration and deceleration of the parts on the conveyor. However, a model with this level of detail is normally not required, and the computational burden of such a model is detrimental to the performance of the simulation.

Arena makes the assumption that each part that is placed on the conveyor moves at a constant speed until it reaches the queue of parts that have accumulated at the end of the conveyor. To define a simple conveyor, we must define three things:

1. two stations,
2. a segment, and
3. a conveyor.

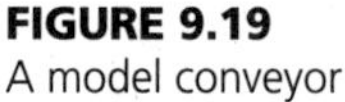

FIGURE 9.19
A model conveyor

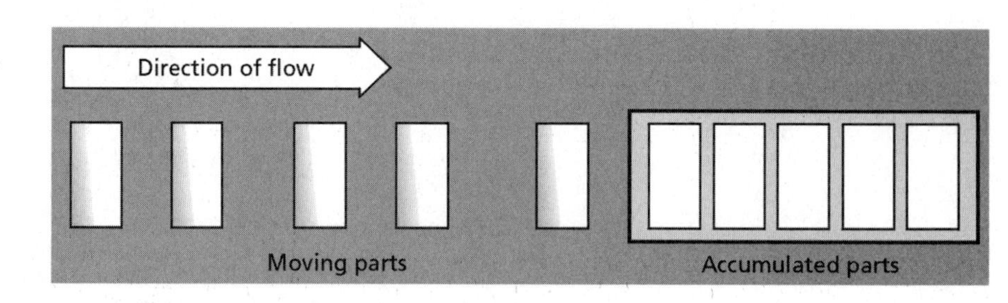

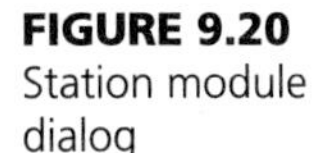

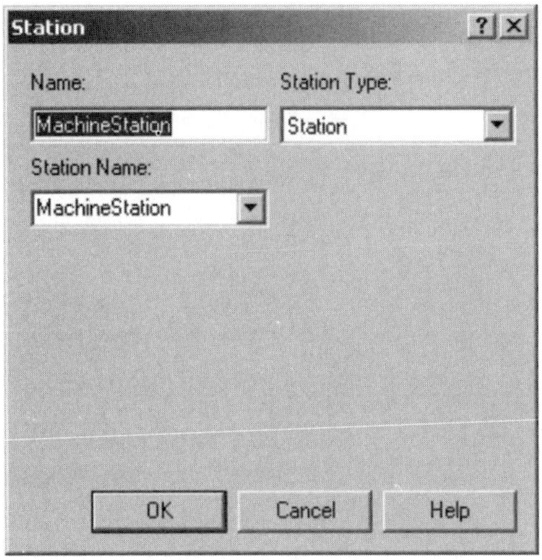

FIGURE 9.20
Station module dialog

A *station* is a logic module that defines a place where system logic, such as loading or unloading a conveyor, changes an entity's picture or where other system state changes can occur. The **Station** module is located on the **Advanced Transfer** panel and is placed on the model window by dragging and dropping as any other module. The dialog for a **Station** module is shown in Figure 9.20. The **Name** attribute is just the name that shows on the module in the model. The **Station Type** can be either **Station** or **Set.** A **Set** is a collection of stations. Here we are just defining a single station. The **Station Name** is the identifier Arena uses to refer to this station, and it is used to define conveyors as we will see shortly.

A *segment* is a construct that connects two or more stations. Segments are defined using the **Segment** module in the spreadsheet window, as shown in Figure 9.21, and the spreadsheet to specify the stations on the segment, as shown in Figure 9.22. Besides a **Beginning Station** and **Next Station,** the segment has a **Length.** The length can be defined in any physical units you choose, but it is constrained by the requirement that an entity on the conveyor must occupy an integral number of units of length. Thus, if entities need three inches, you cannot define the unit of length to be a foot because that would require entities to occupy .25 units. In this case, the unit of length should be inches.

The **Conveyor** module is defined using the **Conveyor** spreadsheet as shown in Figure 9.23. Each conveyor has a name that is used in reports. The physical layout of

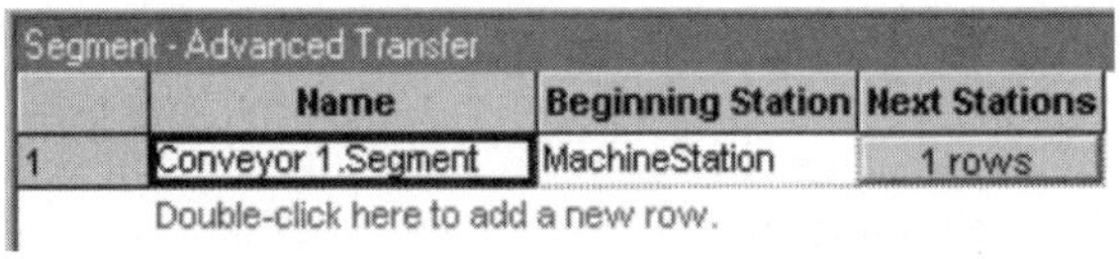

FIGURE 9.21
Spreadsheet for defining segments

Segment - Advanced Transfer

	Name	Beginning Station	Next Stations
1	Conveyor 1.Segment	MachineStation	1 rows

Double-click here to add a new row.

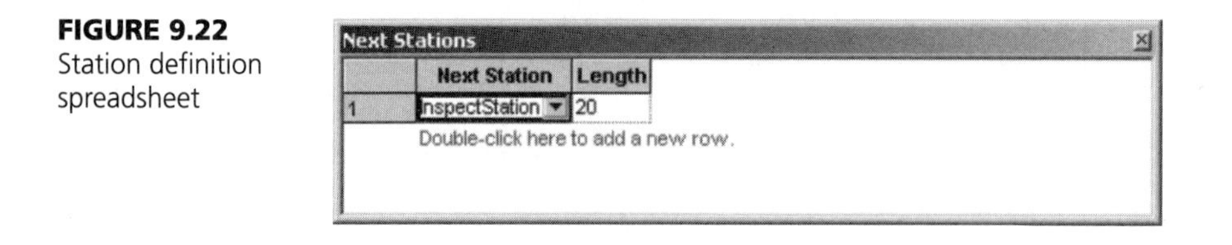

Next Stations

	Next Station	Length
1	InspectStation	20

Double-click here to add a new row.

FIGURE 9.22 Station definition spreadsheet

FIGURE 9.23 Conveyor module spreadsheet

Conveyor - Advanced Transfer

	Name	Segment Name	Type	Velocity	Units	Cell Size	Max Cells	Accumulation	Initial	Report
1	Conveyor 1	Conveyor 1.Segment	Accumulating	1.	Per Minute	1	1	1	Active	☑

Double-click here to add a new row.

a conveyor is defined by a segment, which is identified by the second parameter, the **Segment Name.** In this case, we specified the segment defined in Figure 9.21. The **Type** parameter can be **Accumulating** or **Non-Accumulating.** The **Velocity** is the number of length units per time unit the conveyor moves. The **Units** parameter specifies the time units used to define velocity. If length units are inches and time units are seconds, then the velocity is in inches per second. **Cell Size** is the number of length units that define a cell on the conveyor. Each item on the conveyor will occupy at least one cell and may occupy more than one cell. The **Max Cells Occupied** parameter specifies the maximum number of cells an item can occupy. On an accumulating conveyor, the number of cells an item can occupy when it is accumulating at the end can be different from the number of cells it occupies when it is moving. Generally, items are spaced farther apart when they are moving and closer or bunched together when they are accumulating. The **Accumulation Length** parameter specifies the number of cells the item occupies when it is accumulating. A conveyor can be active, meaning that it is operating, or inactive, meaning that it is not working. The **Initial Status** parameter specifies the status when the simulation starts. The status will remain the same unless it is explicitly changed. The last parameter is a check box to specify whether the conveyor statistics will be reported.

9.3.4 Building the Model in Steps: Version 1

Rather than trying to proceed directly from a blank model window to a finished manufacturing model, we will build the model in several phases. In the first phase, we will create version 1, which only includes the arrival and machining processes and the conveyor to the inspection station. Our objective here is to show how to incorporate an accumulating conveyor into a model. The model is shown in Figure 9.24. We have placed an **Arrival** module and a **Process** module to provide the parts arrivals and machining operations. We have defined two resources, `Machine 1`

FIGURE 9.24 Version 1 of the manufacturing model

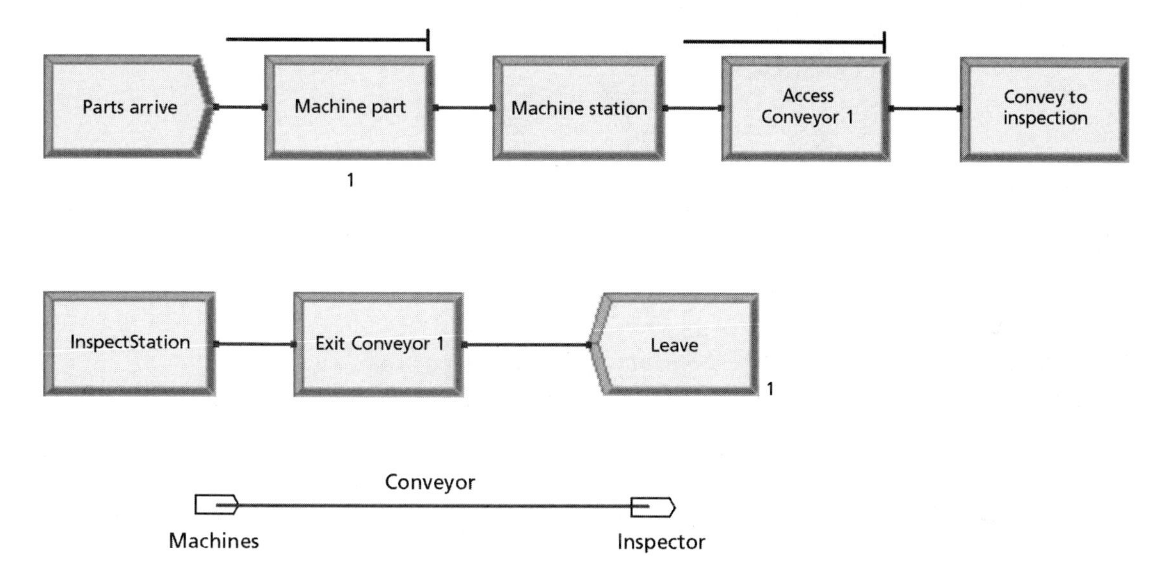

and `Machine 2`, that are used in the `Machine Parts` process. After the `Machine Parts` process, we have placed a **Station** module `Machine station` to define the start of `Conveyor 1`, which moves parts from the machining operation to the inspection station. In the second row of modules, we have placed a **Station** module `InspectStation` to define the end of the conveyor. When a conveyor is used in a model, **Station** modules must be used to define the points in the module logic where the item enters and exits the conveyor.

The procedure for moving an item on a conveyor is:

1. access the conveyor,
2. convey the item to its destination, and
3. exit the conveyor.

These three operations are represented in the model by three modules: the **Access, Convey,** and **Exit** modules found on the **Advanced Transfer** panel.

Figure 9.25 shows the dialog for the `Access Conveyor 1` module. In this dialog, we specify the conveyor that will be used. It is useful if the conveyor has already been defined—that is, if the data in Figures 9.21, 9.22, and 9.23 have already been entered. Entities that cannot enter the conveyor are held in a queue or other list until the conveyor has a place for them. This dialog provides an input to specify the type of structure that holds waiting entities; if they are held in a queue, the dialog provides a place for the queue's name. Note that the module has a queue graphic above it to hold the entities that are waiting.

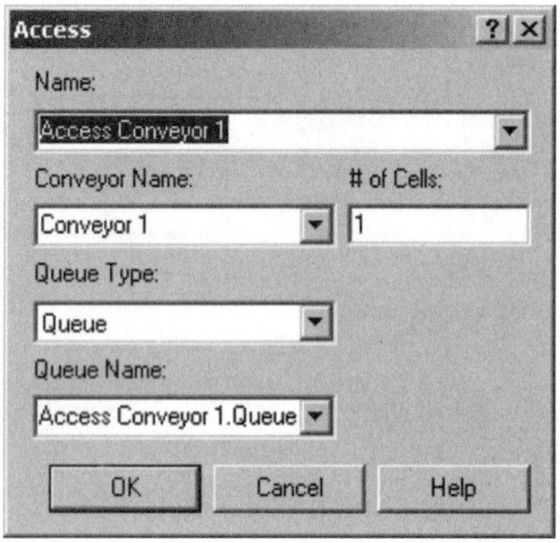

FIGURE 9.25
Dialog for Access module

The dialog for the **Convey** module in Figure 9.24 is shown in Figure 9.26. Here we specify which conveyor is used to convey the item. Normally, this is the same conveyor as in the access dialog. We also specify the destination type and identity. In our case, the destination is a **Station** and the specific station is `InspectStation`. The destination could also be the next station in a sequence or given by an expression. Arena's online help explains these options. In this case, because the conveyor moves the item to a station, the next module in the simulation logic will be that station. In Figure 9.24, there is no connector between the **Convey** module and the **Station** module for `InspectStation`. Arena implements a delay equal to the travel time, and at the end of this delay the item appears in the **Station** module. In terms of the simulation logic, stations are just placeholders, similar to labels in programming languages. Once the entity enters a station, it moves immediately to the next module.

In version 1, the next module is an **Exit** module. The dialog for this module is shown in Figure 9.27. All the **Exit** module needs to know is which conveyor to exit

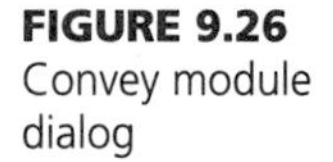

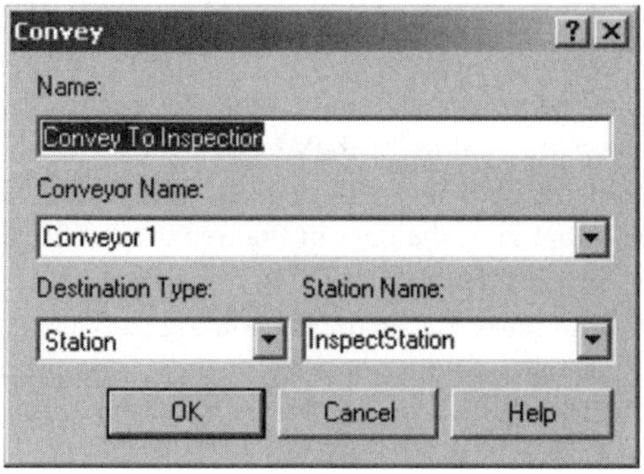

FIGURE 9.26
Convey module dialog

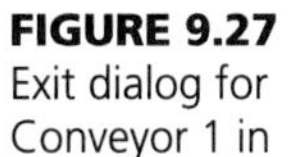

FIGURE 9.27 Exit dialog for Conveyor 1 in version 1

and how many cells to release. Normally, the number of cells to release is the same as the number taken when the item entered the conveyor. All of the model logic related to moving the entities on the conveyor and accumulating them at the end of the conveyor is automatically handled by Arena.

This completes version 1, which, of course, is not a complete model. It is a good idea to develop a complex model by creating a series of increasingly complex working versions as we are doing here.

9.3.5 Modeling Blocking

A workstation can become *blocked* after finishing processing on a part, if the conveyor ahead of it is full, thus preventing the workstation from off-loading the part so the machine or person can process the next part. In many models, the capacity of conveyors is so large that this seldom occurs, so a model that ignores this possibility is a good approximation to the real system. In other models, this can and does happen frequently, so we must include it in the model. We will assume for this model that any workstation that cannot off-load a part cannot begin work on the next part.

To model the blocking behavior, consider the logic we need to incorporate in the model. A part has finished the machining operation and now must be placed on the conveyor to go to the inspection station. However, `Conveyor 1` is full, so the part remains on the machine until a position is available on the conveyor. To implement this logic, our strategy will be to create a new resource, `Conv1Space`, that represents the number of positions on `Conveyor 1`. Before a part releases the machine it is using, it will seize 1 unit of `Conv1Space`. After a unit of `Conv1Space` has been seized, it will release the machine and proceed to access the conveyor and convey to the next station. In this way, the part will not be removed from the machine until it can be placed on the conveyor. Figure 9.28 shows the model with this blocking included.

FIGURE 9.28 Version 1 of model with blocking for full conveyor

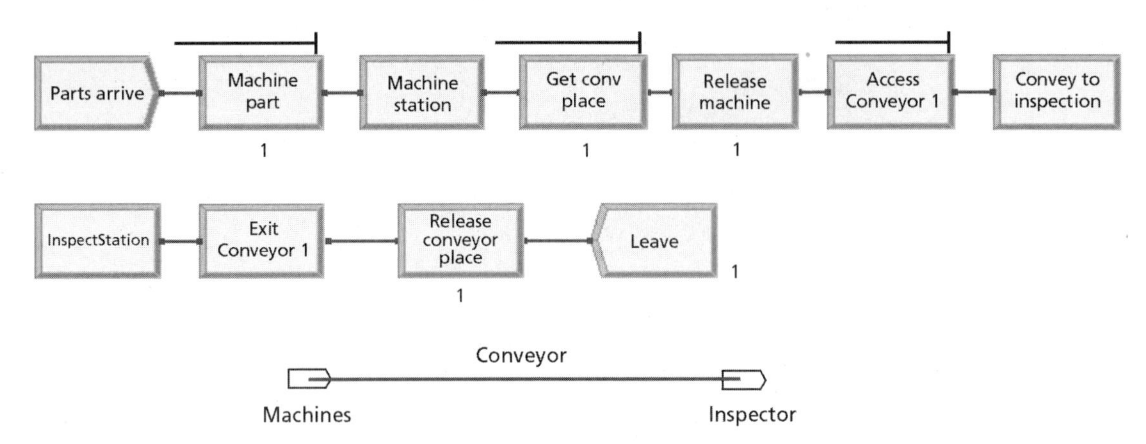

First, we need to change **Process** module `Machine Part 1` to **Seize, Delay** rather than **Seize, Delay, Release** because we do not want to release the machine immediately at the end of this process. We also added an attribute, `Machine Number`, to save the index of the machine that was seized. We will need this attribute when we release the machine. Figure 9.29 shows the dialog. Then add the new resource `Conv1Space` with an appropriate number of units to represent the number of spaces on the conveyor. Initially, we set the number of units to 2 so we could force the blocking to occur in order to test if the logic is working correctly. Later we can set the number of units to the true value.

FIGURE 9.29 Dialogs for new process to machine part

Before the Access Conveyor 1 module, we added two new **Process** modules: Get Conv Place and Release Machine. The first of these is a **Seize, Delay** process that seizes 1 unit of Conv1Space. The second is a **Delay, Release** module that releases 1 unit of Machine 1 or Machine 2. Here we specify the attribute Machine Number to identify the index of the machine that was seized. In both of these dialogs, the time is a constant, 0. Now the part has one unit of a resource that measures available space on Conveyor 1. However, it does not actually have a place on the conveyor. The **Access** module actually takes a place on the conveyor for the part.

One final change must be made to the model: After the part arrives at the inspection station, it is holding one unit of the resource Conv1Space and must release this unit so the accounting for space on Conveyor 1 will be accurate. To accomplish this, we added a **Process** module Release Conveyor Place after the Exit Conveyor 1 module. This process is a **Delay, Release** module, and the time is also 0.

This model can be run to test the blocking operation. You will note that the queues in modules Machine Parts and Get Conv Place will show parts waiting. In the first of these modules, parts will be waiting for an available machine. In the second, parts will be waiting for the conveyor. The module Access Conveyor 1 also has a queue, but it will be empty. Actually, you will occasionally see one part waiting for the cell in the conveyor to arrive, but this will be brief and this queue will never have more than one part waiting.

Blocking was added to version 1 to demonstrate how it works. For versions 2 and 3, we will omit blocking. You can easily add it if blocking must be included.

9.3.6 Adding the Rework Loop: Version 2

Version 2 adds the inspection station and rework loop of the conveyor. Before adding these elements to version 1, let's examine in detail what we need to add and develop a strategy to add the new elements. After the part exits Conveyor 1, we need to add another process to perform the inspection. When the inspection is finished, we must decide whether the part passed inspection or not. If it did not pass inspection, it will enter Conveyor 2 to return to the machines to be reworked. For now, we will let either machine do the rework. Later, we will modify the model so that all rework is done by Machine 1. If the part passes inspection, it is placed on Conveyor 3 to proceed to a station where it will enter the loop conveyor. The model is shown in Figure 9.30.

First, we can add the new resources we need. We have two identical inspectors at the inspection station, so we will add a new resource, Inspector, with two units. We also need to add two conveyors. One conveyor, which we will call Conveyor 2, carries parts from the inspection station back to the machines for rework. Its starting station, InspectStation, already exists. We will create a station, Rework, to denote the end of Conveyor 2. This station physically corresponds to the machines but logically needs to be a new station because parts that arrive at this

FIGURE 9.30 Version 2 of the manufacturing model with rework loop

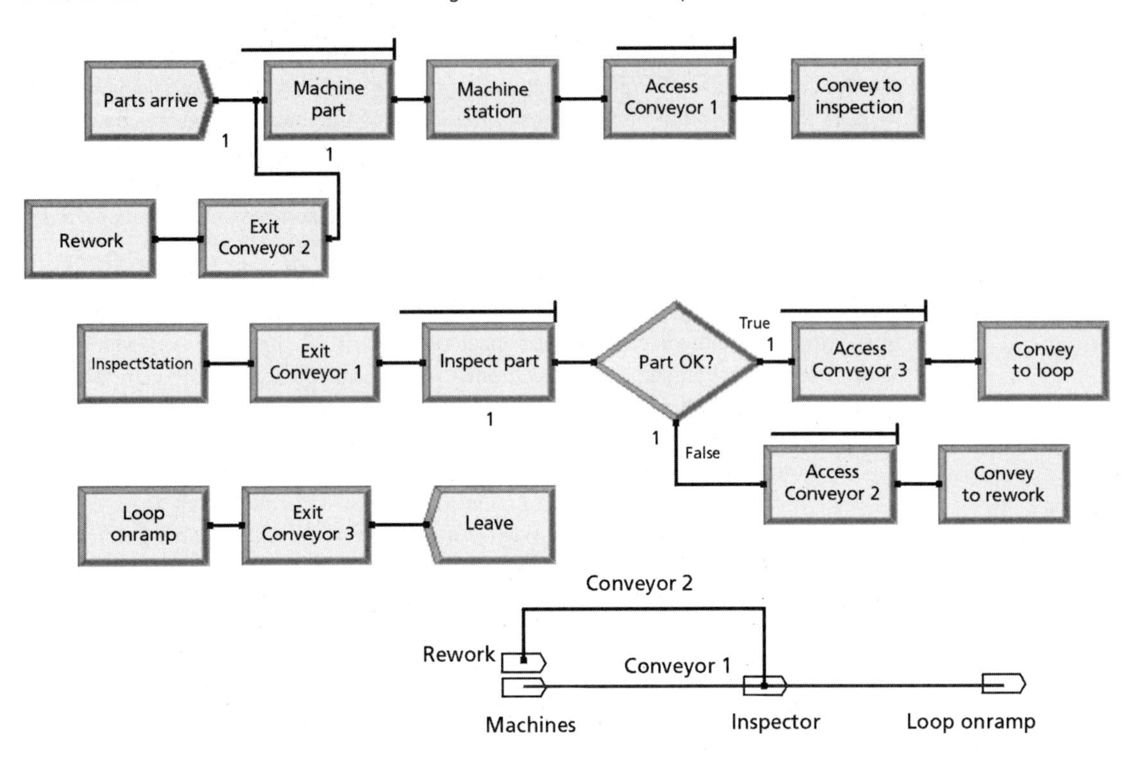

station will be on `Conveyor 2` and must exit this conveyor before they can begin rework. The other conveyor is `Conveyor 3`, which carries parts from the inspection station to the on-ramp of the loop conveyor. For this conveyor, we need a station to represent the on-ramp to the loop conveyor. For now, you can place these two station modules on the model window in a temporary location just to define the stations. Later, you can move them to their appropriate place in the model logic.

After creating the new stations, we can create segments in the **Segment** spreadsheet to represent the conveyor segments and define the two new conveyors in the **Conveyor** module spreadsheet. Now that the two new conveyors have been defined, we can proceed to add the new pieces to version 1 to complete version 2. Start by detaching the **Dispose** module `Leave` from `Exit Conveyor 1` so we can insert more modules between them. Then, we add the **Process** module `Inspect Part` and the **Decide** module `Part OK?` to implement the inspection process. After each of the outputs passes to the `Part OK?` module, we have an **Access–Convey** pair of modules to access the appropriate conveyor and convey the part. Each of the **Station** modules `Rework` and `LoopOnRamp` is followed by an **Exit** module to remove the part from the conveyor.

After exiting `Conveyor 2`, each part attempts to seize a machine and do the machining operation, so we simply need to connect the `Exit Conveyor 2` module

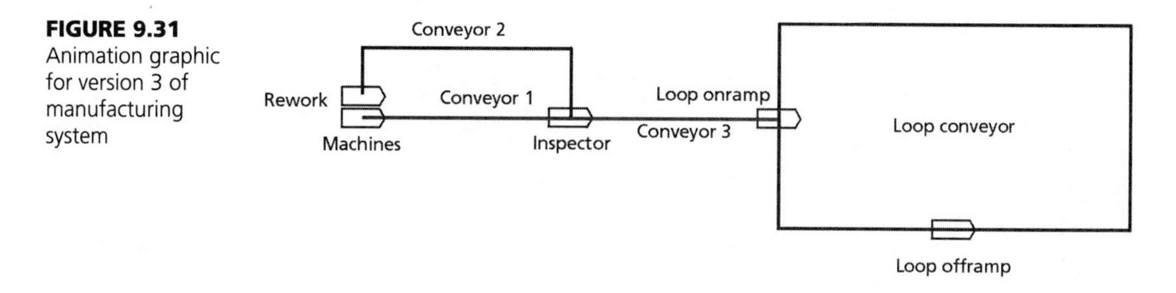

FIGURE 9.31 Animation graphic for version 3 of manufacturing system

to the `Machine Part` module to implement this logic. Since we are not modeling the loop conveyor in this version, we will just dispose of the part after exiting `Conveyor 3` at the loop on-ramp. Note that we must exit the conveyor before disposing an entity. Arena will issue an error message if we attempt to dispose an entity that is on a conveyor.

In version 2, we have not introduced any new concepts or modules but simply extended version 1 by adding more of the same types of elements. We also extended the conveyor animation by adding the two new stations and conveyor segments.

9.3.7 Adding the Loop Conveyor

Version 3 adds the loop conveyor and restricts parts undergoing rework to use only machine 1. Figure 9.31 shows the graphic for the system animation, including the loop conveyor. We must add a station for the off-ramp of the loop conveyor and a segment to represent the conveyor. This segment consists of two parts: One part connects the loop on-ramp to the off-ramp; the other connects the off-ramp to the on-ramp. Together, these two pieces form the loop conveyor. Figure 9.32 shows the segment spreadsheet for version 3 with the two parts of the segment and their lengths.

The new segment can now be used to define the loop conveyor. Figure 9.33 shows the conveyor spreadsheet for version 3. The loop conveyor is defined similarly to the

FIGURE 9.32 Segment spreadsheet for version 3

Next Stations

	Next Station	Length
1	Loop Off Ramp	20
2	LoopOnRamp	35

Double-click here to add a new row.

Segment - Advanced Transfer

	Name	Beginning Station	Next Stations
1	Conveyor 1.Segment	MachineStation	1 rows
2	Conveyor 2.Segment	InspectStation	1 rows
3	Conveyor 3.Segment	InspectStation	1 rows
4	Loop.Segment	LoopOnRamp	2 rows

Double-click here to add a new row.

FIGURE 9.33 Conveyor spreadsheet for version 3

Conveyor - Advanced Transfer

	Name	Segment Name	Type	Velocity	Units	Cell Size	Max Cells Occupied	Accumulation Length
1	Conveyor 1	Conveyor 1.Segment	Accumulating	2.	Per Minute	1	1	1
2	Conveyor 2	Conveyor 2.Segment	Accumulating	2	Per Minute	1	1	1
3	Conveyor 3	Conveyor 3.Segment	Accumulating	2	Per Minute	1	1	1
4	Loop	Loop.Segment	Non-Accumulating	1.5	Per Minute	1	1	1

Double-click here to add a new row.

first three conveyors, except that the **Type** parameter is **Non-Accumulating.** Arena uses the definition of the stations and segments to determine that the conveyor is a loop and to determine a path from each station to each other station or back to itself. Thus, if we convey a part from the loop off-ramp around to the on-ramp, we do not need to tell Arena what path to take.

Figure 9.34 shows the model for version 3. To add the loop conveyor and assembly operation to the model, we moved the **Dispose** module `Leave` and added modules to access the loop conveyor after the `LoopOnRamp` station module. Note that we first access the loop conveyor, then exit `Conveyor 3`. This makes sure the part is always on a conveyor. If we did these operations in reverse order, there could be a time when the part has exited conveyor 3 at the loop on-ramp but not yet been placed on the loop conveyor. Once on the loop conveyor, a **Convey** module is used as before to move it to the next station, the loop off-ramp.

When the part reaches the loop off-ramp, it can exit the conveyor only if the input buffer at the assembly process is not full. We need to place a **Decide** module at this point to test the number of parts in the `Assemble` process queue and, if there is room in the buffer, exit the conveyor; otherwise, convey again to the loop off-ramp. The dialog for the **Decide** module `Assemble Buffer Full?` is shown in Figure 9.35. The input buffer to the `Assemble` process is named `Assemble.Queue` by Arena. So the test for this **Decide** module is `NQ(Assemble.Queue) < 4` to let the parts exit the conveyor if the number of parts in the input buffer is less than 4.

To complete this version of the model, we also need to make changes to force parts being reworked to use `Machine 1` only. This was effected by adding a new process, `Rework Part`, when parts exit `Conveyor 2` at the `Rework` station; see Figure 9.34. Figure 9.36 shows the dialog for this **Process** module. This process is identical to the process `Machine Part`, except that `Machine 1` is the only resource seized. Now there are two buffers of parts waiting for `Machine 1` and `Machine 2` because there are two processes that use these two resources. Thus, if you want to know statistics on the total number of parts waiting for any of these resources, you will need to combine the data for these two queues.

FIGURE 9.34 Model for version 3 of manufacturing system

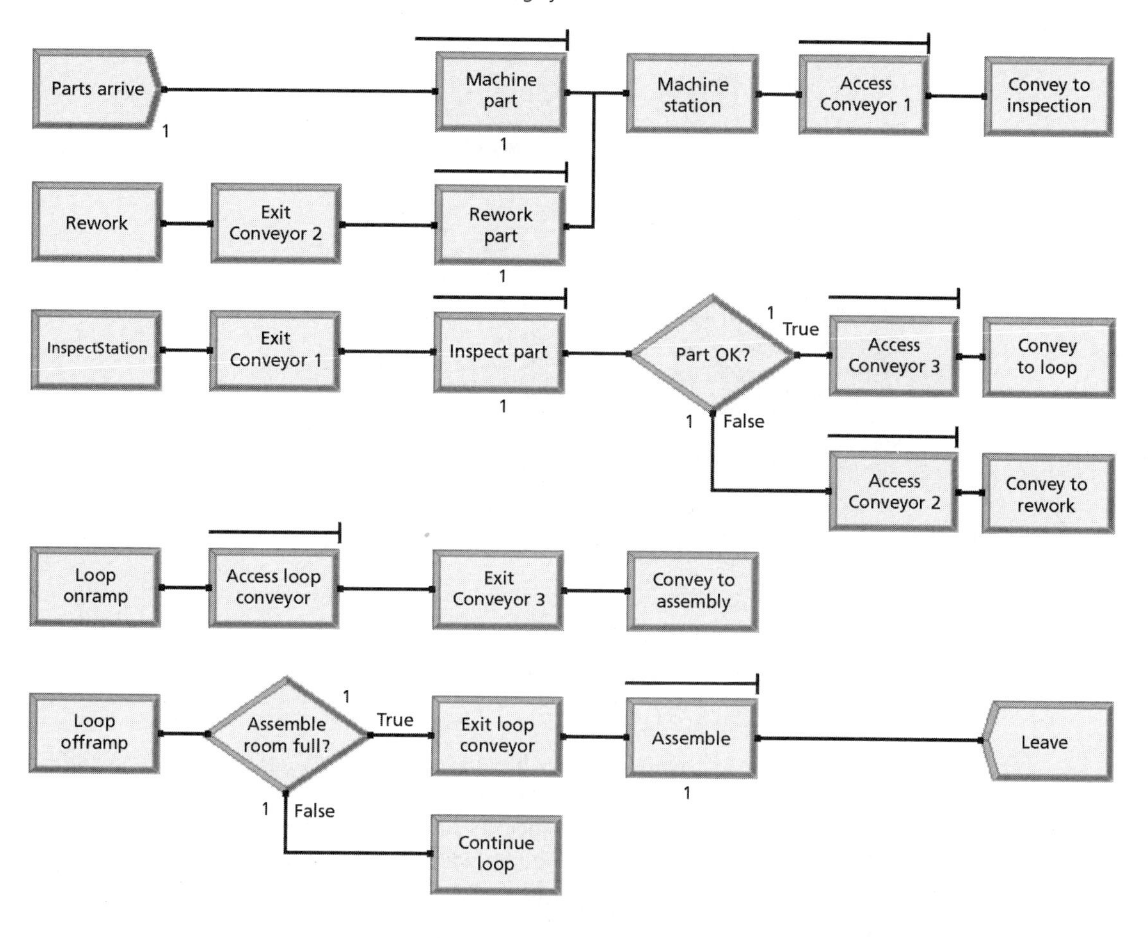

FIGURE 9.35 Decide module dialog for version 3

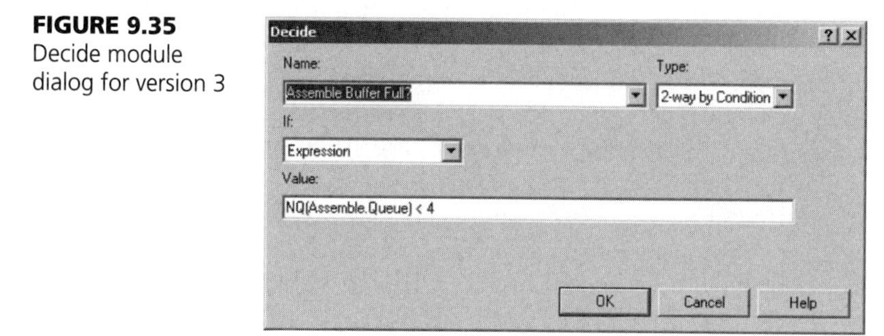

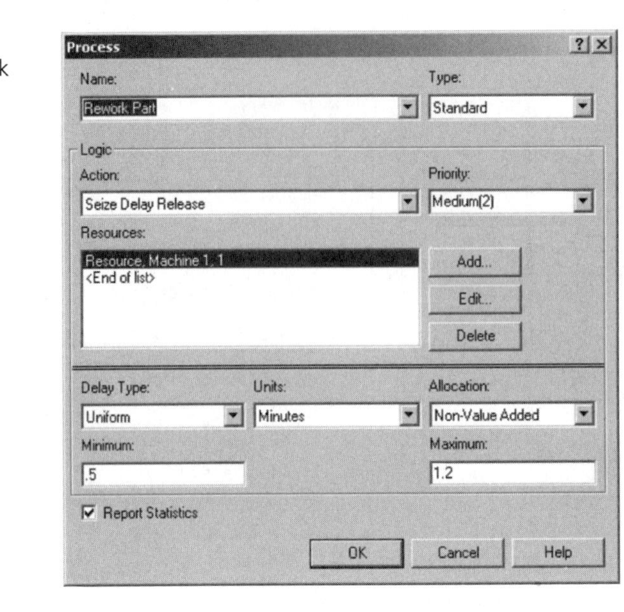

FIGURE 9.36
Dialog for rework part process

9.3.8 Special Characteristics: Transporters and Batch Processing

Manufacturing and other systems often have special characteristics such as intelligent transporters and batch processing that require additional modeling logic and capabilities. As an example, consider the flexible manufacturing system in Figure 9.37. This system consists of a loading dock, four machine centers (each with its own input and output buffers), a shipping dock, transporters, a system of tracks or paths for the transporters to follow, and a storage area for idle transporters. This system has two characteristics often found in modern manufacturing systems: transporters and flexible manufacturing cells. The transporters may be called automated guided vehicles (AGVs) because they can have robotic capabilities and can be programmed to go to a specific location. Thus, AGVs can have a certain level of intelligence and can respond to changing system states. For example, an AGV can change its route to pick up a finished part, wait at a location if a part will be finished soon, and respond to an order from a manufacturing cell to pick up a part. Flexible manufacturing cells are work centers that can be programmed to do different tasks. One part might need drilling, but the next part might need to be milled. These are especially useful in electronics manufacturing because a single machine is able to build and test a variety of circuit boards.

Models that involve intelligent transporters and intelligent machines are much more complex than simple models with simple rules of operation. Consider, for example, the system in Figure 9.37. The simplest operation rule might be that the transporters each follow a fixed route at the same speed around the loops and are able to

FIGURE 9.37
Example of a flexible manufacturing system using AGVs

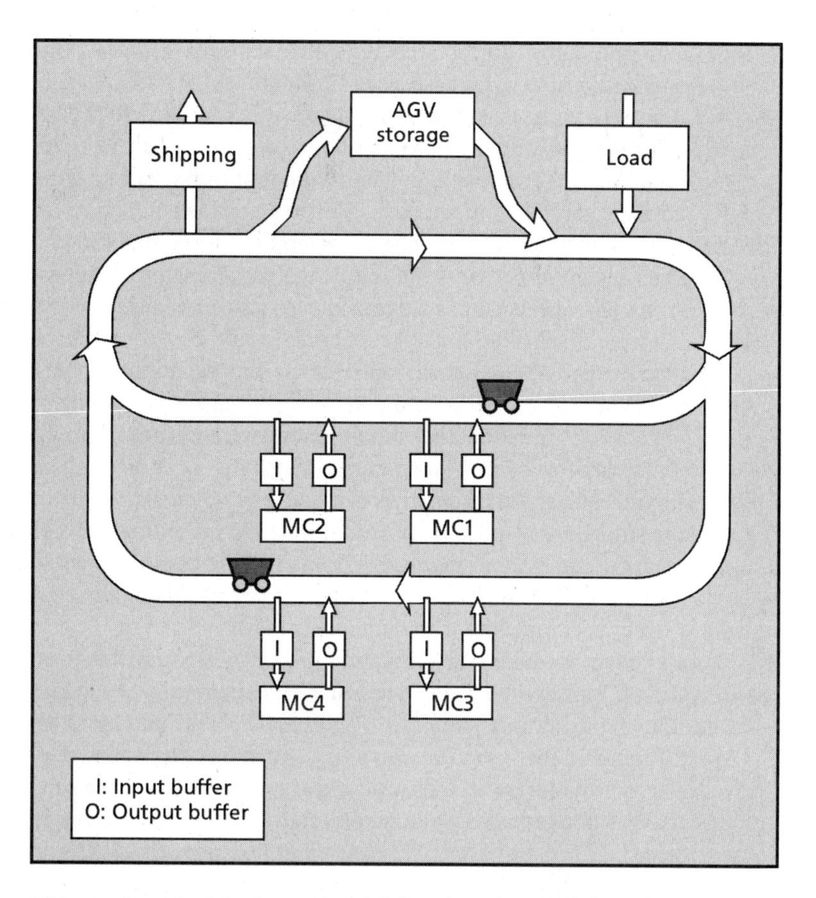

When an item is picked up at Load, it is assigned to MC1, MC2, MC3, or MC4. The AGV then goes to the appropriate MC.

AGVs do not stop at a pickup or drop-off point unless there is something to pick up or there is room to drop off the item.

instantaneously pick up or drop off parts in the loading, machining, and shipping locations. This would allow the transporters to schedule their arrivals autonomously because the transporters would never pass one another. Thus, when a transporter arrives at a location, it schedules its arrival at the next location from its own list of stops around the loop. This model would be especially easy to program using the process view because each transporter would have its own process and the processes only interact through moving the parts within the system. However, this would not be a realistic model because it is not able to account for the time to pick up and drop off the parts or the flexibility we want to have in scheduling the transporters.

Consider the changes required to allow the transporters to adjust their routes and schedules to more efficiently move the parts through the system. Suppose that

(1) input buffers to each machine have capacity 3, and (2) processing of each part on any machine requires exactly 12.3 minutes. In addition, suppose that the completion time for each part being processed is available to the transporters. Now consider the transporter's options if it arrives with a part at an unloading station to an input buffer for a machine when the input buffer is full. The transporter can wait until the part being processed is finished and then unload its cargo and proceed, or it can proceed to go around the loop another time and possibly deposit the part at another input buffer that is not full. If the part being processed is almost finished, then it would make sense for the transporter to wait. However, if the remaining processing time is substantial, the best choice would be to continue on the loop and to unload the part at another station. Transporters cannot pass one another (if they are on the same track), so this would keep the transporters moving on the path and reduce congestion. To model this situation, every transporter must know not only the time remaining for all parts in process but also the location of all other transporters at all times. Thus, if a transporter is stopped, any other transporters that might go through the stopped transporter's position must also stop themselves. Moreover, when the stopped transporter begins moving again, the other stopped transporters must start moving also. It is easy to see that the logic in this model is an order of magnitude greater in complexity than the models considered so far.

Many commercial simulation packages for manufacturing have skeleton logic for AGVs, flexible tools, and other characteristics of modern manufacturing systems built into the simulation engine. The user must provide the structure of the system and the logic for the rules governing the system. This is quite a bit of detail to provide, but it does relieve the user from the burden of controlling the interactions between AGVs and other parts of the system that interact in predefined ways.

If one is modeling a system such as this using one of the worldviews presented in Chapter 5, the activity view is especially useful. The activity view works well in this situation because there are many conditional events that can occur and each must be checked after each scheduled event. Recall that the activity view divides events into scheduled events (B-events) and conditional events (C-events). The clock moves when a B-event is executed; after the scheduled event, the conditions for each C-event are checked and all conditional events that are enabled are then executed. Actions such as stopping a transporter because another transporter is stopped and changing the path of a transporter would normally be implemented as conditional events.

Arena includes modules to model transporters, which are devices that move entities from one place to another in a system. Arena's view of transporters can be divided into two categories: *free-path transporters* and *guided transporters.* Both transporters move entities from one location to another, but free-path transporters move from one station to the next unconstrained by geometry or the congestion of the system. Guided transporters are constrained to defined paths and stations along those paths.

The modules in Table 9.4 are available to implement the logic for guided transporters.

TABLE 9.4 Arena modules to control transporters

Module Name	Operation
Transporter	Define a transporter.
Halt	Change the status of a transporter to inactive.
Activate	Change the status of a transporter to active.
Request	Assign a transporter to an entity and move it to the entity's location.
Allocate	Assign a transporter to an entity but do not move it.
Move	Move a transporter to another station without moving the entity.
Transport	Move the entity and transporter to another station.
Free	Release the transporter.

These modules allow the user to define transporters with multiple units. The transporters can be active or inactive, but only active transporters can move entities from one station to another. When an entity uses a transporter, either a **Request** or an **Allocate** module must be used to assign the transporter to the entity. These modules can select a specific transporter, or they can use one of several rules to select the transporter. The available rules are *cyclical, random, preferred order, largest distance,* and *smallest distance.* We then think of the entity as controlling the transporter. Once the assignment is made, a **Move** module can be used to move the transporter without the entity to another station, or the **Transport** module can be used to move the entity and the transporter to a destination station. Of course, these can be used many times to move transporters to multiple locations. Finally, a **Free** module is used to release the transporter and make it available for another entity to use.

Arena's built-in facilities for transporters can handle many modeling problems that arise with AGVs, but if the decision logic used by an AGV uses more complex rules that involve the entire system state, then these rules must be provided and programmed by the user, or a modeling approach that might not be obvious may need to be employed. For example, a "control" entity can be created that takes control of the transporter and moves it from station to station. Using this approach, the **Pickup** and **Dropoff** modules in the **Advanced Process** panel can be used to move other entities in the system. You can find out more about modeling with transporters in the Arena online help system and in Kelton, Sadowski, and Sadowski (2002), section 7.3.

9.3.9 A Manufacturing Decision Support System

We started this section by discussing how a simulation of a manufacturing system can serve as a decision support system by allowing management to evaluate alternative system designs. Here we will use the model developed in section 9.3.7 to evaluate two proposed system changes. First we will establish an experimental design for this model.

TABLE 9.5 Initial costs in the manufacturing model

Entity or Resource	Cost Category	Amount ($)
Part	Holding/hour	2.00
Machine 1	Busy/hour	6.00
Machine 1	Idle/hour	6.00
Machine 1	Per use	1.10
Machine 2	Busy/hour	6.00
Machine 2	Idle/hour	6.00
Machine 2	Per use	1.10
Inspector	Busy/hour	8.00
Inspector	Idle/hour	10.00
Assembler	Busy/hour	9.00
Assembler	Idle/hour	13.00

The system operates continuously 24 hours per day. We will establish a day as the length of a replication and run the model for 10 days. Parts arrive every 5 minutes, and the mean number of parts per arrival instance is 3.4. Thus, the arrival rate for parts is 3.4/5 = .68 per minute, and the number of parts to arrive and be processed per day is approximately 1440(.68) = 980. This is a reasonable sample size for each replication.

The costs for this model are given in Table 9.5. Any costs not shown in this table are zero, including those for purchasing, operating, and maintaining the equipment as well as managing the manufacturing system.

When we ran the model using the initial system configuration, we found that the average total waiting time for parts was 9.46 minutes, but 7.30 minutes of this was spent in the queue waiting to access `Conveyor 3`. Therefore, an upgrade to this conveyor would be a good candidate to improve the flow of parts and thus reduce costs. An upgrade is available that will increase the conveyor speed from two cells per minute to four cells per minute. The improvement to the conveyor costs $10,000 but must be amortized over 200 days. Therefore, to be profitable, the improvement must decrease the daily operating cost by at least $50. Another improvement could come from replacing machines 1 and 2 with more reliable machines. The new machines produce only 5 percent of defective parts that need to be reworked (the original machine produced 20 percent defectives), but they are more expensive to operate, costing $6.35 per hour whether they are busy or idle.

Four scenarios were run to evaluate these two improvements. These scenarios are presented in Table 9.6 and the results are shown in Table 9.7.

In each scenario, the estimates are taken from the Category Overview page of the Arena report. In each column, the **Total Entity Cost** is the sum of the four category costs. The **Other** category was omitted because its cost was zero. The **Total Resource Cost** is the sum of the three resource costs. However, the **Total System Cost** is not the sum of the **Total Entity Cost** and **Total Resource Cost** because the **Busy**

TABLE 9.6 Four scenarios for manufacturing model

Scenario	Description
1	Original system—no improvements
2	Faster conveyor 3
3	Improved machines 1 and 2
4	Improved machines 1 and 2, and faster conveyor 3

TABLE 9.7 Results of four scenarios

Cost Category	Scenario 1 ($)	Scenario 2 ($)	Scenario 3 ($)	Scenario 4 ($)
Non-value added	423.88	424.69	91.31	87.37
Transportation	2046.23	1613.59	1828.95	1372.86
Value added	2227.83	2235.45	2192.00	2189.37
Wait	455.13	159.66	542.57	125.54
Total entity	5153.07	4433.30	4654.82	3775.15
Busy	543.19	545.31	495.86	494.68
Idle	401.61	398.72	473.46	474.81
Usage	1965.37	1971.20	1660.01	1645.95
Total resource	2910.17	2915.23	2629.33	2624.45
Total system	5554.68	4832.12	5128.28	4249.96

Cost and **Usage Cost** were also included in the entity cost categories. Thus, the **Total System Cost** is the **Total Entity Cost,** plus resource **Idle Cost.**

From Table 9.7, we see that adding the faster conveyor reduced total system costs by approximately $723, which is far more than the $50 needed to justify the conveyor improvements. In scenario 3, we see that improvements in the machines resulted in a reduction of approximately $426 in total system costs, so this improvement appears justified, despite the increased costs for the two machines. Scenario 4 shows that we are justified to implement both improvements, resulting in a reduction of $5,554.68 – $4,249.96 – $50 = $1,254.72. Recall that the $50 is the daily cost of the conveyor improvement.

9.4 TRANSPORTATION SYSTEMS

Transportation systems involve the movement of people or items from one location to another. Most of the models in this book involve transportation systems in some sense because they model the arrival and processing of entities. In particular, the manufacturing models we just discussed have a large component that concerns the transportation of parts and finished goods into, within, and from a manufacturing facility. This transportation can be performed by forklifts, cranes, conveyors, trucks, and other means. Usually, parts and finished products are moved as quickly as possible to maximize the production rate.

In this section, we examine a transportation model that is not a manufacturing system. There are many transportation systems in our society that we deal with daily either directly or indirectly. These include:

- airport systems from the passenger's perspective (parking, terminal check-in, boarding), the planes' perspective (landing, taxiing, unloading and loading passengers, changing crews), and the airlines' perspective (crew scheduling, aircraft maintenance);
- mass transit systems, including rapid rail systems and buses;
- highways, including computer-monitored and controlled intelligent vehicle highway systems (IVHS), bridges, and toll roads;
- ferries;
- trucking networks, including terminals where cargo are loaded and sorted;
- rail and maritime freight-shipping systems (also including terminals), which are concerned with moving goods from one place to another;
- package delivery, which is a big business and offers good examples of transportation systems that have been designed to move goods quickly and efficiently.

This list shows how ubiquitous and important transportation systems are in the modern economy.

Most of these transportation systems have much in common with the systems studied earlier. They involve entities that move through a network of routes, queueing at times as they go and using resources to facilitate their movement. Thus, they can be modeled and simulated using the same basic techniques as the earlier applications. However, transportation models frequently have two characteristics that distinguish them from many other models:

1. They involve a schedule, and
2. they involve the coordination of various facilities and entities within the system to move the goods efficiently.

Most public mass-transit systems, airlines, and shipping companies have schedules that provide planned times when each vehicle will stop at each location to pick up and discharge passengers and goods. Even highway systems use a schedule in the sense that traffic volumes are higher during morning and evening rush hours; if the highway uses signals or other means to manage traffic differently during more congested times, these controls must normally operate on a schedule.

The coordination characteristic is a little more difficult to see for transportation systems. All nontrivial systems involve some form of coordination between entities.

However, in transportation systems, this seems to be a larger issue. As an example, consider an airport terminal where passengers are arriving and checking in while planes are arriving to discharge passengers and load passengers that have checked in. For the system to operate smoothly and efficiently, the movements of airplanes must be coordinated with those of the passengers (as well as the crew, the weather, and plane repair and maintenance). The arrival and departure schedule is a major part of this coordination effort, but when environmental factors force operations off the schedule, other mechanisms must be in place to coordinate these activities. For example, the flight can be cancelled or rescheduled, or passengers can be placed on another flight. For the model, well-defined policies must be provided to implement this coordination. Obviously, these can greatly complicate the model.

The model we will discuss in this section involves terminal operations for a package delivery service. Management wants a model of this system because it believes that packages are taking too long to be processed at the terminal and the terminal in general is not operating efficiently. An efficiently operating terminal would have trucks to immediately start unloading at a dock when they arrive and delivery vans would also not have to wait to load their packages and begin delivering them. Workers would be busy most of the time, and the system would be as inexpensive as possible to operate. Obviously, all of these criteria cannot be met simultaneously, but a simulation lets management evaluate the trade-offs.

Figure 9.38 is a diagram of the terminal operations. The model of terminal operations is described as follows: Large trucks unload at the two loading docks. Arrivals of these trucks are scheduled each half hour between 7 A.M. and 8:30 A.M. However, due to uncertain traffic conditions, they can be early or late. Their actual arrival time is a uniformly distributed random variable between 5 minutes before their scheduled arrival time and 10 minutes after the scheduled arrival time. In addition, unscheduled trucks arrive with interarrival times that are exponentially distrib-

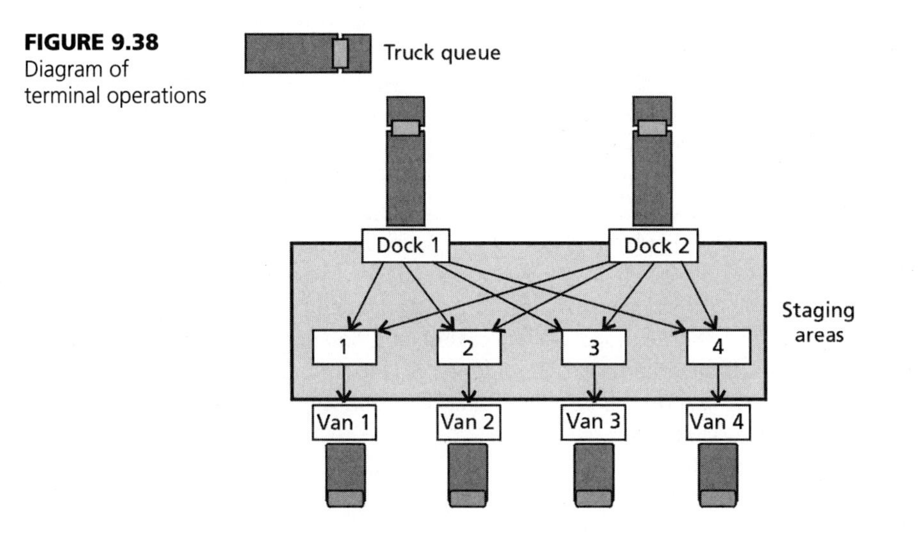

FIGURE 9.38 Diagram of terminal operations

uted with mean 12 hours. If both docks are occupied, arriving trucks queue first-come, first-served to await unloading. The queue is assumed to have infinite capacity, and trucks will wait as long as necessary for unloading. The number of packages in each truckload is represented by a normal random variable with mean 200 and standard deviation 30, truncated to an integer.

The terminal has 10 workers available for either unloading trucks or loading vans. We assume that each worker can unload any truck or load any van. Workers unload trucks at a rate of 80 packages per hour and load packages on a van at a rate of 100 packages per hour. The total unload rate is the product of the number of workers unloading and the unload rate per worker. So if three workers are unloading a truck, packages are being moved into the terminal from that truck at rate $80 \times 3 = 240$ packages per hour. Similar rates obtain for loading packages onto vans. At least two workers are required to unload a truck, and at most four workers can simultaneously unload any single truck.

On the other side of the terminal, delivery vans are loaded. The service owns four vans, each of which can hold a random number of packages represented by a truncated normally distributed random variable with mean 100 and standard deviation 10. The capacity is a random variable because the sizes and shapes of packages are random, and returning vans will frequently contain some packages that could not be delivered on that trip. As packages are unloaded from trucks, they are sorted for each of the four routes in the city: A, B, C, and D. Each package has the following probabilities of going to an address on each route: .25, .30, .20, and .25.

All vans are scheduled to arrive at the terminal at 9 A.M. for loading the day's first round of deliveries. Each van has its own dock, so there is no need for vans to queue waiting for a loading dock. For the 10 workers in the terminal, unloading trucks takes priority over loading vans, although this priority is not preemptive—that is, if a worker is loading a van when a truck arrives, she continues to load the van until loading is completed. When a truck finishes unloading, the workers are assigned to help unload the other truck if one is at the dock and being unloaded. Otherwise, the workers are assigned to load vans if they are present and being loaded or ready to be loaded. All vans have the same priority, and workers will be assigned to vans on a first-come, first-served basis. Thus, if trucks or vans are present and not being serviced by the maximum number of workers, then workers who become idle will immediately start working on those vehicles. If a truck, for example, has two workers unloading and another worker starts helping to unload, the time to unload the truck will change because the number of workers has changed. If no truck can be unloaded and no van can be loaded, the workers become idle and will wait until either a van arrives or a truck arrives. When a van arrives, it will be loaded if at least 20 packages are waiting to be loaded for its route and there is at least one idle worker. At most, two workers can load a van. When the van is finished loading, the workers will be assigned first to unloading a truck if the maximum number of workers is not unloading the truck, then to loading another van, and then to idle—in that priority order.

Once a van is loaded, it will leave the terminal to deliver the packages on its route. The time to deliver is uniformly distributed between one and two hours. After completing its route, the van will return to the terminal to pick up additional packages and deliver them. This process continues until either the number of packages waiting to be delivered is less than 20 or the time of day is after 4 P.M. If the van arrives before 4 P.M., it will wait; if more than 20 packages are available to be delivered to its route, then it will load and deliver these packages. Any vans that are waiting at the terminal at 4 P.M. and are not being loaded are sent home and scheduled to arrive at 9 A.M. the next day for that day's deliveries. If a van returns to the terminal after 4 P.M., it is sent home until 9 A.M. the next day.

This model has the characteristics we mentioned for a transportation system. Trucks, vans, and workers move packages from trucks through the terminal to vans and delivery destinations. The system has a schedule: Trucks are scheduled to arrive with packages in the mornings, and vans are scheduled to arrive in the mornings to load and deliver the packages and to leave at 4 P.M. or on their first return after 4 P.M. It also has resources—workers and docks—that constrain the system operations.

9.4.1 Phase 1: The Truck Unloading Process

We will develop this model in two phases. In phase 1, we will create version 1, which is a simple model that omits much of the detail; in phase 2, we will add details that make the model more faithful to the description.

First, note that this model has two processes operating: trucks arriving and unloading, and vans arriving and loading. The only interactions between these two processes are that they share resources (workers) and that the van loading process is constrained by the truck unloading process because packages unloaded from the trucks are loaded into the vans. We will start by creating the truck unloading process.

Both processes share workers, so we first define the resource, `Worker`, that represents the 10 workers. Also, there are two docks at which trucks can unload. We declare two resources, `Dock 1` and `Dock 2`, to represent these docks. Figure 9.39 shows the resource spreadsheet. After defining the two dock resources, we define a set, `Dock Set`, that consists of these two docks. You might be wondering why we defined `Worker` to be a resource with 10 units but `Dock Set` to be a resource set with two resources. We did this because later we will define a variable array to hold

FIGURE 9.39 Resources for package depot model

Resource - Basic Process

	Name	Type	Capacity	Busy / Hour	Idle / Hour	Per Use	StateSet Name	Failures	Report Statistics
1	Worker	Fixed Capacity	10	0.0	0.0	0.0		0 rows	☑
2	Dock 1	Fixed Capacity	1	0.0	0.0	0.0		0 rows	☑
3	Dock 2	Fixed Capacity	1	0.0	0.0	0.0		0 rows	☑

Double-click here to add a new row.

the number of packages to be unloaded at each dock. A truck will arrive and request a member of the set `Dock Set`. We will use the index of the dock in the set that was seized to index the element of the array associated with this dock. This provides a rather natural way to associate an array and a set of resources, and it is useful in this application.

The Truck Arrival Process

Now we are ready to create the truck arrival and unloading process. The version 1 model is shown in Figure 9.40, with the truck arrival and unloading process on the top and the van arrival, loading, and delivery process below.

There are two parallel arrival processes for trucks: scheduled arrivals and unscheduled arrivals. We use two **Create** modules to represent these arrival processes. Figure 9.41 shows the dialog for the **Create** module for scheduled truck arrivals. These arrivals start at 8.00 hours, or 8 A.M., occur every .5 hours, and end when four trucks have arrived. Figure 9.42 shows the dialog for unscheduled truck arrivals. These occur with a random delay having an exponential distribution with mean 4 hours, starting with the first delay after 7 A.M. At most, five unscheduled arrivals occur per day.

After the truck arrives, the number of packages is sampled in an **Assign** module and the truck requests a dock from `Dock Set`. After a dock has been seized, the truck is ready to unload. In version 1, we will unload the truck in a way that is a little different from the model description, but we will replace this portion of the model when it is refined later. At this point, we don't want to be distracted by the complexities of unloading trucks or loading vans. For now, the process is to first request two workers in **Process** module `Get 2 Workers`. Since the truck cannot be unloaded unless two workers are available, we must request two workers. However, if three or four workers are available, we will also use them. In the **Assign** module `Workers Ready to Unload`, we compute the number of workers that are available with the expression

```
MN(2,MR(Worker)-NR(Worker))
```

`MR(Worker)` is the total number of workers, and `NR(Worker)` is the number of workers who have been seized by processes. Thus, the difference is the number of workers available. If this is more than two, we will limit it to two because we cannot use more than four workers total and we have already seized two workers. We also compute the total number of workers that will be unloading the truck by adding 2 to this number. We will need this number when we release the workers after finishing the unload process.

Figure 9.43 shows the dialog for the **Process** module `Unload Truck`. In this module, we seize the additional workers (if any) and delay the unload time. Because the workers can unload at a particular rate per hour, contained in variable UnloadRate, the time will be given by the expression at the bottom of the dialog. We do not release the workers in this module because we do not have the option of releasing more workers than are seized. We could make the action Seize, Delay,

FIGURE 9.40 Phase 1 package depot model: (a) truck arrival and unloading process; (b) van arrival, loading, and package delivery process

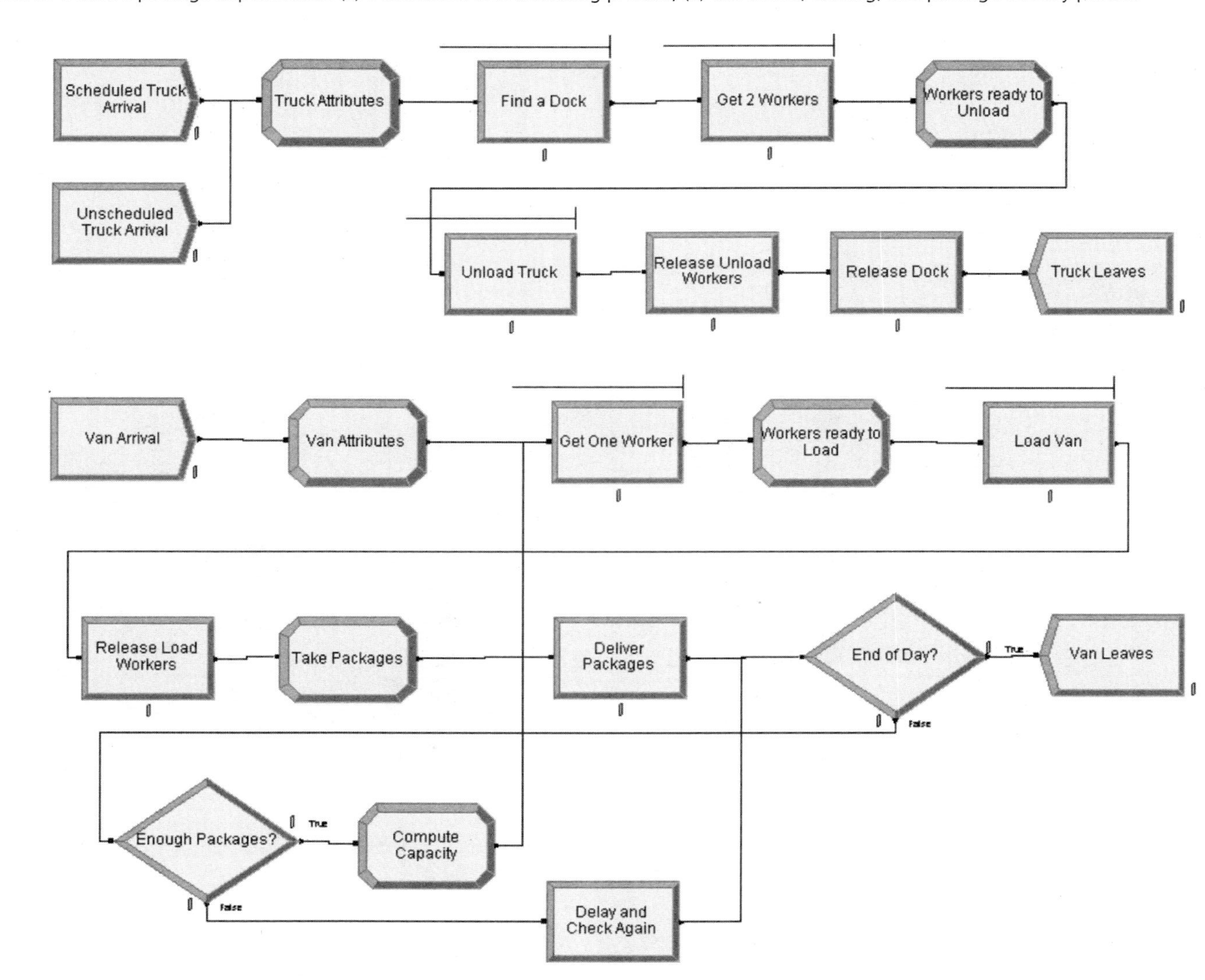

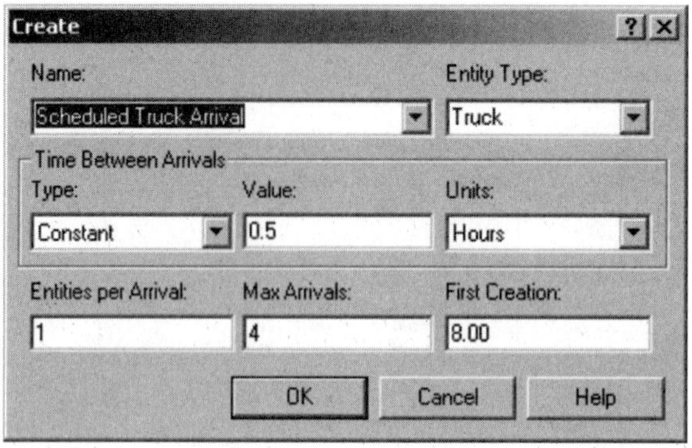

FIGURE 9.41
Dialog for scheduled truck arrivals Create module

Create
Name: Scheduled Truck Arrival
Entity Type: Truck
Time Between Arrivals
Type: Constant
Value: 0.5
Units: Hours
Entities per Arrival: 1
Max Arrivals: 4
First Creation: 8.00
OK Cancel Help

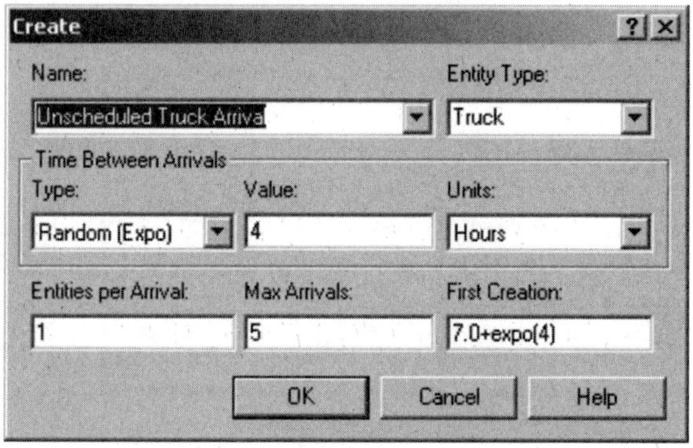

FIGURE 9.42
Dialog for unscheduled truck arrivals Create module

Create
Name: Unscheduled Truck Arrival
Entity Type: Truck
Time Between Arrivals
Type: Random (Expo)
Value: 4
Units: Hours
Entities per Arrival: 1
Max Arrivals: 5
First Creation: 7.0+expo(4)
OK Cancel Help

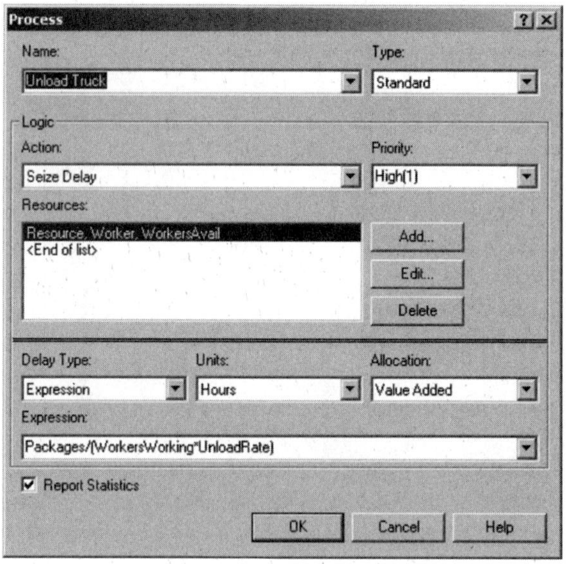

FIGURE 9.43
Process dialog for unloading trucks

Process
Name: Unload Truck
Type: Standard
Logic
Action: Seize Delay
Priority: High(1)
Resources:
Resource, Worker, WorkersAvail
<End of list>
Add... Edit... Delete
Delay Type: Expression
Units: Hours
Allocation: Value Added
Expression: Packages/(WorkersWorking*UnloadRate)
Report Statistics
OK Cancel Help

Release to release the number of workers seized in this module and then release two workers in the next **Process** module. Instead, we use a separate **Process** module to release all workers seized. Both approaches will work.

Finally, after the truck has finished unloading, we release the dock and dispose of the truck. In the **Process** module `Release Dock`, a 5-minute delay accounts for the time to pull away from the dock and allow the next truck to begin docking.

In a later version, we will replace the **Process** modules where we seize workers and delay the unload time. You might ask why this model is not faithful to the description above. We will discuss this point a little later, but give it some thought for now and see if you can answer the question before we get to that section.

The Van Loading Process

The bottom process in Figure 9.40 represents the vans arriving, loading, delivering packages, and returning for another load. In the **Create** module `Van Arrival`, four arrivals are scheduled at exactly 9.00. Thus, all four vans arrive simultaneously at 9 A.M. The next module, `Assign Attributes`, assigns the attribute `Capacity` to establish the van's capacity, the variable `VanCount` to count the vans, and the attribute `VanNum = VanCount` to give each van a unique number between 1 and 4. At this point, all four vans exist and have attributes `Capacity` and `VanNum`.

The next four modules—`Get One Worker`, `Workers Ready to Load`, `Load Van`, and `Release Load Workers`—operate similarly to the four analogous modules in the truck unloading process to load the vans using one or two workers. In the truck unloading process, two workers are initially requested, and up to two more workers will be seized if they are available. In this case, one worker is initially requested and another will be seized if available. Otherwise, the processes are the same.

After loading the van, the next module, `Take Packages`, removes the packages from the inventory waiting for the truck. The number of packages remaining is zero if the number of packages to be loaded is less than the van's capacity and the difference between the number of packages and the van's capacity otherwise. The expression `MX(RouteInv(VanNum)-Capacity,0)` does this calculation neatly.

Now the van experiences a delay while the packages are being delivered. We assume that the length of time to drive the route has a distribution that is uniformly distributed between one and two hours and does not depend on the number of packages on the van. After the delivery is completed, a **Decide** module, `End of Day?`, determines if the van should end the day or go back for another load. If the time is after 4 P.M. (or 16.0 hours), the van ends the day. Since we plan for each replication to consist of one 24-hour day, we can dispose of the van after 4 P.M.

If the time is before 4 P.M. when the van completes delivery, it returns to the depot and attempts to start loading another load. Recall that the van will not start loading unless there are at least 20 packages waiting. Thus, the van's process enters a **Decide** module, `Enough Packages?`, to see if at least that many packages are waiting. If so, then the new capacity is computed again and the van goes to the module `Get 1 Worker` to start the loading process. If the number of packages waiting is fewer than 20, then the van enters a delay of 10 minutes and checks again for the end of day.

9.4.2 Phase 2: Refining the Model

Version 1 of the model failed to implement the intended model for two reasons. First, the packages are taken from the truck at the *end* of the unloading period. When a van arrives after delivering packages, it checks to see if 20 packages are ready for loading. In the intended model, packages are unloading continuously during the unloading period, so 20 packages will usually be available sometime before the end of the unloading period. This behavior is not captured in version 1. Second, all workers are seized at the start of the unloading process for trucks or at the start of the loading process for vans. If a worker becomes idle during an unloading process or during a loading process, then the version 1 model does not allow that worker to begin work on a truck or van that is in process. If that worker could begin work at this time, the end of the process would occur sooner because the number of workers has increased. Although we cannot predict exactly how important these behaviors are in the model, our intuition says that they are important and must be accommodated.

The modification that needs to be made involves replacing the truck unloading process and the van loading process with processes that are much more complex. First, we will deal with the unloading process; the loading process will work similarly. The unloading process will be replaced with a submodel, as shown in Figure 9.44. Rather than treating a load as a single work process, our approach is to unload the truck one or two packages at a time. The submodel is shown in Figure 9.45.

Arena provides two ways to create a submodel. First, you can convert a **Process** module into a submodel by specifying submodel in the type field as we did in section 8.4. A second means is to select **Submodel** from the **Object** menu then **Add**

FIGURE 9.44 Dialog for Process module unload truck (truck arrival and unloading)

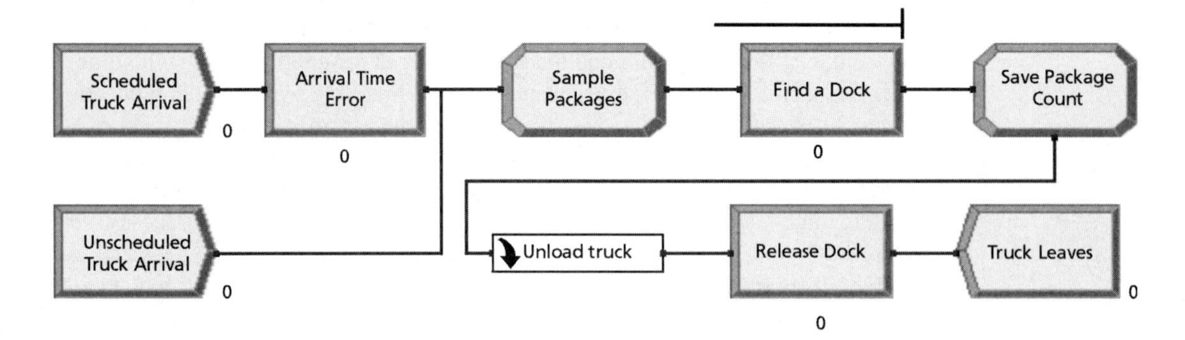

FIGURE 9.45 Truck unloading subprocess model

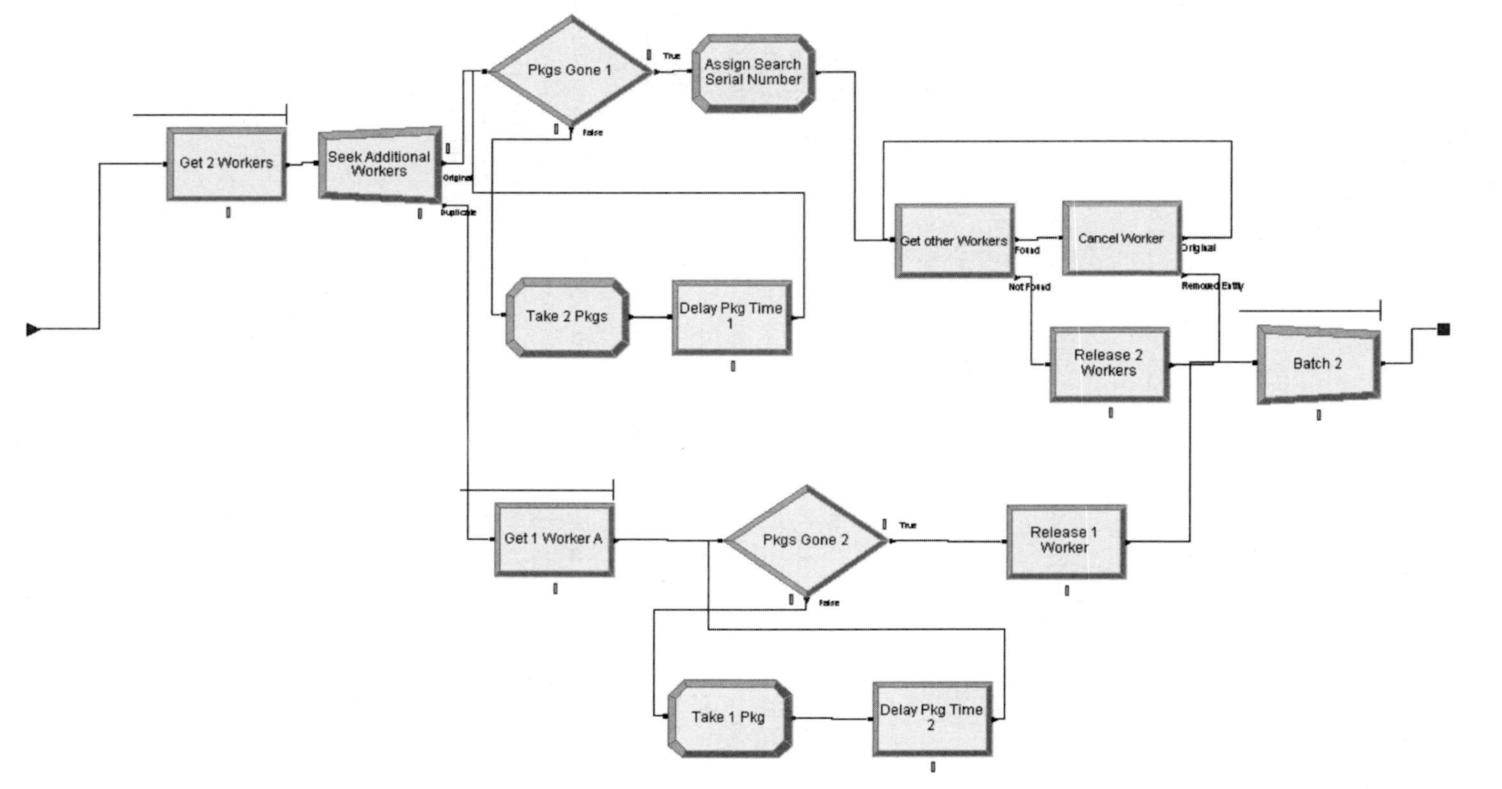

Submodel. This second type of submodel is somewhat more flexible than the first because it allows multiple inputs and outputs. This is the type of submodel we will use in the current model.

Before we can implement this revision of the unloading process, we must create variables to hold the number of packages in each truck, the number waiting to be loaded into each van, and the number loaded in each van. To handle this need, we created three array variables: `PkgCount`, which has two elements to store the number of packages on each truck; `RouteInv`, which has four elements to store the number of packages in the depot destined for each van route; and `VanInv`, which has four elements to store the number of packages loaded on each van. All elements of these arrays are initialized to zero.

The attribute `DockNum` was specified in the **Process** module `Find a Dock` to store the number of the dock that is seized. This value will be either 1 or 2 because there are only two docks. The **Assign** module `Save Package Count` just before the submodel to unload the truck stores the value of the attribute `Packages`, which contains the number of packages in the truck, in the variable `PkgCount(DockNum)`. In this way we are able to use `DockNum` to index the array `PkgCount`. Note that we cannot use the attribute `Packages` to track the number of packages remaining in the truck. Can you explain why?

Now let's look at the submodel to unload the truck. In Figure 9.45, we first request two workers. Once two workers are seized, we separate the process into three processes in the `Seek Additional Workers` module. The original process uses the two workers to unload the truck two packages at a time. The other processes will request one worker each and, if a worker is seized, each will unload the truck one package at a time. Using this approach, the requests for additional workers will be active until the truck is unloaded, so if one or two workers become available during the unloading process, they will be seized and participate in the unloading process. As you might imagine, it is possible that the truck could finish unloading before one or more of the additional workers becomes available. If this happens, the request for the worker is still active, so it must be canceled when the truck is empty.

In the process of unloading the truck, we must check to see if the truck is empty. Each time packages are removed from the truck, we will reduce the number of packages on the truck; thus, the signal to stop the unloading process is that `PkgCount(DockNum)` is zero or less. The **Decide** module `Pkgs Gone 1` does this. If there are more packages in the truck, then the original entity first removes two packages using the **Assign** module `Take 2 Pkgs` and enters a **Process** module `Delay Pkg Time 1`, which represents the delay while two workers remove two packages from the truck. The delay to remove one package is .75 min. Then another check is made to see if there are additional packages to remove.

The duplicated entities go through a process that is almost identical to the process for the original entity. Their process is shown in Figure 9.46. The only difference is that there may be two entities going through the process, each representing one additional worker, and each process takes just one package during each loop through the **Decide** module `Pkgs Gone 2`. When all packages are gone—that is, when `PkgCount(DockNum)` $\leqslant$ 0—the worker is released immediately.

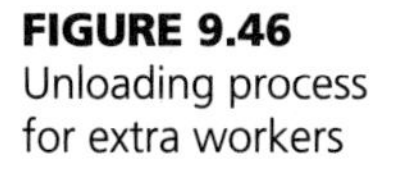
FIGURE 9.46
Unloading process for extra workers

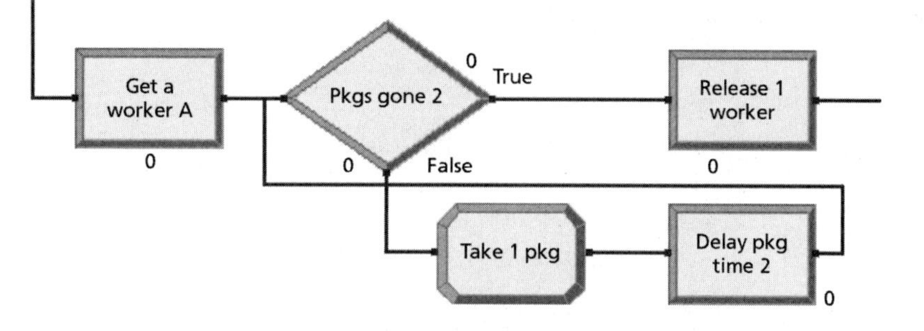

As mentioned earlier, it is possible for the duplicated entities to be waiting for workers when the truck finishes unloading. In this case, the workers are no longer needed for this truck, and the entity instances waiting for the workers must be removed from the queue in the **Process** module `Get 1 Worker A` and disposed. This responsibility is taken by the original entity and is done in the **Process** fragment shown in Figure 9.47. The procedure here is to search the queue to find all unfilled requests for workers and cancel the requests. First, we use the **Assign** module `Assign Search Serial Number` to assign the serial number of the current entity to a variable, `SearchNum`, to use as a search criterion. In the **Separate** module, all duplicated entities have the same serial number. Therefore, any entity that has the same serial number as the current entity is being requested to unload the same truck and is thus no longer needed.

The next module is a **Search** module located in the **Advanced Process** panel. The dialog for this module is shown in Figure 9.48. In this module, we search from the front to the back of the queue for any entity whose serial number has the same value as this variable `SearchNum`. The **Search** module has two outputs: one for when the search is successful and one for when the search fails to find an entity. If an entity that satisfies the search criterion is found, then the next module in the process is a **Remove** module (also on the **Advanced Process** panel) named `Cancel Worker`. This module simply removes an entity from a queue and sends it to a designated module. This module also has two outputs: one for the original entity and one for the entity that is removed from a queue. The top output loops back to the

FIGURE 9.47
Process module fragment for unload truck submodel

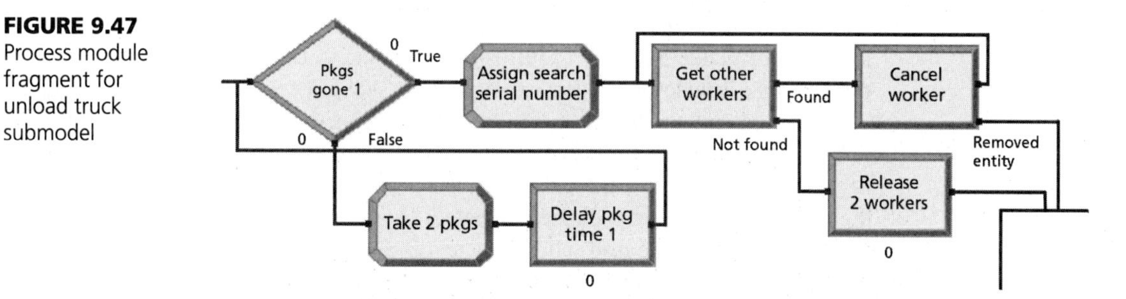

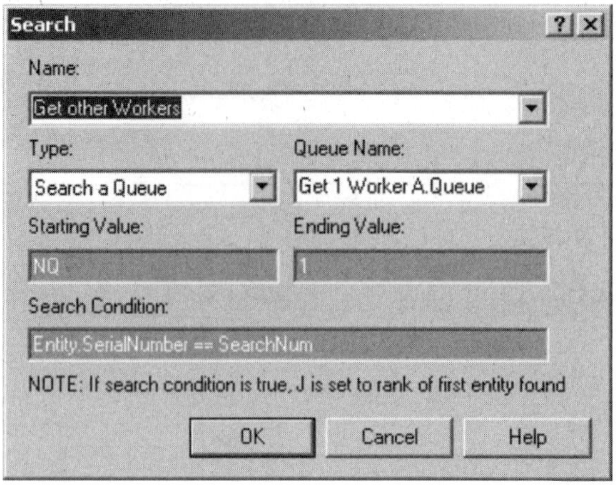

FIGURE 9.48 Search module for finding other worker requests

Search module `Get Other Workers`. The second output connects to the **Batch** module that recombines the entities that were duplicated.

You can see that the loop to search the queue and cancel requests for workers will continue until all entities in the queue are removed. Once all unmet requests for workers have been identified and canceled, the original process releases its two workers, and all entities are batched again to form one entity that represents the original truck

9.4.3 Phase 3: Data Collection and Experimental Design

Now that the model is implemented and represents the intended system, we are ready to tell it the data to collect. The first step in this process is to reexamine the performance measures that are important to management. The initial model description early in section 9.4 indicates the following performance measures:

1. mean time in system per package,
2. mean time trucks wait to start unloading,
3. mean time to load vans (including waiting time), and
4. mean cost per hour.

Each performance measure requires specific data to be collected and analyzed to produce an estimate of the desired parameter. We will consider them one at a time.

If we had modeled this system from the perspective of each package that flows through the depot, it would be easy to estimate the mean time in system per package. In that case, we would just record the arrival time for each package and record the total time when the package is deleted. However, packages were represented as a variable that is increased when they arrive and decreased when they are moved. Actually, we have three variables: `PkgCount` is an array with two elements containing the numbers of packages to be unloaded at each dock, `RouteInv` is an array with four elements that stores the numbers of packages in the terminal waiting to be loaded on each of the four delivery vans, and `VanInv` is an array with four elements that stores the numbers of packages already loaded on each of the four delivery vans. If we set each `VanInv` element to zero when the van leaves to deliver packages, then the total number of packages in the depot is the sum of all elements of these three variables.

We can use these variables to estimate the mean number of packages in the depot. Little's Law ($L = \lambda W$) gives us a means to estimate mean waiting time for packages (W) given the mean number of packages in the depot (L) and the arrival rate of packages (λ). Because the expected number of truck arrivals per day is six (four scheduled and two unscheduled), and each arrival has an average of 200 packages, the arrival rate is 6(200) = 1200 packages per day or 50 packages per hour. Thus, if our estimate of L is $\hat{L}$, our estimate of W is $\hat{W} = \hat{L}/\lambda$, where $\lambda = 50$ per hour. We must also note the assumptions of Little's Law: The system is one in which entities flow into it, spend some time in the system, and flow out. No other assumptions are made (other than technical assumptions that do not concern us here). In particular, we assume that the system neither creates nor destroys entities within it.

The mean time trucks wait to start unloading and the mean loading time for vans can be accommodated easily with Arena's standard data-collection and analysis facilities. Since trucks are modeled as entities, we can just mark the arrival time and the time at which the truck starts to unload and then tally the difference between these two times to compute the truck waiting times. For the mean loading time for vans, we can also mark the time the van arrives to start loading and the time the van leaves for its route and then tally the difference as the loading time.

During the operation of the model, a truck can be in one of three states: arrived and waiting for a dock, at the dock and waiting for enough workers to unload, and unloading at the dock. To represent this information, the model uses a variable for each dock to indicate its status, as well as a queue to hold the trucks waiting for a dock. The queue cannot be represented simply as a variable because each truck generates the number of packages it contains on arrival, and this attribute must be stored with the truck in the queue. Similarly, each van can be in one of four states: not present, present but not loading, loading, and delivering. A van is not present if it has not yet arrived for the day or has left after 4 P.M. Since there are four vans and each has its own loading dock, vans do not have to queue for a dock. A van may be present but not loading if it has arrived at 9 A.M. or has returned from a trip but lacks enough workers or packages to load the van. Once the van is finished loading, it delivers the packages and returns to the loading dock after a period of time. During this time, it is in the last state: delivering. To represent these states, each van has a variable that contains the current state of the van.

The appropriate performance measure for this system depends on the objectives for the system. It seems reasonable that management wishes to move the packages

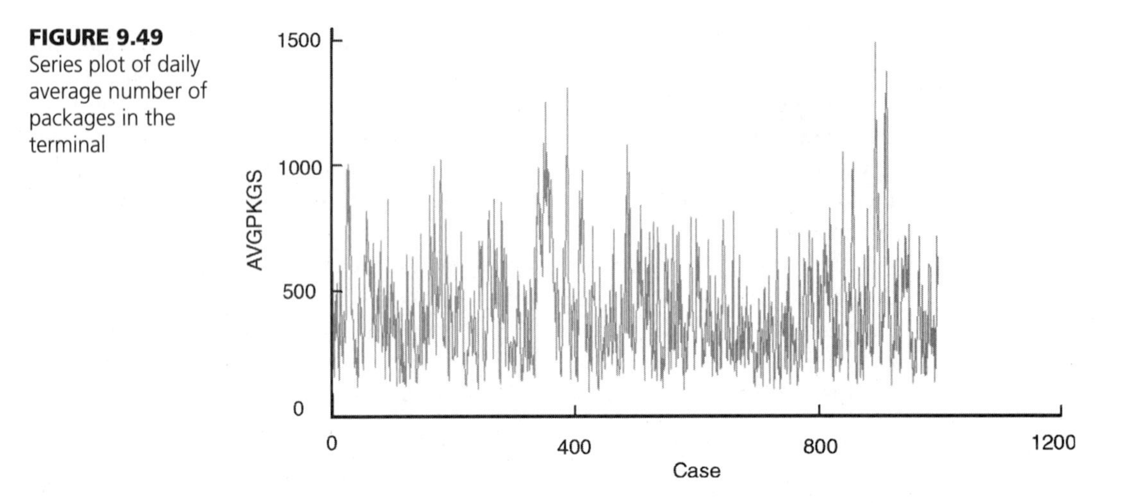

FIGURE 9.49 Series plot of daily average number of packages in the terminal

through this terminal as quickly as possible. Thus, the mean processing time per package in the terminal is a good measure of performance for the system. However, because the packages are not modeled as discrete entities, the mean waiting time cannot be estimated directly from waiting time data. Instead, it can be estimated indirectly from the mean number of packages in the system. A package will be assumed to be "in the system" if it is in a truck waiting in the queue, in a truck being unloaded at the dock, in the staging area for a van, or on a van waiting to leave for delivery. Thus, packages enter the system when a truck arrives, and they leave the system when a van departs. Let $N(t)$ denote the number of packages in the system at time t. Then it is easy to see that because of the daily schedule, $N(t)$ is generally low before 6:30 A.M., increases during the morning hours, and decreases in the late morning and early afternoon hours. Since there are unscheduled deliveries that can occur anytime, $N(t)$ can increase at any time when an unscheduled truck arrives.

The strategy for collecting and analyzing data from this simulation will be as follows: The unit of data will be the average number of packages in the system over one day (12 midnight to 12 midnight). Let $X_1, X_2, \ldots$ be the average number of packages for days 1, 2, ... Generally, these observations will be autocorrelated, so the batch means approach will be applied to them. Because the number of packages that stay in the system more than 24 hours is quite small, the autocorrelation function for these observations would be expected to die off rather quickly.

The simulation was run for 1000 days, and the time average of the number of packages in the system each day was computed. Figure 9.49 is a time series plot of the average number of packages in the system during each of the 1000 days. There does not appear to be an initial transient period. However, Figure 9.50, which is a plot of the autocorrelation function of these observations, shows significant autocorrelation in the daily averages. To compute a reliable confidence interval for the mean, a batch size of 20 was used, producing 50 batches of observations. Figure 9.51 shows the 50 batch means, and Figure 9.52 shows the autocorrelation function for the batch means.

FIGURE 9.50 Autocorrelation function plot for average number of packages in the terminal (original unbatched data)

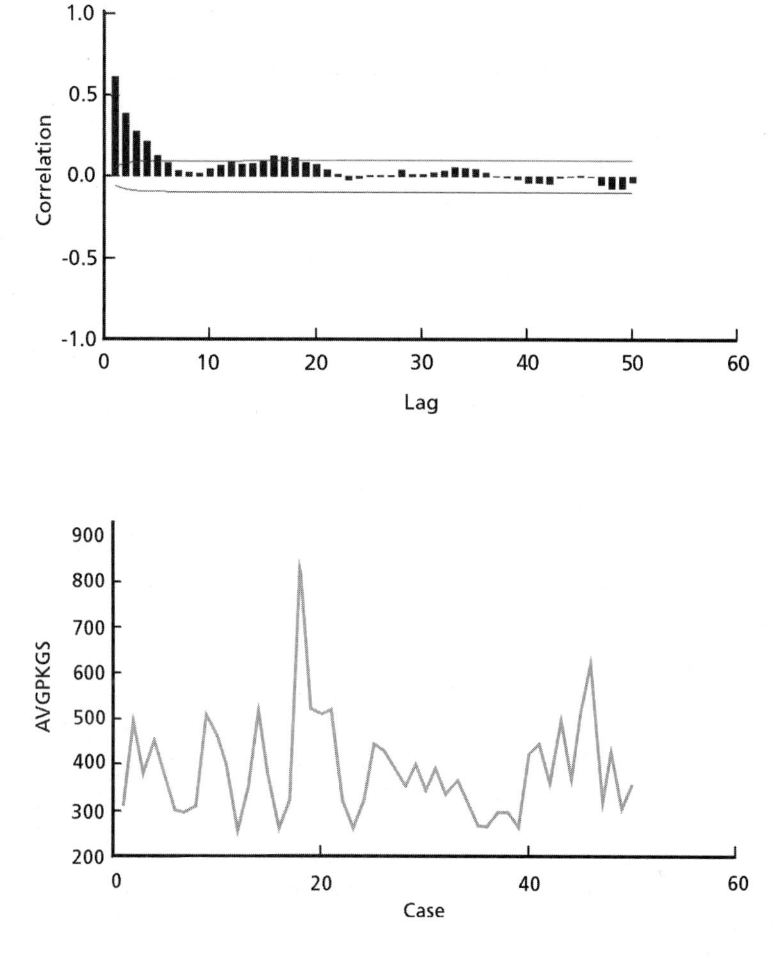

FIGURE 9.51 Batch means for batches of 20 observations

FIGURE 9.52 Autocorrelation function for batch means (batch size 20)

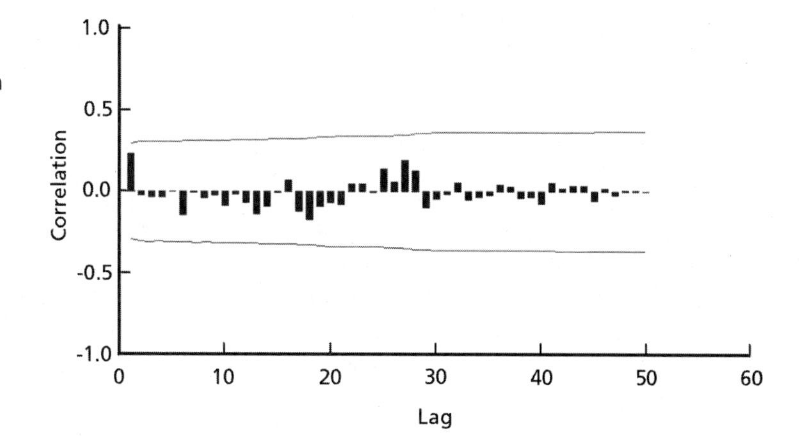

TABLE 9.8

	AVGPKGS
Number of batches	50
Minimum	251.993
Maximum	832.244
Sum	19,421.223
Mean	388.424
95% confidence interval (upper)	419.408
95% confidence interval (lower)	357.441
Standard error	15.418
Standard devation	109.021
Variance	11,885.595

9.5 SUMMARY

In this chapter, we have examined some more realistic models of service systems that a simulation analyst might encounter in practice. A service system is any system that provides some sort of "service" to a stream of customers. In today's "service economy," these systems are ubiquitous on both the personal level and in business offices, manufacturing, transportation, health care, government, and other aspects of life that we deal with directly and indirectly on a daily basis. Organizations consider it a high priority for their service systems to operate efficiently to provide short waiting times, high throughput, and low cost. Thus, it is significant that simulation is a useful tool to evaluate the performance of such systems.

In the simplest queueing system, there are servers and queues to hold customers waiting for service. The system can also be characterized by its arrival process, numbers of queues, queue behavior, numbers of servers, and server characteristics. Arrival streams can have independent interarrival times or be dependent in some way. Customers can arrive singly or in batches of a fixed or random size. If not allowed to join the system, customers can leave permanently or they can retry—that is, return after a period of time. After joining the queue, customers can renege. If the system has multiple queues, customers can jockey to move to a shorter queue.

Service can be provided to customers individually or in batches, and service times can be independent or correlated to model dependence on the order of service to customers. The queue discipline can be FIFO (first-in, first-out), LIFO (last-in, last-out), random, or priority, among other possibilities. The system can have a single server or many servers, and the servers can be identical or different. If service is delivered on a priority basis, the service can be preemptive or nonpreemptive, and preempted customers can resume service, leave the system, or be served by another server.

Most realistic service systems consist of a collection of simpler queueing systems in a network configuration. A *network* is a collection of nodes connected by arcs. The interpretation of the nodes and arcs depends on the model the network represents. If the model is that of a service system, then the nodes represent simple queueing systems and the arcs represent possible paths for customers when service at a node is completed. The simplest type of queueing network is a *tandem* network in which all customers proceed directly from the first service center to the second. In more complex models, the rules for moving from node to node must be included in the model. Queueing networks are natural to model using Arena because the Arena flowchart is designed to represent them.

Performance measures for service systems can reflect the customer's interests or the system management's interests. Customers usually want to minimize waiting time or minimize queue lengths. These objectives are actually the same since they are related by Little's Law: $L = \lambda W$. Management is frequently interested in maximizing resource utilization or minimizing mean system cost per unit time. Most models that are developed to help decision makers choose among alternative system designs or policies must have a cost model embedded in order to compute estimates of the cost of the system under alternative decisions. Arena includes a rather elaborate cost model that includes both fixed (i.e., per use) and variable (i.e., per unit time) costs for entities and resources. Costs can be allocated to one of five categories: *waiting, value added, non-value added, transfer,* and *other.* Arena allocates all costs accumulated while an entity is in a queue to the waiting category. The allocation of all other costs can be controlled by the user.

Manufacturing systems are an important part of our economy and an important class of service systems that can be effectively modeled and analyzed using simulation. Manufacturing systems have servers, queues, and customers, but these are generally machining or assembly centers, buffers for subassemblies, and parts, respectively. In addition, manufacturing systems have conveyors, transporters, and other types of equipment to physically move parts and unfinished and finished products in the system. Conveyors can be classified as *accumulating* or *nonaccumulating.* Transporters can also have many different characteristics such as their capacity and route, and they can have intelligence. Conveyors and transporters add an additional dimension of complexity and difficulty to the modeling process. We demonstrated how to model manufacturing systems using Arena by building a sample model that includes conveyors, inspection, and rework. This model included an animation of the system showing the movement of parts and subassemblies on conveyors; we also demonstrated how to build a model that has a separate animation.

Transportation systems include airports, docks, highways, mass-transit systems, railroads, and trucking whose objective is to move people, packages, or other materials from one place to another. These systems also have specific characteristics that call for more advanced modeling techniques. Transportation systems usually include a schedule to coordinate the movement of people, transporters, packages, and other elements of the system. These models usually have more complex logic that represents the interaction of the entities in the system.

These ideas were illustrated by developing an Arena model of a package delivery system that includes scheduled and unscheduled truck arrivals and scheduled van arrivals. In this model, packages are delivered to the depot by trucks and delivered to the customers by vans. At the depot, packages are sorted for delivery on one of four routes serviced by vans. The packages were not modeled as individual entities but were represented as a variable, resulting in a simpler and more efficient model. The model also included logic for ending van deliveries at the end of the day. One complicating characteristic of this model was the requirement that workers could be reassigned to an ongoing task when they completed work on a task. This was accomplished using more complex modeling techniques in the submodels. Both the manufacturing and transportation models in this chapter were developed in a stepwise fashion to illustrate the process of building a more complex model by iterating through several simpler models and refining the model to introduce new features.

We cannot hope to discuss every type of model that would need to be developed. Indeed, we can hardly imagine the types of models that will be developed in the future. Simulation can be used to model processes in such disparate areas as communication systems and health care. Each area will have its own special modeling challenges. In this chapter, we have demonstrated techniques that are useful in many models and have hopefully helped you develop your knowledge and skill so that you feel comfortable exploring and learning new techniques.

References

CNN. 2001. Available online via <http://www.cnn.com/2001/ALLPOLITICS/07/31/election.laws/index.html>.

Kelton, W. D., R. P. Sadowski, and D. A. Sadowski. 2002. *Simulation with Arena.* Boston: McGraw-Hill.

Problems

In some of the following problems, needed distributions or parameters (or both) may not be provided. In these cases, please provide reasonable values.

9.1 Consider a model of a barbershop with one barber. The time required for the barber to cut a customer's hair is uniformly distributed between 10 and 20 minutes. Customer arrivals are Poisson with mean interarrival time of 20 minutes. The shop has five chairs for waiting customers. If a customer arrives to find the shop full, he returns after 10 minutes with probability .3 and leaves without returning with probability .7. The shop opens at 7:30 A.M. and serves all customers who arrive before 6 P.M. Develop an Arena simulation of this system and use it to estimate (1) the proportion of arriving customers who get a haircut and (2) the mean waiting time for customers. If the barber adds another chair, how will these two parameters change?

9.2 Change the model in problem 9.1 so that customers will return to the shop no more than one time. How does this affect the two parameters when the shop has five and six chairs?

9.3 Change the model in problem 9.1 to have two identical barbers. Customers will select the first available barber. The distribution of time to cut hair is uniformly distributed between 10 and 20 minutes for each barber. Arrivals are Poisson with mean interarrival time of 10 minutes. The probability of a retrial is .3. Compute the parameters—the proportion of customers getting a haircut and the mean waiting time—using five, six, seven, and eight chairs in the shop. How do these parameters compare to those computed in problem 9.1?

9.4 Develop an Arena simulation of the barbershop with two barbers, Sam and Dave, who are not identical. Sam's service time is uniformly distributed between 10 and 20 minutes. Dave's service time is uniformly distributed between 15 and 30 minutes. Customer arrivals are Poisson at rate 5 per hour. Sixty percent of arriving customers prefer Sam, and 40 percent prefer Dave. Customers will only allow their preferred barber to cut their hair. The shop has six chairs, and 30 percent of arriving customers who find the shop full will return after a delay of 10 minutes. The shop opens at 7:30 A.M. and serves all customers who arrive before 6 P.M. In addition to the proportion of arriving customers who get a haircut and the mean waiting time, estimate the utilization of each barber.

9.5 Consider the model in problem 9.4, but assume that some customers have no preference for their barber. Suppose that 50 percent of arriving customers prefer Sam, 20 percent prefer Dave, and 30 percent have no preference. When customers arrive, they are served in strict first-come, first-served (FCFS) order if possible. If a customer prefers a particular barber, then he will wait until his preferred barber is available. But if, for example, a customer with no preference is at the front of the queue when Sam becomes free, he will receive service from Sam even if there is a customer behind him who prefers Sam. Develop an Arena simulation of this model and use it to estimate the parameters described in problem 9.4.

9.6 Consider the following modification to problem 9.5: A customer's preference can change from a specific barber to "none" if he waits long enough. For each customer, after waiting a uniformly distributed length of time between 30 minutes and 60 minutes, his preference will become "none." Develop an Arena simulation and use it to estimate the mean waiting time for customers and barber utilization for each barber. How do the estimates compare to the values computed in problem 9.5?

9.7 Modify the barbershop model in problem 9.2 so that a haircut involves three stages: washing, cutting, and styling. Each barber does all three stages for a customer. The service times are uniform (3, 6), uniform (5, 12), and uniform (2, 6) for the three stages, respectively. All service times are in minutes. Twenty-five percent of customers skip the washing stage, and 40 percent skip the styling stage. All customers have their hair cut. The shop opens at 7:30 A.M. and serves all customers who arrive before 6 P.M. Develop an Arena simulation of this system and use it to estimate the proportion of arriving customers who get a haircut and the mean waiting time when the shop has five and six chairs.

9.8 Suppose that the barbershop in problem 9.2 now takes appointments. Appointments are scheduled every 15 minutes, and 35 percent of each day's appointments are filled. The overall arrival rate is the same as in problem 9.2 but an average of 35 percent of customers arrive for an appointment. The remainder arrive according to a Poisson process with an appropriate arrival rate to make the expected number of arrivals per day the same as in problem 9.2. Customers who have an appointment will be served at the first opportunity after their appointment time. Customers who do not have an appointment will wait until the barber is free for their haircut. Develop an Arena simulation of this system and use it to estimate the proportion of arriving customers who get a haircut and the mean waiting time when the shop has five and six chairs.

9.9 A realistic model of a barbershop must include the morning, lunch, and afternoon breaks the barber takes. Assume that the barber takes a 15-minute break at 10 A.M. or the first time after 10 when he finishes the customer whose hair he is cutting at 10. He takes a 15-minute break at 3 P.M. using the same rule. At noon, he takes a 30-minute break for lunch using the same rule. These breaks are not long, so customers will wait until he returns if they are waiting for a haircut. The customer waiting time during the barber's break is included in mean waiting time calculations, but the time the barber is on break is not included in barber utilization computations. Develop an Arena simulation of this model and use it to estimate the proportion of arriving customers receiving a haircut, the mean waiting time for customers, and the utilization of the server.

TABLE 9.9

Node	Service Time Distribution
2	Triangular (10, 25, 100)
3	Uniform (4, 20)
4	Triangular (20, 60, 300)
5	Uniform (2, 5)
6	Triangular (60, 180, 300)
7	Triangular (60, 300, 600)
8	Uniform (4, 10)

9.10 Develop an Arena simulation of the election poll model in Figure 9.13. Nodes 2 through 8 represent places where voters might wait for service. The service time distributions for these nodes are given in Table 9.9 (time units are seconds). Ten percent of voters are not on the roll and must go to node 4 to determine their eligibility to vote. Of those, half are eligible and half must leave without voting. Fifteen percent of voters are handicapped. For this model, assume all queues have infinite waiting room. Use the Arena simulation to estimate the mean queue length for all queues, the mean waiting time in the queue at node 1, and the mean time in the system.

9.11 Using the election poll system described in problem 9.10, assume the queue capacity at node 6 (voting in the voting booth) is 1. If that queue is full, node 5 is blocked. How does this affect the mean queue lengths and mean time in the system?

9.12 Modify the model in problem 9.11 to make the queue capacity 3 at node 3. If the queue at node 3 is full, node 2 will be blocked. How does this change affect the mean queue lengths and the mean time in the system?

9.13 In the barbershop model in problem 9.4, barbers are paid $15 per haircut. Sam is paid $25 per hour and Dave is paid $20 per hour. The cost of the barbershop (rent, insurance, advertising, supplies, and other expenses) is $200 per day. Estimate the expected profit per day for problem 9.4 using these parameters.

9.14 Consider the following model of manufacturing operations for a custom manufacturer of kitchen and other household cabinets: The carcasses (basic cabinets) and doors are made on separate production lines. Thus, when an order is received, it is separated into an order for the carcasses and another for the doors.

TABLE 9.10

Doors	Probability
0	.10
1	.30
2	.40
4	.20

TABLE 9.11

Step	Distribution
Cut parts	Uniform (5.0, 8.0)
Assembly	Uniform (2.5, 6.0)
Finish	Uniform (1.6, 2.8)

TABLE 9.12

Step	Distribution
Assembly	Triangular (1.5, 2.5, 4.0)
Finish	Uniform (1.4, 1.7)

Table 9.10 shows the distribution of the number of doors per case.

Each carcass goes through three steps in manufacturing: cut parts, assemble the case, and finish the case. The shop has three production lines for doors. Table 9.11 gives the distributions of process times for the three steps.

Because the parts for doors are standard and are precut, doors only go through two manufacturing operations: assemble the door and finish the door. The shop has two production lines for doors; the distributions of process times for each door for the two steps are given in Table 9.12.

The craftsman assembles all doors for a cabinet as a batch. For the model, assume that the time to complete the batch is computed by multiplying the sampled process time by the number of doors in the batch.

Completed carcasses are placed on an accumulating conveyor and moved to an assembly area. Finished doors are also placed on a separate accumulating conveyor for movement to the assembly area. In the assembly area, doors are assembled to the cases and the finished products are packaged for shipment. The time to complete the final assembly is computed from $T_1 + N \times T_2$ where T_1 is uniformly distributed between 1.6 and 2.3 minutes, N is the number of

TABLE 9.13

Orders	Probability
0	.20
1	.45
2	.25
3	.10

doors on the cabinet, and T_2 is triangular with parameters 1.2, 1.8, and 3.0.

The number of orders released to the shop per day has the distribution shown in Table 9.13. Orders are released at the start of each day. The shop works from 7 A.M. until 6 P.M. each day.

Develop an Arena simulation of this system and use it to estimate the mean time to produce a cabinet and the mean number of cabinets in production.

9.15 Modify the model in problem 9.14 to include a transporter that moves cabinets from the finish room to the final assembly area. A transporter can move one cabinet, and the time required to move a cabinet from the finish room to the assembly area has a gamma distribution with parameters 3 and .4.

9.16 The following is a description of the operations of a medical clinic. The clinic has one doctor, two nurses, two nurse's aides and one receptionist. The building has a waiting room, six examination rooms, a laboratory, and a pharmacy. The lab has one lab technician, and the pharmacy has one pharmacist. The clinic is a walk-in clinic for urgent care, so patients do not have appointments. Patient arrivals are a Poisson process with rate eight per hour. Arriving patients must register with the receptionist. After registration, the patient waits until an aide takes him to an examining room. Once in the examining room, the patient will wait for a nurse to do a preliminary assessment and take a history or update the notes on the patient's condition. Nurses are permitted to treat certain mild conditions such as sore throats or mild infections that apply to 20 percent of patients. For the other 80 percent, the nurse will then tell the doctor that the patient is ready to be seen. The doctor then examines the patient. The doctor orders lab tests for 60 percent of patients.

The procedure for lab tests is that the nurse draws blood and the specimen is then sent to the lab where the lab technician does the testing. Lab tests are done in FCFS order. The patient waits for the lab test to be completed. After the test, the doctor will see the patient again in 50 percent of cases. In the other 50 percent of cases, the nurse will discuss the results of the lab tests with the patient. After the return visit by the doctor or nurse, the patient will leave the examining room.

TABLE 9.14

Process	Distribution
Registration	Uniform (1.0, 2.0)
Aide to exam room	Uniform (.2, 1.3)
Nurse assessment	Triangular (2.0, 4.0, 10.0)
Doctor examination	Triangular (4.4, 6.1, 12.2)
Lab test	Uniform (3.4, 13.9)
Doctor second visit	Triangular (3.2, 4.5, 9.6)
Nurse second visit	Triangular (2.4, 6.1, 10.3)
Prescription filling	Uniform (2.3, 5.9)

A prescription is written for 80 percent of patients. If the patient has a prescription, she will take it to the pharmacy for filling. Prescriptions are filled in FCFS order. After receiving the prescription, the patient leaves the clinic.

Distributions for various process times in this model are given in Table 9.14 (all times are in minutes). The clinic operates from 8 A.M. to 4 P.M. daily. All patients who arrive before 4 P.M. will be seen the day of arrival.

Develop an Arena simulation of this model and use it to estimate the utilization of the doctor, the utilization of the nurse, and the mean total waiting time of the patients while in the clinic. (Total waiting time includes the time waiting for an aide, doctor, nurse, lab test, or prescription.)

9.17 For the clinic in problem 9.16, add the feature that patients can have two priority categories: normal and urgent. Urgent patients have higher priority than normal patients when served by the aide, doctor, or nurse but the same priority in the lab and pharmacy. Priority is not preemptive, however. Thirty percent of patients are urgent, and the rest are normal. Use the Arena simulation to estimate the same parameters as in problem 9.16.

9.18 Add the following cost model to the clinic in problem 9.16. The clinic receives \$50 for each patient visit. The costs incurred by the clinic are \$20 per hour for each nurse, \$12 per hour for each aide, \$7 per hour for the receptionist, \$100 per hour for the doctor,

$40 per hour for the lab (and lab technician), and $50 per hour for the pharmacy (and pharmacist). Use the Arena simulation to estimate the expected profit (or loss) per week for the clinic.

9.19 The following is a model of a fast-food restaurant. The restaurant is open from 6 A.M. to 10 P.M. It has two stations where arriving customers order food and pay for the order. An order will include a random number of sandwiches, a random number of orders of french fries, a random number of salads, and a random number of drinks. Some orders will include none of some of these items but all orders include at least one item. When the order is taken, it is displayed via a monitor in the food preparation area where the food is prepared and assembled. Payment is made immediately after placing the order. The restaurant has five employees preparing food. Two assemble hamburgers and other sandwiches to order. Two more employees prepare salads to order. One employee manages preparation of french fries and gets the drinks for orders. French fries are cooked in a batch that will serve between 12 and 20 (uniformly distributed) orders of fries. The time to cook and prepare a batch of fries has a uniform distribution between 5.5 and 6.8 minutes. When the employee notices that the number of remaining orders of fries is 5 or fewer, she begins cooking a new batch. This occurs a short but random length of time after the number of orders remaining drops below 5. The completed order is delivered to the customer, and she leaves the order station.

Develop an Arena simulation of this restaurant. Provide reasonable distributions and parameter values where needed. Use the model to estimate the mean waiting time for customers and the mean number of orders pending.

9.20 Modify the model in problem 9.19 to allow the servers to take new orders while waiting for previous orders to be filled. In addition, customer arrivals vary with the time of day. Table 9.15 gives the arrival rates.

Use this model to estimate the mean number of orders pending and the mean waiting time for customers.

9.21 Modify the model in problem 9.19 to include a cost model. This cost model should include value added and non-value added costs. Costs should be allocated to all processes and subjective costs should be allocated to customer waiting time.

TABLE 9.15

Period	Arrival Rate
6 A.M. – 9 A.M.	44
9 A.M. – 11 A.M.	16
11 A.M. – 1:30 P.M.	59
1:30 P.M. – 5 P.M.	12
5 P.M. – 7:30 P.M.	54
7:30 P.M. – 10 P.M.	8

9.22 Consider the following modification to the package depot model in section 9.4.2. Packages are presorted before being placed on trucks so that the packages are stored in containers with between 5 and 15 packages per container. All packages in a container are for the same delivery route; therefore, the group of packages can be handled by a single worker and delivered to the appropriate van's staging location. When handling containers, workers can unload at a rate of 40 containers per hour. The number of containers per truck is a normally distributed random variable with mean 20 and standard deviation 3, truncated to an integer. To handle the containers, which can be quite heavy, a nonaccumulating conveyor has been installed from each truck dock to the van staging area. Containers move at a speed of 1.2 feet per second on the conveyor and require 8 feet when moving but only 4 feet when accumulated on the end. The length of each conveyor is 32 feet.

Modify the Arena simulation of the package depot model to include these changes. How do the performance measures compare to those presented in section 9.4.3?

9.23 Add the following cost model to the package depot model in section 9.4. The terminal costs $67 per hour to operate while it is open. Workers are paid $12.10 per hour; vans (with drivers) cost $32.16 per hour. When they are waiting in the terminal, the time is considered non-value added, but when they are on their delivery route, the time is considered value added.

9.24 One model of an urgent care facility involves patients who require service from a random number of medical personnel (nurses, doctors, and other technicians), depending on the nature and severity of their illness or injury. Patient arrivals are a Poisson process with rate 20 per day. Patients are first examined by a

TABLE 9.16

Number	Probability
1	.30
2	.40
3	.15
4	.10
5	.05

triage nurse, which takes between 1.5 and 10 minutes. Then they are treated by a random number of personnel. Table 9.16 gives the distribution of the number of medical staff needed to treat a randomly selected patient.

The clinic has seven staff members. Treatment time has a triangular distribution with parameters 10, 30, and 80 minutes, regardless of the number of people treating the patient. Develop an Arena simulation of this model and use it to estimate the mean waiting time for patients. The waiting time is the time between arrival and the start of treatment.

9.25 Modify the model in problem 9.24 so that patients also have priorities. Let the number of personnel needed by a patient denote that patient's priority. Patients who need more personnel to treat them have a more serious condition and thus higher priority. In this version, patients are treated in priority order, nonpreemptively. Use the Arena simulation to estimate the mean waiting time for patients and compare the results with the value computed in problem 9.24.

10

Simulation in Practice

This chapter seeks to put the material in the rest of the book into the context of the real world. Simulation practice in the real world is not as clean and easy as the relatively simple problems in this book have been. Real-world problems are messy, and the analyst must develop skills and judgment to deal with them. We open with an overview of the major themes in the chapter—that the simulation analyst must be a generalist, a *systems* analyst. The analyst must also have skills to manage the project and the team that works on the project. Effective communication with the client is a requirement, and problems with data availability and data collection will often have a direct impact on the model. The analyst must also manage the model—its development, use, and deployment. This chapter concludes with practical advice for the person who plans to be a simulation analyst.

The first nine chapters of this book have been devoted to teaching the concepts and methods of simulation. These chapters have presented the technical aspects of simulation: creating an appropriate model, implementing the model, input data analysis, planning and making simulation runs, statistical analysis of output data, and interpreting the results. These are all important methodologies, but in practice simulation involves much more than just these technical aspects.

Simulation projects are special projects outside the normal course of system operations. They are initiated in response to some problem that suggests a change needs to be made to the system. Each simulation project has a *sponsor*, the person who requests the project. We will use the term *client* to denote the person for whom the project is being undertaken and who has a stake in the successful outcome of the project. The client doesn't normally know simulation methodology. She just knows there is a problem and some analysis is needed to determine the proper decisions. When the project begins, you do not know that simulation is the proper tool to solve the problem. So part of the project involves a systems analysis and a determination of the appropriate tools and methodologies to use. For this reason, we will often refer to the person doing the analysis as a *systems analyst* rather than a *simulation analyst*.

Among those who have applied simulation to solve real problems in business, there is much agreement on the what is needed for success. The works by Musselman (1994) and Sadowski and Grabau (1999) provide excellent advice concerning how to successfully run a simulation project. The suggestions in this chapter are based on their recommendations as well as the experience of the authors.

10.1 A SIMULATION PROJECT

Before we get into the details of applying simulation in the real world, we will consider a project that involves modeling an emergency department (ED). We will use the term *ED project* to refer to this project. Although this project takes place in a health care system, the experiences and lessons apply to any system in manufacturing, logistics, government, or other area. We will use this project to illustrate the considerations and issues that can be encountered in a real project.

The hospital administration was in the process of building a new urgent care center (UCC) adjacent to the emergency department. Hospital management expected that during its hours of operation, most walk-in patients would present to the UCC, and the most serious of these cases would be referred by the triage nurse to the emergency department. All patients arriving by ambulance and those who arrive when the UCC is closed would be treated in the emergency department. Management was concerned about the capacity of the entire system and wanted guidance regarding optimal staffing schedules and ways to minimize patient waiting time. Although they presented no numerical evidence, they felt that the current emergency department was sometimes overstaffed and underutilized, and sometimes understaffed and overutilized. They also felt that patients sometimes had to endure unnecessarily long waits. The volatility of patient arrivals was the primary characteristic of the system that gave them problems with staffing and resource allocation decisions.

10.1.1 Project Origination and Systems Analysis

A consulting group consisting of a physician, a systems analyst who specializes in simulation, and two staff analysts was retained by hospital management. Initially, management failed to inform the medical director of the emergency department about this project or get his cooperation. When the consultants needed access to the physical facilities and the personnel, cooperation was difficult because management had not convinced the medical staff (physicians and nurses) of the need for problem analysis or the benefits that could result from the project. Moreover, the employees were skeptical of the reasons for bringing in outside consultants and felt their jobs might be in jeopardy. The team met with the medical director and explained to him the benefits to be derived from the work as well as to assure him that the work posed no risk to him or his employees. Moreover, the consultants requested that a team consisting of themselves and some key ED personnel be formed to do the analysis.

Once access and cooperation were gained, the team needed to understand how the current system functioned. Interviews were conducted with ED personnel to document the flow of patients through the department. Because the ED personnel were not trained in discrete-event simulation, the interviewer had to spend time discussing the possible scenarios involving patient arrival, initial examination, diagnosis, treatment, and discharge. Nurses and physicians who were intimately familiar with the system often were found to overlook routine parts of the procedures, and the interviewer had to prompt them to include these activities. For example, a registration clerk must register all patients. If the patient is critically ill, relatives or others can register the patient, or the registration clerk can come to the patient's bedside. Otherwise, the patient must go to a desk where the registration clerk enters the patient's data into a computer. This process is routine for all patients and the medical staff are not directly involved, so they tended to ignore the delays due to registration.

10.1.2 Data Collection

Data were also needed to establish portions of the model. For example, how many times does the physician visit a patient? Does the physician visit once except on rare occasions? Or does she visit two or three or more times? If she visits multiple times, what distribution should be sampled to select the number of visits? As another example, the order of activities for a patient depends on the patient's initial diagnosis. If the patient has been seriously injured in an automobile crash or he is suspected to have had a heart attack, then the initial interview and examination by a triage nurse will be dispensed with and treatment will begin immediately by a physician. Data concerning the various initial diagnoses and treatment routines for them would be helpful to determine which activities to include in the model. It was clear that there would not be enough time to develop a highly detailed model, so patients with rare conditions that require special treatments—for example, a patient presenting with an extremely contagious disease—would not be included in the model.

Members of the team were given access to all ED patient records or charts, provided they signed a confidentiality agreement. Because patient records are required by law to be confidential, the hospital had to carefully monitor the records that were examined. The consequences of releasing patient data, even information that might have been considered unimportant, included the possibility of a lawsuit for the consultant and the hospital. The team found that while the charts contained useful information about patient arrival times, the patient's initial diagnosis on arrival, and the times when nurses and physicians first interacted with the patient, they did not include critical data about the lengths of time for examination and treatment. Data kept on patient charts are recorded for medical and legal requirements, not for system modeling needs. Data stored on the hospital's information system provided information about the number of arrivals per hour, but other data on X-ray and lab times could not be accessed because they were stored on incompatible computer systems that were actually operated by other physician groups who worked in the hospital under contract.

Some data regarding how to classify the patients into groups by severity of condition were available on the patients' charts, but medical expertise was required to interpret the data, and the information generally did not provide direct data about the order of the visits and examinations. For the model, the analyst needed to define categories that described how the patient was managed in the ED. The model was not concerned about the medical details, only how the patient entered, flowed through, and exited the ED. Some data that could be useful for this were available on the hospital's information system but could not be retrieved in a timely manner because the system was old and the job of matching diagnosis codes to categories was too involved to be done with the available time and resources.

Some data had to be collected manually to estimate lengths of times for examination and treatment by nurses, doctors, and other technical personnel; times to complete registration information; the number of bedside visits to the patient by the doctor; and other operational measures for the system. Over a period of approximately one week, members of the team observed approximately 50 patients. Considerable effort went into collecting this small set of observations, and it included only two patients who were critically ill. Thus, estimates of the proportion of critical patients would have a very large variance and estimates of the treatment time distribution could not be made with any degree of certainty.

When the data-collection procedure was being set up, the team had to agree on the definition of the start and the end of a physician's visit to a patient. Does the visit begin when the physician starts to go to the patient's room or when the physician actually enters the door and begins interviewing the patient? Similar decisions had to be made for nurses' visits to patients. The decision was made to include travel time in all measurements, and thus travel would not be a separate activity in the model.

The model that was constructed was much simpler than it could have been for two reasons. First, the model was built just to answer the questions and concerns posed by management. Second, while a more elaborate and more accurate model could have been built, data were not available to provide parameter estimates for a more detailed model and sufficient time was not available to collect these data. For example, the model could have sampled the diagnosis code from a historical distribution and even based the sample on the time of day and day of week. Patients who presented with flu symptoms could be examined once by a physician and discharged; patients who presented with asthma could be examined and treated several times until the physician decided to discharge the patient to the hospital or to home. (This level of detail was not necessary for the model, but it was felt that having reports that included medical details would help sell the model to the medical staff and the administration.) However, data were simply not available to determine the appropriate distributions or the relationship between examination and treatment paths and times and the diagnosis code. Data were the limiting factors in the model design.

10.1.3 The Preliminary Model

Once a model had been proposed and a graphical representation had been created and presented to the members of the team, agreement was almost immediate. With current simulation software, model development was quick and painless.

The preliminary model was presented to hospital management, including the medical director of the emergency room. A lengthy discussion followed regarding what specific numerical measures should be used to judge the performance of the system. For example, the administration stated that it wanted to minimize patient waiting time. However, patient waiting times vary considerably. Should the mean patient waiting time be used? Or should some percentile such as the 90th be used? The consultants spent considerable time explaining the alternatives to the management and working to arrive at a consensus. Additional discussions addressed whether to present measurements of physician utilization (the percentage of time the doctors were busy) and whether these measurements were even meaningful since certain physician activities, such as charting, were ignored in the model. The charting activity takes much of the physician's time and normally should have been part of the model. However, the decision was made to not include this activity because it would be hard to measure since the persons involved would know they were being timed and it is often done in conjunction with other activities. For example, the physician often will enter data or comments in the chart while interviewing and examining the patient. The team decided to include physician utilization estimates but plot them on a graph with the understanding that the maximum should be 70 percent because the other 30 percent of the time they were charting or doing other required duties such as phoning to follow up on patients, calling in prescriptions, and so on.

Before the model could be used, it had to be validated. Data on patient waiting times and the number of patients waiting were used for validation. However, current estimates of these performance measures were not available, although some raw data were available on the hospital's information system. The consultants obtained the data from the database and computed estimates of these parameters. The data used to compute parameter estimates and fit distributions for the model were now several months old, and the hospital had made a point of noting that the workload on the emergency room was seasonal and appeared to have an increasing trend. The data used had been collected in January, but the validation month was May. The team was concerned about the seasonal effects. The mix of illnesses and trauma is different in May because more people are active in warm weather. They worried that this difference could cause the model to be invalid. Nevertheless, the simulation was run and estimates computed from the output. Mean waiting time from the model was 20 percent greater than the actual mean for the validation period. Similar results were found for the mean number of patients waiting in the ED. Although the consultants wished the estimates were closer, it was concluded that 20 percent was sufficiently close to conclude that the model was valid.

During the project, the consultants communicated regularly with management. Communication took the form of periodic reports to inform management of the progress that was being made and sometimes as informal conferences to resolve problems.

10.1.4 Project Conclusion

A final presentation was made to the hospital administration. A graphical version of the model was shown along with an animation, but most of the time was spent discussing the assumptions behind the model, the results obtained from the simulation runs, and the recommendations to the hospital administration. The consultants concluded that the 90th percentile of patient waiting time could be cut almost in half if patients were more carefully screened in the urgent care facility and the average time to return labs and X-rays were reduced by 30 percent. Because these systems were not under the hospital's administration, specific recommendations regarding how to do this were not made, but this recommendation would become part of an upcoming contract renewal for the groups that operate these systems. The model also predicted that overutilization of nurses and excessive waits by patients on weekends could be avoided by adding another nurse on each of two shifts. The model showed this to be a more efficient solution than adding another physician. Among the recommendations was a strong endorsement of the idea that the hospital's information systems be upgraded to automatically collect input data for the model such as the length of time the physician sees a patient and the number of times the physician visits each patient. Specific recommendations regarding how these data could be easily and cost-effectively collected and merged with the current database were also made. Finally, the consultants suggested that systems could be put into place to allow the hospital to access the simulation models over the Internet. This would allow the hospital and the consultants to improve their efficiency and reduce costs.

The administration was pleased with the results of the project, felt that the information provided was useful, and decided to continue to use the model periodically as the patient load increased to determine the best changes to make to adjust to a changed environment. The consultants provided documentation and training for the model, and they helped the information systems personnel design a system to automatically collect the input data needed for the model. The data-collection system was estimated to cost the hospital $15,000 per year, but the administration judged that the cost was justified by the savings from the use of the model. The data-collection system would guarantee a large quantity of input data, and it would guarantee that the data were recent and relevant to the model.

The purpose of this example is to show that the tasks involved in developing the model and running the simulation constitute just a small part of the overall simulation modeling activity. Problems with all of the following issues had to be addressed:

- establishing the project scope and boundaries,
- obtaining support from top management,
- communicating with management and employees,
- collecting and analyzing data,
- dealing with lack of data,
- interacting with information systems,
- interpreting available data,
- choosing performance measures,
- validating the model,
- implementing the model and the recommendations, and
- presenting the results.

The resolution of all of these issues required sound judgment, experience, and creativity.

10.2 THE SCIENCE OF SIMULATION PROJECT MANAGEMENT: PROBLEM ANALYSIS AND SOLUTION PROCESS

The systems analysis process generally should follow a well-defined sequence of steps: problem formulation, project planning, system analysis, model creation, data collection and analysis, model building, model validation, experimentation, reporting, and implementation. It is not always the case that these steps will occur in exactly this order, but each step is important and must be included.

10.2.1 Problem Formulation and Statement of Objectives

Each project is unique. The system is unique, and often the decision problem is unique. So it is important to understand exactly why an analysis is requested. The best way to approach problem formulation is to start by asking, Exactly what decisions are contemplated? Who will be making them? What are the choices? The answers to these questions will provide the core information to establish the objectives of the project.

For example, in the ED model, management wanted to know whether the new system consisting of the emergency department and the urgent care facility would be adequate to handle the anticipated patient load and what numbers of physicians, nurses, and registration clerks on each shift would provide optimal performance of the two departments. The decisions would be made by the vice president for clinical services, who is in charge of the emergency department as well as other parts of the hospital. Other possible decisions included locating a portion of the laboratory in the ED, and making greater use of portable X-ray equipment to avoid the delays in sending patients to the radiology department.

The problem formulation and statement of objectives provides information useful to defining the scope of the project and therefore the model. It also should include information about the limitations of the model and the resources available for the project. For example, the ED model did not include detailed models of the radiology or laboratory departments because they were operated by different organizations, data were not available, time was not available to include them, and this level of detail was judged not critical to the model.

10.2.2 Project Plan and Schedule

The project plan consists of breaking the project into phases and putting them in some logical order. Once the decision problem and objectives have been identified, the scope of the project can be defined. It is important that each project have a well-defined and reasonable scope so resources can be allocated to each phase. In the ED project, the major phases were (1) data collection from patient charts, (2) system analysis and model conceptualization, (3) model building and testing, and (4) experimental runs and report generation. Each phase was further defined and milestones were established so the team would know when the phase was completed.

Each major phase was then divided into subphases. For example, system analysis and model conceptualization consisted of (1) interviews with key ED personnel, (2) observation of ED operations, (3) preliminary data collection for measurements that could not be obtained from patient charts, (4) preliminary model design, and (5) model review by other team members. At finer levels of detail, milestones may or may not be specified. These subphases could be further divided into subsubphases, and so forth, as necessary.

Each phase and subphase should have a schedule so the project manager will know how much time to devote to it. This is often difficult to do because it is hard to know exactly how long each phase will take to complete. The schedule suggests guidelines, but judgment and experience are needed to provide good estimates of time and other resource requirements. If a phase misses its deadlines by a short time, there is no cause for alarm, but if it appears that it will miss its deadline by a significant time, the project manager will need to review the scope of the phase and either revise the tasks in the phase or revise the schedule (or possibly both).

10.2.3 System Analysis

At some point in the project, the team must carefully look at the system and understand exactly how it works. This is the system analysis step. It actually begins in the problem formulation stage because some understanding of system operation is necessary to define the decision problems. For example, most people understand how an ED works—patients arrive with medical problems, are examined and treated, and then leave the ED. With this level of understanding, it is clear that an inefficiently operated department can lead to excessive patient waiting time. However, this level of understanding is not sufficient to develop a model. The modeler needs to know exactly what happens to patients once they arrive at the ED. What protocols are applied to determine which patients receive immediate treatment and what stages do the patients go through as they are registered, examined, tested, treated, and discharged?

System analysis usually involves observing system operation, talking to key personnel, and collecting data. It can be a time-consuming process, so it should be done with a keen awareness of the scope of the model. It is not desirable to develop a more detailed system description than is necessary for the model.

10.2.4 Model Conceptualization and Formulation

Once the system analysis is completed, a model can be formulated. Novice modelers tend to start with modeling as soon as they acquire information on the system structure. However, this urge should be resisted. The idea of this stage is to develop a concept of the simulation model that is consistent with the stated objectives. Several alternative approaches to modeling and simulation should be considered and evaluated. It is here that the modeler first establishes the scope and level of detail for the model.

It is important to make good choices about the basic structure of the model because errors at this stage can be fundamental and correcting them can require that the model be rebuilt. So take time at this point to review your system analysis and develop a concept for the model. Consider alternative modeling techniques and alternative structures for the model. What are the options? Which option is best? How much detail should be included in the model? What is the appropriate scope for the model—that is, what parts of the system should be modeled explicitly and what parts should be omitted or considered implicitly in the model?

In reality, most modelers are familiar with only a few simulation software packages, and they are usually highly skilled at using only one package. The modeling methodology of the software normally dictates the modeling techniques that will be used in the current model, and general-purpose simulation packages such as Arena are powerful enough to handle most problems. If a similar system has been modeled before, then you might be able to reuse or modify code from a previous model for this project. In some cases, specialized software is available for the system under study. When this is the case, you will spend more for the software but save time on development if the time to learn how to use the software is not excessive. The idea here is that there are many options and it is wise to invest time considering them.

10.2.5 Data Collection and Analysis

Most organizations keep large quantities of data. However, simulation modeling almost always requires the collection of additional data. Data are kept for customer records and performance reports but not with the level of detail needed for simulation modeling. Data can be classified as *input* data and *performance* data. Input data are observations on fundamental system activities such as interarrival times, service times, and routing choices. Analysis of input data supplies distributions and parameters for stochastic variates that the model will need in order to run. Performance data are observations on such system characteristics as customer waiting times and resource status (in use versus idle). Performance data are used by management to estimate how well the system is operating. Simulation models use input data to produce predicted performance data. The important concept here is that organizations frequently keep performance data because managers use them to track how well the system is working. They usually do not keep input data, so it must be collected manually. In the ED project, each patient's time in the system (performance data) could be computed from arrival and discharge times on the patient's chart, and each patient's time to first examination by a physician (another performance measure) could also be computed from chart entries. Data on arrival rates (input data) could be extracted from hospital databases, but data on lengths of time for patient examination and laboratory tests (also input data) were not available and had to be collected by hand.

The team usually must collect additional data. Data collection can be expensive and time-consuming. Therefore, what data are collected should be carefully considered. The scope and level of detail of the model and likely additions to the model will determine what data should be collected. It is often less expensive to increase the variables to be collected in the first effort than to repeat the data collection later when additional data are needed.

Sometimes you will need to collect data to document the current system condition. For example, if hospital administration feels that patients are waiting too long in the ED, you might need to collect some patient waiting times and compute the performance measures to see if the perception is based on only extreme values or there are systematic delays that could be reduced. Current system performance data are also needed for model validation.

10.2.6 Model Building

Once the detailed structure of the system is known, you are ready to actually build the model. With current simulation software such as Arena, model building is done interactively. It is important to keep in mind that the model you build will probably be revised multiple times during and after this project. Therefore, the structure of the model should be established to facilitate model modifications. Modular and hierarchical structures can be especially helpful here.

It is important that you document your model, not only so you can remember model details and reasons you chose the particular model structure, but also so the client will be able to maintain and update the model as necessary. Most interactive simulation packages have some facility for model documentation. You should learn these and use and augment them as needed.

After the model-building effort, you will have a model that you can show to other team members and to management. It is important to get their review and approval. They can frequently spot problems in the model because they are much more familiar with system operation. This is also a good place to review the objectives of the project and reaffirm the level of detail in the model. Decision makers must have confidence in the model and its predictions in order to use the model.

10.2.7 Model Validation

The model must be shown to be a valid system representation to establish management's confidence in the model predictions. Model validation means ensuring that the model behaves in approximately the same way as the real system. Techniques for validation involve comparing the output of the model with the corresponding output of the real system. For example, the distribution of patient waiting times in the ED model could be compared with the distribution of patient waiting times in the real system. If the two distributions are not significantly different, then the model would be declared valid. Another technique for model validation is the Turing test in which experts are shown reports from the model and from the real system and asked to tell which is which. If they cannot identify the model output from the real system report, then the model is considered valid.

If a model is found to be invalid, it may be that the model is not detailed enough, that model logic is not correct, or that important parameter estimates are wrong. Judgment and experience are required to quickly identify the reason and correct it.

The validation process is often time-consuming, tedious, and risky because you risk finding that the model cannot be validated and therefore must be modified. It is important that you start validation as early in the project as possible in order to have time to modify and then revalidate the model.

10.2.8 Experimentation and Analysis

Once you have a valid model implemented and ready to use, you can make simulation runs to evaluate system designs or decisions. The term *experimental design* refers to the selection of input parameter values and other conditions, decisions, or policies that you can specify in the model. If your objective is to evaluate a few specific decisions, then the experimental design is simple. For example, suppose that mean patient waiting time is too long and you are considering two alternatives: (1) adding another physician to each weekday evening shift and (2) adding two additional nurses to each weekday evening shift. Then you can set up these two staff schedules and run the model with each to see which does a better job in reducing mean patient waiting time.

If your objective is to determine which staff schedule—including all physicians, nurses, and registration clerks—is optimal, then you have many thousands of possible runs in your experimental design. Since you do not have enough time to make all of the runs, you must use a systematic and rational approach to select which schedules you will test. Often it is useful to make several runs over the entire range of possible input values and then refine the grid in subsequent runs.

When doing the experimentation, you will need to determine the length of the runs. Longer runs yield more precise estimates of system performance but they take more time to complete. Typically, you do not know how long a run needs to be to give estimates with a desired precision, so some experimentation is needed to determine optimal run length.

10.2.9 Reports and Presentations

Once you have made the simulation runs, analyzed the output, and prepared your recommendations, you will need to write a report that documents your findings and make a presentation so you can explain your recommendations. The success or failure of a simulation project can hinge on the quality of the presentation. If you fail to be convincing in your recommendations, if you are so sure of your conclusions that you lose credibility, or if you present your results in unintelligible tables and charts, then management will not be inclined to use your recommendations or to pay you to do another analysis. Your reports and presentation must be clear, focused, and complete. You should clearly identify your key recommendations and put them up front. Supporting numerical evidence should be presented graphically, and tables should be used minimally, if at all. Error estimates should be presented for all parameters that are estimated from the model. Your recommendations should also include any caveats and a reminder of the major assumptions in the model. If the assumptions do not hold, then your results may not be valid. The decision makers should be reminded of this.

10.2.10 Implementation

Simulation projects are expensive, and the resulting model is an asset for which the organization has paid. It would be foolish to discard a model that has proved useful. Thus, the model needs to be documented; if the model will be used by other personnel, they will need to be trained to operate the model and understand the output. This process can consume a lot of time, and the project budget should include funds to pay for this activity.

Sometimes, management will want to implement the model so it can be used routinely. When this is the case, you might need to modify the user interface and the interface that supplies parameters to the model. Generally, the model will need to access a database where current system data and parameter estimates are stored. The process of implementing a model for routine use can be quite extensive. For example, the director of the ED creates a new staff schedule every six weeks. He said that he would like to have the model available to evaluate the schedule when he makes the periodic changes. Before this could be done, data-collection procedures would have to be put in place to collect and make available patient arrival times and examination and treatment times. In addition, the user interface to the model would need to be modified to make it more intuitive and other changes would be needed to produce the reports wanted by the ED director so they could be exported to his favorite spreadsheet program.

Finally, implementation can also refer to the implementation of the recommendations that result from the model. As the system modeler, you understand the model results better than anybody, so it is helpful for you to remain involved in the implementation of these results to make sure that they are interpreted correctly and the implementation is faithful to the model recommendations.

10.3 THE ART OF PROJECT MANAGEMENT

Project management is both a science and an art. Most of the ideas in section 10.2 fall on the science side. Other aspects are much less scientific. Project management involves working with people and understanding their motivation and incentives (Verma and Thamhain, 1996). A successful project deals not only with data collection and analysis, scheduling, programming, experimental design, and other tasks, but also with managing the human interface. Understanding the political environment of the project and managing the project team effectively are critical to a successful project.

10.3.1 Problem Identification and Formulation

The first step in any systems analysis project is to determine the exact problem. Why is a system analysis and simulation model being considered? Normally, a problem will present some symptom to the system manager. In the ED scenario, the hospital received a large number of complaints from patients about excessively long waiting times. Management was also under pressure from the hospital's owners to reduce costs and justify building the UCC. The analyst and other team members must understand the circumstances and decision requirements that are driving the project.

It is important to understand the organizational structure of the people in the system. When interviewing system managers, make sure that you determine exactly who makes the decisions and who manages the system. In the ED, the medical director manages the system and establishes staff work schedules, but the hospital administration makes decisions about acquiring new equipment, building new facilities, and hiring staff. Because anticipated decisions involved both staff schedules and use of new equipment, it was important to learn the points of view of both managers.

Virtually all problems are unique. Most emergency departments operate basically in the same way, but in the project discussed in section 10.2, the emergency department has an attached urgent care facility. Other EDs make use of health professionals such as physicians' assistants or nurse practitioners. Those options are considered politically unacceptable in the current system. It is important to identify the unique characteristics of the system and problem. These might or might not be critical to the model.

10.3.2 Understand the Client's Objectives

Part of the problem formulation process is to discern the objectives of decision makers. Often, management's understanding of its objectives is different from the analyst's, and it is usually stated in imprecise or nonspecific terms. For example, management might say that it wants to improve patient satisfaction. Exactly what does this mean? Encourage the client to provide a clear, concise, and concrete statement of the objectives. Does she want to improve patient satisfaction by decreasing patient waiting times? Would it be acceptable to reduce average waiting times at the expense of increasing the waits of a few patients? Is management primarily concerned about the extreme waiting times or is it concerned about the average waiting time? A concrete statement would be something like: Management wants 85 percent of patients to wait no more than 45 minutes in the emergency department (exclusive of triage, registration, examination, treatment, and other service times). Sometimes, managers have not thought specifically about these ideas, so you as analyst must question and probe to see what they are really concerned about. Objectives should be clear and measurable.

Problems may be stated in highly personal terms. For example, an employee might say, "The reason we are so slow is that Jack Smith over there is holding everybody up because he is so meticulous," or "The real problem is that the people in the laboratory do not do a very good job, so we often have to send additional specimens for testing." These concerns might or might not be valid. However, it is dangerous to get involved in internal political issues, especially personnel issues. You need the cooperation and confidence of all employees in order to understand system operation, and if they feel that giving you too much information might jeopardize their jobs, they will always hold back. In the ED project, some physicians stated that others were slow in their work and, in fact, hinted that some of the other physicians were not fully competent so they would often "lend a hand" because they were concerned about the quality of care. The analysts listened to this information but did not share it with anyone. When developing the model and discussing the problems with management, all ED physicians were considered identical and interchangeable in their speed and work characteristics. It is critical that the analysts be objective and impartial in any ongoing political issues.

10.3.3 Collect Data to Verify and Document Problems

Sources of data include existing internal databases and records, external datasets, and special surveys and data collections. Although data in the most well known internal sources are often not useful, it is helpful to search carefully for other internal data. Sometimes engineering or other departments keep data that are useful and measure appropriate phenomena. It is not uncommon for one department to be unaware that another collects certain data. This can be discovered by making inquiries and asking probing questions. For example in a hospital, the registration system automatically records the time a registration activity begins. On busy days, the registration clerks serve many patients without a break. Thus, the times between registration starts can be used to measure the registration activity times since the periods do not include an idle time. Many external datasets are available from the government and other public sources. These data can sometimes be used to provide certain parameters. In the ED project in section 10.1, public health data were used to provide the probabilities of various levels of severity for arriving patients.

If existing internal and external data are insufficient, then the only alternative is to collect the data manually. This process normally involves considerable planning, several people, and significant time. Thus, it is important to determine carefully how much data you might need. The data-collection procedure could involve manual observation and recording, as was done in the ED project in section 10.1, or the team could set up a system to automatically collect the data. The latter approach will usually be more expensive for a small dataset, but it could be economically justified if a large amount of data is required or if data will need to be collected on an ongoing basis in order to use the model periodically.

Perceived problems are sometimes not the actual problems. It is important to use data to verify that the problems exist. As an example, suppose that the director of the ED has stated that patients are waiting too long and waiting times are increasing. This perception is based on the fact that he received a few angry letters from patients a few weeks ago and more angry letters last week. After collecting data and investigating, we find that a few weeks ago, one of the ED physicians was out of work for the week for personal reasons, and last week, there was an unusually large number of patient visits for no explainable reason. Data collected showed that patient waiting times have followed a stable distribution for the past year. Thus, while management might want to explore ways to reduce patient waits or innovative ways to handle unusually large patient volumes, controlling increasing patient waiting times is not a valid reason for the simulation model.

The systems analyst can serve an important role as an outside observer. Often, persons working with the system have difficulty seeing system operation as a whole. It is their job to deal directly with details, and they find it unnatural to think about the high-level performance of the system. In the ED project, all employees who were on the analysis team were trained as health care workers. Their training focused on solving individual problems for individual patients, and they had some trouble thinking in terms of the entire ED and UCC as a single system with patients as one component of the system. This is a matter of their training and mind-set. You, as an outside observer, can help them see the system from a broader perspective.

The result of the problem identification and formulation activity should be a written document that describes and documents the problem. After managers have reviewed this document, they should sign off on it so you as the analyst have a written authorization to proceed with the project. Without a written record, management could say later that the problem you solved or the model you built was not really the problem with which it was concerned.

10.3.4 Consider Alternative Models

Some problems do not require a simulation model. If the ED were experiencing long delays at the triage nurse and little congestion elsewhere, then a simple queueing model could determine how many triage nurses are needed to handle the load. In other cases, another modeling approach is appropriate. Suppose the problem is with a dialysis clinic where patients are scheduled for dialysis treatment and all parameters are known with relative certainty. If management wants to establish the best work schedule for the employees and make sure there is always enough staff to handle the workload, then an optimization model is a better approach. Simulation models are expensive and time-consuming to build, so it is important to determine that the problem justifies this level of effort. Approach the problem identification task with an open mind and an understanding that this might not be a simulation problem.

10.3.5 Project Planning and Management

Careful planning is the key to success. Simulation projects are usually large and involve a core team of people and considerable resources. To make sure that the project is completed successfully on time and on budget, it must be carefully managed.

Three resources must be managed: people, time, and money. Your goal is to use all of these resources efficiently to complete the task as quickly and reliably as possible. If you finish the project but present the recommendations too late to influence the decision, then clearly the project cannot be called a success. If you complete the project on time but use significantly more resources than anticipated, then your credibility will be hurt and future projects will be jeopardized. So you have a great incentive to manage the project well.

Techniques for simulation project management are similar to those for managing any other project. Project management is a topic about which much has been written. See Fleming and Koppelman (1996) and Verma (1997). The project plan should include a financial plan as well as a time schedule. Simulation projects may cost anywhere from tens of thousands to millions of dollars. When this much money is devoted to a project, planning is necessary to make sure the money is allocated efficiently. The financial plan is also an important means to sell the project to the sponsor based on its financial objectives. You should be able to show that the expected benefits exceed the cost of the project. Most managers' objectives are stated in terms of money, so your presentation should also be made in terms of money.

Most organizations have political issues that concern simulation projects. These issues should be dealt with directly when planning the project. For example, certain key executives who are not directly involved in managing the system under study might be interviewed and briefed because they are influential and might be future clients. When the plan is created, think about all people who could have an interest in the project and how they might be considered in the plan.

The first step in project planning is to establish the major tasks involved and estimate the time and other resources needed for each of them. The tasks in section 10.2 can serve as a guide. You might want to combine some of these steps, or you might need to subdivide some. Normally, you should have no more than 6 to 10 major tasks. If you have more than 10, you are probably thinking of too much detail at this point.

Second, identify the major constraints and deadlines involved. Usually, the sponsor of the project will have a date at which she plans to make a decision. This will provide a guideline for completing the project and reporting your recommendations.

Third, identify the people who will make up the project team. Each person should have a specific reason for being on the team and a specific job he or she will do. Assign duties to each person and discuss these duties so that each team member understands his or her own responsibilities.

Fourth, establish a budget for the project. If the project is internal to an organization and all team members are employees of the organization, the budget might seem superfluous. If the project is conducted by a consulting organization, the budget is important to make sure that the project has been priced correctly. Indeed, if the consulting firm proposed a price to do the work, a budget was most likely already established.

Fifth, create a project schedule so that you will have a planned time when each task will begin and end. A Gantt chart is a useful graphical tool to display the schedule and show the relationship between tasks. Within this schedule, establish important milestones. For example, one such milestone is the point when the first draft of the model has been completed and is ready to be presented to the team for approval. At this time, the team will need to meet to review the model and either approve it or suggest changes. Obviously, another important milestone is the time when the final presentation is made to management. Normally, a project will have several milestones like these.

Once you have a project plan, it should be communicated to all members of the team and to the project sponsor. It is important to keep the project moving according to the plan. For this reason, you will need to spend time reviewing the plan to ensure that it is realistic. Any project has a lot of uncertainties, so the plan must include assumptions. Therefore, target dates and resources must be set up with slack time to allow for unexpected events. This involves a lot of judgment that comes from experience and some guesswork. Remember that if all estimates for completion dates and resources are set at optimistic levels, they will likely be missed and the effect will be cumulative. As each task falls behind schedule, it will force the following tasks to also fall further behind.

The project plan should be reviewed regularly, and any signs that it is falling behind should be taken seriously. Sometimes team members will be assigned project duties in addition to their regular work duties. This can result in the project duties receiving the lowest priority and being done last. It is difficult to say how to handle a problem like this in general, but some possibilities are to continue to sell the project to team members so they understand the importance of the project and to see if other, less busy employees can be assigned to the team or assume work responsibilities for team members. Finally, keep the team members informed of milestones and deadlines that are coming up.

10.3.6 Team Management

Most simulation projects have to be undertaken by a team. Few people have the expertise that a collection of people do and, except for the simplest projects, none has the time to do all tasks involved. Typically, the team needs a simulation modeler and one or more people who are intimately familiar with the system to be modeled. The team may also need software developers; a database manager; a writer; experts in accounting, finance, engineering, management, or other specialized areas; and one or more representatives of the project sponsor. When you have a group of people working together, it is important to manage their activities and interactions so they work together efficiently.

The project team needs to have a designated project manager with the responsibility and authority to lead the team. A project manager with responsibility but not authority is ineffective. This manager should possess the same characteristics as any leader: a good communicator, an innovator, a motivator. The project manager's primary responsibility is to organize the tasks and delegate them to team members. If there are problems in collecting data or getting cooperation from employees, then the solution usually must come from the project manager. This person's responsibilities include monitoring the project and detecting schedule slippages or other problems in the execution of the schedule.

Because the simulation expert usually must communicate with all members of the team and understands the simulation technology as well as most other aspects of the problem, it is often the case that she is tapped to be the project manager. To be an effective manager, however, this person must know much more than simulation. She must understand the perspective of each team member as well as the sponsor. Her point of view must be that of a problem solver rather than just a simulation technician.

10.3.7 Communicating with Clients

Poor communication is the biggest single reason for project failure. As a simulation modeler, it is tempting to just think in terms of the model and simulation methodology, but it is important to resist this temptation. Remember that the client has a problem—a system problem, not a simulation problem—that he has hired you to solve. He doesn't really care whether you develop a simulation model, an optimization model, or a differential equation model. His concern is with obtaining enough information to make good decisions, solve the problem, and achieve his objectives. If you use modeling terms such as *random variates, events, activities,* and so forth, you neither contribute to his understanding nor increase his confidence in you. When the ED project was under way, the term *simulation* was never used. Instead, the client was told that we would develop a *computer* model that mimics the behavior of the ED and allows us to predict its behavior when changes are made to the system. This was a true statement and included all of the information the client needed. If the client asks about details of the model, then you can provide as many details as needed to satisfy the client. It is important to remember that the client is concerned with the system and the problem, not the methodology used to build the model.

Communication is a vital part of the process of problem solving, and much effort is required to do it well. When the project begins, you will need to communicate with the client and others to understand the nature of the problem and how the system operates. You will also communicate with the client to explain how you can help solve the problem and what resources you will need. Effective communication at this stage can save time and effort by avoiding incorrect assumptions and repeated tasks.

Communication involves listening to the client. You must keep an open mind throughout the project and always view the problem and the system from the client's perspective. At the same time, it is important to manage the client's expectations. Deal with unrealistic ideas or expectations as soon as you know about them. In the ED project, the client questioned whether she could use the model to evaluate her tactical decisions. For example, could she interactively run the model and change schedules and other decisions as the model runs? The project manager explained to her that this model was not set up to allow interactive operation. He also explained that the performance measures might not be meaningful if the simulation is interrupted at arbitrary times and changes are made in the system. Once these caveats were explained, she was comfortable with using the model to evaluate policy alternatives but not as an interactive tool.

Part of managing the client's expectations also concerns preparing her for problems. Every project must deal with problems, so the client should be told to expect them. In this vein, let the client know the project plan and provide milestones so she can track the progress of the project. Like anyone else, the client is uncomfortable with uncertainty. You should remove any uncertainty you can.

You will need to be professional and somewhat formal in communicating with the client. Regular project reviews and reports to the project team and the client communicate and document the project status and any problems that have occurred. These reports also keep the client involved and keep the project a high priority, all of which contribute to project success.

10.4 DATA COLLECTION

Most simulation modelers do not think much about data collection until they are involved in a real-life project and then realize that obtaining and analyzing data is the most time-consuming and critical part of the project. Sometimes they have too little data and sometimes too many data are available. The former situation—too little data—is more common. We pointed out earlier that organizations keep huge amounts of data, but they keep them for purposes other than modeling. Data are collected for legal, accounting, operational, and managerial reasons. Available data often consist of aggregate or summarized quantities such as the number of items in inventory or the number of late shipments last week and do not contain the kind of detail needed as input for a simulation model.

10.4.1 Coordinating Model Development and Data Collection

You must have parameter estimates and therefore collect data before you can use the model. However, you can begin model development before data are collected. Indeed, the model tells you what parameters are needed, and therefore what data you need to collect. So the model must be formulated before data are collected. Because data collection and analysis can take a long time, it should be started as early as possible. Therefore, it is critical to establish your data needs as early as possible in the project, and it is equally critical to be clear and complete in your specification. Once you start data collection, you do not want to have to repeat it to include variables that were overlooked.

10.4.2 The Amount and Availability of Data

Sometimes data are not available for certain system activities or even entire systems. In the ED project, the UCC had not yet opened. Therefore, data for arrival rates, service times, distributions of diagnoses, and other system parameters did not exist. In this case, the analysts assumed many of the parameters such as the distribution of time to triage a patient would be the same as in the ED. They also found a database at a government Web site that provided diagnosis data. In this case, data were available from some source for all system parameters even though data could not be collected in the urgent care center. When no data are available for a system, you have three basic choices: Find a similar system somewhere and use data from that system, find a public data source, or use judgmental values for the parameters. Although there is some risk in each option, the third is most risky because many managers are not particularly good at guessing numerical values even though they are intimately familiar with the operation of the system.

When data are available, it is often the case as in the ED study that the time and expense of collecting data is so much that only a tiny sample can be collected. In your statistics courses, you learned that you must have more than 1000 observations to estimate a probability to within plus or minus 3 percent of its true value. Clearly, if your sample consists of only 50 or 100 observations, you will have considerable error associated with the parameter estimates. Models typically have dozens or hundreds of parameters, so these errors can have a large influence on the model. It is not easy to deal with this problem. Obviously, the best solution is to collect a large sample for every parameter. When this cannot be done, consider each parameter and judge how critical it is to know its value with great precision. Data-collection efforts can be redirected to collect observations for those parameters that are more critical. Once the model has been built and validated, it is a good idea to make simulation runs to determine how sensitive the model is to changes in parameter values. This can usually tell you where additional sampling effort should be directed.

10.4.3 Data Quality

Do not assume all observations are valid. Challenge the data. In another study involving a hospital, the analyst developed a model of the flow of patients through the hospital. Part of the model involved the process of cleaning and preparing patient rooms after a patient was discharged. The hospital had a new computer system that stored data recording the times when each room was vacated, when the hospital staff began cleaning the room, and when the cleaning was completed and the room ready for a new patient. This system utilized the hospital's phone system. Staff dialed a special phone number and, when the computer answered, they punched in special codes to designate what activity was beginning or ending. The analyst was excited to learn that this data existed because this is exactly the type of detail data she needed to develop part of the model. However, no employee at the hospital knew how to dump the data to a diskette or other medium in order to transfer it to the analyst's computer. And once the data were obtained, she spent considerable time processing it to compute the lengths of time between the start and finish of the cleaning operation. Finally, when she plotted the histogram of the cleaning times, she found that approximately 40 percent of them were between 1 minute and 3 minutes. These data were clearly invalid because the minimum length of time to clean a room is 12 to 15 minutes. Further investigation found that the night staff were cleaning several rooms, then going to the phone and punching in all codes at one time. They had not been instructed on how to operate the system or did not understand the importance of operating the system exactly as instructed. The analyst decided to discard all observations that were less than 10 minutes. She realized that some of the remaining data might still be invalid, but she decided that it was better to have some data that included possibly invalid observations than no data at all. Unfortunately, this example is not atypical. It is common to find that data have been contaminated or have not been recorded correctly. In that case, you have to make pragmatic decisions.

10.5 MODEL MANAGEMENT

A simulation model goes through a life cycle that we can characterize by four stages:

1. The model is born when you first conceptualize the model and goes through its infancy as your concept becomes more definite.
2. It grows up as you refine it, implement it on the computer, and validate it.
3. It matures as it is used and the input parameters are further refined.
4. Finally, it dies as the systems it models are replaced, the people who sponsored the project move on, and it is no longer useful.

Model management is a useful part of each stage and can greatly enhance the life and usefulness of the model.

10.5.1 Model Complexity

Simulation models for real systems are quite complex. The models in this textbook have been selected to illustrate important concepts in simulation modeling, but models for real systems are at least one order of magnitude larger. They may have at least 10 times as many types of entities or 10 times as many events or activities. Arena models may have 10 times as many modules. The ED model was developed using Inprise Delphi and a simulation library. The simulation program totals approximately 5000 lines of user-written code, not including the simulation library. Actual models developed using Arena have hundreds and sometimes thousands of modules.

Real models also have many more parameters, and user interfaces must be much more complex to handle the parameters and interaction with users who may be managers. The ED model had approximately 990 input parameters. Some of these were staff schedules that specified the number of workers on each shift for an entire week. With three shifts per day and seven days in a week, this is 21 parameters for the week. Because there are three types of workers (physicians, nurses, and registration clerks) and two units (ED and UCC), the total number of parameters needed to specify staff schedules is 126. Most of the parameters are numerical values computed from data. For example, they include arrival rates for each unit for each hour during the week (2 units × 168 hours = 336 values total), distributions for initial diagnosis and severity, parameters for the examination times, and others. A complex input form was created to give the user control over the input parameters, but if the user had to enter all of the parameters for each run, the model would be unusable. So parameters could be stored on a file and retrieved, allowing quick input. An even better approach is to store parameters in a database.

Parameter estimates are computed from the input data that are collected. These observations should be saved in a database where they are easily retrieved, identified, and augmented as new data become available. A great deal of effort is required to fit distributions and estimate parameters, so some effort is justified in organizing and maintaining the data as well as the estimates.

The output to simulation models of real systems usually consists of hundreds of parameter estimates for performance measures that provide details that managers require to understand the system's response to changes. As a result, output reports are more complex and must be organized carefully. An organized approach to model development and management is needed to handle this level of complexity. Without an organized approach, your progress will be slow, errors and omissions will be made, and delays will occur.

As you develop the model, keep an open mind. The model will evolve as additional understanding and insight accumulates. Learn to revise your thinking and assumptions throughout the project. Remember that the project is a learning experience for you. You should think differently about it at the end than you did at the start.

10.5.2 Hierarchical Model Building

The ED model is also quite complex. The model was designed using the event view and has 27 distinct activities with approximately 50 events, including shift change events. Within many events, the logic was rather complex because the model had to sample many variates and make many decisions based on the values of these and other variables. The model developer is faced with the same problems as other persons such as architects, engineers, and programmers in managing large, complex projects. In order for its complexity to be managed, the model must be divided into manageable pieces. The best way to do this is to use a hierarchical approach and divide the entire model into two or more pieces—that is, submodels—and define the relationship between them. Then, for each submodel, repeat the process: Divide the submodel into two or more parts and define the relationship between the parts. This process continues until all submodels are small enough to be developed as a single unit.

Figure 10.1 is a simplified process diagram of the model used in the ED project. This model can first be divided into the ED model (submodel 1) and the UCC model (submodel 2). The only connection between them is that the UCC can transfer patients to the ED, so combining these two submodels was rather simple. Within the ED model are four natural submodels: The process of treating patients arriving by ambulance prior to the MD Assessment 1 activity (submodel 1.1), the process of treating walk-in patients before being examined by a physician (submodel 1.2), the examination and treatment process (submodel 1.3) and the discharge or admit activity (submodel 1.4). These processes all had relatively simple connections. The UCC submodel can be divided into two natural pieces: The process of examination and assessment prior to seeing a physician (submodel 2.1) and the process of examination and treatment by the physician (submodel 2.2). Most models have a natural structure that supports hierarchical model building.

For larger models that require a team of people to develop, hierarchical model management works well with a team approach because it allows you to divide a model into pieces and assign each piece to a team member. Note that submodels 1.2 and 2.1 above were so similar that they could easily have been assigned to a single modeler. Many simulation software tools such as Arena have built-in support for hierarchical model structures, making it that much more natural to use a hierarchical model.

10.5.3 Start with a Simple Model

When developing the model, do not get lost in model details. As you gain experience, you will develop judgment for the level of detail required to support the project goals. Our recommendation is to err on the side of simplicity. When you are defining the model, it is difficult to know how much detail is required. In the ED project, should we include technicians such as respiratory therapists, phlebotomists, and emergency medical technicians in the model? These people are definitely present in the system, so their omission does make the model less realistic. However, we decided to leave these details out of the model because the decisions under consideration did not involve them. If your model does not have sufficient detail, then it will be revealed in the validation process. But if your model has too much detail, you will not know it. However, you will have spent more time and other resources developing the model and it will require more time to run and maintain.

FIGURE 10.1 Emergency department patient flow and urgent care center model

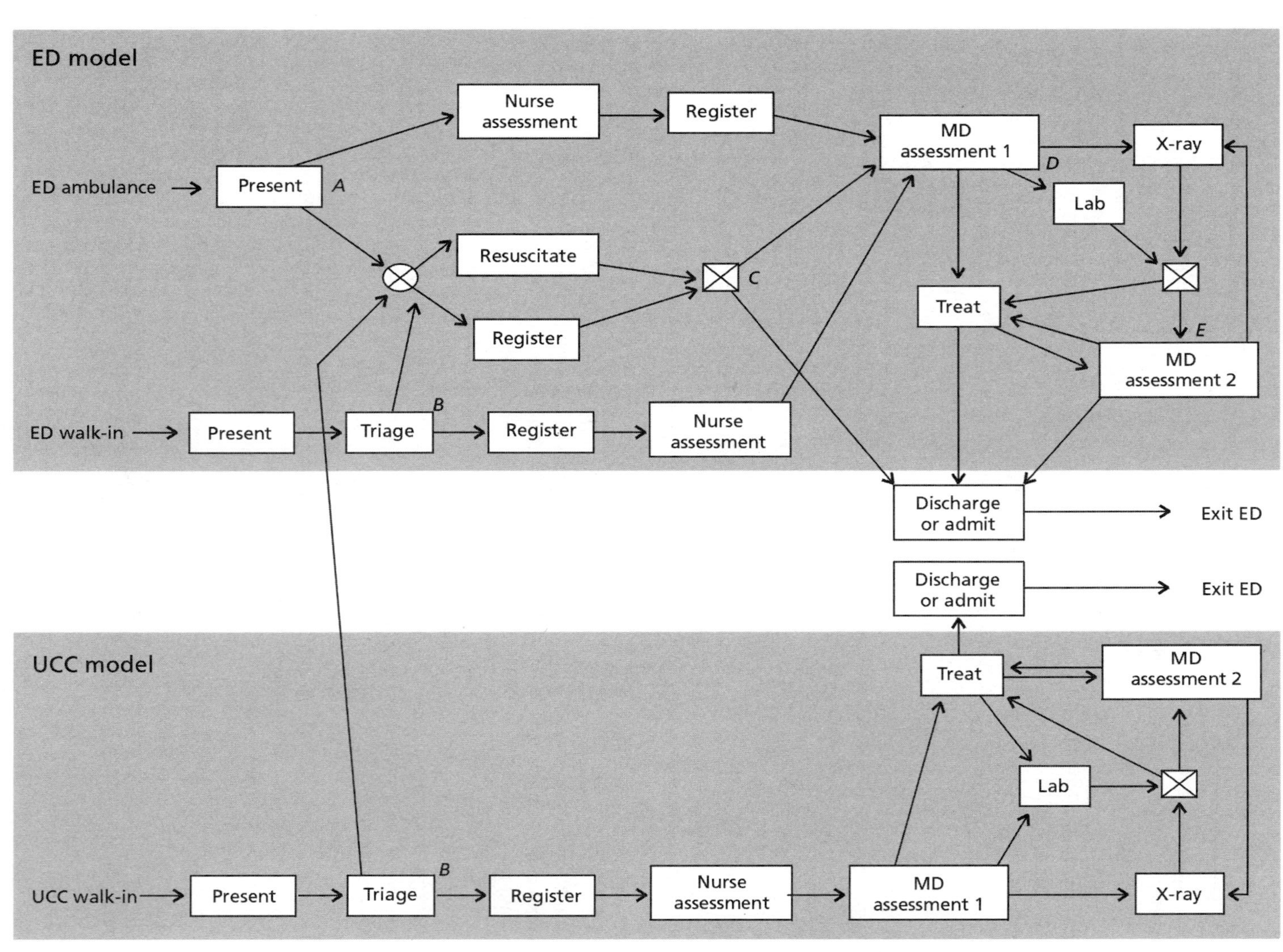

Thus, it is always better to start with the simplest model you think will work and then add detail as needed.

Similarly, define the scope of the model and stick to it. Musselman (1994) suggests that you start with the objectives and work back to the model. Use the objectives to develop a list of questions you want the model to answer and then create a list of performance measures the model should provide to answer the questions. Next establish the scope of the model. All systems interact with the external environment. When developing the ED model, we had to determine what parts of the environment would be modeled in detail and what parts would be modeled implicitly. Two systems under consideration were the lab and radiology departments. Patients experienced delays waiting for lab and radiology reports. This delay resulted from the patients' specimens or the patients themselves entering another queueing system in another department. However, none of the decision alternatives under consideration directly involved these departments, so the scope of the model was established to exclude the detail of these departments. Other reasons for excluding them were that not enough time or personnel resources were available to model them, and if results showed that the model needed to be expanded, then they could have been added as separate modules. Indeed, the hospital administration considered this a potential direction for expanding the model. Another discussion involving the scope of the model concerned the patient population. One team member questioned whether the model should include a submodel that allowed changes in community sociological conditions to influence the arrival process of patients. For example, if drug use increased in the community, it would result in more patients presenting for treatment of drug overdoses and related conditions. The team concluded that this portion of the model would have required more resources than were available, and the effects could be modeled directly by changing the distribution of initial diagnosis. Every model will include many considerations like this, and each must be carefully considered. It is not a good idea to create a general-purpose model designed to answer any questions that might arise in the future. Instead, it is best to create a focused model to answer the specific questions under consideration. The decisions about model scope should also be documented so future users of the model will know why decisions were made to include or exclude parts of the system.

10.6 SUMMARY

In this chapter, we have discussed many ideas concerning how to have a successful simulation project. A lot of experience actually conducting simulation projects is required to develop the skills and expertise to run a successful project most of the time. In this section, we will review and summarize the main recommendations.

- Make every project a learning experience. Because every model and every project is unique, you will have to innovate in every project. Only through experience will you develop the knowledge to innovate successfully.
- Listen to the client. She has the problem and she knows the system. Only the client can tell you the objectives of the project.
- Be professional. This advice is rather general and abstract. It is important to treat every aspect of the project seriously. Professionalism contributes to client confidence in your work, and it encourages cooperation by all members of the team.
- Set clear, specific, and measurable objectives. Work with the client to establish objectives that everyone understands and that can be measured so you know when they have been met.
- Set reasonable limits on the project and the model. It is better to complete a simple model on time than to leave a complex model unfinished or finish it too late to be useful.
- Communicate often. Communication is the key to effective project management, and you cannot communicate too much. It should be both oral and written as well as embedded in the model software.
- Make the model as simple as possible. If you start with a simple model, you maintain your modeling freedom because you can add detail and make changes as needed. If you start with a complex model, then you will find it hard to justify making changes. Complex models also have many more places for errors to hide.
- Start data collection as early as possible. Data collection will take a lot of the team's time and other resources, so it should be started early to avoid delaying the project.
- Develop confidence in your model and its predictions. If you do not have confidence in your model, then you will find it hard to sell the model to the client.
- Keep all members of the team as well as the client involved. If the client or team members do not stay involved, then they will lose interest in the project and will not give their tasks the highest priority.
- Prepare your final recommendations carefully and deliver them forcefully. The ultimate success of this or any project lies in the extent to which the system is changed and the benefits of the changes are achieved. If you do not sell the results of your modeling efforts well, then the client could choose to not implement the changes you advocate. In that case, the project cannot be considered a success.

References

Fleming, Q. W., and J. M. Koppelman. 1996. *Earned value project management.* Newtown Square, Pennsylvania: Project Management Institute Publications.

Musselman, K. 1994. Guidelines for simulation project success. In *Proceedings of the 1994 Winter Simulation Conference*, ed. J. Tew, S. Manivannan, D. Sadowski, and A. Seila. Lake Buena Vista, Florida.

Sadowski, R., and M. Grabau. 1999. Tips for successful practice of simulation. In *Proceedings of the 1999 Winter Simulation Conference*, ed. P. Farrington, H. Nembhard, D. Sturrock, and G. Evans. Phoenix, Arizona.

Verma, J. K. 1997. *Organizing projects for success.* Newtown Square, Pennsylvania: Project Management Institute Publications.

Verma, V. K., and H. J. Thamhain. 1996. *Human resource skills for the project manager: The human aspects of project management.* Newtown Square, Pennsylvania: Project Management Institute Publications.

Yeates, D., and J. Cadle. 1996. *Project management for information systems.* 2d ed. New York: Pittman Publishing.

Appendix

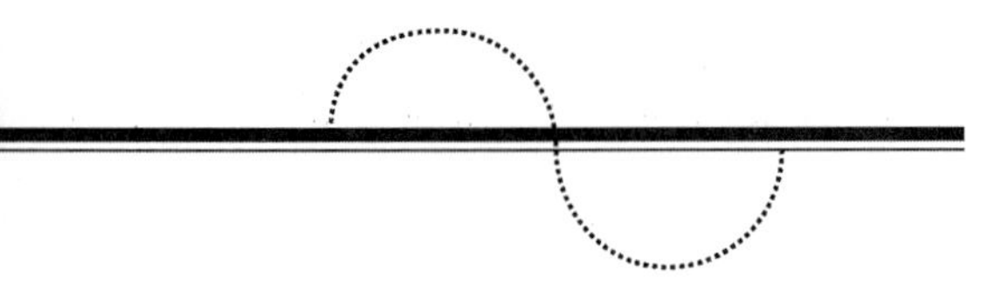

Review of Probability and Statistics

In this appendix, we provide a brief review of some elementary concepts of probability and statistics. Our presentation will be brief and include just the basic statistical ideas used in this book. For a more in-depth review, see Albright, Winston, and Zappe (2002), Hogg and Craig (1995), or Hayter (2002).

A.1 PROBABILITY CONCEPTS: SOME DEFINITIONS

The probability of an *event* is a number between zero and 1 (both inclusive) that indicates the likelihood of that event.

An event is a set of one of more *outcomes* of an experiment.

An *experiment,* in the probability context, is any condition or situation that can be replicated essentially under the same conditions.

The *complement* of an event A, denoted $\overline{A}$, contains all the sample points in the sample space that are not included in A.

The *union* of two events, denoted by $A \cup B$, is an event that contains all the sample points (outcomes) that are *either* in A *or* in B or in *both.*

The *intersection* of two events, denoted by $A \cap B$, is an event containing all the outcomes that are common to *both* events.

The definition of both the union and the intersection of events can be extended to more than two events.

Two or more events are said to be *mutually exclusive* if their intersection is a null set (i.e., does not contain any sample points or outcomes).

Two or more events are said to be *exhaustive* if their union is the universal set (i.e., contains the entire sample space).

Note that a set A and its complement $\overline{A}$ are mutually exclusive and exhaustive events.

A.1.1 Probability Rules

1. Probability of the complement of an event:

$$P(\overline{A}) = 1 - P(A)$$

2. Probability of the union of the events:

$$P(A \cup B) = P(A) + P(B) - P(A \cap B)$$

(See Figure A.1.) Note that if A and B are mutually exclusive, $P(A \cap B) = 0$ and $P(A \cup B) = P(A) + P(B)$.

3. *Conditional probability:* Probability of the event A given that B occurs,

$$P(A \mid B) = \frac{P(A \cap B)}{P(B)}$$

4. *Independence:* If A and B are independent, then $P(A \cap B) = P(A) \bullet P(B)$ and $P(A \mid B) = P(A)$. Also, if $P(A \cap B) = P(A) \bullet P(B)$, then A and B are independent.

FIGURE A.1

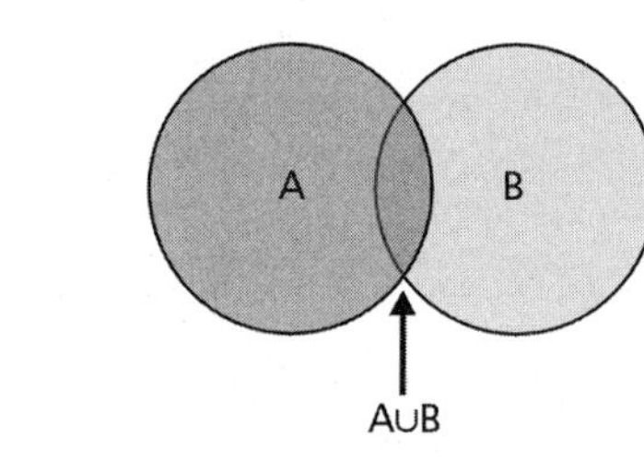

Example

We will explain the above probability concepts with the following example. Consider an experiment in which we repeatedly draw balls from an urn. The urn contains 4 red balls and 6 green balls. Among the red balls, 1 has stripes and 3 have dots. Among the green balls, 2 have stripes and 4 have dots. You pick one ball at random and define the following events:

R: a red ball is drawn
G: a green ball is drawn
S: a striped ball is drawn
D: a ball with dots is drawn

The following probabilities are clear from the above experiment:

$P(R) = 4/10$
$P(G) = 6/10$
$P(S) = 3/10$
$P(D) = 7/10$

Also, $P(R \cap D) = 3/10$, $P(R \cap S = 1/10$, $P(G \cap D) = 4/10$, and $P(G \cap S) = 2/10$.

Note that $R = \bar{G}$ and $G = \bar{R}$, and therefore R and G are mutually exclusive and exhaustive; similarly, S and D are complementary events.

The probability that a red *or* a dotted ball is drawn:

$$P(R \cup D) = P(R) + P(D) - P(R \cap D)$$
$$= 4/10 = 7/10 - 3/10 = 8/10$$

The probability that a green *or* a striped ball is drawn:

$$P(G \cup S) = P(G) + P(S) - (P(G \cap S) = 6/10 + 3/10 - 2/10 = 7/10$$

Similarly, we can calculate the probabilities of the events $R \cup S$ and $G \cup D$.

Suppose we draw a ball and observe its color to be red. The following conditional probabilities can be calculated:

$$P(S \mid R) = \frac{P(S \cap R)}{P(R)} = \frac{1/10}{4/10} = 1/4$$

$$P(D \mid R) = \frac{P(D \cap R)}{P(R)} = \frac{3/10}{4/10} = 3/4$$

Are the events R and D independent? In other words, does the event R affect the probability of the event D or vice versa?

Is $P(R \cap D) = P(R) \bullet P(D)$?

$$P(R \cap D) = 3/10. \ P(R) \bullet P(D) = (4/10) \bullet (7/10)$$

The events R and D are *not* independent because these are not equal.

Bayes's Theorem

Conditional probabilities can be viewed as revising the prior probabilities based on some sample information. Bayes's theorem formalizes or provides a mechanism to calculate the revised (posterior) probabilities. Formally, if $A_1, A_2, \ldots A_k$ are k mutually exclusive and exhaustive events in the sample space, the posterior probability of any event A_i given that the event B has been observed is given by

$$P(A_i \mid B) = \frac{P(A_i \cap B)}{P(B)}$$

$$= \frac{P(A_i)\, P(B \mid A_i)}{P(A_1)\, P(B \mid (A_1) + P(A_2)\, P(B \mid A_2) + \ldots + P(A_k)\, P(B \mid A_k)}$$

With prior probabilities $P(A_1), P(A_2), \ldots, P(A_k)$ and the conditional probabilities $P(B_i \mid A)$, the above equation can be used to calculate the posterior probabilities of the events $A_1, A_2, \ldots, A_k$.

A.2 RANDOM VARIABLES, PROBABILITY DISTRIBUTIONS, AND EXPECTATIONS

A *random variable* assigns a *numerical value* for every outcome of the experiment. Random variables can be either discrete or continuous. A discrete random variable can only assume one of a finite or countably infinite set of values. A continuous random variable can assume any value in an interval of values. Once an experiment, its outcomes, and the random variable of interest have been clearly defined, we can obtain (calculate) the probabilities of the random variables taking on a specific value—or a range of values in the continuous case. This results in a *probability distribution.*

A.2.1 Discrete Random Variables

Consider the experiment of throwing two dice. Let the random variable, X, be the sum of the points on the two dice. It is easily seen that X is a discrete random variable because it can take on only the values 2, 3, 4, . . . 12. Also, it is easy to show that the probability distribution, also called the *probability mass function* (pmf) for discrete distributions, is as shown in Table A.1.

The *cumulative distribution function* (cdf) of a random variable, denoted by $F(x)$, is given by

$$F(x) = P(X \leq x), \quad -\infty < x < \infty$$

TABLE A.1

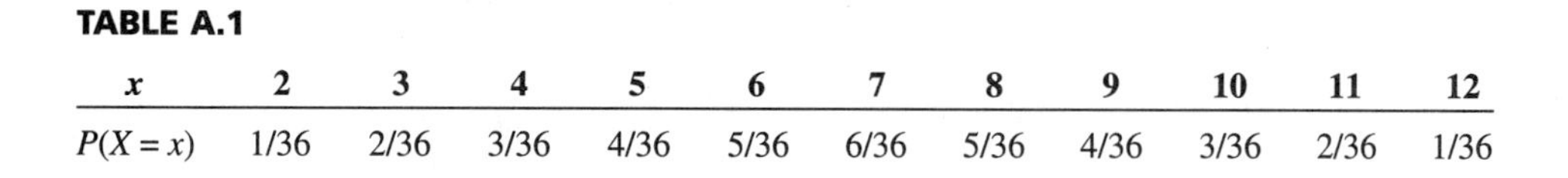

x	2	3	4	5	6	7	8	9	10	11	12
$P(X = x)$	1/36	2/36	3/36	4/36	5/36	6/36	5/36	4/36	3/36	2/36	1/36

TABLE A.2

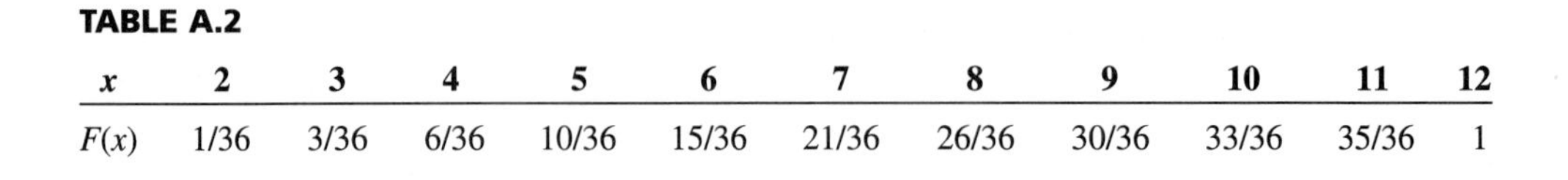

x	2	3	4	5	6	7	8	9	10	11	12
$F(x)$	1/36	3/36	6/36	10/36	15/36	21/36	26/36	30/36	33/36	35/36	1

For discrete random variables, the cdf is a step function with steps equal to the probabilities at each value the random variable can assume.

The function

$$F(x) = P(X \leqslant x), \; x = x_i$$

where x_i are the values that a discrete random variable can assume is called the *cumulative mass function* (cmf). The cmf for X above is shown in Table A.2, and the cdf is shown in Figure A.2.

The *expected value,* also called the *mean* and denoted by μ, for a discrete random variable is given by

$$E(X) = \mu = \sum x \bullet P(X = x)$$

where the sum is taken over all values x that the random variable can assume.

The rth moment is given by

$$E(X^r) = \sum x^r \, P(X = x)$$

In particular, the *variance,* denoted by σ^2, is given by

$$\sigma^2 = E(X - \mu)^2 = E(X^2) - \mu^2$$

Higher standardized moments such as skewness (α_3) and kurtosis (α_4) are given by

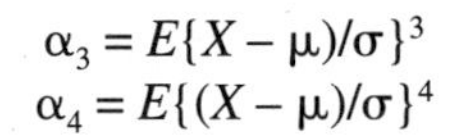

$$\alpha_3 = E\{X - \mu)/\sigma\}^3$$
$$\alpha_4 = E\{(X - \mu)/\sigma\}^4$$

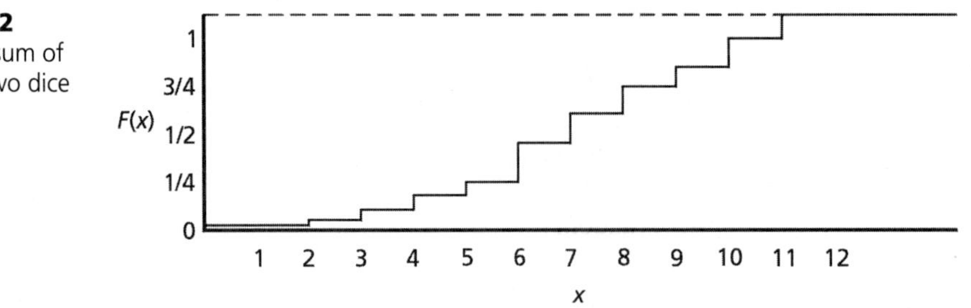

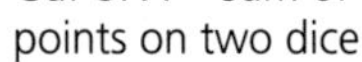
FIGURE A.2

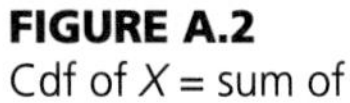
Cdf of X = sum of points on two dice

A.2.2 Some Discrete Distributions

Some of the most commonly used discrete distributions are the *Bernoulli, hypergeometric, binomial, Poisson,* and *geometric.* The pmf's, cmf's, and moments for some of these distributions are given below, along with Excel functions to compute probabilities and percentiles, and Arena functions to sample from the distribution.

Bernoulli Distribution

The Bernoulli random variable can take on only one of two values: zero or 1. Since there are only two possible outcomes of the experiment, it is used to model a random selection between two choices:

pmf
$$p(x) = \begin{cases} 1-p & x=0 \\ p & x=1 \\ 0 & \text{otherwise} \end{cases}$$

cmf
$$F(x) = \begin{cases} 0 & x<0 \\ 1-p & 0 \leqslant x < 1 \\ 1 & x \geqslant 1 \end{cases}$$

parameters p

mean p

variance $p(1-p)$

skewness $\dfrac{1-2p}{p^2(1-p)^2}$

kurtosis $\dfrac{1-3p+3p^2}{p^3(1-p)^3}$

Excel function None

Arena function None

Hypergeometric Distribution

The hypergeometric distribution is a model for sampling without replacement. Suppose there is a finite population of N items, such as 20 finished goods, and some number D ($D \leqslant N$) of the items fall into a group that is of interest—for example, 5 defective items. A random sample of size n is selected from this population without replacement; and the random variable of interest, X, is the number of items in the sample that belong to the class of interest—defectives, in our example.

pmf $$p(x) = \begin{cases} \dfrac{\binom{D}{x}\binom{N-D}{n-x}}{\binom{N}{n}} & x = 0, 1, 2, \ldots, \text{Min}(n, D) \\ 0 \text{ otherwise} \end{cases}$$

where

$$\binom{N}{n} = \frac{N!}{n!\,(N-n)!}$$

cdf no closed form

parameters N, D, n

mean $$n\left(\frac{D}{N}\right)$$

variance $$n = \left(\frac{D}{N}\right)\left(1 - \frac{D}{N}\right)\left(\frac{N-n}{N-1}\right)$$

skewness $$\frac{(N-2D)(N-1)^{1/2}(N-2n)}{[ND(N-D)(N-n)]^{1/2}(N-2)}$$

kurtosis $$\frac{3(N-1)(N+6)}{(N-2)(N-3)} + \frac{(N-1)N(N+1)}{(N-n)(N-2)(N-3)}$$

$$= \left\{1 - \frac{6N}{N+1}\left(pq + \frac{n\,(N-n)}{N^2}\right)\right\}\frac{1}{npq}$$

where $p = \dfrac{D}{N}$

and $q = 1 - p$

Excel function HYPGEODIST

Arena function None

Binomial Distribution

The binomial distribution is a model for sampling with replacement. Suppose (1) an experiment consists of n trials; (2) each trial has one of two outcomes, arbitrarily named *success* (S) and *failure* (F); (3) the probability of success on any trial is a constant (p); and (4) the trials are independent. Such trials are called Bernoulli trials, and any experiment that satisfies the above three assumptions is called a *binomial experiment.* It is easy to see that a binomial random variable is the sum of n independent Bernoulli random variables, each having parameter p. The random variable X, the number of successes on the n trials, is a binomial random variable.

pmf	$p(x) = \binom{n}{x} p^x (1-p)^{n-x} \quad x = 0, 1, 2, \ldots, n$ 0 otherwise
cmf	$F(x)$ no closed form solution
parameters	$n, p \ (q = 1 - p)$
mean	np
variance	npq
skewness	$(q - p)(npq)^{-0.5}$
kurtosis	$3 + \dfrac{1 - 6pq}{npq}$
Excel functions	BINOMDIST, CRITBINOM
Arena function	None

Poisson Distribution

Let the number of arrivals at a toll booth in a unit of time, such as five minutes, be represented by the random variable X. Let λ be the mean number of arrivals per minute. Under some assumptions [(1) the probability that exactly one arrival will occur in an interval Δt is approximately $\lambda \, \Delta t$; (2) the probability that two or more arrivals occur in the interval $\rightarrow 0$ as $\Delta t \rightarrow 0$], the random variable X follows the Poisson distribution.

A Poisson distribution approximates a binomial distribution when the number of trials, n, is very large and the probability parameter, p, is very small. It is often used to model the number of events that occur when there is a large number of opportunities for an event but a small probability that the event occurs on each opportunity.

pmf	$p(x) = \begin{cases} \dfrac{e^{-\lambda} \lambda^x}{x!} & x = 0, 1, 2, \ldots, \\ 0 \text{ elsewhere} & \end{cases}$
cmf	$F(x) =$ no closed form solution
parameters	$\lambda > 0$ If approximating a binomial distribution, $\lambda = np$.
mean	λ
variance	λ
skewness	$\dfrac{1}{\sqrt{\lambda}}$
kurtosis	$3 + \dfrac{1}{\lambda}$

Excel function	POISSON
Arena function	None

Geometric Distribution

In the case of the Bernoulli trials (see binomial distribution), let X denote the number of trials before the first success; then X is said to be a *geometric random variable.*

pmf	$p(x) = \begin{cases} p(1-p)^x \\ 0 \text{ elsewhere} \end{cases}$	$x = 0, 1, 2, \ldots,$
cmf	$F(x) = \begin{cases} 0 & x < 0 \\ 1-(1-p)^{[x]+1} \end{cases}$	$x \geq 0$; [ ] is an integer function
parameters	$0 \leq p \leq 1$	
mean	$\frac{1-p}{p}$	
variance	$\frac{1-p}{p^2}$	
skewness	$\frac{2-p}{\sqrt{1-p}}$	
kurtosis	$\frac{p^2-9p+9}{1-p}$	
Excel function	None	
Arena function	None	

A.3 CONTINUOUS RANDOM VARIABLES

A probability distribution defined for a continuous random variable is called a *probability density function* (pdf) and denoted by $f(x)$. In the case of continuous random variables, it only makes sense to specify the probability that the random variable is in a range. In fact, the probability that a continuous random variable takes a specific value, x, is zero; that is, $P(X = x) = 0$. The cumulative distribution function denoted by $F(x)$ is given by

$$F(x) = P(X \leqslant x) = \int_{-\infty}^{x} f(x)\,dx$$

The probability that the random variable X lies between a and b is given by

$$P(a \leqslant X \leqslant b) = \int_{a}^{b} f(x)dx = \int_{-\infty}^{b} f(x)dx - \int_{-\infty}^{a} f(x)dx = F(b) - F(a)$$

The expected value is given by

$$E(X) = \mu = \int_{-\infty}^{\infty} x\, f(x)dx$$

and

$$E(x^r) = \mu'_r = \int_{-\infty}^{\infty} x^r\, f(x)\, dx$$

The expressions for variance, skewness, and kurtosis are similar to those for the discrete case.

A.3.1 Some Continuous Distributions

Some of the commonly used continuous distributions are uniform, triangular, exponential, normal, Student's *t*, lognormal, gamma, and Weibull. The expressions for the pdf's, cdf's, and moments along with Excel and Arena sampling functions for some of these distributions are given below.

Uniform Distribution

The uniform distribution is used to model an activity known to take a value between two limits, a and b, but when nothing else is known about the likelihood of values in this interval. If $a = 0$ and $b = 1$, then this distribution becomes the uniform distribution on (0, 1), which is used as the basis to sample from other distributions.

pdf
$$f(x) = \begin{cases} \dfrac{1}{b-a} & a \leqslant x \leqslant b \\ 0 & \text{elsewhere} \end{cases}$$

cdf
$$F(x) = \left\{\begin{array}{ll} 0 & x \leqslant a \\ \dfrac{x-a}{b-a} & a \leqslant x \leqslant b \\ 1 & x \geqslant b \end{array}\right\} \quad a \leqslant b$$

parameters	a, b
mean	$\frac{a+b}{2}$
variance	$\frac{(b-a)^2}{12}$
skewness	0
kurtosis	1.8
Excel function	None
Arena function	UNIF

Triangular Distribution

The triangular distribution is useful to model a variable X that is known to assume values in a specific interval, $[a, b]$; in addition, there is a value c in this interval such that the likelihood of observing a value x increases as x is closer to c but nothing more is known about the distribution of X. The minimum, maximum, and most likely values of X are a, b, and c, respectively.

pdf

$$f(x) = \begin{cases} \dfrac{2(x-a)}{(b-a)(c-a)} & a \leqslant x \leqslant c \\ \dfrac{2(b-x)}{(b-a)(b-c)} & c < x \leqslant b \\ 0 & \text{elsewhere} \end{cases}$$

cdf

$$F(x) = \begin{cases} 0 & x \leqslant a \\ \dfrac{(x-a)^2}{(b-a)(c-a)} & a \leqslant x \leqslant c \\ 1 - \dfrac{(b-x)^2}{(b-a)(b-c)} & c \leqslant x \leqslant b \\ 1 & x \geqslant b \end{cases}$$

parameters	a, b, c
mean	$\frac{a+b+c}{3}$
variance	$= \frac{1}{18}(a^2 + b^2 + c^2 - ab - ac - bc)$

skewness	$-\frac{8\sqrt{6}}{45}\left[\frac{(b+c-2a)(a+c-2b)(a+b-2c)}{(b-a)^3}\right]$
kurtosis	$\frac{12}{5}$
Excel function	None
Arena function	TRIA

Two special cases are the symmetric triangular distribution, where

$$c = \frac{a+b}{2}$$

and the right triangular distribution where $c = a$ or $c = b$.

Symmetric Triangular Distribution

pdf

$$f(x) = \begin{cases} \frac{4(x-a)}{(b-a)^2} & a \leqslant x \leqslant \frac{a+b}{2} \\ \frac{4(b-x)}{(b-a^2)} & \frac{a+b}{2} \leqslant x \leqslant b \\ 0 & \text{elsewhere} \end{cases}$$

cdf

$$F(x) = \begin{cases} 0 & x \leqslant a \\ \frac{2(x-a)^2}{(b-a)^2} & a \leqslant x \leqslant \frac{a+b}{2} \\ 1 - \frac{2(b-x)^2}{(b-a)^2} & \frac{a+b}{2} \leqslant x \leqslant b \\ 1 & x \geqslant b \end{cases}$$

parameters	$a, b \quad a < b$
mean	$\frac{a+b}{2}$
variance	$\frac{(b-a)^2}{2}$
skewness	0
kurtosis	$\frac{12}{5}$

Right Triangular Distribution with $c = a$

pdf $$f(x) = \begin{cases} \dfrac{2(b-x)}{(b-a)^2} & a \leqslant x \leqslant b \\ 0 & \text{elsewhere} \end{cases}$$

cdf $$F(x) = \begin{cases} 0 & x \leqslant a \\ 1 - \dfrac{(b-x)^2}{(b-a)^2} & a \leqslant x \leqslant b \\ 1 & x \geqslant a \end{cases}$$

parameters $a, b \quad a < b$

mean $\dfrac{2a+b}{3}$

variance $\dfrac{(b-a)^2}{18}$

skewness $\dfrac{16\sqrt{6}}{45}$

kurtosis $\dfrac{12}{5}$

Normal Distribution

The normal distribution is appropriate to represent many random phenomena due to the central limit theorem. This theorem states that the average of a large number of independent observations from the same population has a distribution that is approximately a normal distribution. In many cases, there is reason to model a variable as the sum of a number of identical components. For example, the amount of fuel sold per day by a fuel depot can be considered the sum of the amounts sold in each transaction.

pdf $$f(x) = \frac{1}{\sqrt{2\pi\sigma^2}} \exp\left[-(x-\mu)/2\sigma^2\right] \qquad -\infty < x < \infty$$

cdf No closed form expression. Tables and software are available to compute cumulative probabilities.

parameters $\mu, \sigma^2 \ (\sigma^2 > 0)$

mean μ

variance	σ^2
skewness	0
kurtosis	3
Excel functions	NORMDIST, NORMINV, NORMSDIST, NORMSINV
Arena function	NORM

Lognormal Distribution

If X has a normal distribution, then e^X has a lognormal distribution. This distribution has been found useful to describe some economic phenomena (e.g., stock price changes) as well as some engineering phenomena (e.g., lifetimes of equipment).

pdf	$f(x) = \begin{cases} \frac{1}{x\sqrt{2\pi\theta^2}} \exp\left\{-\frac{(\ln x - \lambda)}{2\theta^2}\right\} & x > 0 \\ 0 & \text{elsewhere} \end{cases}$
cdf	No closed form expression but can be calculated from the standard normal df.
parameters	λ, θ^2 ($\theta^2 > 0$)
mean	$e^{\lambda + \frac{\theta^2}{2}}$
variance	$w(w-1)e^{2\lambda}$ where $w = e^{\theta^2}$
skewness	$\sqrt{(w-1)}\,(w+2)$
kurtosis	$3 + (w-1)(w^3 + 3w^2 + 6w + 6)$
Excel functions	LOGNORMDIST, LOGINV
Arena function	LOGN

Exponential Distribution

The exponential distribution is frequently used to describe random variables in queueing, service, and reliability systems. If arrivals in a queueing or service system form a Poisson process, then the interarrival times have an exponential distribution. The Poisson process is a model for a completely random sequence of events in which each event has an equal probability of occurring in a small interval of time,

Δt. The exponential distribution is also a special case of the gamma and Weibull distributions.

pdf	$f(x) = \begin{cases} \frac{1}{\beta} \cdot e^{-x/\beta} & x > 0 \\ 0 & \text{elsewhere} \end{cases}$
cdf	$F(x) = 0 \quad x > 0$ $= 1 - e^{-x/\beta} \quad x \geqslant 0$
parameters	$\beta > 0$
mean	β
variance	β^2
skewness	2
kurtosis	9
Excel function	EXPONDIST
Arena function	EXPO

Gamma Distributions

The gamma distribution has been used to model service times in queueing and service systems, equipment lifetimes, times between purchases of nondurable consumer goods, and machine repair times, among other phenomena. There are several special cases. If $\alpha = 1$, then it is an exponential distribution with mean β. If α is an integer, then it is an Erlang distribution, which is the distribution of the sum of α independent exponential random variables, each with mean β. If $\beta = 2$, then the gamma distribution is a χ^2 distribution with 2α degrees of freedom. As $\alpha \rightarrow$ to ∞, it approaches to a normal distribution.

pdf	$f(x) = \begin{cases} \frac{x^{\alpha-1} e^{-x/\beta}}{\beta\alpha\, \Gamma(\alpha)} & x > 0 \\ 0 & \text{elsewhere} \end{cases}$
cdf	Does not exist in closed form except for special cases.
parameters	$\alpha, \beta \quad \alpha > 0, \beta > 0$
mean	$\alpha\beta$

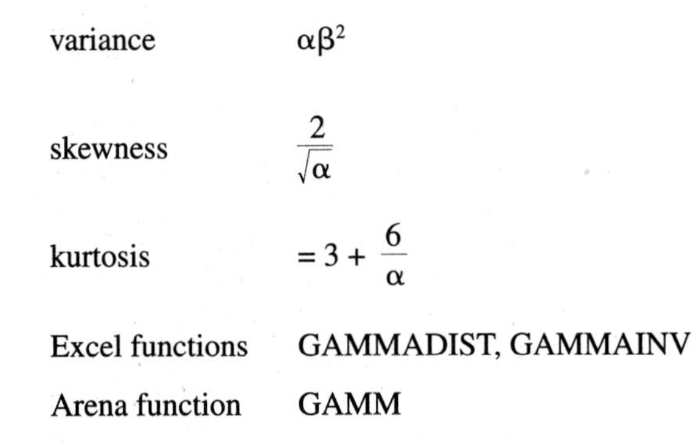

variance	$\alpha\beta^2$
skewness	$\dfrac{2}{\sqrt{\alpha}}$
kurtosis	$= 3 + \dfrac{6}{\alpha}$
Excel functions	GAMMADIST, GAMMAINV
Arena function	GAMM

Weibull Distribution

The Weibull distribution has been found to fit lifetime data in many reliability studies. If $\alpha = 1$, then it is an exponential distribution. If $\alpha = 2$, then it is a Raleigh distribution.

pdf
$$f(x) = \begin{cases} \dfrac{\alpha\, x^{\alpha-1}\, e^{-(x/\beta)^{\alpha}}}{\beta^{\alpha}} & x > 0 \\ 0 & \text{elsewhere} \end{cases}$$

cdf
$$F(x) = \begin{cases} 0 & x < 0 \\ 1 - e^{-(x/\beta)^{\alpha}} & x \geqslant 0 \end{cases}$$

parameters α, β $\qquad \alpha > 0, \beta > 0$

mean
$$\beta\Gamma\left(1 + \frac{1}{\alpha}\right)$$

variance
$$\beta^2\left\{\Gamma\left(1 + \frac{2}{\alpha}\right) - \left[\Gamma\left(1 + \frac{1}{\alpha}\right)\right]^2\right\}$$

skewness
$$\frac{\Gamma\left(1 + \frac{3}{\alpha}\right) - 3\Gamma\left(1 + \frac{2}{\alpha}\right)\Gamma\left(1 + \frac{1}{\alpha}\right) + 2\left[\Gamma\left(1 + \frac{1}{\alpha}\right)\right]^3}{\left\{\Gamma\left(1 + \frac{2}{\alpha}\right) - \left[\Gamma\left(1 + \frac{1}{\alpha}\right)\right]^2\right\}^{3/2}}$$

kurtosis
$$\frac{\Gamma\left(1 + \frac{4}{\alpha}\right) - 4\Gamma\left(1 + \frac{3}{\alpha}\right)\Gamma\left(1 + \frac{1}{\alpha}\right) + 6\Gamma\left(1 + \frac{2}{\alpha}\right)\left[\Gamma\left(1 + \frac{1}{\alpha}\right)\right]^2 - 3\left[\Gamma\left(1 + \frac{1}{\alpha}\right)\right]^4}{\left\{\Gamma\left(1 + \frac{2}{\alpha}\right) - \left[\Gamma\left(1 + \frac{2}{\alpha}\right)\right]^2\right\}^2}$$

Excel function WEIBULL

Arena function WEIB

Beta Distribution

The beta distribution models random variables whose values are between zero and 1. It is useful for modeling probabilities or proportions that are not known. The ratio of X to $(X + Y)$, where X and Y are independent gamma random variables, has a beta distribution, so it is an appropriate model for a proportion in many cases. With a suitable transformation, it can be used to model a variable that is known to be between limits $a < b$. Some studies have found the beta distribution to be useful in modeling activity times such as in a PERT network.

pdf $$f(x) = \begin{cases} \dfrac{\Gamma(\alpha+\beta)}{\Gamma(\alpha)\Gamma(\beta)} x^{\alpha-1}(1-x)^{(\beta-1)} & 0 < x < 1 \\ 0 & \text{elsewhere} \end{cases}$$

cdf Does not exist in closed form.

parameters α, β $\alpha > 0, \beta > 0$

mean $$\frac{\alpha}{\alpha+\beta}$$

variance $$\frac{\alpha\beta}{(\alpha+\beta)^2(\alpha+\beta+1)}$$

skewness $$\frac{2(\beta-\alpha)(\alpha+\beta+1)^{1/2}}{(\alpha\beta)^{1/2}(\alpha+\beta+2)}$$

kurtosis $$\frac{3(\alpha+\beta+1)[2(\alpha+\beta)^2+(\alpha\beta)(\alpha+\beta-6)]}{\alpha\beta(\alpha+\beta+2)(\alpha+\beta+3)}$$

Excel functions BETADIST, BETAINV

Arena function BETA

A.4 INFERENTIAL ANALYSIS

The basic idea of statistical inference is to "estimate" some characteristics or parameters of the distribution of a random variable of interest using observations of that random variable. Examples of random variables of interest are the individual incomes of people in a community, state, or country; the heights of people; company sales; the lifetimes of tires produced by a company; the amount of carbon monoxide in the emissions of automobiles from a specific company; and the defectives produced by an assembly line. Examples of parameters of random variables that we may want to estimate are the mean, median, variance, and 90th percentile. If the random variable of interest has normal distribution, we may be interested in estimating its mean (μ) and its variance (σ^2). If the number of defectives in a production process has a binomial distribution, we may be interested in estimating the proportion that are defective, p.

Generally, statistical influence involves taking a *random sample* of size n, denoted by $(X_1, X_2, \ldots, X_n)$, from the population. We calculate some quantities such as the sample mean, $\bar{X}$, and sample variance, s^2.

$$\bar{X} = \sum \frac{X_i}{n}$$

and

$$s^2 = \sum \frac{(X_i - \bar{X})^2}{n} = \frac{\sum X_i^2 - \frac{\left(\sum X_i\right)^2}{n}}{n}$$

These quantities calculated from the sample are called *sample statistics.* Inference generally concerns computing point and interval estimates (confidence intervals) for the unknown parameter(s) and testing hypotheses concerning the values of the unknown parameter(s).

A.4.1 Sampling Distribution of a Statistic

One fundamental concept in probability and statistics is that sample statistics are random variables and thus have their own distributions. It is intuitive that these distributions (of the sample statistics) are functions of the parent population (the type and their parameters) and the sample size. A pictorial representation is shown in Figure A.3. In the figure, θ is the parameter (or characteristics) of interest from the parent population and $\hat{\theta}$ is a sample statistic that is a point estimator for θ. The two most important statistics are the sample mean

FIGURE A.3

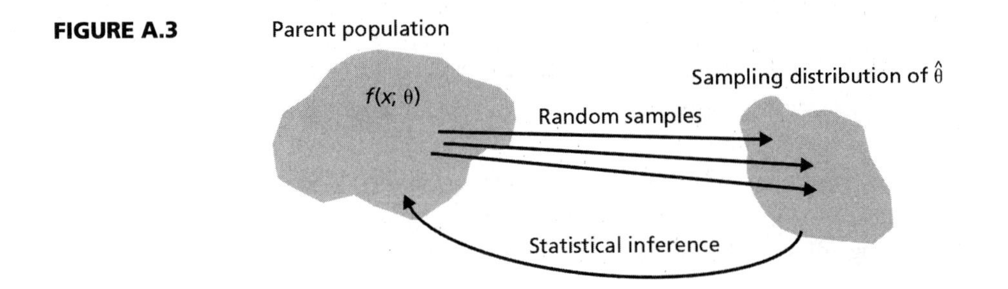

$$\overline{X} = \frac{1}{n}\sum_{i=1}^{n} X_i$$

and the sample variance

$$s^2 = \frac{1}{n-1}\sum_{i=1}^{n}(X_i - \overline{X})^2$$

Excel provides the functions AVERAGE and STDEV to compute the sample mean and sample standard deviation.

Sampling Distribution of $\overline{X}$

Let X be the random variable with mean μ_X and variance σ_X^2. If $X_1, X_2, \ldots, X_n$ is a random sample of size n from this parent population, then the sample mean

$$\overline{X} = \frac{\sum X_i}{n}$$

is a random variable.

The mean of $\overline{X}$, denoted by $\mu_{\overline{X}}$, is μ_X; and the variance of $\overline{X}$, denoted by $\sigma_{\overline{X}}^2$, is $\sigma_{\overline{X}}^2 = \sigma_X^2/n$. If the parent population variance σ_X^2 is not known, then the sample variance s^2, where

$$s^2 = \frac{1}{n-1}\sum (X_i - \overline{X})^2$$

is used as an estimator of σ_X^2. The probability distribution of $\overline{X}$ depends on (1) whether the parent population is normally distributed, (2) whether the population variance σ_x^2 is known or unknown, and (3) whether the sample size is large ($n \geqslant 30$) or small.

The mean and variance of the sampling distribution of $\overline{X}$ are given by $E(\overline{X}) = \mu_{\overline{X}} = \mu_X$ and

$$\text{var}(\overline{X}) = \sigma_{\overline{X}}^2 = \frac{\sigma_X^2}{n}$$

The distribution of $\dfrac{\overline{X} - \mu_X}{\sigma_X}$ is

1. *unknown* if parent is non-normal and n is small ($n < 30$);
2. *t-distribution* with $n - 1$ degrees of freedom if parent is normal, σ^2 is unknown, and n is small;
3. *standard normal* if parent is normal and σ^2 is known; and
4. *approximately standard normal* if n is large (due to the central limit theorem), regardless of the distribution of the parent population.

Sampling Distribution of s^2

If the parent population is normal and if s^2 is the sample variance, then

$$\frac{(n-1)s^2}{\sigma^2}$$

is a χ^2 (chi-square) distribution with $(n - 1)$ degrees of freedom.

A.4.2 Examples: Sampling Distribution of $\overline{X}$

Example 1

An automatic machine that is used to fill cans of soup has the following characteristics: mean filling weight $\mu = 15.9$ ounces and $\sigma = .5$ ounces.

1. Show the sampling distribution of $\overline{X}$, where $\overline{X}$ is the sample mean for 40 cans selected randomly by a quality control inspector.
2. What is the probability of finding a sample of 40 cans with a mean, $\overline{X}$, greater than 16 ounces?

Solution The parent population is not known to be normal. However, the sample size is large ($n > 50$). Because of the central limit theorem,

1. $\overline{X}$ is normal with mean $\mu X = 15.9$ ounces and standard deviation

$$\sigma_{\overline{X}} = \frac{\sigma_X}{\sqrt{n}} = \frac{.5}{\sqrt{40}} = .079$$

(See Figure A.4.)

2. $P(\overline{X} \geqslant 16) = ?$

$$Z = \frac{16 - 15.9}{.079} = 1.27$$

(See Figure A.5.)

$$P(Z \geqslant 1.27) = 1 - F(1.27) = 1 - .8980 = .102$$

FIGURE A.4
Distribution of $\overline{X}$

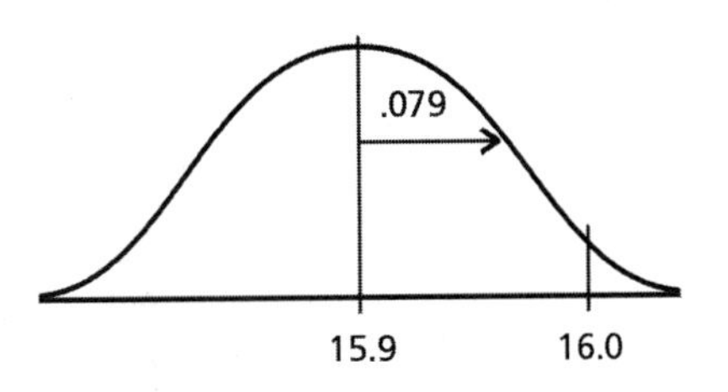

FIGURE A.5 Distribution of $Z = \frac{\overline{X} \quad \mu_X}{\sigma_{\overline{X}}}$

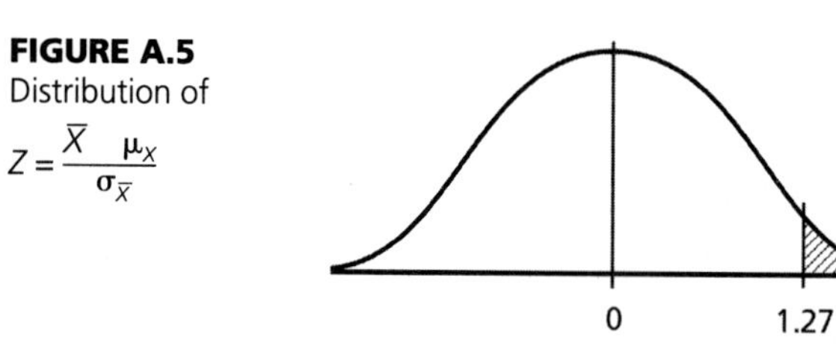

$F(z)$ is the cumulative standard normal distribution.

Example 2

Assume that an economist estimates that the number of gallons of gasoline used monthly by each automobile in the United States is a normally distributed random variable with mean $\mu = 50$ and unknown variance.

1. Suppose that a sample of nine observations yields a sample variance of $s^2 = 36$. What is the probability that $\overline{X}$ lies between 45 and 55?

Solution X is normal with $\mu = 50$ and σ unknown. Also, $n = 9$ (small), $s^2 = 36$. The distribution of

$$\frac{\overline{X} - \mu}{s_{\overline{X}}}$$

is a t-distribution (df = 8) (see Figure A.6):

$$s_{\overline{X}}^2 = \frac{s^2}{n} = \frac{36}{9} = 4$$

$$P(45 \leqslant \overline{X} \leqslant 54) = ?$$

$$= P\left(\frac{45 - 50}{2} \leqslant \frac{\overline{X} - \mu}{s_{\overline{X}}} \leqslant \frac{54 - 50}{2}\right)$$

$$= P\,(-2.5 \leqslant t \leqslant 2.0)$$

FIGURE A.6 Distribution of $t = \frac{\overline{X} \quad \mu_X}{s_{\overline{X}}}$

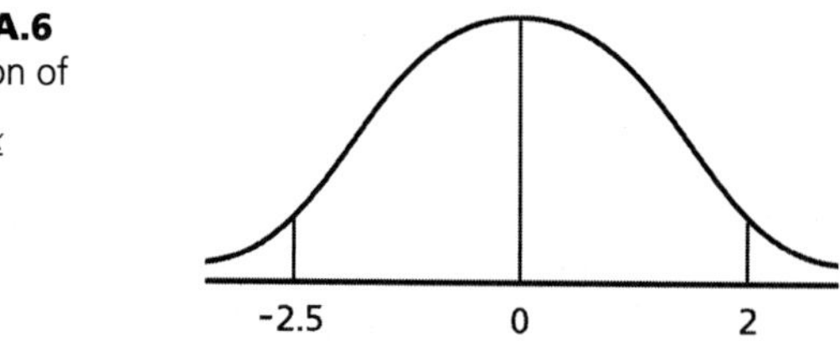

$$F(2) - F(-2.5) = .960 - .018 = .942$$

In Excel, the function TDIST can be used to compute this probability. The appropriate function would be

$$= 1 - \text{TDIST}\ (2.0, 8, 1) - \text{TDIST}\ (2.5, 8, 1)$$

TDIST ( ) computes the probability in the upper tail of Student's t-distribution if the last parameter is 1.

A.4.3 Estimation

Point Estimation

If X is a random variable with probability distribution $f(x; \theta)$, where θ is an unknown parameter and $X_1, X_2, \ldots, X_n$ is a random sample of size n from the distribution $f(x; \theta)$, then the statistic $\hat{\theta} = h\ (X_1, X_2, \ldots, X_n)$ corresponding to θ is called the *estimator* of θ. Note that $\hat{\theta}$ is a random variable (in the previous section, $\overline{X}$ was a random variable). After the sample has been selected, $\hat{\theta}$ takes on a particular numerical value and is called the *point estimate* of θ.

Example

Suppose that the random variable X is normally distributed with unknown mean μ and known variance σ^2. The sample mean $\overline{X}$ is a point estimator of μ—that is, $\hat{\mu} = \overline{X}$. If a random sample of size $n = 5$ has been obtained from this population given by $X_1 = 4.6$, $X_2 = 5.9$, $X_3 = 3.7$, $X_4 = 7.2$, and $X_5 = 5.5$, then

$$\overline{X} = \frac{4.6 + 5.9 + 3.7 + 7.2 + 5.5}{5} = 5.4$$

is a point estimate of the unknown parameter μ.

If the population variance is unknown, then the sample variance

$$s^2 = \frac{1}{n-1}\left[\sum (X_i - \overline{X})^2\right]$$

is a point estimator of the unknown variance σ^2, and the numerical value $s^2 = 1.46$ is a point estimate of the variance s^2.

Properties of Estimators

Unbiasedness Good estimators in some sense are "close" to the true value of the unknown parameters. Formally, an estimator $\hat{\theta}$ is said to be an *unbiased estimator* of an unknown parameter θ if $E(\hat{\theta}) = \theta$. Thus, if the "average" of all values of $\hat{\theta}$ in the sampling distribution is equal to θ, then $\hat{\theta}$ is unbiased; that is, the mean of the sampling distribution of $\hat{\theta}$ is equal to θ. It can be shown that $\overline{X}$ and s^2 are unbiased estimators of the mean μ and variance σ^2.

Efficiency The mean square error of $\hat{\theta}$ is defined as

$$\text{MSE}(\hat{\theta}) = E(\hat{\theta} - \theta)^2$$

If $\hat{\theta}$ is unbiased, $E(\hat{\theta}) = \theta$ and $\text{MSE}(\hat{\theta})$ is the variance of $\hat{\theta}$. If $\hat{\theta}_1$ and $\hat{\theta}_2$ are two estimators of the parameter θ, then the relative efficiency of $\hat{\theta}_2$ to $\hat{\theta}_1$ is defined as

$$\frac{\text{MSE}\,(\hat{\theta}_1)}{\text{MSE}\,(\hat{\theta}_2)}$$

If this relative efficiency is less than 1, then we can conclude that $\hat{\theta}_1$ is a more efficient estimator of θ than $\hat{\theta}_2$. An estimator θ^* whose mean square error is less than or equal to any other estimator $\hat{\theta}$ is called an *optimal estimator* of θ.

Consistency Another way to define the closeness of an estimator $\hat{\theta}$ to the parameter θ is in terms of consistency. If $\hat{\theta}_n$ is an estimator of θ based on a random sample of size n, then we say that $\hat{\theta}_n$ is consistent for θ if

$$\lim_{n \to \infty} P(|\hat{\theta}_n - \theta| < \epsilon) = 1$$

for any value ϵ. Estimators whose mean squared error (or variance, if the estimator is unbiased) tend to zero as $n \to \infty$ are consistent. Note that $\overline{X}$ is a consistent estimator of the mean of the normal distribution [$\sigma^2_{\overline{X}} = \sigma^2/n \to 0$ as $n \to \infty$].

A.4.4 Confidence Intervals

Point estimates of parameters carry little information and do not tell how close they are to the true parameter value. In this section, we will discuss confidence intervals for the population mean μ. Confidence intervals (CIs) are interval estimates of the parameters or characteristics of interest. CIs are a function of three factors: (1) the level of confidence, (2) the sample size, and (3) the population variance.

The *lower confidence limit* (LCL), A, and the *upper confidence limit* (UCL), B, are computed from the data so that $100(1 - \alpha)$ percent of all possible samples of size n would produce values of A and B such that $A \leq \theta \leq B$. Note that A and B will vary from sample to sample, whereas θ is a fixed but unknown constant. Thus, the confidence statement for a 95-percent confidence interval [A, B] for parameter θ is "the probability is approximately .95 that the interval [A, B] includes the (unknown) value θ."

If the confidence coefficient of a confidence interval [A, B] is $100(1 - \alpha)$, then the probability that the interval does *not* include θ is α.

Based on the sampling distribution of $\overline{X}$ discussed in section 4.1, the 100(1 – α)-percent CI for the mean μ takes one of the following forms (let σ_x^2 be the parent population variance and s^2 be the sample variance):

$\overline{X} \pm Z_{\alpha/2}\frac{\sigma}{\sqrt{n}}$ if the parent population is normal and σ_x^2 is known

$\overline{X} \pm Z_{\alpha/2}\frac{s}{\sqrt{n}}$ if the parent population is non-normal, σ_x^2 is unknown, and the sample size is large

$\overline{X} \pm t_{\alpha/2,\, n-1}\frac{s}{\sqrt{n}}$ if the parent population is normal, σ_x^2 is unknown, and the sample size is small

Excel functions NORMSINV and TINV can be used to compute $Z_{\alpha/2}$ and $t_{\alpha/2,\, n-1}$, respectively. Excel also provides the function CONFIDENCE to compute $Z_{\alpha/2}\sigma/\sqrt{n}$. The half-width of the confidence interval for μ, $Z_{\alpha/2}\sigma/\sqrt{n}$, is affected by three parameters:

1. the sample size (larger sample sizes result in narrower CIs);
2. the population standard deviation, σ (higher standard deviations result in wider CIs); and
3. the level of confidence, 1 – α (higher values of confidence levels—lower values of α—result in wider CIs).

Confidence Interval Example 1

A sample of 64 customers at Ron and Ted's service station shows a sample mean of 13.6 gallons of gasoline purchased per customer. If the population standard deviation is 3.0 gallons, what is the 95-percent confidence interval estimate for the mean number of gallons purchased per customer?

Solution The distribution of X is unknown, μ is unknown, $\sigma = 3.0$. The sample mean $\bar{x} = 13.6$, $\alpha = .05$, and $n = 64$. Because of large sample size and the central limit theorem, the distribution of $\overline{X}$ is approximately normal.

The 95-percent CI or μ is given by $\bar{x} \pm Z_{\alpha/2}\sigma/\sqrt{n}$. Substituting the appropriate values results in a CI of

$$\left[13.6 \pm 1.96\left(\frac{3.0}{8}\right)\right] = (12.865,\ 14.335)$$

Confidence Interval Example 2

In testing a new production method, 18 employees were randomly selected and asked to try the new method. The sample mean production rate for the 18 employees was 80 parts per hour. The sample standard deviation was 10 parts per hour. Provide a 90-percent confidence interval estimate for the mean production rate for the new method. (Assume that the production rate is normally distributed.)

Solution The distribution of X is normal, and variance is unknown. The sample size is small ($n = 18$); the sample mean, $\bar{x}$, is 80; and the sample standard deviation, s, is 10.

The distribution of $\bar{X}$ is t with 17 degrees of freedom.

The 90-percent confidence interval for μ is given by $\bar{X} \pm t_{.05,\,17}\, s/\sqrt{n}$. Substituting the appropriate values for $\bar{x}$, t, s, and n results in a CI of

$$80 \pm 1.74\left(\frac{10}{\sqrt{18}}\right) = (5.899,\ 14.101)$$

A4.5 Sample Size Determination

In many situations, the decision maker does not know what sample size should be used. He or she may want to specify (1) the width of the CI (denoted by W) or (2) the maximum difference (or "error") between the estimate of the population mean and the true (unknown) population mean, denoted by $D = W/2$.

The required sample size can be determined from the expression for half-width of the CI:

$$Z_{\alpha/2} \frac{\sigma}{\sqrt{n}} = D$$

$$n = \frac{Z^2_{\alpha/2}\,\sigma^2}{D^2} = 4Z^2_{\alpha} \cdot \frac{\sigma^2}{W^2}$$

If σ is not known, the appropriate distribution of $\bar{X}$ may be t-distribution. Because we do not know the sample standard deviation or the degrees of freedom in advance (sample size is not known), the process of determining the sample size becomes iterative.

Example

A national survey research firm has data that indicate the interview time for a consumer opinion study has a standard deviation of 6 minutes. How large a sample should be taken if the firm wants a .98 probability of estimating the population mean interview time to within 2 minutes or less.

Solution

$$\alpha = .02,\ \sigma = 6,\ D = 2,\ Z_{.01} = 2.33$$

Using the above formula,

$$n = (2.33)^2 \frac{6^2}{2^2} = 48.86 = 49$$

A.4.6 Hypothesis Testing

Another way to make inferences about the population is through hypothesis testing. Consider a criminal case. The U.S. judicial system assumes that a defendant is innocent and it is up to the prosecution to prove he or she is guilty of a crime. A trial takes place. Based on the evidence presented during the trial, the jury (or judge) decides (produces a verdict) whether or not the defendant is guilty. The above situation can be formulated as a "hypothesis-testing" problem.

Null hypothesis (H_0): The defendant is not guilty.
Alternate hypothesis (H_a): The defendant is guilty.

In general, the null hypothesis is assumed to be true. In order for us to accept the alternate hypothesis, strong and convincing evidence must be presented.

The evidence produced during the trial is the "sample information," and it is used to make inferences about the true *state of nature* (i.e., the innocence or guilt of the defendant). Obviously, whenever we make a decision based on sample information, the decision process is subject to errors. In hypothesis testing, there are two kinds of error: type I and type II (see Table A.3).

The type I error is the error of accepting H_a when H_0 is true. The type II error is the error of accepting H_0 when H_0 is false. Let α denote the probability of a type I error and β denote the probability of a type II error. The value of α is usually chosen to be .05. Also $(1 - \beta)$ is called the *power of the test.*

General Procedures for Hypothesis Testing

The steps for hypothesis testing are as follows:

1. Formulate the hypotheses. Specify α.
2. Obtain a random sample. Calculate the appropriate summary statistics (e.g., $\bar{x}$, s).
3. Using the summary statistics from step 2, calculate the appropriate test statistic.

TABLE A.3 True state of nature

		H_0 True	**H_a True**
Decision	Accept H_0	No error	Type II error (β)
	Accept H_a	Type I error (α)	No error

4. Compare the test statistic from step 3 with the appropriate percentile of the sampling distribution and make the decision (reject or do not reject H_0).

Three Forms of Hypothesis Testing for μ

Hypothesis testing for μ takes one of these three forms:

1. H_0: $\mu = \mu_0$
H_a: $\mu \neq \mu_0$

2. H_0: $\mu \geqslant \mu_0$
H_a: $\mu < \mu_0$

3. H_0: $\mu \leqslant \mu_0$
H_a: $\mu > \mu_0$

1. Procedure for Two-Sided Hypothesis Testing for μ

H_0: $\mu = \mu_0$
H_a: $\mu \neq \mu_0$

a. From the sample, compute $\bar{x}$ and s; choose an appropriate value for α.

b. Calculate $z_{\text{statistic}} = \dfrac{\bar{x} - \mu_0}{\sigma\sqrt{n}}$.

c. If $|z_{\text{statistic}}| > z_{\alpha/2}$, reject H_0.

***p*-Value** The probability that the test statistic (Z or t) takes on a value more extreme than the calculated value of the test statistic is called the *p-value* and is calculated as

$p = P(Z > Z_{\text{statistic}})$ if H_a is $\mu > \mu_0$ or
$p = P(Z < Z_{\text{statistic}})$ if H_a is $\mu < \mu_0$ or
$p = 2P(Z > |Z_{\text{statistic}}|)$ in the case of two-sided hypothesis testing.

Excel provides the function ZTEST to compute the p-value for a two-sided hypothesis. If the p-value is less than the specified α value, then we reject the null hypothesis.

2. Procedure for One-Sided Hypothesis Testing for μ

H_0: $\mu \geqslant \mu_0$
H_a: $\mu < \mu_0$

a. From the sample, obtain $\bar{x}$ and s; choose α.

b. Calculate $z_{\text{statistic}} = \dfrac{\bar{x} - \mu_0}{\sigma\sqrt{n}}$.

c. If $|Z_{\text{statistic}}| < -z_{\alpha}$, accept H_a.

In this case, the p-value is calculated as

$$p\text{-value} = P(Z < Z_{\text{statistic}})$$

Using the Excel function ZTEST, compute

$$\frac{\text{ZTEST (array, } \mu_0)}{2}$$

where *array* is the range of cells containing the data.

3. *Procedure for Testing the Other One-Sided Hypothesis*

$H_0: \mu \leqslant \mu_0$
$H_a: \mu > \mu_0$

a. From the sample, obtain $\bar{x}$ and s; determine (given) α.

b. Calculate $z_{\text{statistic}} = \dfrac{\bar{x} - \mu_0}{\sigma / \sqrt{n}}$.

c. If $Z_{\text{statistic}} > z_\alpha$, accept H_a.

$$p\text{-value} = P(Z > Z_{\text{statistic}})$$

If the t-distribution is the most appropriate distribution for $\bar{X}$, we will replace Z with the t-distribution with $(n - 1)$ degrees of freedom in all three forms of hypothesis testing above. Also note that the population standard deviation σ is replaced by the sample standard deviation s if σ^2 is unknown.

A.4.7 Examples: Hypothesis Testing

Example 1

An automobile assembly-line operation has a scheduled mean completion time of 2.2 minutes. Because of the effect of completion time on both earlier and later assembly operations, it is important to maintain the 2.2-minute standard. A random sample of 45 times shows a sample mean completion time of 2.28 minutes, with a sample standard deviation of .20 minutes. Use a .02 level of significance and test whether or not the operation is meeting its 2.2-minute standard.

$H_0: \mu = 2.2$ $\quad n = 45$
$H_a: \mu \neq 2.2$ $\quad \bar{x} = 2.28$
$s = .20$
$\alpha = .02$

Since n is large, we can use normal distribution approximation. The test statistic,

$$Z_{\text{statistic}} = \frac{\bar{x} - \mu_0}{s/\sqrt{n}} = \frac{2.28 - 2.2}{\frac{.2}{\sqrt{45}}} = 2.68$$

$$Z_{.01} = 2.33$$

Since $Z_{\text{statistic}} > Z_{\alpha/2}$, reject H_0.
The p-value $= 2\{P(Z > Z_{\text{statistic}})\} = 2(.0037) = .0074$

Example 2

A long-distance trucking firm believes that its mean weekly loss due to damaged shipments is \$2000 or less. A sample of 15 weeks of operations shows a sample mean weekly loss of \$2200, with a sample standard deviation of \$500. Use a .05 level of significance and test the trucking firm's claim that the mean weekly loss is \$2000 or less.

H_0: $\mu \leq 2000$ $\quad n = 15$
H_a: $\mu > 2000$ $\quad \bar{x} = 2200$
$s = 500$
$\alpha = .05$

A t-distribution with 14 df is appropriate (σ is unknown and the sample size is small; assume that the parent is normal):

$$t_{\text{statistic}} = \frac{2200 - 2000}{\frac{500}{\sqrt{15}}} = 1.55$$

$$t_{.025,\,14} = 2.145$$

Since $t_{\text{statistic}} \leq 2.145$, we do not have a sufficient reason to reject H_0:

$$p\text{-value} = P(t_{14} > 1.55) \approx 1 - .92 = .08$$

Example 3

New tires manufactured by a company in Findlay, Ohio, are designed to provide a mean of at least 28,000 miles of wear. Tests with 20 tires show a sample mean of 27,500 miles with a sample standard deviation of 1000 miles. Use a .01 level of significance and test for whether or not there is sufficient evidence to reject the claim of a mean of at least 28,000 miles.

Solution

H_0: $\mu \geqslant 28{,}000$ $\quad n = 20$
H_a: $\mu < 28{,}000$ $\quad \bar{x} = 27{,}500$
$s = 1000$
$\alpha = .01$

A t-distribution with 19 df is appropriate (assume the parent is normal, σ is unknown, and n is small):

$$t_{\text{statistic}} = \frac{27{,}500 - 28{,}000}{\frac{1000}{\sqrt{20}}} = -2.24$$

Since $t_{\text{statistic}} > -t_{.01,\,19} = -2.539$, we do not have a sufficient reason to reject H_0 (accept H_a).

A.4.8 Testing on the Difference of Two (Population) Means

We can test for the difference between two population means just as in the case of one sample mean. In this case, the two-sided hypothesis would be of the form

H_0: $\mu_1 = \mu_2$
H_a: $\mu_1 \neq \mu_2$

which also implies the following hypothesis test:

H_0: $\mu_1 - \mu_2 = 0$
H_a: $\mu_1 - \mu_2 \neq 0$

The hypothesis-testing procedure will be identical to the case of a single mean except for the calculation of the test statistic. We consider two cases.

Population Variances σ_1^2 and σ_2^2 Are Known

If the parent populations are assumed to be normal and their variances σ_1^2 and σ_2^2 are assumed to be known, then the test statistic is normal and calculated as

$$Z_{\text{statistic}} = \frac{\bar{x}_1 - \bar{x}_2}{\sqrt{\left(\frac{\sigma_1^2}{n_1}\right) + \left(\frac{\sigma_2^2}{n_2}\right)}}$$

where $\bar{x}_1$ is the mean of the sample (size n_1) from population 1 and $\bar{x}_2$ is the mean of the sample (size n_2) from population 2. Then, as in the one-sample case, the $|Z_{statistic}|$ is compared with $Z_{\alpha/2}$, and we reject the null hypothesis if $|Z_{statistic}| > Z_{\alpha/2}$.

Example Suppose we want to test the equality of mean starting salaries of graduates from two large universities. A random sample of size $n_1 = 100$ from one university yields $\bar{x}_1 = 28{,}250$, while a random sample of size $n_2 = 60$ from the other university yields $\bar{x}_2 = 28{,}150$. Let us assume that the variances are $\sigma_1^2 = 40{,}000$ and $\sigma_2^2 = 32{,}400$. Assume $\sigma = .05$. The null and alternate hypotheses will be

$$H_0\colon \mu_1 - \mu_2 = 0$$
$$H_a\colon \mu_1 - \mu_2 \neq 0$$

The Z-test statistic value will be

$$Z_{statistic} = \frac{28{,}250 - 28{,}150}{\sqrt{\left(\frac{40{,}000}{100}\right) + \left(\frac{32{,}000}{60}\right)}} = 3.273$$

This value of $Z_{statistic}$ exceeds the tabulated Z-value of 1.96. Thus, the null hypothesis must be rejected in favor of the alternate hypothesis.

Population Variances (σ_1^2 and σ_2^2) Are Not Known but Assumed Equal

If the population variances σ_1^2 and σ_2^2 are not known and are assumed to be equal, the t-test is more appropriate. Let $\bar{x}_1$ and s_1^2 be the mean and variance of sample (size n_1) from population 1 and let $\bar{x}_2$ and s_2^2 be the corresponding quantities of sample (size n_2) from population 2. Then the test statistic

$$t_{statistic} = \frac{\bar{x}_1 - \bar{x}_2}{\sqrt{\left[\frac{(n_1 - 1)s_1^2 + (n_2 - 1)s_2^2}{(n_1 + n_2 - 2)}\right]\left(\frac{n_1 + n_2}{n_1 n_2}\right)}}$$

and is compared with a t-distribution with $(n_1 + n_2 - 2)$ degrees of freedom.

References

Albright, S. C., W. Winston, and C. Zappe. 2002. *Managerial statistics.* Pacific Grove, California: Brooks/Cole.

Hogg, R. V., and A. T. Craig. 1995. Introduction to mathematical statistics. Englewood Cliffs, New Jersey: Prentice Hall.

Hayter, A. J. 2002. *Probability and statistics for engineers and scientists* 2d ed. Pacific Grove, California: Brooks/Cole.

Index

Access module, 351, 352
accumulating, 350
accumulating conveyor, 346
Accumulating Length, 350
actions, 217
activate, 225
Activate module, 363
active state, 212, 216
ActiveX, 250
activity, 215, 283
activity cycle diagram, 281, 283
activity routine, 217, 218
activity view, 210, 216, 236
actual parameter, 179
A-D GOF, 109
Advance Transfer Panel, 310, 351, 363
advertising, 144
after-tax cash flow, 31, 63
aggregated symbol, 303
AGV. See automated guided vehicle
Albright, S. C., 418, 448
Alexopoulos, C., 150, 127
algorithm, 160
Alla, H., 285, 323
Allocate module, 363
alternate hypothesis, 443
American option, 89
Anderson-Darling test, 109
Animate Transfer Toolbar, 311
animation, 280, 309
arc, 175
Arena, 8, 246, 249
arrival event, 214
arrival process, 329
arrival rate, 119
Asian option, 93
Assign module, 263
ATCF, 31, 63
atomic symbol, 303
attribute, 189, 212, 265

autocorrelated, 126
  interarrival times, 329
  service times, 330
autocorrelation, 380
automated guided vehicle (AGV), 360
automatic connection, 251

balk, 330
bank lobby model, 264
Banks, J., 8, 17, 42, 47, 73, 108
Basic Process Panel, 250
batch means, 126, 171, 380
Batch module, 263, 279, 290
Bayes's theorem, 420
begin service event, 213
Bell, P. C., 277, 247, 248, 275
Beninga, S., 78, 88
Bernoulli
  distribution, 423
  random variable, 42
  trial, 424
BestFit, 106
beta distribution, 434
BETADIST, 434
BETAINV, 434
B-event, 221
BINOMDIST, 425
binomial distribution, 424
binomial experiment, 424
binomial random variable, 48
blocking, 333, 353
breakdown event, 214
Bright, G., 248, 277
buffer, 346
buffer storage, 347
built-in variable, 268
Buxton, J. N., 217, 242

C, 70, 246
Cadle, J., 417
call option, 89
Carson, J. S., 17, 73
cdf. *See* cumulative distribution function
cell size, 350
central limit theorem, 430
Ceric, V., 280, 323
certainty equivalent, 5
C-event, 221
child, 303
Chi-sq GOF, 109
chi-square, 74
  distribution, 437
  goodness of fit, 109
Chokshi, R., 7, 17
CINEMA, 248
claims, 67
Clementson, A. T., 222, 242
client, 391
client objectives, 403
clock, 159
CNN, 338, 384
colored Petri net, 288
communication, 408
complement, 419
conditional decision rule, 284
conditional event, 178, 214
conditional probability, 419
CONFIDENCE, 441
confidence interval, 29, 440
  width, 440
confidence limit, 440
CONNECT option, 252
connector animation, 310
consistent estimator, 440
consumable, 332
contingent contract, 88
continuous random variable, 421
continuous simulation, 10
Convey module, 351, 352
conveyor, 346, 348
Conveyor module, 349
Conway, R. W., 150, 127
copy center model, 291
corporate planning model, 143
correlated variates, 99, 101
correlation matrix, 99, 100
cost function, 50
cost parameter, 341
Cox, D. R., 120, 121, 124, 150
Craig, A. T., 108, 418, 448
Crane, M. A., 136, 150
Create module, 251, 252, 263
CRITBINOM, 425
cumulative density function, 426
cumulative distribution function (cdf), 421
cumulative mass function, 421
customer, 327, 328
cycle, 137

cycle time, 261

database, 402
data collection, 392, 398, 399, 404
data quality, 410
David, R., 285, 323
Decide module, 263, 289
decision model, 5
derivative, 88
derivative security, 88
Design/CPN, 248
deterministic model, 26
difference of two means, 447
directed arc, 285
discrete-event dynamic system, 210
discrete-event simulation, 10
discrete-event system, 154, 157
discrete random variable, 42, 45, 421
Dispose module, 251, 252, 263
distribution parameters, 60
documentation, 11, 14
drift parameter, 79, 99
Drop-off module, 363
dynamic, 10
dynamic model, 93
dynamic model description, 156
dynamic simulation, 118
dynamic system simulation, 10

ECSL/CAPS, 220
ED Project, 391
efficient estimator, 440
election system, 337
emergency department, 391
empirical distribution, 42
end condition, 118
end service event, 214
entity, 189, 210, 262
entity attribute, 212
entity life cycle, 283
Entity module, 257
Entity.picture, 257
Entity.Serial Number, 290, 298
Erlang distribution, 63
estimator, 439
European option, 89
Evans, J. B., 287, 323
Evans, J. R., 27, 73
event, 10, 157, 213, 419
  canceling, 187
  graph, 175, 281, 282
  list, 158
  notice, 158
  parameter, 178, 179, 180, 181
  routine, 158
  view, 210, 233
Examples library, 250
excess cost, 50
exercise price, 89
exhaustive, 419
Exit module, 351, 352
expected value, 422, 427
experiment, 419
experimental design, 13, 378, 401
expiration date, 89
EXPONDIST, 431
exponential distribution, 431
exponential random variable, 42, 44

facility, 210, 332
facility-based animation, 310
FCFS, 330
Ferrin, D. M., 7, 17
Fiddy, E. J., 248, 277
financial model, 78
Fishman, G. S., 9, 17, 42, 47, 73, 127, 136, 150
fitting distribution(s), 60, 104
fixed cost, 342
fixed time advance, 118
flat model, 303
Fleming, Q. W., 406, 417
flexible manufacturing, 360
formal parameter, 179, 180
FORTRAN, 70, 246
Free module, 363
free-path transporter, 362
fuel depot model, 304
future, 88
future event list, 158

gaming simulation, 9
gamma distribution, 432
GAMMADIST, 432
GAMMAINV, 47, 432
Gantt chart, 406
Gatersleben, M., 7, 17
GE, 99
generic graph, 281

generic simulation language, 160
geometric distribution, 426
geometric random walk, 79
GI/G/1, 120
GI/M/1, 120
Giron, G., 7, 17
Gordon, K. J., 323
Gordon, R. F., 303, 323
GPSS, 124, 246
GPSS/H, 124, 248
Grabau, M., 391, 417
graphical user interface (GUI), 129, 136, 247
Gray, P., 144, 150
Grosz, D., 7, 17
group service, 330
GUI. *See* graphical user interface
guided transporter, 362

Halt module, 363, 129
Hayter, A. J., 129, 418, 448
HD, 99
hedging, 91
Heyman, D. P., 181, 205
hierarchical model, 412
  building, 250
hierarchical modeling, 279, 303
Hillier, F. S., 4, 7, 17, 49, 73, 120, 121, 123, 150
histogram, 86
HLOOKUP, 45, 47
HOCUS, 222, 248
Hogg, R. V., 108, 418, 448
holding cost, 342
Hull, J. C., 78, 88
Hurrion, R. D., 248, 275, 277
hypergeometric distribution, 423
HYPGEODIST, 424
hypothesis test, 443

Iglehart, D. L., 136, 150
immediate transition, 287
implementation, 13, 402
independence, 419
independent replication(s), 55, 128
inference, 434
INFORMS College on Simulation, 7
initial status, 350
initial transient, 123, 126
initiate, 226
input analyzer, 250
input buffer, 347
input distribution, 11
input parameter, 5
inputs, 83
inspection station, 347
insurance, 66
interarrival time, 120
intersection, 419
interval estimate, 440
inventory, 48
inverse transform, 47
iterations, 83

Java, 246
Jensen, K., 288, 323
Jenshing, J. S., 7, 17
jockeying, 330

Karlin, S., 120, 121, 124, 150
Kelton, W. D., 8, 9, 17, 42, 47, 73, 108, 106, 150, 127, 128, 250, 384
Kemeny, J. G., 129, 150
Kolmogorov-Smirnov test, 109, 111
Koppelman, J. M., 406, 417
Kotcher, R. C., 7, 17
K-S GOF, 109
Kurose, J. F., 323
kurtosis, 422

Laski, J. G., 217, 242
Latin hypercube sampling, 82
Law, A. M., 9, 17, 42, 47, 73, 106, 108, 126, 127, 128, 150
LCFS, 330
Lieberman, G. J., 4, 7, 17, 49, 73, 120, 121, 123, 150
Lindley, D. V., 119, 150
Lindley's formula, 119
list, 190
Little, J. D. C., 181, 205
Little's Law, 339, 379
load station, 346
LOGINV, 431
lognormal distribution, 431
LOGNORMDIST, 431
lookup table, 45
loop conveyor, 348, 357
loss ratio, 67

LPT, 330
Luenberger, D. G., 78, 88
Luna, J. J., 303, 323

management science, 2
Mandelkern, D., 247, 277
manufacturing system, 231, 345
market share, 144
marking, 285
Markov, 120
    chain, 129, 131
    process, 130
    property, 130
MathCAD, 70
Mathematica, 70
MATLAB, 70
Max Cells Occupied, 350
maximum likelihood, 108
McNair, E. A., 323
mean, 422
mean recurrence time, 136
mean time between failures, 136
MedModel, 248
Members window, 266
memoryless, 130
message-processing model, 191
M/G/1, 120
Miller, M. J., 7, 17
M/M/1, 120, 123
model complexity, 411
model formulation, 398
model management, 411
module, 250
Monte Carlo sampling, 82
Move module, 363
moving average, 123
multiple servers, 332
Murata, T., 286, 303, 323
Musselman, K., 391, 415, 417
mutually exclusive, 419

Named view, 320
Navigate panel, 320
Naylor, T. H., 150, 144
Nelson, B. L., 17, 73
network, 337
Nicol, D. M., 17, 73
node, 175
nonaccumulating, 350
nonaccumulating conveyor, 346
nonconsumable, 332
nonidentical servers, 333
nonlinear cost, 50
nonpreemptive service, 331
non-value added cost, 341
non-value added time, 341
normal distribution, 430
normal random variable, 42, 44
NORMDIST, 431
NORMINV, 431
NORMSDIST, 431
NORMSINV, 431
null hypothesis, 443

objectives, 11
office building, 31
O'Keefe, R. M., 247, 248, 275, 277
Olson, D., 27, 73
operations research, 2
OPTIK, 248
option, 89
order quantity, 49
other cost, 341
outcome, 418, 419
output data, 29
output parameter, 6
outputs, 83

parameter, 5, 434
parent, 303
parts, 346
Pascal, 70, 246
passivate, 225
passive state, 212, 216
Paul, R. J., 280, 323
pdf. *See* probability density function
performance data, 399
performance measures, 339, 399
period, 118
permanent entity, 190, 210
Petri net, 281
Pickup module, 363
Pidd, M., 8, 17, 42, 47, 73, 217, 242
place, 285
pmf. *See* probability mass function
point estimation, 439

point-and-click, 247
POISSON, 426
Poisson distribution, 425
Poisson process, 329
poll, 338
Poole, T. G., 222, 242
Pooley, R. J., 280, 323
population, 440
posterior probability, 420
power of the test, 443
p–p plot, 108
prediction interval, 29
preemptive nonresume, 331
preemptive resume, 331
preemptive service, 331
premiums, 67
present value, 60, 89
prior probability, 420
priorities, 284
probabilistic model, 4
probability, 418, 419, 420
    decision rule, 284
    distribution, 421
probability density function (pdf), 426
probability mass function (pmf), 421
problem formulation, 397
problem identification, 403
process, 215, 225
process graph, 281, 289
process interaction, 228
Process module, 251, 252, 263, 289
process view, 210, 225, 236
project management, 390, 396, 405
project parameters, 257
project plan, 397
project report, 401
project schedule, 397
project sponsor, 391
ProModel, 248, 277
Proof Animation, 248
put option, 89
p-value, 109, 444

q–q plot, 108
queue, 190, 263, 283, 284, 329
queue discipline, 330
queueing network, 337
queueing system, 120

RAND, 42, 47
random number generator, 55
random sample, 435
random variable, 421
random variate, 41
ratio estimator, 139
Record module, 264
recurrent, 133
recurrent state, 133
regeneration cycle, 137
regeneration point, 137
regenerative method, 136
regenerative process, 137
Reisig, W., 285, 286, 323
Remove module, 377
reneging, 330
renewal process, 329, 330
reorder point, 304
replication, 22, 27, 83
Request module, 363
resource, 210, 262, 332
Resource module, 256
Resource set, 333
Resource spreadsheet, 256
resource usage cost, 342
resource utilization, 340
retrial, 330
right triangular distribution, 430
risk, 33
@RISK, 78
risk analysis, 60
RiskCorrmat, 102
RiskLogistic, 113
RiskNormal, 80, 81, 82
roller conveyor, 347
Ross, S. M., 120, 121, 124, 150
Route module, 310
Route tool, 312
*r*th moment, 422
Run menu, 259
run setup, 257

Sadowski, D. A., 17, 384, 277
Sadowski, R. P., 8, 17, 277, 384, 391, 417
sample correlation, 100
sample mean, 435
sample statistics, 435
sample variance, 436
sampling distribution, 435

Sargent, R. G., 175, 205
schedule, 158
Schedule module, 266
Schedule spreadsheet, 266
scheduled event, 214
Schruben, L. W., 175, 205, 277
Search module, 377
seed, 55
SEE_WHY, 248
segment, 349
Seila, A. F., 127, 150
sensitivity, 31
sensitivity analysis, 32, 52
Separate module, 263, 279, 290
sequence, 317
server, 329
server utilization, 339
service rate, 119
service system, 327
service time, 120
ServiceModel, 248
set, 190
Set module, 266
set of resources, 263
Set spreadsheet, 266
shortage cost, 50
SIGMA for Windows, 248
SIMAN, 246, 248, 250
SIMFACTORY, 246, 248
SIMGRAPHICS, 248
Simon, H. A., 303, 323
Simplan Systems, Inc., 144, 150
SIMSCRIPT II.5, 246
SIMULA, 246
simulate, 163
simulation, 8
  clock, 159
  model, 8
single-server queue, 161, 198, 226
skewness, 422
SLAM II, 248
SLAMSYSTEM, 248
SMARTS Library, 250
Smith, W. L., 120, 121, 124, 150
SNAP option, 252
Snell, J. L., 129, 150
Sobel, M. J., 181, 205
Society for Computer Simulation, 7
special sale promotion, 23
SPT, 330
state of nature, 443
state space, 130
state variable, 118, 284
static model description, 156
static simulation, 10, 21
static system description, 213, 409
station, 310, 312, 349
Station module, 310, 349
station sequence, 317
stationary, 117, 126
statistical inference, 434
stochastic model, 4
stock price model, 79
strike price, 89
submodel, 303, 304, 313, 380
subsystem, 3
suspend, 226
symmetric triangular distribution, 429
system, 2
system attributes, 190
system dynamics, 213
system state, 118, 156, 159, 190, 213
systems analysis, 11, 392
systems analyst, 391
Systems Modeling, 277, 250
Szymankiewicz, J. Z., 222, 242

Tam, C. M., 7, 17
tandem queue, 333
Taylor, H. M., 120, 121, 124, 150
t-distribution, 436
team management, 407
temporary entity, 189, 210
terminal, 366
testhead, 217
Thamhain, H. J., 402, 417
three-phase approach, 220
threshold, 142
timed transition, 286, 287
timing routine, 159
TINV, 441
Tocher, K. D., 220, 242, 283, 323
token, 285
tolerance interval, 29
top-down design, 192
Torn, A. A., 286, 323
transfer cost, 341
transfer time, 341

transient, 117, 123, 133
  period, 135
  state, 133
transition, 285
transition firing, 285
transition probability matrix, 132
transport equipment, 346
transportation, 365
transporter, 346, 360
Transport module, 363
triangular distribution, 428
triangular random variable, 42, 43
trigger level, 304
truck, 367, 368, 369
truncation point, 126
type I error, 443
type II error, 443

unbiased estimator, 439
uniform distribution, 427
uniform random variable, 42, 43
union, 419
unload station, 346
urgent care center, 391

validation, 11, 400
value added cost, 261, 341
value added time, 341
van, 367, 368, 373
van der Weij, S., 7, 17
variable, 263, 268
variable time advance, 118
variance, 422
VBA, 250
velocity, 350
verification, 11, 246
Verma, J. K., 402, 406, 417
Vincent, S., 106
Visio, 250
visual interactive simulation, 246, 247, 281
VLOOKUP, 45, 47
volatility, 79, 99

wait cost, 341
wait time, 341
waiting state, 212
WEIBULL, 433
Weibull distribution, 433
Welch, P., 126, 150
what-if, 33
Winston, W. L., 8, 17, 27, 73, 418, 448
Winter Simulation Conference, 7, 8
Wirth, N., 192, 205
work in progress, 261
workstation, 346
worldview, 210

XCELL, 246, 248

Yeates, K., 417

Zappe, C., 418, 448
Zheng, S. X., 7, 17
ZTEST, 444, 445